Flow and Combustion in Advanced Gas Turbine Combustors

FLUID MECHANICS AND ITS APPLICATIONS

Volume 102

Aims and Scope of the Series

The purpose of this series is to focus on subjects in which fluid mechanics plays a fundamental role.

As well as the more traditional applications of aeronautics, hydraulics, heat and mass transfer etc., books will be published dealing with topics which are currently in a state of rapid development, such as turbulence, suspensions and multiphase fluids, super and hypersonic flows and numerical modeling techniques.

It is a widely held view that it is the interdisciplinary subjects that will receive intense scientific attention, bringing them to the forefront of technological advancement. Fluids have the ability to transport matter and its properties as well as to transmit force, therefore fluid mechanics is a subject that is particularly open to cross fertilization with other sciences and disciplines of engineering. The subject of fluid mechanics will be highly relevant in domains such as chemical, metallurgical, biological and ecological engineering. This series is particularly open to such new multidisciplinary domains.

The median level of presentation is the first year graduate student. Some texts are monographs defining the current state of a field; others are accessible to final year undergraduates; but essentially the emphasis is on readability and clarity.

For further volumes:
http://www.springer.com/series/5980

Johannes Janicka • Amsini Sadiki • Michael Schäfer
Christof Heeger

Flow and Combustion in Advanced Gas Turbine Combustors

Springer

Prof. Dr.-Ing. Johannes Janicka
Department of Mechanical Engineering
Institute of Energy and Power Plant Technology
Technische Universität Darmstadt
Darmstadt
Germany

Prof. Dr. Amsini Sadiki
Department of Mechanical Engineering
Institute of Energy and Power Plant Technology
Darmstadt University of Technology
Darmstadt
Germany

Prof. Dr. Michael Schäfer
Department of Mechanical Engineering
Institute of Numerical Methods in Mechanical Engineering
Technische Universität Darmstadt
Darmstadt
Germany

Dr.-Ing. Christof Heeger
Center of Smart Interfaces
Technische Universität Darmstadt
Darmstadt
Germany

ISSN 0926-5112
ISBN 978-94-007-5319-8 ISBN 978-94-007-5320-4 (eBook)
DOI 10.1007/978-94-007-5320-4
Springer Dordrecht Heidelberg New York London

Library of Congress Control Number: 2012945627

Printed on acid-free paper

Springer is part of Springer Science+Business Media (www.springer.com)

Preface

Worldwide over 80% of the energy production is basically achieved by combustion of fossil fuels. This dominant role of combustion in energy conversion will not change substantially despite recent efforts to promote renewable energy in the next decades. Meanwhile combustion will remain a key process in traditional energy conversion systems either in transportation or the energy production sector. Emerging technology fields, such as micro-scale power generation, material synthesis, biofuel applications, novel hybrid systems etc., still depend on chemically reacting flow processes. With regard to both, the environmental sustainability and operating efficiency demands, modern combustion research has to face two main objectives: the optimization of combustion efficiency and the reduction of pollutants in various technological applications. Focus is put here on issues in stationary and aircraft gas turbine combustors since their applications have dramatically transformed and will continue to impact society behavior as one of the driving development factors.

A gas turbine combustor is a rather sophisticated system comprising complex confined configurations in which heat and mass transfer as well as combustion, radiation and multiphase flow phenomena play a central role. Efforts to increase the efficiency of modern gas turbines on the one hand and to comply with the current emission regulations on the other hand have already achieved substantial improvements and are strongly pursued further. Novel concepts have been developed and driven to a certain degree of maturity. However, further developments with respect to reliability, emissions, efficiency, profitability and a rapid implementation of these attractive concepts suffer essentially from still incomplete knowledge of the underlying technical and scientific mechanisms, such as

- Turbulent flows and mixing processes
- Liquid fuel breakup, atomization, evaporation and combustion of sprays in a turbulent environment
- Chemical kinetics of complex fuels, soot formation
- Droplet interaction during evaporation and combustion
- Interaction between turbulence, evaporating droplet and reaction
- Heat transfer by convection, soot and gas radiation
- Interaction between combustion chamber and compressor or turbine

All these issues are very complex. Due to the demanding requirements, e.g. stable combustion, low emission, low mechanical and thermal loads on the wall, compatibility with compressor and turbine etc., a reliable development of modern engines can only be achieved by close interaction between advanced numerical simulations, optimization techniques and some well-designed validation experiments. The latter allow for a deep and reliable understanding of the physicochemical mechanisms and interactions in a complementary way.

This book reflects the outcome of the research activities within the interdisciplinary Collaborative Research Center 568 "Flow and combustion in future gas turbine combustion chambers" funded by the Deutsche Forschungsgemeinschaft for more than a decade (2001–2011) at the Technische Universität Darmstadt and the University of Heidelberg in Germany. This compilation aims firstly at reporting on the newest achievements and improvements from the various research activities of the Center, having received considerable recognition worldwide. It aims secondly at providing an opportunity for researchers and interested workers to quickly capture the state of the art. These are current developments, solutions and procedures in the areas of flow and combustion modeling, unsteady simulation methods, optimization strategies, multiphase flow characterization, experimental validation methods and measurement techniques.

The Collaborative Research Center 568 pursued to develop a completely integrated modeling and numerical simulation of the complex interacting processes. By incorporating turbulent heat and mass transport, single or multi-phase flow phenomena, chemical reactions/combustion and radiation, the development and design process of advanced gas turbine chamber concepts can be substantially supported. Thereby the investigators had in mind that a rapid monitoring and numerical prediction of real system dynamics requires software that combines reasonable costs of computations and sufficient accuracy of results based on physically consistent models and appropriate numerical methods. For that purpose the CFD code package FASTEST has been adopted as a common software platform. A large number of research projects were successfully completed in different specific project areas. These were defined according to the geometrical arrangement of the combustion chamber:

- Injection systems and fuel-air mixture preparation
- Combustion modeling techniques
- Component interactions and near wall thermo-fluid-dynamical processes
- Numerical techniques, data management and optimization strategies

The main achievements in these project areas are reported in sixteen chapters compiled in the five parts of this book. The common feature of all the contributions of the first part is the treatment of liquid fuel injection, subsequent liquid breakup, atomization and spray evolution under gas turbine conditions.

The chapters of the second part dwell on experimental, theoretical and numerical description of combustion in gas turbine combustors. They include the development of methods involving reduced reaction mechanisms for complex fuels, soot modeling and the formulation of appropriate turbulence-chemistry interaction models for non-premixed, partially premixed and premixed combustion. In particular, innovative high-speed measurement techniques allow time-resolved characterization of flame stability and extinction behavior.

Part three is concerned with the integration of the combustion chamber boundaries and the interactions with neighboring components, such as compressor and resulting fluid mechanical processes.

An essential prerequisite for an efficient overall integral model for the simulation of combustion chambers is, on the one hand, the efficiency of sub-models, and on the other hand, the development and efficient integration of numerical methods. The contributions dealing with this topic constitute the content of part four of this book.

In the last funding period of the Center, a number of projects have been formulated in which the methodological knowledge and its procedures had to be transferred to industry in the framework of a partnership between industry and university. The related contributions are assembled in part five of this book.

The individual chapters feature the endeavor accomplished in the advanced combustion modeling along with newest developments and applications of URANS and LES methods as well as related coupling strategies, like Hybrid URANS-LES strategies. They propose methods to treat low emission concepts, computational combustion, chemical kinetics and innovative numerical methods and optimization techniques. Model assessment was mainly supported by the design and investigation of targeted validation experiments and resulting experimental data. The central outcome in this respect comes from combustion-LES particularly being recognized as an attractive approach for combustor simulation due to its demonstrated superiority over classical RANS. This particularly holds in terms of the prediction of mixture formation. In particular, the following achievements need to be outlined:

(i) A complete modeling of specific phenomena in combustion chambers was achieved based on an overall modeling and combustion-LES technique. For the gaseous combustion, the description relies on the detailed tabulated chemistry following the "Flamelet-Generated Manifold" (FGM) approach. In particular, the extension of the FGM for describing pollutants such as NO_x and CO, the transported filtered density function method (Transported FDF)) and the "Artificially Thickened Flame" (ATF) model have been implemented into the common CFD FASTEST code, and validated.

(ii) Physically consistent modeling of two-phase flow processes in the air-blast atomizer nozzle has been accomplished. Particular emphasis was placed on a reliable description of spray droplet-wall interaction phenomena and liquid film-wall heat transfer. Analysis of the overall spray micro-processes including spray combustion has been carried out based on a thermodynamically consistent modeling integrating the FGM tabulated method for combustion description.

(iii) The quality assessment methods for LES developed for simple configurations have been extended to complex geometries following a dynamic quality control technique. For this purpose, an efficient computation of the adjunct flow equations was necessary along with efficient methods of model reduction such as Proper Orthogonal Decomposition (POD) and Centroidal Voronoi Tessellations (CVT). These promising approaches have been developed and integrated into the FASTEST code. Their applicability has been demonstrated in simulations of gas turbine combustors.

(iv) Adequate optimization algorithms have been developed that allow for reaching the proximity of the technical optimum with a relatively small number of complex simulations. A successful evaluation and testing of these methods have been achieved based on a few, complex flow LES calculations.

(v) Among the most interesting technological developments of the last three years, the development of high-speed lasers and high-speed detection systems must be quoted. These techniques enable to provide completely new insights into the time-dependent structure of turbulent combustion systems.

The participants involved in the Collaborative Research Centre 568 are members of ten institutes of the departments of mechanical engineering and mathematics of the Technische Universität Darmstadt and of the Interdisciplinary Center for Scientific Computing (Interdiziplinäre Wissenschaftliches Rechnen IWR) of the University of Heidelberg. They are:

- Institute of Energy and Power Plant Technology
- Chair of Fluid Mechanics and Aerodynamics
- Chair of Fluid Dynamics
- Institute of Reactive Flows and Diagnostics
- Interdisciplinary Center for Scientific Computing
- Technical Thermodynamics
- Turbomachines (2001–2004)
- Gas turbines and Propulsion Techniques (2001–2008)
- Numerical Mathematics
- Institute of Numerical Methods in Mechanical Engineering

The members of the collaborative research Center 568 acknowledge the financial support by the Deutsche Forschungsgemeinschaft over the four funding periods (2001–2011). In particular, they highly appreciate the advice and critical review of Prof. Friedrich, Prof. Pfitzner, Prof. Sattelmayer, Prof. Peters, Prof. Renz, Prof. Kneer, Prof. Leipertz and Prof. Sommerfeld as chairmen of the evaluation committee. The success of this Center was also made possible by the DFG headquarters in Bonn, led by Dr. Lachenmeier, Dr. Lenz and Dr. Küster. The members of the Center thank especially this team. Last not least, the efficient collaboration of the office staff of the institutes involved in this Center is highly appreciated.

Darmstadt

J. Janicka
A. Sadiki
M. Schäfer
C. Heeger

Contents

Part I
Injection Systems and Mixture Formation

In modern low emission aero-engine combustors, spray combustion is used extensively. It follows three preliminary steps: injection of liquid fuel, vaporization of the droplets, and mixing of the oxidant and the fuel. The characteristics of the fuel spray, such as droplet size and velocity distributions, have a great influence on the efficiency and the emissions of the engines. Future compliance with stricter emission regulations demands a deeper physical understanding of the fragmentation processes and the resulting characteristics of the final spray. Presently, the primary atomization used in gas turbine combustors, either using empirical models or using a direct numerical simulation, is inadequately predicted to be of use in optimization studies. Therefore empirical data are generally obtained on a case-by-case basis, in an attempt to capture the most important dependencies of the droplet velocity and size distribution. Without doubt there is a deficit in this area, in particular for atomizers operated at elevated pressures. Even for simple pressure atomizers, the influence of chamber pressure appears ambiguous. For airblast atomizers there exist no reliable models for predicting the resulting spray.

The focus of Part I is to investigate the effect of the main operating parameters on the atomization process of an airblast atomizer. In particular, the flow inside the nozzle is determinant. Primary spray formation in the pressure swirl injector is followed by drop breakup and evaporation during flight, spray impact onto the wall of the atomizer, formation of a liquid film on a wall surface, its breakup and the final or secondary atomization into the chamber. Under real gas turbine conditions these phenomena are influenced also by the swirling hot air flow through the primary swirler and high ambient pressure and wall temperature inside the atomizer.

The investigations presented in Chap. 1 focus on the spray atomization, transport and impact on a solid substrate under cross-flow conditions, as used in airblast atomizers with prefilmers for aero engines and gas turbines. It includes both theoretical analyses and experimental characterization of the spray. The phenomena are observed using a high-speed video system and the spray is characterized using the phase Doppler technique. The governing mechanisms of drop formation, wall collision and aerodynamic breakup are identified. Especially, an atomization model is developed, which accounts for primary atomization, wall film formation and aerodynamic breakup.

Chapter 2 deals with the liquid film evolution on a wall surface, especially with shear-driven liquid film flows as occurring in several locations of fuel preparation systems, e.g. inside air-driven atomizers or in Lean Pre-mixing Pre-vaporizing (LPP) combustion chambers of modern gas turbines. The fundamentals of gravity-driven as well as air-driven film flow and evaporation on unstructured and micro-structured wall surfaces are investigated experimentally and numerically.

The objective of Chap. 3 is to develop and validate a thermodynamically consistent spray module for Large Eddy Simulation that allows describing accurately the essential processes featuring spray combustion in gas turbine combustion chambers. Besides the injection of liquid fuel, these include the turbulent droplet dispersion, the vaporization of the droplets and mixture formation and the subsequent spray combustion.

Chapter 1
Primary Atomization in an Airblast Gas Turbine Atomizer

L. Opfer, I.V. Roisman, and C. Tropea

Abstract This study focuses on the spray atomization, transport and impact on a solid substrate under cross-flow conditions, as used in airblast atomizers with prefilmers for aero engines and gas turbines. The phenomena are observed using a high-speed video system and the spray is characterized using the phase Doppler technique. The governing mechanisms of drop formation, wall collision and aerodynamic breakup are identified. It is shown that three different mechanisms are mainly responsible for the formation of single drops from the bulk liquid. These are: primary atomization, breakup of the liquid wall film and further aerodynamic breakup of droplets. Finally, an atomization model is developed, which accounts for primary atomization, wall film formation and aerodynamic breakup. The model predicts the distribution of the drop diameters and velocities in the generated spray. The agreement between the model predictions and the experimental data is very good.

Keywords Atomization • Airblast atomizer • Spray impact • Aerodynamic breakup

1.1 Introduction

The study in the framework of the project A1 is motivated by the fragmentation phenomena in prefilming airblast atomizers. Current airblast atomizers provide good atomization performance over a wide span of operating parameters and relatively

L. Opfer (✉) • I.V. Roisman • C. Tropea
Institute of Fluid Mechanics and Aerodynamics, Mechanical Engineering,
Technische Universität Darmstadt, Petersenstr. 30, 64287 Darmstadt, Germany
e-mail: lopfer@sla.tu-darmstadt.de

I.V. Roisman • C. Tropea
Center of Smart Interfaces, Mechanical Engineering, Technische Universität Darmstadt,
Petersenstr. 32, 64287 Darmstadt, Germany
e-mail: roisman@sla.tu-darmstadt.de

J. Janicka et al., *Flow and Combustion in Advanced Gas Turbine Combustors*,
Fluid Mechanics and Its Applications 102, DOI 10.1007/978-94-007-5320-4_1,

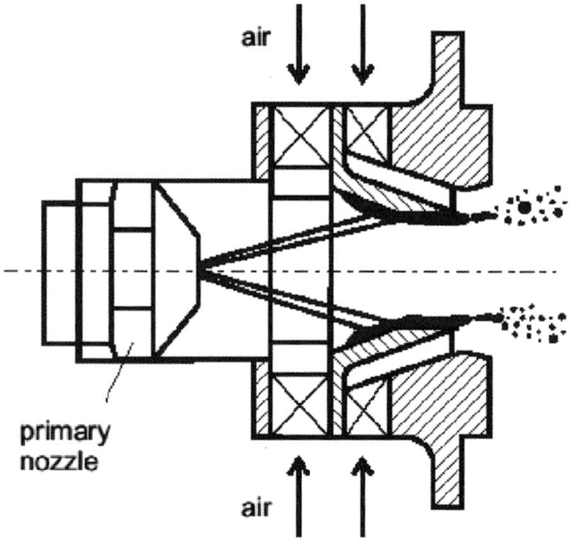

Fig. 1.1 Cross-section of an airblast atomizer

little soot formation [10]. The characteristics of the fuel spray, such as droplet size distribution and droplet velocities, have a great influence on the efficiency and the emissions of aero engines; however, stricter emission regulations require an optimization of the combustion process and the fuel preparation inside the engine.

This requires a deeper physical understanding of the fragmentation processes inside the airblast atomizer and the resulting characteristics of the final airblast spray.

The basic design of an atomizer for aero-engines is shown in Fig. 1.1. The liquid is injected through a hollow cone nozzle and impacts on the annular prefilmer. The impacting liquid generates a liquid film on the wall which is fragmented by a strong tangential airflow. This atomization process comprises several phenomena such as primary atomization of the hollow cone spray, spray impact and splashing, generation of a liquid wall film on the prefilmer, shear stripping from the liquid wall film, film breakup at the edge of the prefilmer and aerodynamic breakup of drops and ligaments. The prediction of the main properties of the generated spray is therefore a rather challenging task, which requires first to determine the dominant mechanisms of atomization. This is one of the main topics of this project.

1.1.1 Spray Generated by Airblast Atomizers, Typical Drop Diameters

The outcome of the different atomization processes inside the airblast atomizer is influenced by many parameters, like the volumetric flow rates of the airflow and the liquid flow (ALR denotes their ratio), velocities of the liquid, U_l, and air, U_a, ambient pressure, material properties of the fluids (surface tension σ, density ϱ, viscosity μ), etc.

One of the most important parameters of a spray is the characteristic diameter of drops. Its value can be presented using the typical length scales:

$$\lambda_{L\sigma} = \frac{\sigma}{\varrho_L U_L^2}, \lambda_{A\sigma} = \frac{\sigma}{\varrho_A U_A^2}, \lambda_\mu = \frac{\mu_L^2}{\sigma \varrho_L}$$

Examples of empirical correlations for the mean drop sizes in the spray generated by plain-jet airblast atomizers are, for example

$$D_{32} = 0.585\lambda_{L\sigma}^{0.5} + 53\lambda_\mu^{0.225} ALR^{-1.5},$$

$$D_{32} = 0.95\frac{\sigma^{0.33}\dot{m}_L^{0.33}}{\rho_L^{0.37}\rho_A^{0.3}U_R}(1 + ALR^{-1})^{1.7} + 0.13\lambda_\mu^{0.5} d_0^{0.5}(1 + ALR - 1)^{1.7},$$

$$D_{32} = 0.022\lambda_{A\sigma}^{0.45}(1 + ALR^{-1})^{0.5} + 0.00143\lambda_\mu^{0.4}(1 + ALR^{-1})^{0.8},$$

$$D_{32} = 0.48d_0\left(\frac{\lambda_{A\sigma}}{d_0}\right)^{0.4}(1 + ALR^{-1})^{0.4} + 0.15d_0\left(\frac{\lambda_\mu}{d_0}\right)^{0.5}(1 + ALR^{-1}),$$

$$D_{min} = 204\lambda_{A\sigma},$$

[11–15], where D_{32} is the Sauter mean diameter. Only the last two expressions are dimensionally correct. Further examples of empirical models can be found in [16]. It is obvious that these models (as all are purely empirical models) are useful only over a restricted range of the operating parameters and cannot be used as a universally reliable prediction tool for spray characterization.

1.1.2 Modeling of Atomization

Several studies describe the breakup of a viscous liquid jet [17–19] or non-Newtonian jet [20] by the development of jet disturbances due to the Kelvin-Helmholtz instability. The most unstable wavelength is determined by the thickness of the vorticity layer in the gas and by the liquid-to-gas density ratio. The development of these Kelvin-Helmholtz waves leads to the appearance of secondary interfacial instabilities and to drop stripping by the Rayleigh-Taylor instability. These instabilities are applicable in the cases of relatively low gas velocity and turbulence levels in the gas and liquid phases.

A commonly used breakup model for numerical simulations is the Taylor analogy breakup model [21] which is based on an analogy between oscillating drops and spring-damper-mass systems. In this approach the aerodynamic load on the drops works as a forcing term, whereas restoring and dissipating terms are given by surface tension and viscosity, respectively. The breakup criterion is defined by a critical deformation of the drop. In [22] the Taylor analogy breakup model is extended to describe the atomization process as a cascade of breakup events. The extended model is then applied to the simulation of high velocity fuel sprays. An adjustable constant has to be introduced into such models, which is tuned to achieve good agreement with experimental results. The main disadvantage of such an approach is that it does not describe the real physics of breakup, types of breakup mode, mechanism of instability, etc.

A useful instrument for the description of the dynamics of a liquid jet is a quasi one-dimensional approach developed in [23]. In [24] this theory has been used to analyze the jet transverse instability and breakup in airflow.

In the case of airblast atomization or high-speed liquid injection into a stagnant dense air the situation becomes much more complicated since the atomization mechanism is influenced by the in-nozzle flow [25, 26]. Such phenomena require alternative modeling approaches.

Several phenomenological models of atomization consider the turbulence in the liquid flow as a driving force of jet disintegration. In [27] the drop size is assumed to be comparable to the typical size of the turbulent eddies. Correspondingly, the liquid wall film starts to disintegrate when the turbulent boundary layer reaches its free surface. Such models cannot be directly applied to the case of airblast atomization of a liquid jet and wall film, in which the turbulent boundary layer is initiated at the free surface by the airflow. Analogous modeling principles have been applied in [28].

The studies using direct numerical simulations (DNS) for high-speed jet atomization [29] can provide some additional information about the atomization process. Nevertheless, such an approach still requires enormous computer resources, especially in the case of simulations of high-Reynolds number turbulent flows.

In the case of extremely high gas velocities, typically required for atomization of viscous liquid jets and jets with suspended particular matter, the atomization can be described using the chaotic disintegration theory [30]. Following this approach the drop size distribution is predicted using percolation theory, which in the asymptotic case, yields the Gamma-type of size distribution observed in numerous experiments [31].

1.1.3 Secondary Drop Breakup

The final size distribution of drops generated by an airblast atomizer is determined by their evaporation, binary collision and aerodynamic breakup. Aerodynamic (secondary) breakup occurs when the relative velocities between the drop and the surrounding gas are high enough. Such conditions are relevant to the atomization in the airblast atomizers, considered in this paper.

The mathematical theory of [32], which considered a random discrete process of breakup of particles, predicts a log-normal distribution of particle sizes. Recent mathematical stochastic models [33] generalize Kolmogorov's theory and predict a fractal distribution of the particle sizes. Such purely mathematical approaches do not consider all physical aspects of atomization.

The aerodynamic breakup of a drop of diameter D and velocity u_l is typically characterized using the Weber and Reynolds numbers, defined as

$$We = \frac{\varrho_g (u_g - u_l)^2 D}{\sigma}, \; Re = \frac{\varrho_g \left(u_g - u_l\right) D}{\mu},$$

where u_g, ϱ_g and μ are the velocity, density and viscosity of the gas, respectively, σ is the surface tension coefficient.

Droplets with $We < 12$ are supposed to be stable whereas droplets with $We > 12$ undergo further breakup [34–36]. At high Weber numbers various regimes of aerodynamic breakup have been observed for low viscosity liquids:

- $12 < We < 80$: bag breakup, multimode breakup [34, 37, 38].
- $80 < We < 350$: boundary layer stripping (or shear break-up), [37, 39, 40] which is alternatively described as stretching/thinning breakup in [41, 42]
- $We > 350$: catastrophic break-up [43].

The outcome of aerodynamic breakup depends on the Weber number. Various empirical models for the average drop diameter exist in the literature for various breakup modes. Among them are:

- For breakup events that occur in the range of $12 < \mathrm{We} < 80$: (bag- and multimode breakup) the correlation of [44] is given in the form

$$D_{32,frag} = 0.32 \left(\frac{\sigma}{u_{rel}^2 \varrho_g} \right) (We(T_{tot} - T_{ini}))^{2/3},$$

where T_{ini} is the initiation time after which a drop is deformed beyond an ellipsoidal shape and T_{tot} is the total time for the complete breakup process. An empirical correlation for the initiation time has been proposed in [34] whereas a correlation for the total breakup time the correlation is given in [45]

- At higher Weber numbers, $80 < \mathrm{We} < 350$, the Sauter mean diameter of the fragments is described by the correlation of Hsiang and Faeth [38]

$$D_{32,frag} = 6.2 \left(\frac{\varrho_l}{\varrho_g} \right)^{1/4} \left(\frac{\mu_l}{\varrho_l D_0 u_{rel}} \right)^{1/2} We \frac{\sigma}{\varrho_g u_{rel}^2},$$

These empirical expressions allow prediction of the typical drop size of the fragments generated by drop aerodynamic breakup of a low viscosity liquid over a wide range of parameters. In the present study they are implemented in a numerical code which predicts the main characteristics of the spray generated by an airblast atomizer.

1.1.4 This Study

The main objective of this study is the investigation of the main mechanisms of the airblast atomization with a prefilmer. It includes an experimental investigation of the atomization of a planar liquid sheet in a cross flow, impact of the spray onto a rigid substrate, and atomization of the liquid film formed on the substrate by spray impact. The atomized spray is characterized using the phase Doppler instrument and high-speed video system. It is shown that at high air velocities the main phenomenon governing the final diameter of the drops in the spray is the secondary (aerodynamic) breakup.

The following main tasks have been performed in the framework of the project:

- Measurement of a spray generated by the MTU-nozzle at various operational parameters and various ambient pressures
- Application of a novel IPI measurement system for characterization of the airblast spray
- Investigation and modeling of natural and forced spray fluctuations
- Investigation and characterization of a liquid film on the prefilmer
- Development of the chaotic disintegration theory in application to the airblast atomizer
- Investigation of the "laboratory" airblast atomizer
- Experimental investigation of single drop aerodynamic breakup, theoretical model of the bag breakup process

One of the significant surprising results of the study is in the fact, that the main atomization mechanism, which determines the final size of the drop in the generated spray, especially at high air velocities, is the aerodynamic drop breakup.

Finally, we propose a model for the spray atomization and transport, which is able to predict the distribution of the drop diameters and velocities in the final spray. The model is validated by comparison with the experimental data. The agreement is very good, especially for higher air velocities.

1.2 Characterization of Airblast Sprays Under Elevated Ambient Pressure Conditions

A pressure chamber as depicted in Fig. 1.2 is employed to characterize the airblast spray under elevated ambient pressure conditions equivalent to those in aircraft combustors. The ambient pressure in the chamber can be adjusted between 1 and 30 bar. The chamber is equipped with transparent fused quartz windows to allow for non-intrusive, optical measurement techniques. In particular, a phase-Doppler instrument is employed to characterize the airblast spray. The particle

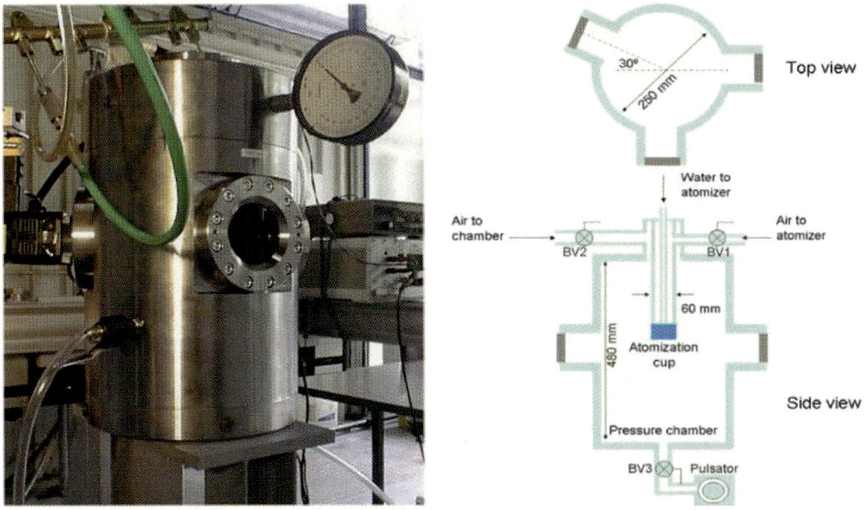

Fig. 1.2 Image (*left*) and scheme (*right*) of the pressure chamber

image velocimetry technique is used to obtain velocity fields of the airblast spray. A high speed camera allowed for time resolved visualizations of the airblast spray formation process inside the pressure chamber.

Figure 1.3a shows the SMD of the droplets at radial positions between 0 and 24 mm and at axial distances ($z = 3, 7, 11, 15$ and 19 mm). The effect of the chamber pressure, or more exact, the air density in the pressure chamber is shown in Fig. 1.3b. The droplet size increases when increasing the air density in the pressure chamber. In this work, the air velocity also decreases when increasing the pressure chamber since the air-mass flow rate is kept constant at different chamber pressure values, which leads to a remarkable increase in the droplet diameter.

Comparing Fig. 1.3b, c shows the opposite effect of the air velocity and the air density on the droplet size. In Fig. 1.3c, the air mass flow rate is increased between 20, 35 and 50 SCMH which causes a decrease in the droplet size as increasing the air-mass flow rate is increased. The air velocity affects also the film thickness on the prefilmer, which plays an important role in determining the droplet diameter after the breakup takes place. The effect of the liquid flow rate on the droplet size distribution is limited, despite the fact that the flow rate increases by double and triple, as illustrated in Fig. 1.3d.

The spray parameters are carefully investigated at various operational conditions, including various ambient pressures. For more details see [4, 5] in the list of own publications. Moreover, the data has been used for the validation of the numerical code for simulation of spray transport ([2, 3] in the list of own publications). This is a cooperation study in the framework of SFB 568.

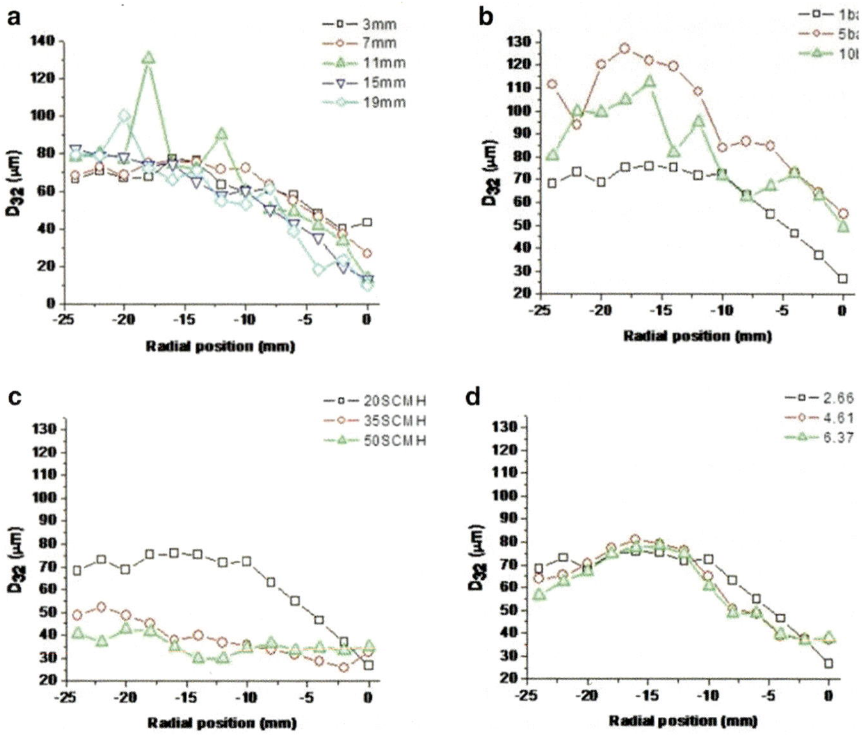

Fig. 1.3 Profiles of the Sauter Mean Diameter of the airblast spray for various operating parameters. (**a**) SMD for increasing distance from the nozzle, (**b**) SMD for increasing chamber pressure, (**c**) SMD for increasing air flow rates and (**d**) SMD for increasing liquid flow rates

Spray fluctuations at various chamber pressures are characterized using two techniques, namely: Proper Orthogonal Decomposition of time-resolved images and spectral analysis of laser Doppler velocity data. It has been found that the airblast spray frequency exhibited a strong dependency on the chamber pressure and the gas-phase flow rate and is totally independent of the liquid phase flow rate. The obtained frequencies from both techniques match each other closely, as shown in Fig. 1.4. The predicted freequency of the spray oscillations

$$f_c = \frac{U_{ax} S}{\pi D_{nozzle}}$$

agrees well with the experimental data. Here U_{ax} is the axial velocity of the air in the nozzle, D_{nozzle} is its diameter, S is the swirl number (see [1] in the list of own publications for more detail).

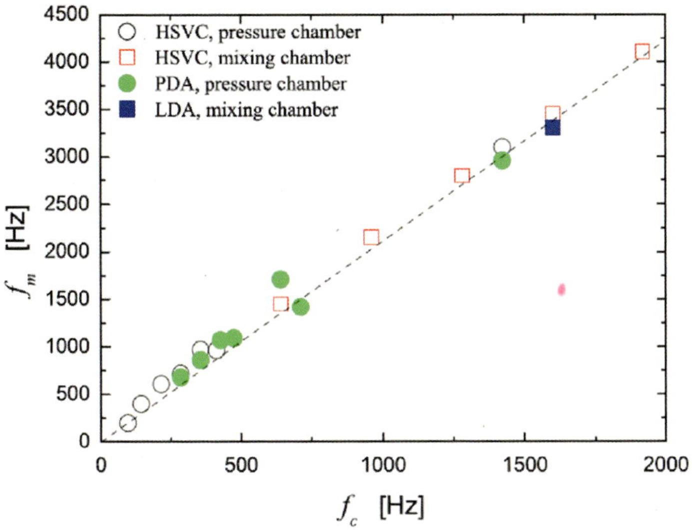

Fig. 1.4 Measured frequency f_m as a function of the calculated characteristic frequency f_c

1.3 Fragmentation Phenomena Inside the Airblast Atomizer

The pressure chamber allows for spray characterization and visualization experiments under high ambient pressure conditions. However, due to the limited optical access into the airblast atomizer, no information about the breakup processes within the atomizer can be revealed in the pressure chamber. For this reason a set of new experiments under atmospheric conditions has been designed and conducted to obtain a deeper insight in atomization phenomena inside the airblast atomizer [6–9].

A generic geometry is chosen to investigate the complex atomization phenomena in prefilming airblast nozzles. This experimental setup represents a two-dimensional laboratory model of an airblast atomizer. The experimental setup is shown in Fig. 1.5.

A flat fan nozzle is used to generate a thin liquid sheet which disintegrates in a cross-flow, generating a primary spray. The spray is directed perpendicular to the impingement plate. The impingement plate is located inside a wind tunnel. The direction of the airflow is parallel to the surface of the impingement plate. The leading and trailing edge of the impingement plate have a blade shape in order to prevent flow separation and the emergence of recirculation areas. The impinging spray forms a liquid film on the substrate which is shear-driven by the cross-flow and breaks up at the atomizing edge at the end of the substrate. The flow in the wind tunnel is driven by a radial blower. The air velocity is controlled by an electronic frequency converter. Before the airflow enters the test section it is decelerated into a settling chamber. Inside the settling chamber the flow is straightened by a

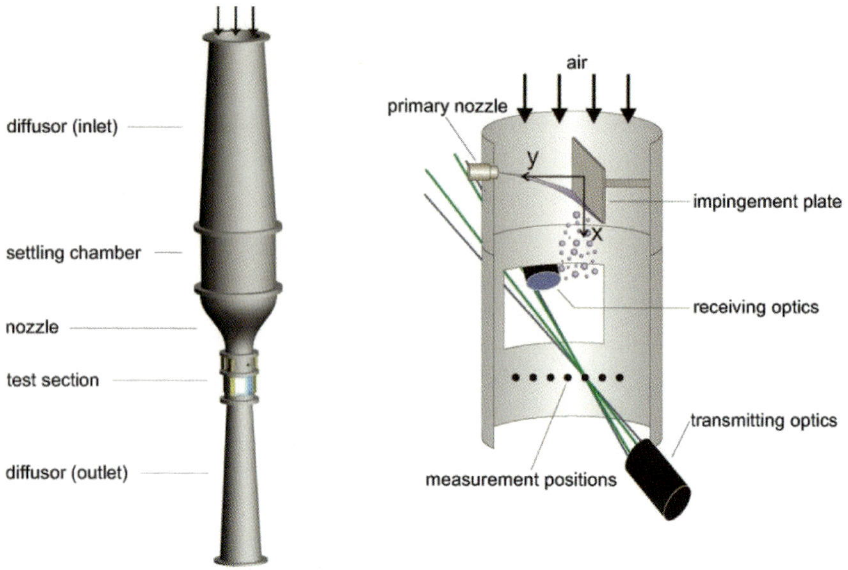

Fig. 1.5 Experimental setup. The complete wind tunnel is shown on the left side, a more detailed sketch of the test section on the right side

honeycomb to suppress vortices below a given length scale and velocity components normal to the main flow direction.

The inlet channel of the test section is equipped with pressure taps for absolute and static pressure. The difference between these is measured by a differential pressure sensor in order to determine the current volume flux. The flat fan nozzle is connected to a pressurized tank in which the bulk liquid is contained. The injection pressure of the primary spray is given by the pressure in the tank. The test section of the wind tunnel is equipped with windows and provides access for optical measurement techniques. A high-speed video system is used for visualizations of spray impact on the wall and breakup of the liquid wall film. A dual-mode phase Doppler instrument is employed for measuring droplet diameters and two velocity components (x- and y-direction) in the spray.

This setup allows for experiments with air velocities up to 125 m/s under atmospheric conditions (1 bar, 293 K). All experiments have been conducted with water as the liquid phase and air as the gas phase.

A *Phantom V12* high-speed camera is used for the shadowgraph visualization. The camera is equipped with a *Nikon* 85 mm lens. It has been operated at a frame rate of 56,000 fps. A mercury vapor lamp is used as a light source.

The airblast spray has been characterized at $x = 60$ mm downstream the edge of the impingement plate. Diameter measurements at $x = 40$ mm and 80 mm yielded the same results which shows that the liquid breakup is completed at the selected measuring positions.

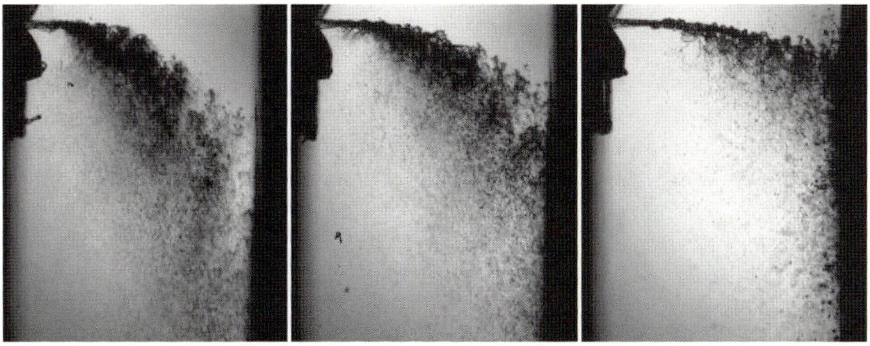

Fig. 1.6 Shadowgraphs of the impacting spray at an injection pressure of 4 bar and crossflow velocities of 50 m/s (*left*), 75 m/s (*middle*) and 100 m/s (*right*). Crossflow direction is from top to bottom

1.3.1 High-Speed Visualization of the Spray Impact Process

Shadowgraphs of spray impact under various cross-flow conditions are shown in Fig. 1.6. The first interaction between liquid and gas phase occurs when the primary liquid sheet is injected into the gas flow. In the near-nozzle region, the gas surrounds the liquid sheet. However, at greater distances from the primary nozzle the airflow destabilizes the liquid sheet. Liquid bags are first formed and then stretched by the airflow until they finally breakup up. In [46] this phenomenon is described for laminar and round liquid jets. For single drops this phenomenon is widely known as bag breakup.

Figure 1.7 depicts the influence of the air velocity on the spray impact and fragmentation process. The shadowgraphs clearly show that primary atomization of the liquid sheet is heavily influenced by the cross-flow. At $u_g = 25\ \mathrm{m/s}$ the liquid reaches the wall as a cluster of connected ligaments. At increasing gas velocities more droplets are stripped from the liquid sheet by the cross-flow. The position at which the primary liquid sheet is completely disintegrated in single fragments moves closer to the orifice. The amount of liquid that impinges on the wall decreases for increasing gas velocities since more liquid fragments are transported along with the gas flow.

The motion and breakup of the liquid wall film is also dominated by the airflow. The film forms large scale ligaments at the end of the impingement plate at low gas velocities. Higher gas velocities result in a continuous decrease of fragment sizes that emerge from the film breakup. It is obvious that the impingement plate has a strong impact on the atomization process at low gas velocities. However, if the gas velocity is increased, the influence of the spray impact process on the final airblast spray decreases. For the presented configuration, the impingement plate does not affect the fragmentation for gas velocities well above 100 m/s.

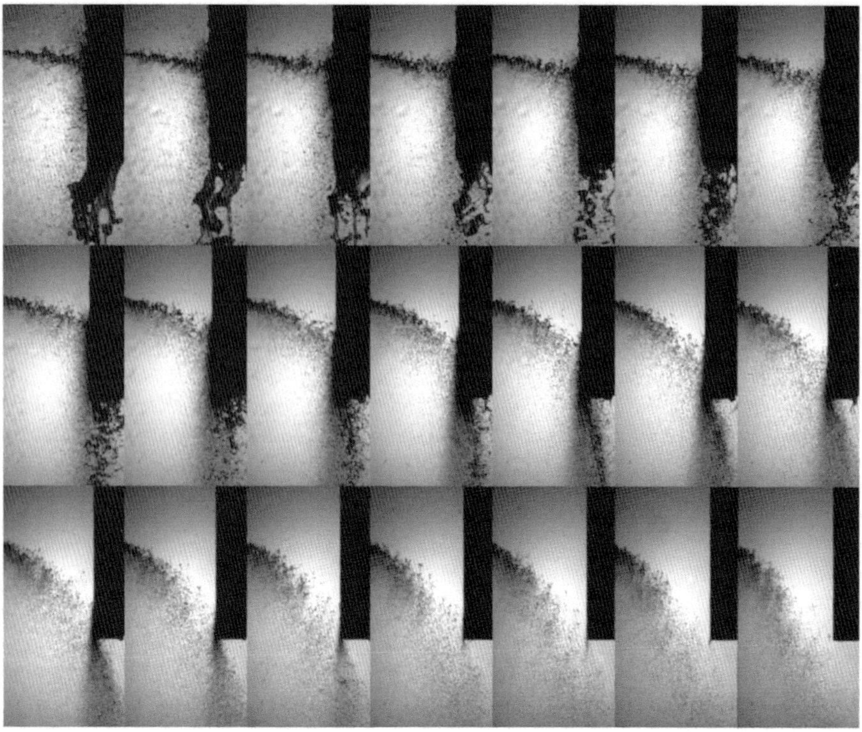

Fig. 1.7 Influence of the air velocity on the impact and fragmentation process for $u_g = 25$ m/s (*top left image*) to $u_g = 125$ m/s (*bottom right image*). The velocity increases from *left* to *right* in steps of 5 m/s. Direction of the airflow is from top to bottom

1.3.2 Characterization of the Liquid Film

A thin liquid film is formed when the pressure driven primary spray impacts on the prefilmer. The thickness of this liquid film gives the dominant length scale for drops and ligaments that originate from the film breakup. The liquid film thickness is typically in the order of $\sim$100 μm and is highly unsteady due to the impacting drops. This makes the measurement of the film thickness a challenging task.

In cooperation with the ETH Zuerich, the most sophisticated liquid film thickness sensor available is used to investigate the spray impact and its outcome. The sensor that has been developed at the Laboratory for Nuclear Energy Systems at the ETH Zuerich measures the conductivity between a pair of electrodes which is a direct measure for the liquid height above the electrodes. These electrodes are mounted flush to the surface of the sensor so that the measurement technique is non-intrusive. The distance between 2 electrode pairs is 2 mm which defines the spatial resolution technique. The sensor is equipped with an array of 64×16 electrode pairs and has

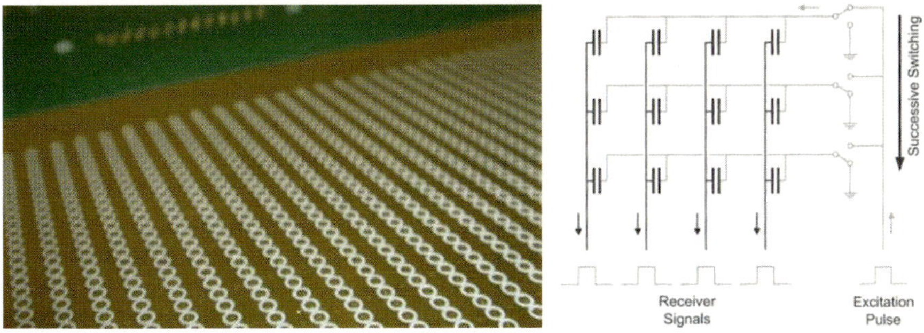

Fig. 1.8 Image (*left*) and scheme (*right*) of the liquid film thickness sensor (Source: [47])

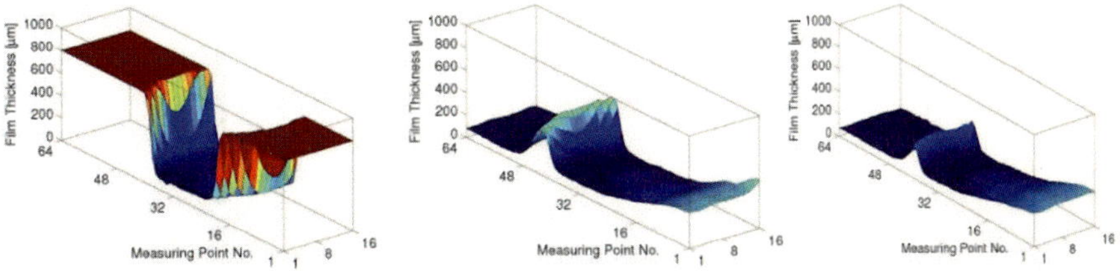

Fig. 1.9 Time-averaged results of the film thickness measurements. Inclination angles are 0°, 12° and 24° from *left* to *right*

a very high temporal resolution of 10 kHz. A detailed description of the sensor and the measurement technique can be found in [47] (Fig. 1.8).

The results of the film thickness measurements for three different inclinations of the sensor are shown in Fig. 1.9. It can be observed that the impacting spray forms a thin liquid film with a mean thickness of 100 μm in the spray impact area for $\alpha = 0°$. The impact area is enclosed by a circular hydraulic jump which increases the mean film thickness to more than 800 μm. The width of the hydraulic jump is about 4 mm.

For $\alpha = 12°$ and $\alpha = 24°$ the liquid is driven to the lower side of the sensor by gravity. The shape of the liquid film is no longer axisymmetric. A hydraulic jump occurs at the upper side of the sensor while the film thickness on the lower side gradually increases. The height of the hydraulic jump decreases with increasing inclination angles.

The results of the film thickness measurements show that the spray impact area is dominated by inertial forces. The film thickness in this area remains constant for all inclination angles. For the $\alpha = 12°$ and $\alpha = 12°$ cases the inertia dominated area is bounded by a hydraulic jump only on the upper side of the sensor.

The measurements with the electrode array sensor revealed unique data about the evolution of the liquid film that originates from an impacting spray. This data was used to obtain a deeper understanding about the complicated spray impact process and is available for the validation of numerical codes.

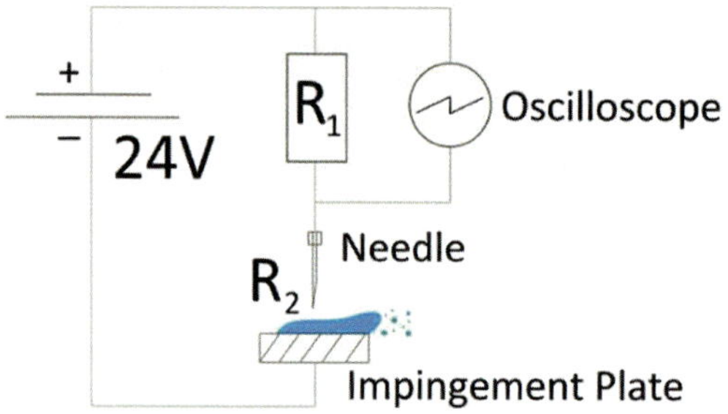

Fig. 1.10 Electrical circuit used for the film thickness measurements under crossflow conditions

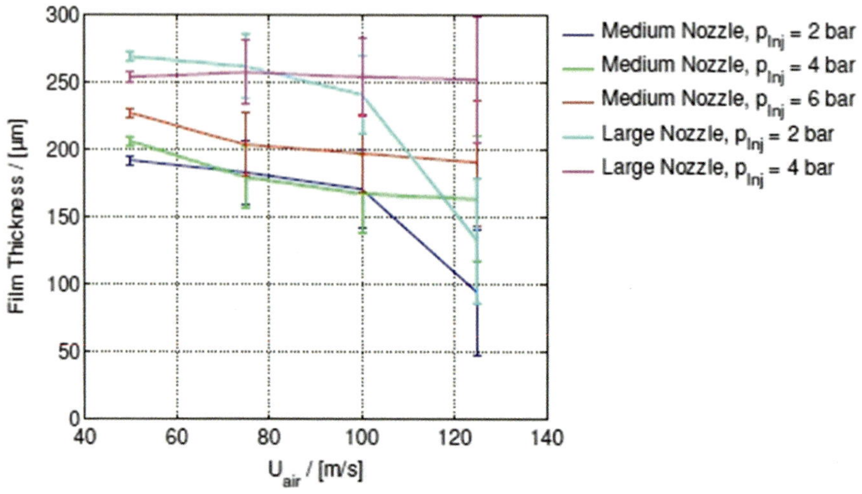

Fig. 1.11 Results of the film thickness measurements under cross-flow conditions

However, due to geometric restrictions of the previously described wind tunnel the electrode array sensor could not be used to investigate the film thickness under cross-flow conditions. Consequently, an alternative method based on the electrical conductivity of the fluid was used for this task. This method employs a single set of electrodes. One electrode is the impingement plate itself while the second electrode has the form of a needle and can be moved towards the wall in fine, well-defined increments. Once the needle dives into the liquid film on the impingement plate, the electrical circuit between the electrodes will be closed and a current or a voltage drop can be measured. For this experimental series an oscilloscope was used to measure a change in voltage. A 24 V DC power supply was used. The electrical circuit is depicted in Fig. 1.10, measurement results are presented in Fig. 1.11.

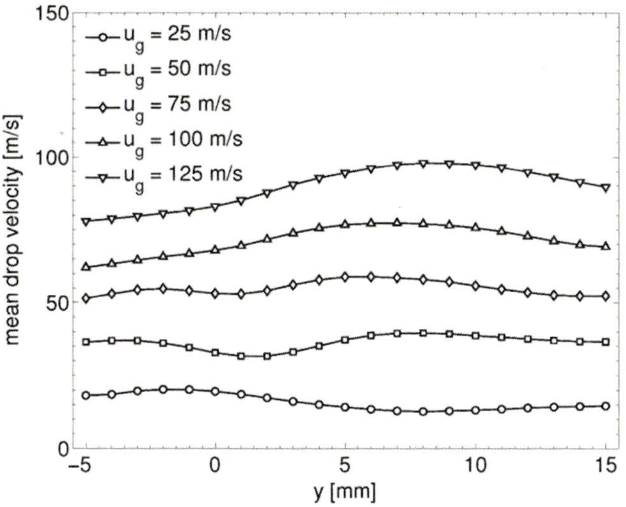

Fig. 1.12 Results of the film thickness measurements under cross-flow conditions

1.3.3 Characterization of the Final Airblast Spray

Mean drop velocities in the final airblast spray are shown in Fig. 1.12. Mean drop velocities are in the range from 60 to 80% of the air velocity. The smaller velocities around $y = 0$ can be related to the wake generated behind the impingement plate.

Figure 1.13 depicts the evolution of the Sauter mean diameter of the final airblast spray for various gas velocities and primary orifice sizes. Each data point in Fig. 1.13 represents a concentration weighted average of all Sauter mean diameters that have been measured over a profile scan from $y = -5$ mm to $y = 15$ mm in steps of 1 mm.

The Sauter mean diameters tend to decrease for higher air velocities, as one would expect. This is due to higher aerodynamic forces to which ligaments and drops are exposed. This leads to a decrease in the maximum stable drop size, which strongly affects the Sauter mean diameters. Interestingly, the differences in mean diameters for the investigated orifice sizes become very small for high gas velocities. This shows how the influence of the primary liquid shape becomes less important when aerodynamic breakup is the most important fragmentation mechanism.

When the drop is heavily accelerated during the breakup process the relative velocity can drop below a critical value so that capillary forces are able to stabilize the drop. Since smaller drops are accelerated faster and the influence of surface tension increases for small diameters the fragmentation of smaller drops is harder to achieve. Considering that, aerodynamic forces play a somehow ambiguous role in the fragmentation process. They are necessary to provide the forces needed for

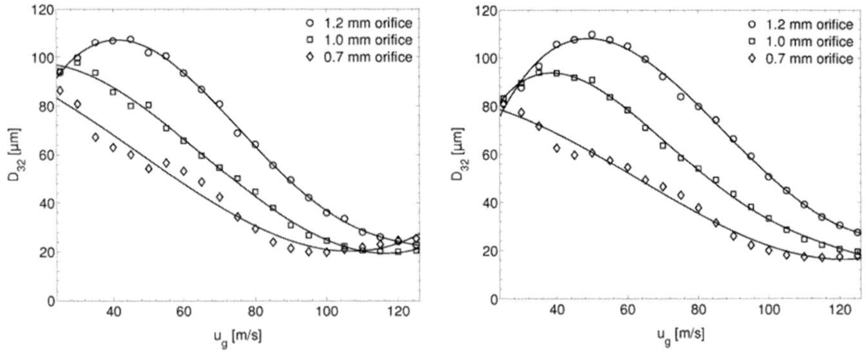

Fig. 1.13 Dependence of the Sauter mean diameter of the final airblast spray on air velocities for various injection pressures and orifice sizes. *Symbols* are experimental data points, *solid lines* are polynomial fits of the data

Fig. 1.14 Shadowgraphs of the final airblast spray at air velocities of 25, 75 and 100 m/s (from *left* to *right*)

fragmentation of the liquid. At the same time the liquid is accelerated by these forces, which reduces the relative velocity and thus the aerodynamic load.

However, for the larger orifice sizes and low gas velocities an opposite trend can be observed. The Sauter mean diameter first increases until it reaches a maximum at gas velocities between 40 and 50 m/s, then decreases. This effect can be related to the presence of large nonspherical ligaments in the airblast spray as depicted in Fig. 1.14. The existence of large ligaments is favored by large orifice sizes and high liquid flow rates (high injection pressures). Since the phase Doppler instrument is unable to detect these fragments, the Sauter mean diameter of the airblast spray is underestimated in that region. Higher air velocities result in an higher aerodynamic load on these ligaments and the maximum stable size is reduced. A greater amount of relatively large ligaments is then dominated by surface tension and adapts to a spherical shape. These drops can now be detected by the phase Doppler instrument which yields an increase of the Sauter mean diameter. For gas velocities above 60 m/s no nonspherical ligaments can be observed in the final airblast spray.

The droplet size distributions of the airblast spray for various cross-flow velocities are depicted in Fig. 1.15. One can easily recognize that larger drops are unstable

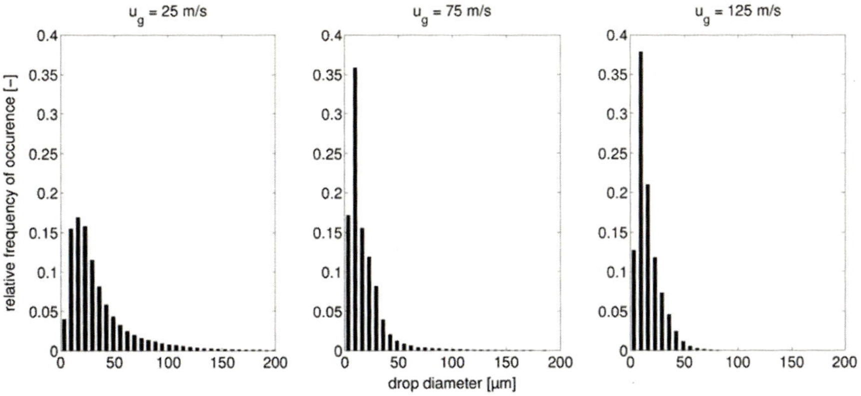

Fig. 1.15 Droplet size distributions of the final airblast spray for various air velocities. The injection pressure is 4 bar, 1.0 mm orifice

at higher air velocities and no longer exist. It is interesting to note that the maxima of all drop size distributions are at about 20 μm. The decrease of mean diameters for higher gas velocities can therefore be related to the absence of large drops while small drops in the range from 5 μm to 20 μm can be found at all operating points.

1.4 Modeling

The atomization of liquids in airblast atomizers is governed by the phenomena that occur at very small time and length scales. The breakup process consists of several main phenomena, such as primary atomization, spray impact on liquid films and aerodynamic breakup. Each of these phenomena involves various mechanisms which are not fully understood up to now. This fact makes the experimental, numerical and theoretical investigation of airblast atomization (as most of the atomization problems) a challenging task.

In this situation it is very important to identify the leading mechanism which determines size and velocity of the majority of the generated drops in the final spray. For this purpose a numerical code has been developed, which is based on a Lagrangian description of drop trajectories and accounts for drop aerodynamic breakup. It consists of three main parts: description of spray transport with drop aerodynamic breakup; description of the initial conditions, namely primary atomization of the liquid sheet; description of the boundary conditions at the wall, namely spray generated at the wall due to spray collision.

1.4.1 Description of Spray Transport Accounting for Aerodynamic Drop Breakup

The code calculates the trajectories of the drops in the spray, predicts the cascade of aerodynamic drop breakup and estimates the distribution of the drop velocities and diameters in the final spray.

The air velocity is assumed uniform and constant. This assumption is valid for the case of airblast atomization since the air-to-liquid mass ratio is very high and the influence of the spray on the velocity field in the gas region can be negligibly small. The spray transport is described in a simplified form. The motion of each drop (which is assumed spherical of diameter D) between its formation and its aerodynamic breakup is calculated using an ordinary differential equation

$$\ddot{x} = \frac{3C_d \varrho_g}{4D\varrho_l}\left(u_g - u_l\right)\left|u_g - u_l\right|$$

where $\boldsymbol{x}$ is the instantaneous position vector of the drop center-of-mass. The drag coefficient C_d is a well-known for a sphere function of the instantaneous Reynolds number, based on the relative velocity of the drop and the surrounding air flow and on the viscosity of the gas [48]. The expressions for the duration of the drop motion and for the average diameter of the fragments are taken from the existing in the literature empirical models. The breakup leads to the appearance of secondary drops. The histories (trajectory and consequent breakup) of each of the secondary drop are calculated independently. The initial velocities of the fragments are assumed equal to the velocity of the "parent" drop at the instant of breakup.

The child droplet size distributions that originate from the breakup of parent drops or ligaments are assumed to fulfill a Gamma distribution, as proposed in [49]. The probability density function of the drop diameters is

$$p\left(\xi\right) = n^n \xi^{n-1}\frac{e^{-n\xi}}{\Gamma(\mathrm{n})}, \quad \xi = \frac{D}{D_{10}}$$

where $n = 4$. It can be shown that for such distribution $D_{10} = \frac{2}{3}D_{32}$.

In order to generate such a distribution numerically a set of random numbers R uniformly distributed between 0 and 1 are first generated. The diameter of drops corresponding to the number R is a root of equation

$$R = 1 - \int_0^{\xi} p\left(\xi\right) d\xi$$

This leads to the transcendent equation

$$\frac{32\xi^3 + 24\xi^2 + 12\xi + 3}{3} \exp\left(-4\xi\right) - R = 0,$$

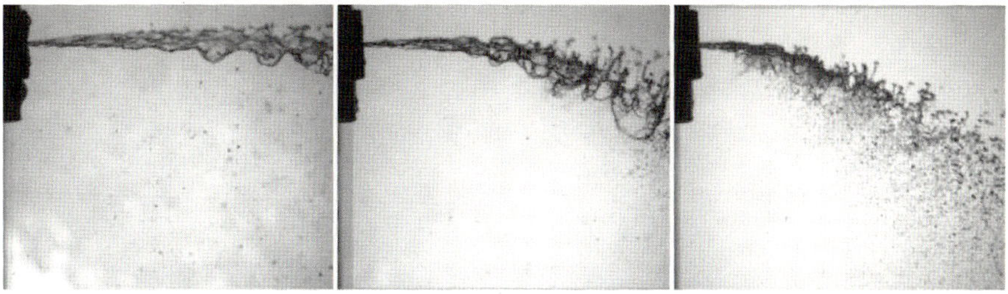

Fig. 1.16 Primary breakup of a liquid sheet under crossflow conditions without wall impact. Air velocities are 10, 25 and 50 m/s from *left* to *right*. Direction of the airflow is downwards

which can be solved numerically for each generated random number R. The corresponding diameters of secondary drops can then be calculated for a known value of D_{10}.

In order to provide reliable statistics for the drop distributions the number of numerically generated drops has to be much higher than the real number of the secondary drops formed by breakup. Therefore, each computed droplet is marked by a "number factor" c that ensures mass conservation during the fragmentation process

$$c_{frag} = \frac{c_{parent} D^3_{parent}}{\sum D^3_{i,frag}}.$$

Introduction of the "number factor" c for each generated drop is rather important in order to calculate the correct distribution of drops obtained by breakup of different parent drops.

The numerical code describes the cascade of aerodynamic drop breakups until the Weber number of the drops reduces to the threshold value $We = 12$.

1.4.2 Initial Conditions: Primary Breakup

The breakup of the liquid sheet without wall impact is depicted in Fig. 1.16. It can be observed that initially small disturbances grow and finally lead to fragmentation of the bulk liquid. In order to describe this breakup process theoretically a Rayleigh-Taylor linear stability analysis is conducted for the liquid sheet that emerges from the flat fan nozzle [50]. This liquid sheet is exposed to the cross-flow. The side of the liquid sheet facing the cross-flow is subjected to the stagnation pressure of the gas flow, whereas the pressure at the opposite side of the liquid film is given by the static pressure.

The Rayleigh-Taylor stability analysis yields a correlation for the mean breakup length, which is given by

$$L_{break} = \frac{U_{liquid}}{\omega^*} ln \left[\frac{h_0 R_0}{\delta_0 (R_0 + L_{break})} \right].$$

Then, the average diameter of the fragment can roughly be estimated from the volume balance:

$$D_{drop} = \left[\frac{6}{\pi} \pi \left(\frac{x h_{break}}{K^*} \right)^2 h_{break} \right]^{1/3} \approx 2.84 K^{*-\frac{1}{3}} h_{break}.$$

1.4.3 Boundary Conditions: Wall Impact and Wall Film Formation

In the next step the trajectory of each fragment is computed from the equation of motion. It is checked whether a drop impacts on the impingement plate or whether it is transported downstream with the cross-flow without impact on the substrate. All impinging drops contribute to the formation of a liquid film on the wall that is shear driven by the airflow.

The motion inside the liquid film is described by the x-component of the Navier–Stokes equation for incompressible flows

$$\varrho \frac{Du_l}{Dt} = \varrho k_x - \frac{\partial p}{\partial x} + \mu \frac{\partial^2 u_l}{\partial x_i \partial x_i}.$$

For quasi-steady thin liquid films it can be assumed that the flow is dominated by viscous forces and that the pressure gradient $\partial p / \partial x$ is negligible. Together with the assumption $\partial^2 u_l / \partial x^2 \ll \partial^2 u_l / \partial y^2$ this yields

$$0 = \frac{\partial^2 u_l}{\partial y^2}$$

Assuming that the liquid film flow is considerably slower than the airflow on its surface the Blasius equation for turbulent boundary layers [48] defines the boundary condition at the liquid/gas interface

$$\tau_{int} = \frac{\nu_g^{1/5} \varrho_g u_g^{9/5}}{x^{1/5}}.$$

The second boundary condition is defined by the no-slip condition at the solid/liquid interface $u_l\ (y = 0) = 0$. The integration of the simplified Navier–Stokes-Equation yields the following expression for the liquid film thickness on the wall

$$h_0 = \sqrt{2\frac{\dot{V}_l \mu_l}{b\tau_{int}}},$$

where b ist the distance between nozzle and substrate. The liquid wall film breaks up at the atomizing edge at the end of the impingement plate. Droplets that are formed from the film are assumed to have a Sauter mean diameter that is in the order of the film thickness h_0. This mean diameter is then used to generate a droplet size distribution as described above.

1.5 Results and Discussion

The procedure of aerodynamic breakup is further applied to all drops that have been created from primary breakup, aerodynamic breakup of parent drops and breakup of the liquid wall film. This is done until all remaining droplets in the spray are stable. This can either be achieved by reduction of diameters through breakup processes or by acceleration by the gas flow which leads to decreasing relative velocities.

The model has been implemented as a set of ordinary differential equations which are solved numerically. For all computations presented here the number of numerically generated secondary droplets for one breakup event is 500 for primary breakup and liquid wall film breakup and 150 for aerodynamic breakup of single fragments. At an air velocity of 125 m/s these settings result in an airblast spray that consists of about 60,000 single droplets.

Figure 1.17 depicts a comparison between the experimental results and the predictions of the model. The experimental results are averaged over all measurement positions for each gas velocity. The proposed model is able to predict the experimental results well. This indicates that the most important breakup phenomena are captured correctly by the model.

In some cases for gas velocities smaller than ~50 m/s the model overpredicts the values of the Sauter mean diameter. As described above, the Sauter mean diameter in this region of gas velocities is larger than the one that has been measured. At the same time, the accuracy of the model decreases for low gas velocities. There are two reasons for this behavior: The model operates with random numbers. Since less breakup events occur at low gas velocities the final airblast spray consists of less droplets. This decreases the statistical confidence level. The submodel for the motion and breakup of the liquid wall film assumes that shear stress at the gas/liquid interface is the dominant force. This might not be true for low gas velocities since inertial forces have to be taken into account. Moreover, at smaller air velocity large non-spherical fragments (as seen in Fig. 1.14a) are rejected by the algorithms used in the phase-Doppler system, which makes the experimental results less reliable.

At higher air velocities the agreement is rather good, which indicates that the atomization mechanisms considered in the model, mainly the secondary drop breakup, are dominant factors in airblast atomitzation process.

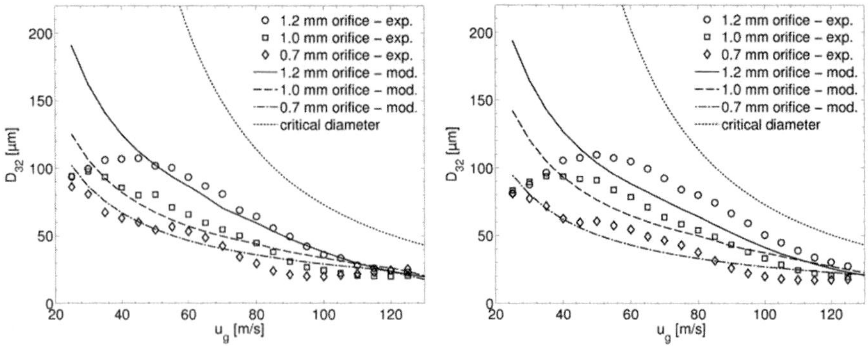

Fig. 1.17 Comparison between experimental results (*symbols*) for the Sauter mean diameter and predictions by the model (*lines*). The *dotted line* shows the evolution of the maximum stable diameter for $We_{crit} = 12$

1.6 Conclusions

The spray formed by an MTU airblast atomizer was characterized under engine conditions in terms of ambient pressure. Droplet Sizes and velocities as well as oscillation frequencies are measured for various operating parameters. A simple model for prediction of the frequencies is proposed.

Then, the impact of a flat fan spray on a wall under cross-flow conditions is investigated being a simple abstraction of an airblast atomizer with impingement plate. The spray impingement on the wall is visualized by a high-speed video system. The breakup phenomena responsible for the evolution of a secondary spray are identified. These are: primary atomization of the liquid sheet that emerges from the flat fan nozzle, subsequent breakup of ligaments and drops due to aerodynamic forces and formation/fragmentation of a liquid wall film. The final airblast spray is then characterized by a phase Doppler instrument for various orifice sizes, injection pressures and cross-flow conditions. It is shown that secondary atomization due to aerodynamic forces is the dominant breakup mechanism at high gas velocities.

A model is developed to predict the mean diameters of the airblast spray. The model consists of several submodels for primary atomization, spray impact and wall film formation/fragmentation and the subsequent breakups of drops due to aerodynamic forces. The proposed model is able to predict the outcome of the fragmentation process for gas velocities greater than ~50 m/s with good accuracy.

The current model provides some advantages over purely empirical correlations. It can easily be enhanced or adapted to different applications. Since it calculates the diameter and velocity of single droplets, the incorporation of an evaporation model should be straightforward. However, to date, the proposed model is only validated for experiments under ambient conditions. The performance of the model under gas turbine conditions still has to be investigated.

1.7 Comparison to Studies Not Related to the SFB 568

While very little literature exists about MTU-type airblast atomizers with spray impact on the prefilmer, a much greater amount of studies has been published about the liquid breakup in standard prefilming airblast atomizers without spray impact inside the nozzle.

A very recent investigation of instabilities that finally lead to the fragmentation of aerodynamically driven annular liquid film can be found in [51]. The authors describe the occurrence of the well-known free shear layer instability and an additional rupturing instability.

In [52] an image analysis algorithm has been applied to determine drop and ligament sizes of the airblast spray very close to the end of the prefilming edge. The mean liquid film thickness on the prefilmer has been measured with an optical sensor, applicable only to smooth films. Their data can be potentially used for improvement of our model for airblast atomization.

Acknowledgements The authors acknowledge the financial support from the German Research Council (DFG) through the SFB568.

References

Project-Related Publications

1. Batarseh, F., Gnirß, M., Roisman, I.V., Tropea, C.: Fluctuations of a spray generated by an airblast atomizer. Exp. Fluids **46**, 1081–1091 (2009)
2. Chrigui, M., Roisman, I.V., Batarseh, F.Z., Sadiki, A., Tropea, C.: Spray generated by an airblast atomizer under elevated ambient pressures. J. Propuls. Power **26**, 1170–1183 (2009)
3. Chrigui, M., Sadiki, A., Batarseh, F., Janika, J., Tropea, C.: Numerical and experimental study of spray produced by an asirblast atomizer under elevated pressure conditions. Proceedings of ASME Turbo Expo 2008: Power for Land, Sea and Air. June 9–13, 2008, Berlin, Germany (2008)
4. Roisman, I.V., Batarseh, F.Z., Tropea, C.: Chaotic disintegration of a liquid wall film: a model of an air-blast atomization. Atomiz. Sprays **20**(10), 837–845 (2010)
5. Roisman, I.V., Batarseh, F.Z., Tropea, C.: Characterization of a spray generated by an airblast atomizer with prefilmer. Atomiz. Sprays **20**(10), 887–903 (2010)
6. Opfer, L., Roisman, I.V., Tropea, C.: High speed visualization of drop and spray impact on rigid walls with cross-flow, poster. In: International Conference on Multiphase Flows, Tampa, USA (2010)
7. Opfer, L., Roisman, I.V., Tropea C.: Spray impact on walls with cross-flow, poster. Workshop on Near Wall Reactive Flows, Seeheim, Germany (2010)
8. Opfer, L., Roisman, I.V., Tropea, C.: Spray Impact on Walls with Cross-flow: Experiments and Modeling. ILASS Europe, Estoril (2011)
9. Opfer, L., Roisman, I.V., Tropea, C.: Laboratory simulations of an airblast atomization: main mechanisms of liquid disintegration and spray characteristics, Exp. Fluids, submitted, March 2012.

Other Publications

10. Lefebvre, A.: Atomization and Sprays. Hemisphere Publishing Corporation, New York (1989)
11. Nukiyama, S., Tanasawa, Y.: Experiments on the atomization of liquids in an airstream. Trans. Soc. Mech. Eng. Jpn. **5**, 62–75 (1939)
12. Lorenzetto, G.E., Lefebvre, A.H.: Measurements of drop size on a plain jet airblast atomizer. AIAA J. **5**, 62–75 (1939)
13. Jasuja, A.K.: Plain-jet airblast atomization of alternative liquid petroleum fuels under high ambient air pressure conditions. ASME Paper 82-GT-32 (1982)
14. Rizk, N.K., Lefebvre, A.H.: Spray characteristics of plain-jet airblast atomizers. J. Eng. Gas Turbines Power **106**(3), 634–638 (1984)
15. Issac, K., Missoum, A., Drallmeier, J., Johnston, A.: Atomization experiments in a coaxial co-flowing mach 1.5 flow. AIAA J. **32**(8), 1640–1646 (1994)
16. Hede, P.D., Bach, P., Jensen, A.D.: Two-fluid spray atomization and pneumatic nozzles for fluid bed coating/agglomeration purposes: a review. Chem. Eng. Sci. **63**(14), 3821–3842 (2008)
17. Lasheras, J.C., Hopfinger, E.J.: Liquid jet instability and atomization in a coaxial gas stream. Ann. Rev. Fluid Mech. **32**(1), 275–308 (2000)
18. Marmottant, P., Villermaux, E.: On spray formation. J. Fluid Mech. **32**, 73–111 (2003)
19. Varga, C.M., Lasheras, J.C., Hopfinger, E.J.: Initial breakup of a small-diameter liquid jet by a high-speed gas stream. J. Fluid Mech. **497**, 405–434 (2003)
20. Aliseda, A., Hopfinger, E.J., Lasheras, J.C., Kremer, D.M., Berchielli, A., Connolly, E.K.: Atomization of viscous and non-Newtonian liquids by a coaxial, high-speed gas jet. Experiments and droplet size modelling. Int. J. Multiphase Flow **34**(2), 161–175 (2008)
21. O'Rourke P.J., Amsden A.A.: The tab method for numerical calculation of spray droplet breakup. SAE Technical Paper 872089 (1987)
22. Tanner, F.X.: Development and validation of a cascade atomization and drop breakup model for high-velocity dense sprays. Atomiz. Sprays **14**(3), 211–242 (2004)
23. Entov, V.M., Yarin, A.L.: Dynamical equations for a liquid jet. Fluid Dyn. **15**(5), 644–649 (1984)
24. Entov, V.M., Yarin, A.L.: The dynamics of thin liquid jets in air. J. Fluid Mech. **140**, 91–111 (1984)
25. Faeth, G.M., Hsiang, L.P., Wu, P.K.: Structure and breakup properties of sprays. Int. J. Multiphase Flow **21**(Supplement), 99–127 (1995)
26. Stahl, M., Damaschke, N., Tropea C.: Experimental investigation of turbulence and cavitation inside a pressure atomizer and optical characterization of the generated spray. In: 10th ICLASS Conference, Kyoto (2006)
27. Dai, Z., Chou, W.H., Faeth, G.M.: Drop formation due to turbulent primary breakup at the free surface of plane wall jets. Phys. Fluids **10**(5), 1147–1157 (1998)
28. Rein, M.: Turbulent open-channel flows: drop-generation and self-aeration. J. Hydraul. Eng. **124**(1), 670–675 (1999)
29. Desjardins, O., Moureau, V., Knudsen, E., Herrmann, M., Pitsch, H.: Conservative Level Set/ghost Fluid Method for Simulating Primary Atomization. ILASS Americas, Toronto (2007)
30. Yarin, A.L.: Free Liquid Jets and Films: Hydrodynamics and Rheology. Longman/Wiley, Harlow/New York (1993)
31. Villermaux, E., Marmottant, P., Duplat, J.: Ligament-mediated spray formation. Phys. Rev. Lett. **92**(7), 074501 (2004)
32. Kolmogorov, A.N.: On the log-normal distribution of particles sizes during breakup process. Dokl. Akad. Nauk. SSSR, pp. 99–101 (1941)
33. Gorochovski, M., Saveliev, V.: Further analyses of Kolmogorov's model of breakup. Phys. Fluids **15**, 184–192 (2003)
34. Hsiang, L.P., Faeth, G.M.: Near-limit drop deformation and secondary breakup. Int. J. Multiphase Flow **18**(5), 635–652 (1992)

35. Guildenbecher, D.R., Lopez-Rivera, C., Sojka, P.E.: Secondary atomization. Exp. Fluids **46**(3), 371–402 (2009)
36. Schmehl R.: Modeling droplet breakup in complex two-phase flows. In: ICLASS Conference, Sorento, Italy (2003)
37. Hinze, J.O.: Fundamentals of the hydrodynamic mechanism of splitting in dispersion processes. AIChE J. **1**(3), 289–295 (1955)
38. Hsiang, L.P., Faeth, G.M.: Drop deformation and breakup due to shock wave and steady disturbances. Int. J. Multiphase Flow **21**(4), 545–560 (1995)
39. Ranger, A.A., Nicholls, J.A.: The aerodynamic shattering of liquid drops. AIAA **7**, 285 (1969)
40. Liu, Z., Reitz, R.D.: An analysis of the distortion and breakup mechanisms of high speed liquid drops. Int. J. Multiphase Flow **23**(4), 631–650 (1997)
41. Snyder, H.E., Reitz, R.D.: Direct droplet production from a liquid film: a new gas-assisted atomization mechanism. J. Fluid Mech. **375**, 363–81 (1998)
42. Lee, C.H., Reitz, R.D.: An experimental study of the effect of gas density on the distortion and breakup mechanism of drops in high speed gas stream. Int. J. Multiphase Flow **26**(2), 229–244 (2000)
43. Hwang, S., Liu, S., Reitz, R.D.: Breakup mechanisms and drag coefficients of high speed vaporizing drops. Atomiz. Sprays **6**(3), 353–376 (1996)
44. Wert, K.: A rationally-based correlation of mean fragment size for drop secondary breakup. Int. J. Multiphase Flow **21**(6), 1063–1071 (1995)
45. Pilch, M., Erdman, C.: Use of breakup time data and velocity history data to predict the maximum size of stable fragments for acceleration-induced breakup of a liquid drop. Int. J. Multiphase Flow **13**(6), 741–757 (1987)
46. Ng, C.L., Sankarakrishnan, R., Sallam, K.A.: Bag breakup of nonturbulent liquid jets in crossflow. Int. J. Multiphase Flow **34**(3), 241–259 (2008)
47. Damsohn, M., Prasser, H.: High-speed liquid film sensor for two-phase flows with high spatial resolution based on electrical conductance. Flow Meas. Instrum. **20**(1), 1–14 (2009)
48. Schlichting, H., Gersten, K., Krause, E., Oertel Jr., H.: Grenzschicht-Theorie, 10th edn. Springer, Berlin Heidelberg (2006)
49. Villermaux, E., Bossa, B.: Single-drop fragmentation determines size distribution of raindrops. Nat. Phys. **5**(9), 697–702 (2009)
50. Chandrasekhar, S.: Hydrodynamik and Hydromagnetic Stability. Dover Publications, New York (1981)
51. Duke, D., Honnery, D., Soria, J.: Experimental investigation of nonlinear instabilities in annular liquid sheets. J. Fluid Mech. **691**, 594–604 (2012)
52. Gepperth, S., Guildenbecher, D., Koch, R., Bauer, H.-J.: Pre-filming Primary Atomization: Experiments and Modeling. ILASS Europe, Brno (2010)

Chapter 2
Experimental and Numerical Investigation of Shear-Driven Film Flow and Film Evaporation

P. Stephan, T. Gambaryan-Roisman, M. Budakli, and J.R. Marati

Abstract Shear-driven liquid film flows can occur in several locations of fuel preparation systems, e.g. inside air-driven atomizers or in Lean Pre-mixing Pre-vaporizing (LPP) combustion chambers of modern gas turbines. In LPP chambers the liquid fuel is primary atomized by a pressure nozzle and sprayed onto a pre-filmer. Fine fuel droplets accumulate at the pre-filmer surface and form a thin liquid film driven by hot compressed air to the inlet section of the combustion chamber. While the thin liquid film is accelerating along the wall, it evaporates and mixes with the hot air. The turbulent air flow induces strong shear forces at the air-liquid interface leading to a destabilization of the liquid film and the development of waves. The hydrodynamics of the wavy film flow govern the heat and mass transport and, hence, the entire fuel preparation process. Hydrodynamics and heat and mass transport strongly depend on the microstructure of the pre-filmer wall surface. In this work, the fundamentals of gravity-driven as well as air-driven film flow and evaporation on unstructured and microstructured wall surfaces have been investigated experimentally and numerically. It has been shown that longitudinal microgrooves have a stabilizing effect on the film flow. Flow regimes leading to a strong increase of evaporation efficiency have been identified. Local film thickness distributions have been measured using high-speed shadowgraphy. Wall temperature distributions have been measured using embedded thermocouples. The measurements have been performed for film Reynolds numbers varying from 225 to 650, for gas Reynolds numbers varying from 10^4 to $7 \cdot 10^4$, and for wall heat fluxes up to 40 W/cm^2. High-speed infrared images have been recorded to visualize local film

P. Stephan (✉) • T. Gambaryan-Roisman • M. Budakli • J.R. Marati
Institut für Technische Thermodynamik, Technische Universität Darmstadt, Petersenstr. 17, 64287 Darmstadt, Germany

P. Stephan • T. Gambaryan-Roisman
Center of Smart Interface, Technische Universität Darmstadt, Petersenstr. 17, 64287 Darmstadt, Germany
e-mail: pstephan@ttd.tu-darmstadt.de

J. Janicka et al., *Flow and Combustion in Advanced Gas Turbine Combustors*, Fluid Mechanics and Its Applications 102, DOI 10.1007/978-94-007-5320-4_2,

break-up and rewetting. Corresponding numerical studies of the gas–liquid flow and heat transfer along a heated wall have been conducted using Computational Fluid Dynamics (CFD). In order to track the moving gas–liquid interface, the volume of fluid (VOF) method has been adopted. Parametric numerical studies have been performed and compared with experimental data.

Keywords Shear-driven thin liquid films • Wavy film flow • Film rupture • CFD • VOF

Nomenclature

A	cross-sectional area [m^2]
a, c	channel geometrical parameters (Fig. 2.6) [m]
d	tube diameter [m]
H	channel width [mm]
h	film thickness [μm]
$\dot{M}$	mass flow rate [kg/s]
n	number of images [–]
p	pressure [bar]
$\dot{q}_{\mathrm{W}}$	wall heat flux [W/cm^2]
t_{W}	wall temperature [°C]
$t_{\mathrm{W,mean}}$	mean wall temperature [°C]
$t_{\mathrm{L,In}}$	liquid inlet temperature [°C]
x, y, z	x, y- and z-axis [m]
Re	Reynolds number [–]

Subscripts

G	gas
Im	image
Int	interface
L	liquid
W	wall

Greek Characters

α	liquid volume fraction [–]
τ	time [s]
τ_{Int}	shear stress [N/m^2]
λ	thermal conductivity [W/(m·K)]]
μ	dynamic viscosity [kg/(m·s)]

2.1 Introduction

Shear-driven liquid film flows can occur in several locations of fuel preparation systems, e.g. inside air-driven atomizers or in of Lean Pre-mixing Pre-vaporizing (LPP) combustion chambers. LPP chambers are used to reduce NO_x emissions in modern gas turbine engines [11]. In LPP chambers, fuel is injected onto the hot wall of the pre-filmer where it forms a thin film which is further driven along the wall by swirled co-current hot compressed gas. Thereby the film partially evaporates and mixes with the gas flow [12]. The characteristics of the wavy thin film influence the atomization/break-up and premixing process and thus the entire fuel preparation process [13, 14].

Numerous experimental and numerical studies aim at the investigation of the parameters influencing the film hydrodynamics and transport phenomena. It has been shown that the thin liquid fuel film flow characteristics are strongly influenced by the pressure gradient and the interfacial shear forces at the gas–liquid interface [1]. Lan et al. [15] used the interferometer film thickness measurement technique to investigate three-dimensional flow characteristics of shear-driven thin films. They observed a decrease of the liquid film thickness with increasing gas velocities at constant liquid mass flow rates. Elsäßer et al. [16] conducted film thickness and velocity measurements of a shear-driven wavy liquid film. They reported a strong change in the characteristics of the film wavyness with increasing gas velocity. Asali and Hanratty [17] investigated the wave formation at high gas velocities theoretically and experimentally. They determined the dependencies of the wave length on gas velocity and on liquid mass flow rate.

Shear-driven evaporating liquid films can be found also in devices for cooling of electronic components [18]. Addressing typical process and material parameters for such applications Kabov et al. [19] found that locally heated and gas-driven liquid films are more stable than gravity-driven (falling) liquid films and thus less tending to film rupture [19].

The liquid film hydrodynamics, transport processes and evaporation can be significantly modified by the wall topography [2, 20, 21]. Several influencing mechanisms of wall microstructure on film flow, heat transfer and evaporation have been identified. Grooves normal to main flow direction of falling or shear-driven films may lead to development of vortices which intensify heat transfer through the mixing of liquid [21]. Grooves parallel to the main flow promote the rivulet flow regime, in which the evaporation rate is significantly intensified due to extremely high evaporation rate in the vicinity of contact lines bounding the rivulets [2, 22]. Apart from that, grooves are known to affect the stability characteristics of liquid film flows [2] and to induce a thermocapillary flow [3, 4]. In the case of film break-up at high heat fluxes and low mass flow rates a certain amount of liquid remains near the troughs of the wall topography. This might lead to a significant increase of the maximal heat flux which can be removed by liquid films flowing along micro-structured walls compared to unstructured walls [5].

Wavy patterns in shear-driven liquid films are very complex and their accurate measurement is a challenging task, especially for films on microstructured surfaces. The parameters that influence the growth of interfacial waves and govern the heat transport have not been identified and verified in detail. In this work, the fundamentals of wavy gravity- and shear-driven film flow and evaporation on structured surfaces are examined and presented in Sect. 2.2. In the following sections experimental and numerical results on the hydrodynamics and heat transfer in the thermally and hydrodynamically developing flow region are presented. The effects of liquid and air mass flow rates and wall heat fluxes on film thickness and on the wall temperature distribution are quantified. Liquid film rupture and rewetting on a micro-grooved surface are visualized using infrared thermography.

2.2 Fundamentals of Film Flow and Evaporation on Structured Surfaces

In this section we consider the gravity- and shear-driven liquid films on surfaces with longitudinal grooves with triangular cross section (see Fig. 2.1). If the wall is only partially wetted, as shown in Fig. 2.1, the liquid evaporation process is governed by the evaporation of an ultrathin liquid film in the transition region (or "micro region") between the macroscopic liquid meniscus and non-evaporating adsorbed film covering the apparently dry zones of the wall surface [6, 22]. Due to very small thermal resistance of the thin liquid film in this micro region, the evaporation rate in this region is extremely high. The intensive evaporation in the micro region is a major reason for the significant heat transfer intensification on structured surfaces in comparison with unstructured surfaces wetted with a continuous liquid film. In order to determine the effect of the liquid and gas flow parameters on heat transfer and evaporation from structured surfaces, a model for the description of gravity- and

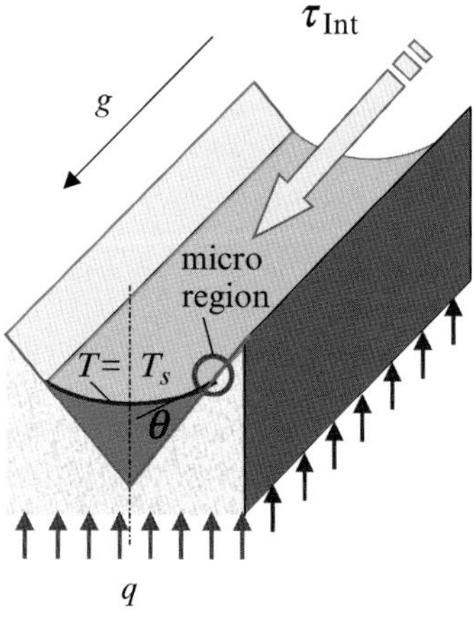

Fig. 2.1 Groove geometry and boundary conditions for modular modelling

shear-driven flow has been developed, in which the complex processes taking place in the micro region have been taken into account as a submodel and integrated into the macroscale model [2, 7].

2.2.1 Modular Modelling of Evaporation of Undisturbed Gravity- and Shear-Driven Liquid Film from structured Surfaces

The position of the undisturbed (waveless) liquid–gas interface depends on the liquid flow rate, the geometrical parameters of the groove and on the shear stress at the liquid–gas interface τ_{Int}, which depends on the gas flow along the wall.

It is assumed that the flow schematically represented in Fig. 2.1 is fully developed, laminar and stationary. In addition it is assumed that the macroscale unidirectional flow of the rivulet is governed by the constant shear stress τ_{Int} at the liquid–gas interface and by gravity. The transverse liquid flow in the micro region is governed by the surface tension and by the disjoining pressure gradient. It is assumed that the interfacial curvature in the direction of the main flow is negligible and that the microscale liquid flow in the micro region has no influence on the macroscopic flow pattern other than through the value of contact angle which depends on the hydrodynamics and transport processes in the micro region. The gas flowing along the wall is pure vapor. The temperature of the liquid–gas interface corresponds to the thermal equilibrium temperature, which is – except from the micro region – the standard saturation temperature T_s. The thermal conductivity of the vapor is assumed to be negligible in comparison with the thermal conductivity of the liquid. A constant heat flux boundary condition is applied at the back side of the wall (see Fig. 2.1). The macroscale liquid velocity distribution in the groove is determined from the Navier–Stokes equation which is reduced to the Poisson equation. The macroscale temperature field is determined from the energy equation, which is reduced to the Laplace equation. The thermohydrodynamic analysis of the micro region [6, 22] leads to functional dependencies of the local integral heat flow in the micro region and the apparent contact angle from the local wall temperature. These functional dependencies are used as sub-models in the macroscopic hydrodynamic and thermodynamic model. The resulting temperature distribution is used for determination of the heat transfer coefficient [2, 7].

Figure 2.2 shows an example of computed heat transfer coefficients in a liquid film flowing under the action of gravity and shear stress over a grooved surface. The computations have been performed for Freon-11 flowing over a copper surface with triangular grooves along the main flow direction (groove width is equal to 0.5 mm, groove angle is equal to $2\varphi = 60°$). The thickness of the copper wall is twice the groove depth. The system pressure is 0.2 MPa and the heat flux 10 kW/m^2. The heat transfer coefficient is plotted versus the mass flow rate per unit film width Γ for three different values of the shear stress. The computational results (solid lines) are compared with experimental data [20] (squares) for the case of a pure falling

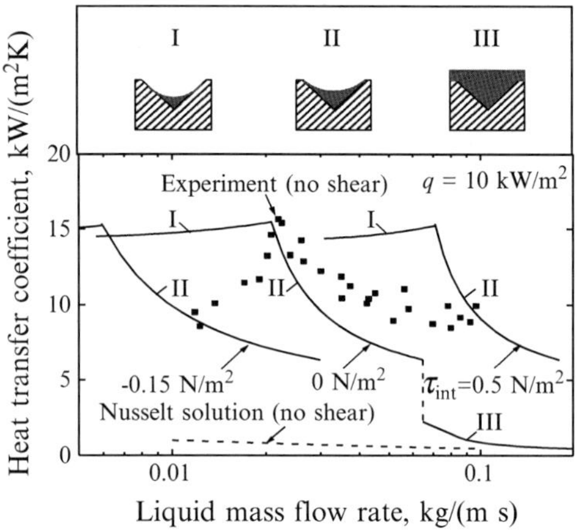

Fig. 2.2 Computed heat transfer coefficient and the experimental data from [20]

film ($\tau_{\mathrm{Int}} = 0$). The dashed line shows the result of the Nusselt theory for a film flow along an unstructured surface. The behavior of the heat transfer coefficient is governed by the wall topography, the mass flow rate of the liquid and the shear stress at the liquid–gas interface. The computed heat transfer coefficients α in the flow regimes I and II significantly exceed the values predicted by the Nusselt theory for unstructured surfaces. In these regimes the liquid mass flow rate is so small that the structure is not fully flooded by the liquid. In flow regime III, where the structure is flooded by a continuous film, the heat transfer coefficient significantly decreases and approaches the values predicted by the Nusselt theory. It can be clearly seen from the figure that the heat transfer coefficient reaches a maximal value at the transition point between the regimes I and II. At this transition point almost the entire area of the groove is covered by liquid, but rivulets are already formed which are bounded by contact lines pinned at the upper edges of the groove with the smallest possible value of the apparent contact angle. The existence of such an optimal mass flow rate at which the heat transfer coefficient reaches a maximal value is a very important result, both, from a scientific and from a practical point of view. It is clearly seen that the optimal liquid mass flow rate is a function of the shear stress at the liquid–gas interface τ_{Int}. The optimal mass flow rate increases with increasing shear stress. However, the maximal heat transfer coefficient is not a direct function of τ_{Int}. Its value depends on the wall topography and on the thermodynamic conditions.

2.2.2 *Effect of the Longitudinal Grooves on the Wavy Structure*

The hydrodynamics of a developed wavy falling and shear-driven film flow has been studied in a dedicated experimental setup. This setup was a preliminary version of the setup described in details in Sect. 2.3, however, with a test channel with a copper tube of 2 m length and 25 mm outer diameter instead of 90 mm length and 19 mm outer diameter as described in Sect. 2.3.

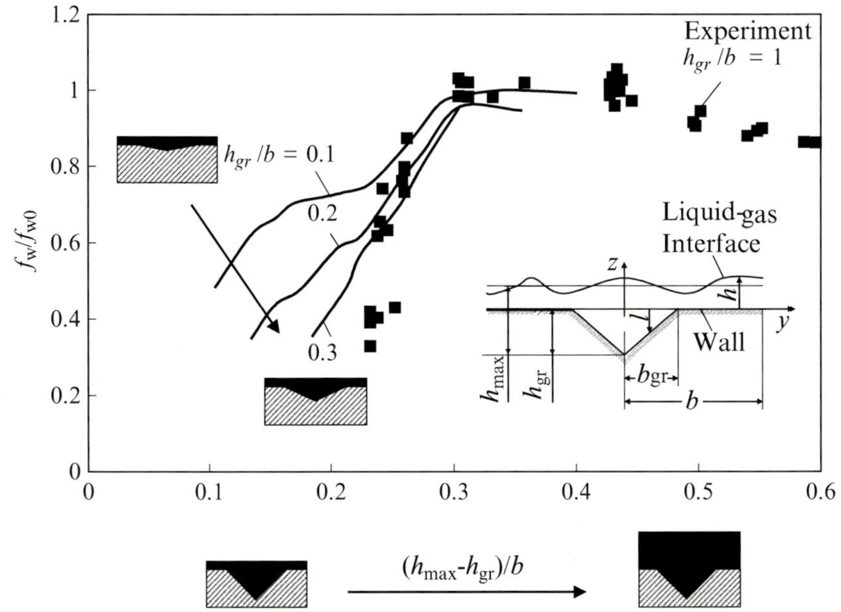

Fig. 2.3 Dominant wave frequencies: the ratio between the wave frequencies on the structured and the unstructured surface. *Solid lines*: results of stability analysis; *squares*: experimental results [8]

Applying the boundary conditions of this experimental investigation a long-wave theory has been developed for the stability analysis of continuously falling films [2] and rivulets [9] flowing over structured surfaces.

The results of the stability analysis for continuous falling films have shown that longitudinal grooves stabilize the film flow and that they tend to decrease the dominant wave frequency and wave propagation velocity. Figure 2.3 shows the effect of the wall topography on the dominant wave frequencies in falling films. The ratio between the dominant wave frequencies at the structured and the unstructured walls is plotted versus a geometrical factor describing the film thickness. The results of the stability analysis are compared with experimental data. The geometrical parameters of the wall topography are shown in an inset. Since the long-wave theory does not allow to analyze wall structures with $h_{gr}/b = 1$, which corresponds to the experimental conditions, three different simulations for $h_{gr}/b = 0.1$, 0.2 and 0.3 have been performed. It can be seen that the theoretical predictions agree with the experimental data both qualitatively and quantitatively. The simplified theory allows predicting the film thickness, at which the dominant wave frequency becomes independent from the wall structure.

2.3 Experimental Investigations

2.3.1 Experimental Setup and Control

The overall setup for the experimental investigation of the hydrodynamics and transport processes in gravity and shear-driven liquid films is schematically shown in Fig. 2.4. The main test section consists of a symmetric experimental channel being built around a copper tube (unstructured or structured) with an outer diameter of 19 mm. This test section is shown in detail in Fig. 2.5. The test liquid (deionized

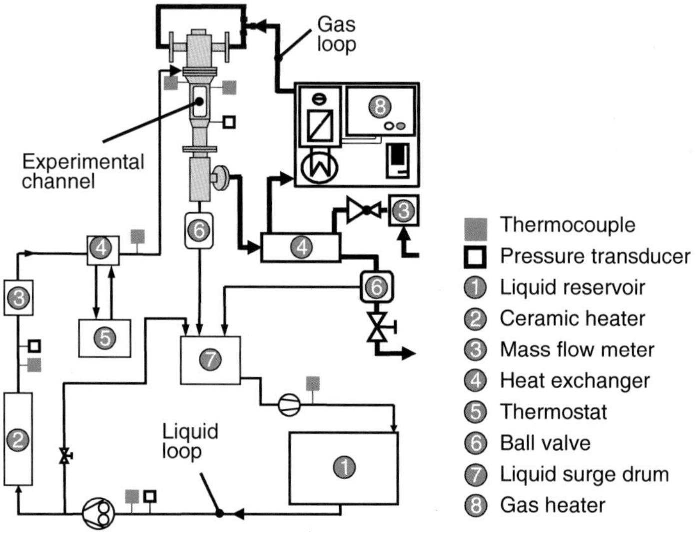

Fig. 2.4 Flow chart of experimental set-up

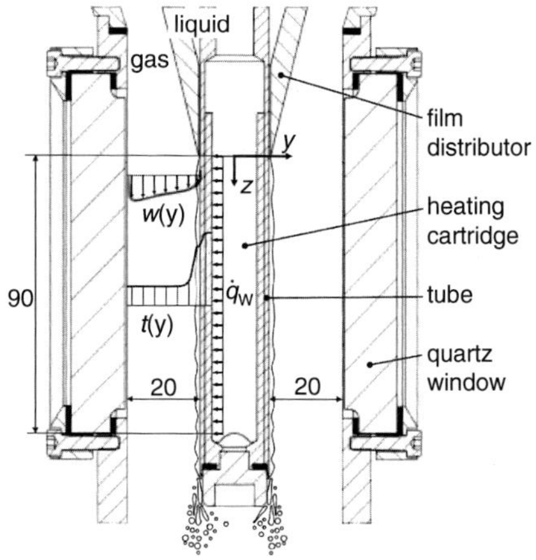

Fig. 2.5 Longitudinal cut of experimental channel

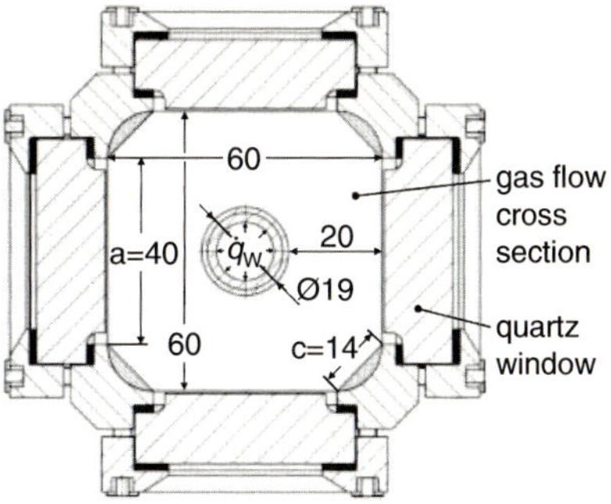

Fig. 2.6 Cross section of experimental channel

water) is circulated in a liquid loop. The liquid enters the test section and flows downwards in a ring-shaped film distributor on the outside surface of the tube. The outer circumference of the experimental channel has been made of stainless steel with 3 quartz glass windows and 1 CaF_2 window for optical observation by shadowgraphy and IR thermography, respectively (see Fig. 2.6). Compressed dried air is supplied in a gas loop. The gas flow enters the experimental channel in co-current direction with the liquid flow as shown in Fig. 2.5.

The gas mass flow rate is controlled by a thermal mass flow meter and a pressure expansion valve. An electrical heater (8) is used to heat up the gas flow to a maximum temperature of $t_G = 450°C$ before entering the test channel. The gas temperature is measured in the gas flow field at the channel inlet, the gas pressure at the channel exit. The liquid is supplied from a reservoir (1) by a frequency controlled variable-speed gear pump. The liquid mass flow rate is measured using a Coriolis mass flow meter. Its inlet temperature is controlled and kept constant using a plate heat exchanger (4) and a high performance thermostat (5). The film distributor creates an annular film with an initial thickness of 425 ± 20 μm distributed homogeneously on the outer surface of the tube that is accelerated by the co-current gas flow further downstream. The liquid inlet temperature is measured using a thermocouple inside the film distributor. In order to quantify the axial wall temperature distribution inside the tube, four thermocouples have been embedded into grooves along the tube at different positions (see Fig. 2.7). Highly conductive soldering material and silver paste were used to minimize the thermal resistances at the embedded thermocouples. The copper tube is heated internally by a cartridge heater with constant homogenous heat flux along a tube length of 90 mm starting from the liquid inlet. A maximum wall heat flux of 45 W/cm^2 $\pm$ 0.1 W/cm^2 can be supplied.

To study the effect of surface topography on hydrodynamics and heat transfer, different tubes with unstructured or structures surfaces can be mounted. The results presented in Sect. 2.3.4 were obtained with a tube with 70 longitudinal micro-grooves having triangular cross section (see Fig. 2.8). Each micro-groove has a depth of 500 ± 40 μm and a mean apex angle of 63°.

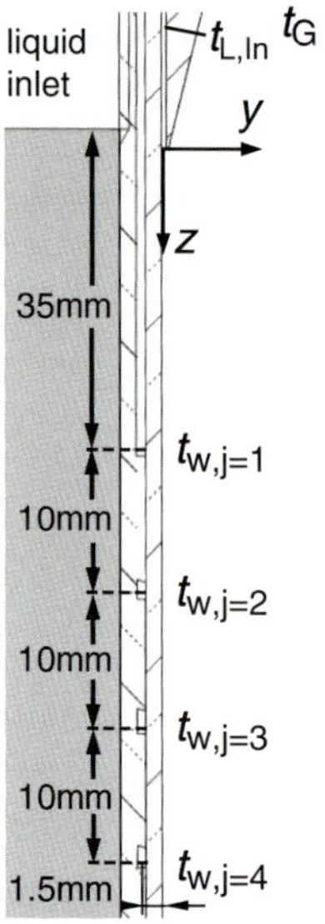

Fig. 2.7 Position of thermocouples

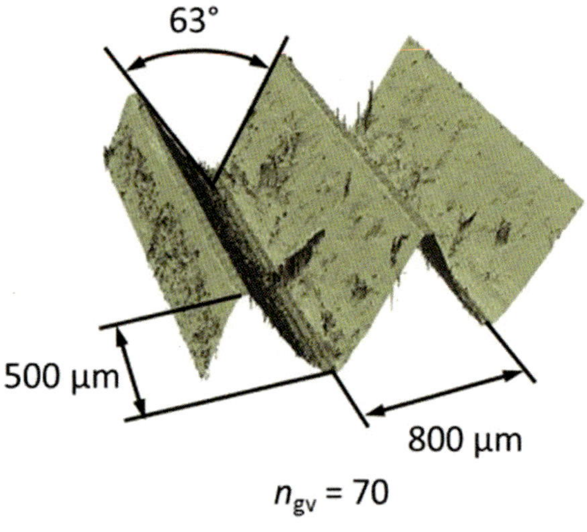

Fig. 2.8 Microscopic image of the micro-structured surface

Table 2.1 Uncertainties of property and process parameter measurements

t	p	$\Delta \dot{M}_G$	$\Delta \dot{M}_L$
±0.3 K	±1% of reading	±1.5% of measurement range	±0.15% of reading

Temperatures, pressures and mass flow rates of the gas and liquid stream were recorded at a rate of 1,000 Hz using an 18-bit National Instruments Data Acquisition System. The uncertainties of the measurements over the full measurement range are listed in Table 2.1.

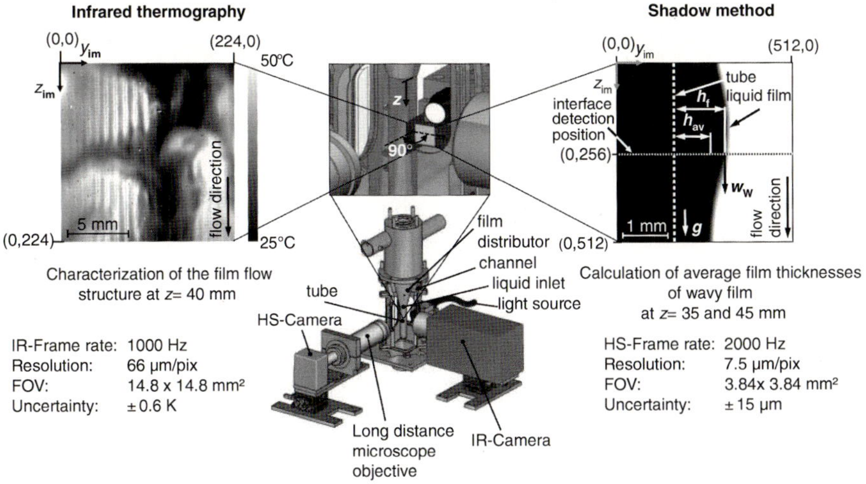

Fig. 2.9 A schematic arrangement of the infrared thermography and shadowgraphy systems. The insets show sample images captured by the IR thermography (*left*) and shadowgraphy (*right*)

2.3.2 Measurement Methods for Film Dynamics

The film dynamics is characterized by the mean film thickness and its waviness. The film thickness development is measured using a confocal chromatic sensor (CHR) technique [1] and additionally a shadowgraphy method. IR thermography is used to visualize local film rupture and rewetting events.

2.3.2.1 Film Thickness Measurements

The shadowgraphy method is schematically shown on the right hand side of Fig. 2.9. A high-speed CMOS camera together with a long distance microscope lens is positioned at an axial distance of $z = 35$ mm and $z = 45$ mm from the liquid inlet using micrometre stages. Prior to the experimental run the CMOS camera is aligned using a needle tip ($AD = 800 \pm 1$ μm) touching the tube surface and backlit by a light source. It is focussed until the shadow of the needle tip is sharply projected. To avoid an under- or over-exposition of the objects in the field of view, the light intensity is varied. At each value of light intensity a snapshot of the needle is logged and subsequently the needle is replaced by a micrometre glass scale with an accuracy of ±3 μm. The field of view is set to 3.84×3.84 mm^2 with 512×512 pixel^2. Using an in-house developed post-processing program, the spatial resolution has been determined to be $\Delta res = 7.5$ μm/pixel.

To evaluate the dynamic film thickness a single reference image of the dry surface is taken. Later, this image is subtracted from each frame taken during the measurement sequences. The film thickness is evaluated along a line passing

Table 2.2 Experimental parameters

Reynolds number liquid	$Re_L = \dot{M}_L/(U_{tu}\mu_L)$	[–]	225 | 375 | 525 | 650
Reynolds number gas	$Re_G = 4\dot{M}_G/(U_{ch}\mu_G)$	[–]	10^4 | $4{\cdot}10^4$ | $7{\cdot}10^4$
Tube circumference	$U_{tu} = d \cdot \pi$	[m]	0.0597
Channel wetted perimeter	$U_{ch} = d \cdot \pi + 4(a+c)$	[m]	0.2757
Dynamic viscosity liquid	μ_L	[kg/(m·s)]	$8.9{\cdot}10^{-4}$
Dynamic viscosity gas	μ_G	[kg/(m·s)]	$18.5{\cdot}10^{-6}$
Wall heat flux	$\dot{q}_W$	[W/cm^2]	20

through the coordinates (0; 256) and (512; 256). The liquid–gas interface is detected based on a threshold value of the grey scale. To quantify the accuracy of the film thickness measurements, the thickness of a wavy falling film has been simultaneously measured using the shadowgraphy method and a confocal chromatic sensor (CHR) [1].

The comparison of the temporal film thickness variation recorded by both techniques showed a very good agreement, and the overall uncertainty of the shadowgraphy measurements could be derived as $\Delta y \pm 15$ μm. For each experimental parameter combination, $n = 2{,}400$ frames were recorded. The average film thickness for each experiment was calculated using the expression $h_m = 1/n\ (\Sigma h_i)$.

2.3.2.2 Observation of Flow Patterns Using Infrared Camera

The IR thermography method is schematically shown on the left hand side of Fig. 2.9. A high-speed mid-wave IR camera with a maximum resolution of 256×256 pixel^2 is used. It allows the observation of the film flow patterns, local film break-up and rewetting on the tube. The camera is positioned at an axial distance $z = 40$ mm from the liquid inlet. Images are taken with a field of view of 14.8×14.8 mm^2 and a resolution of $\Delta res = 66$ μm.

2.3.3 Experimental Results and Discussion

The experimental parameters for the investigations of film flow and heat transfer on the unstructured tube are listed in Table 2.2. In these experiments the inlet temperatures of gas and liquid have been maintained constant at 25°C.

2.3.3.1 Average Film Thickness

The average film thickness at $z = 35$ mm on the unheated unstructured wall is shown in Fig. 2.10 as a function of Re_L and Re_G. By increasing Re_L from 225 to 650 the average film thickness h_m increases from 490 to 620 μm at $Re_G = 10^4$.

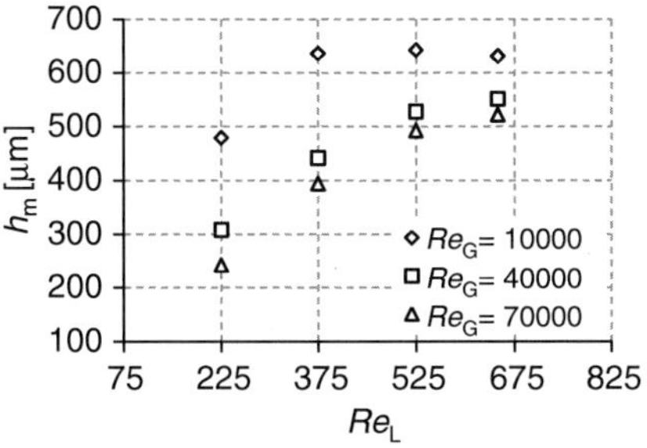

Fig. 2.10 Average film thickness on unheated unstructured surface at $z = 35$ mm

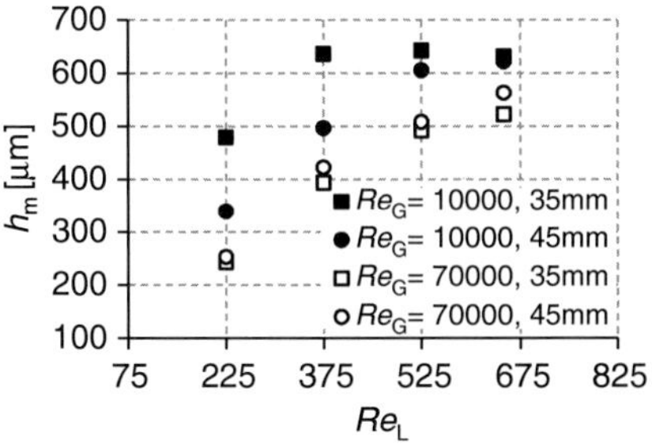

Fig. 2.11 Average film thickness on unheated unstructured surface at $z = 35$ mm and $z = 45$ mm

With increasing Re_G up to $7 \cdot 10^4$, the mean film thickness decreases significantly for $Re_L = 225$–650. This can be attributed to the stronger shear forces acting at the liquid–gas interface resulting in an acceleration of the film and its thinning. However, the dependency of the film thickness on Re_G becomes weaker with increased Re_L. The average film thicknesses for varying Re_L, Re_G and z are shown in Fig. 2.11. At $Re_G = 10^4$ and increasing Re_L, the mean film thicknesses are growing at both axial positions $z = 35$ mm and $z = 45$ mm, though the film thicknesses at $z = 45$ mm is lower than that at $z = 35$ mm. This effect can be explained by acceleration of the liquid by the interfacial shear forces in the hydrodynamically developing flow region, which leads to the film thinning.

2.3.3.2 Wall Temperature Distribution

Figure 2.12 shows the axial wall temperature distribution in a liquid film flowing along an unstructured tube for $\dot{q}_W = 20$ W/cm^2, $Re_G = 10^4$ and varying Re_L. The local wall temperature decreases with increasing Re_L. This effect can be explained by the enhancement of convective heat transfer due to increase of the liquid flow velocity. Since the flowing liquid film is heated up by the hot surface, the wall temperature increases in the streamwise direction. An increase of Re_G from 10^4 to $7 \cdot 10^4$ adds up to a diminishment of t_W (Fig. 2.13). The heat transfer from the heated surface to the liquid film is enhanced, since the flow velocity in the film increases with increasing of the interfacial shear forces. It can be seen that the effect of the liquid Reynolds number on heat transfer is much stronger than the effect of

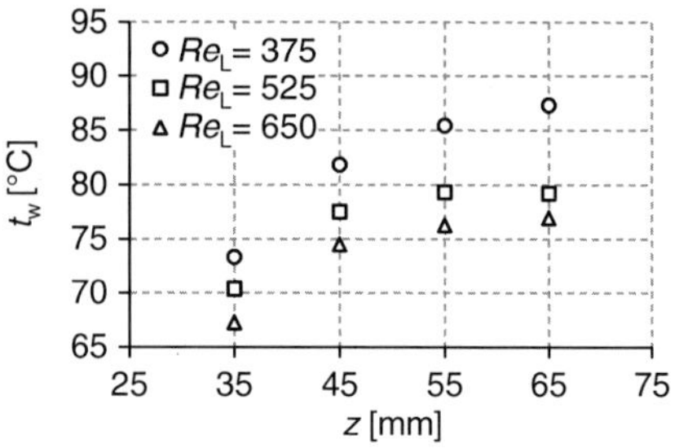

Fig. 2.12 Axial wall temperature distribution at constant $Re_G = 10^4$ and $\dot{q}_W = 20$ W/cm^2 at varying Re_L

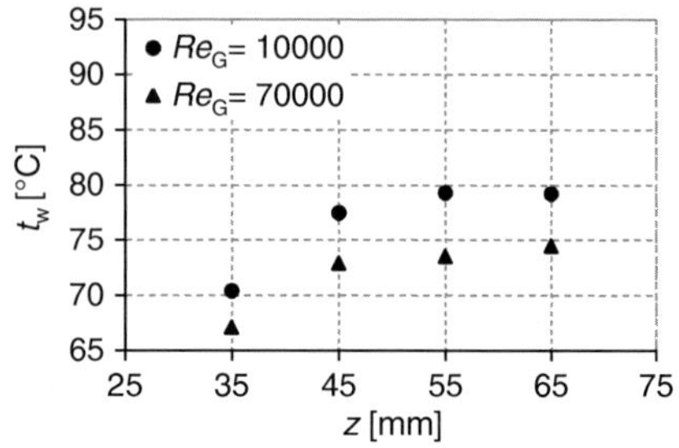

Fig. 2.13 Axial wall temperature distribution at constant $Re_L = 525$ and $\dot{q}_W = 20$ W/cm^2 at varying of Re_G

the gas Reynolds number. A comparable heat transfer enhancement can be achieved by increasing of the gas Reynolds number by a factor of 7 (see Fig. 2.13) and by increasing of the liquid Reynolds number from 525 to 650.

2.3.4 *Film Break-up and Rewetting on a Micro-grooved Tube*

The cooling effect of flowing liquid films can be significantly affected by break-up phenomena [5, 19]. In this work the film break-up and rewetting phenomena have been observed on a micro-grooved tube at a wall heat flux of 40 W/cm^2. In Fig. 2.14 the film flow patterns are shown exemplarily out of three sequences (a_i, b_i, c_i) for constant Re_L, $\dot{q}_W$ and varying Re_G for different time instances. At $Re_G = 10^4$, sequence a_i does not show any considerable wave formation. However, in image a_2 within the domain approximately confined by the coordinates x, $z = (7–9, 0–14)$ mm, distinctive stripe-like patterns with increased temperature take place. These patterns could indicate the existence of a rather thin liquid film where the liquid–gas interface nearly approaches the grooves crest. Starting from about x, $z = (10, 0–4)$ mm hot spots can be seen. At higher values of Re_G the width of the hot spot decreases (see Fig. 2.14b$_2$). This can be explained by the fact that the interfacial shear forces stabilize the liquid film flow and promote a homogeneous distribution of the liquid over the tube perimeter.

At $Re_G = 7 \cdot 10^4$ local film rupture can be observed from the infrared images. In sequence c_i longitudinally advancing waves along the micro-grooves, which appear due to higher shear forces, can be distinguished. The crests of the grooves appear as

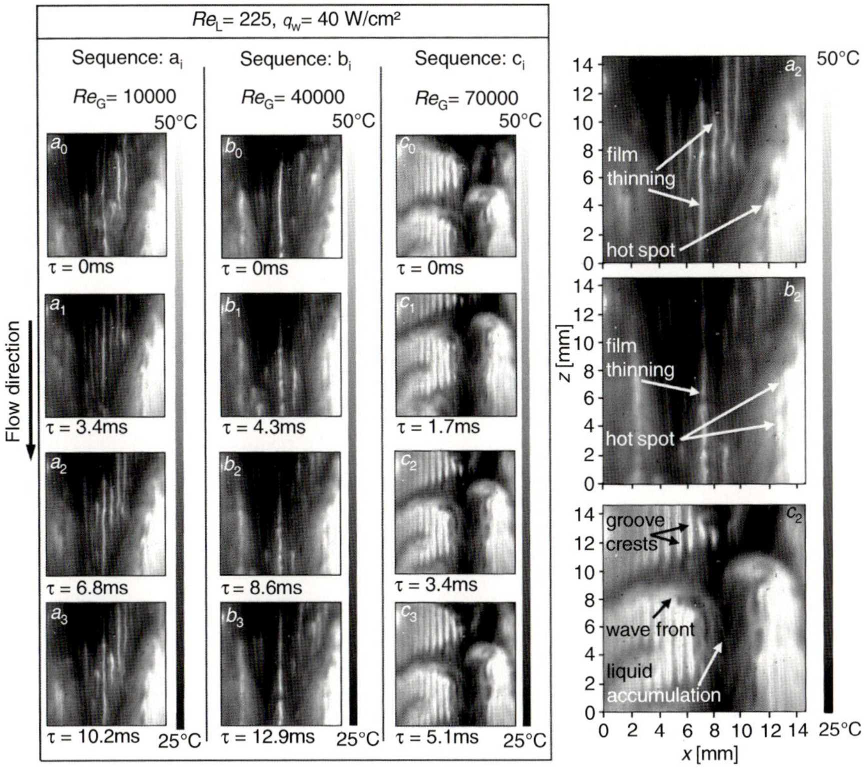

Fig. 2.14 Film break-up and rewetting phenomena on micro-structured surface at fixed $Re_L = 225$, $\dot{q}_W = 40$ W/cm^2 and varying values for Re_G

very thin white vertical stripes contrasting to the hot liquid streams in the vicinity of the groove crests, which appear as grey stripes. We presume that the groove crests in this region are dry. These crests are rewetted as the next wave passes along this location. In Fig. 2.14c$_2$ the wave front can be observed as a very sharp boundary. The wave tail is represented as a notable domain with rake-like liquid peaks. Besides the dewetting-rewetting process observed, an accumulation of liquid mass occurs in the center of the field of view.

2.4 Numerical Investigations

The aim of this numerical study is to understand and describe the hydrodynamic and thermodynamic transport phenomena in wavy thin liquid films sheared by a turbulent gas flow. In order to study the influence of the strong shear forces at the gas–liquid interface on heat transport, computational fluid dynamic (CFD) tools have been adopted. The finite volume technique in OpenFOAM has been employed for investigation of the two-phase flow problem. The numerical method is based on volume of fluid (VOF) method coupled with a continuum surface force (CSF) model, as well as the low Reynolds k-ε turbulence model and the pressure-implicit scheme with splitting of operators (PISO) for pressure–velocity coupling. The numerical results are compared quantitatively with the existing in-house measured data.

2.4.1 Volume of Fluid Method

Three-dimensional (3D) transient simulations of a gas–liquid flow have been performed using the volume of fluid (VOF) method introduced by Hirt and Nichols [23] for tracking the interface between two immiscible fluids. This VOF method enables the fractional volume calculation by introducing a function α to the flow calculations, such that its value is unity for the liquid phase and zero for the gas phase. The tracking of the interface is performed in the cells where the volume fraction α is between 0 and 1. Thus, the interface between the two phases can be tracked by solving the volume fraction continuity equation:

$$\frac{\partial \alpha}{\partial \tau} + \bar{u}_i \frac{\partial \alpha}{\partial x_i} = 0 \tag{2.1}$$

In the two-phase flow system, the properties appearing in the momentum and energy equations are determined by the distribution of phases in each control volume. If the volume fraction of the liquid phase is known, the values of density, viscosity, specific heat capacity and thermal conductivity are calculated by the following expression:

$$\phi = \alpha\phi_{\mathrm{L}} + (1-\alpha)\phi_{\mathrm{G}} \tag{2.2}$$

where ϕ stands for each physical property of fluid named above, the subscript L denotes the liquid phase, and subscript G is used for the gas phase.

2.4.2 Governing Equations

The VOF method is adapted to model two immiscible fluids by solving a single set of governing equations and track the volume fraction of each of the fluids in the computational domain. Assuming that the working fluids are Newtonian incompressible fluids, the governing equations describing the thermo-hydraulic phenomena occurring in two-phase flows are described by continuity, momentum and energy equations, respectively:

$$\frac{\partial \bar{u}_i}{\partial x_i} = 0, \tag{2.3}$$

$$\frac{\partial \rho \bar{u}_i}{\partial \tau} + \frac{\partial}{\partial x_i}\left[\rho \bar{u}_i \bar{u}_j - (\mu + \mu_t)\left(\frac{\partial \bar{u}_i}{\partial x_j} + \frac{\partial \bar{u}_j}{\partial x_i}\right)\right] = -\frac{\partial p}{\partial x_i} + \rho g_i + F_s, \tag{2.4}$$

$$\frac{\partial \rho C_P \bar{T}}{\partial \tau} + \frac{\partial}{\partial x_i}(\rho \bar{u}_i \bar{T}) = \frac{\partial}{\partial x_i}\left[(\lambda + \lambda_t)\left(\frac{\partial \bar{T}}{\partial x_i}\right)\right], \tag{2.5}$$

where u_i and x_i denote the velocity component and the coordinate in the direction i ($i = 1$–3), respectively, τ is the time, p, g and F_s denote, the pressure, the gravitational acceleration and the external body force per unit volume, respectively. The physical properties, ρ, μ, λ, C_p are density, viscosity, thermal conductivity and heat capacity, respectively. The Eqs. (2.4) and (2.5) contain turbulent dynamic viscosity μ_t and turbulent thermal conductivity λ_t, which should be described by an appropriate turbulence model (see Sect. 2.4.3). The surface tension plays an important role in the shear-driven thin liquid film flow. This surface tension creates a pressure jump across the interface. The effect of the surface tension is described using the continuum surface force (CSF) model proposed by Brackbill et al. [24]. The extra body force F_s appearing in the momentum Eq. (2.4) is represented by the following expression:

$$F_s = \sigma_s \kappa(x) \hat{n}, \tag{2.6}$$

where σ_s is the surface tension, κ is the interface curvature and $\hat{n}$ is the normal vector, which are calculated from the following expressions:

$$\kappa(x) = \nabla \cdot \hat{n} = \frac{1}{|\hat{n}|}\left[\left(\frac{\hat{n}}{|\hat{n}|} \cdot \nabla\right)|\hat{n}| - (\nabla \cdot \hat{n})\right], \tag{2.7}$$

$$where \quad \hat{n} = \frac{\nabla \alpha}{|\nabla \alpha|} \tag{2.8}$$

2.4.3 Turbulence Modeling

A direct numerical simulation (DNS) of the thin liquid film flow driven by a turbulent gas stream is not feasible as it is computationally very expensive. The traditional way to reduce the computational efforts in a turbulent two-phase flow simulation is to time-average the Navier–Stokes equations, yielding the Reynolds averaged Navier–Stokes (RANS) equations. In general, for the computations with the empirically derived wall functions in standard k-ε turbulence models, the first grid point should be placed outside the viscous sublayer (i.e. $y^+ > 11$). This may not be feasible in thin liquid film flows, since the liquid film may completely be located in the viscous sublayer.

Table 2.3 Turbulent model constant values

$C_{1\varepsilon}$	$C_{2\varepsilon}$	C_{μ}	σ_k	σ_ε
1.44	1.92	0.09	1	1.3

Therefore, this type of mesh is not sufficient to characterize thin liquid film which is in the range between 200 and 600 μm. However, an alternative approach to the use of wall-functions is to use low Reynolds k-ε turbulence model, which uses "damping functions" near the wall. This model requires a fine elements extending through the viscous sublayer near to the wall ($y^+ < 1$). A measurement for the sufficient grid resolution is the non-dimensional distance of the first grid centroid from the wall, defined as:

$$y^+ = \frac{\rho C_\mu^{0.25} k^{0.5} y}{\mu}. \tag{2.9}$$

The turbulence viscosity (μ_t) in momentum Eq. (2.4) is defined as follows:

$$\mu_t = \rho C_\mu f_\mu \frac{k^2}{\varepsilon}. \tag{2.10}$$

Turbulent kinetic energy k and turbulent dissipation rate ε are determined from the following equations:

$$\frac{\partial(\rho k)}{\partial \tau} + \frac{\partial\left(\rho \bar{u}_j k\right)}{\partial x_j} = \frac{\partial}{\partial x_j}\left[\left(\mu + \frac{\mu_t}{\sigma_k}\right)\frac{\partial k}{\partial x_j}\right] + \mu_t\left(\frac{\partial \bar{u}_i}{\partial x_j} + \frac{\partial \bar{u}_j}{\partial x_i}\right)\frac{\partial \bar{u}_i}{\partial x_j} - \rho\varepsilon, \tag{2.11}$$

$$\begin{aligned}\frac{\partial \rho\varepsilon}{\partial \tau} + \frac{\partial\left(\rho \bar{u}_j \varepsilon\right)}{\partial x_j} &= \frac{\partial}{\partial x_j}\left[\left(\mu + \frac{\mu_t}{\sigma_\varepsilon}\right)\frac{\partial \varepsilon}{\partial x_j}\right] \\ &\quad + C_{1\varepsilon} f_1 \frac{\varepsilon}{k}\mu_t\left(\frac{\partial \bar{u}_i}{\partial x_j} + \frac{\partial \bar{u}_j}{\partial x_i}\right)\frac{\partial \bar{u}_i}{\partial x_j} - f_2 C_{2\varepsilon}\rho\frac{\varepsilon^2}{k}.\end{aligned} \tag{2.12}$$

The turbulence model constants are listed in Table 2.3. The closure of the turbulence model is achieved by prescribing the wall damping functions f_1, f_2 and f_μ. These wall damping functions presented by Lam and Bremhorst [25] are provided in the following equations:

$$f_1 = 1 + \left(0.05/f_\mu\right)^3, \tag{2.13}$$

$$f_2 = 1 - \exp\left[-\left(\rho k^2/\mu\varepsilon\right)^2\right], \tag{2.14}$$

$$f_\mu = \left[1 - \exp\left(-0.0165\rho k^{1/2} y/\mu\right)\right]^2\left(1 + 20.5\mu\varepsilon/\rho k^2\right). \tag{2.15}$$

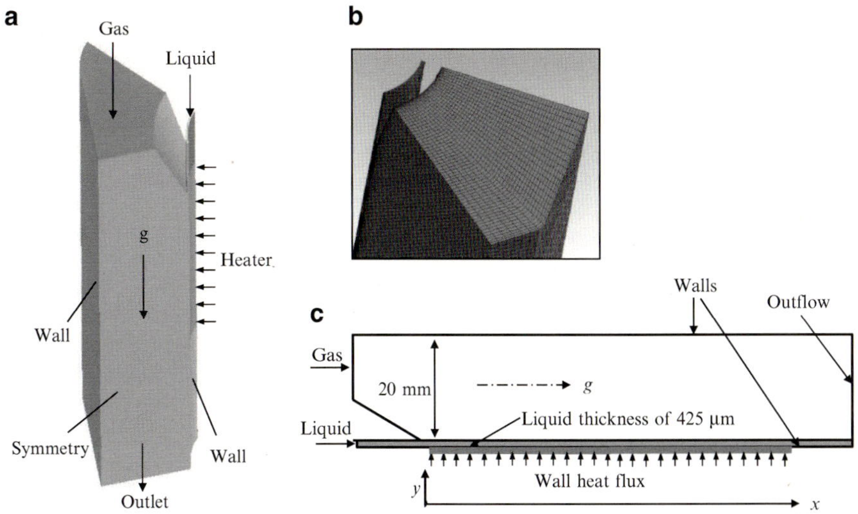

Fig. 2.15 (**a**) 45° sector three-dimensional domain; (**b**) computational grid; (**c**) two-dimensional computational domain

2.4.4 Computational Grid and Boundary Conditions

The computational domain for three-dimensional simulations is shown in Fig. 2.15a. This domain corresponds to one-eighth of the real geometry, whereas the symmetry properties of the test section have been exploited in order to reduce computational time. The corresponding mesh is illustrated in Fig. 2.15b.

Figure 2.15c shows boundary conditions in two-dimensional computational domain used for parametric study. The geometry is created and meshed in ANSYS pre-processor ICEMCFD and exported into OpenFOAM [26]. The mesh density depends on near-wall modeling strategy adopted for resolution of the problem under turbulent flow conditions, and is determined by the y^+ characteristic parameter (Eq. 2.9). The first grid point near the wall is to be placed such that y^+ is of the order less than unity. Several mesh sizes were evaluated from $13{\cdot}10^4$ to approximately $5{\cdot}10^5$ hexahedral elements for different inlet Reynolds numbers of gas and liquid.

Let the positive x-direction be aligned with gravity and the y-direction perpendicular to the wall. The two inlets 'Liquid' and 'Gas' represent the inlet for the water and the air, respectively. At the channel inlet, the uniform velocities and constant temperature for liquid and gas have been posed as boundary conditions. At the outlet boundary, zero-gradients of all variables and constant value for pressure have been imposed. No-slip conditions at the outer solid walls are considered. No boundary conditions are imposed at gas–liquid interface which is not a computational boundary in VOF method. Additionally, it is assumed that heat is transferred only through the "Heater" wall, while the rest of the walls are considered adiabatic.

The turbulent kinetic energy (k) and dissipation rate (ε) at the inlet have been prescribed based on the following equations:

$$k = 1.5(|\bar{u}|I)^2 \quad \text{and} \quad \varepsilon = \frac{C_\mu^{0.75} k^{1.5}}{l}, \tag{2.16}$$

where $I = 10\%$ is the turbulent intensity, and the characteristic length scale l has been set equal to 7% of the hydraulic diameter. The turbulent Prandtl number Pr_t is defined as the ratio of turbulent viscosity to the coefficient of turbulent thermal diffusivity and is assumed constant value equal to 0.9 [27]. In order to reduce the computational time, the computational domain is considered to be initially partially filled with the water at a height of 425 μm and remaining region is filled with air. All the computational elements in the domain are set to have a uniform initialization with the pressure of 0 Pa and temperature of 297 K for all simulations. In the numerical simulations, air and water have been chosen as working fluids, which correspond to the in-house experiments. The thermo-physical properties of water and air at 24°C are used in computations. The PISO (Pressure Implicit with Splitting of Operators) algorithm is used for pressure–velocity coupling. Based on CFL (Courant-Friedrichs-Lewy) stability criterion with CFL <0.5, the time step was set to be around 10^{-5} s. On an average, this corresponds approximately to 50 h of CPU time for 0.3 s simulation time in 2 × Quad core processor machine with 2 × 3.16 GHz speed and 16 × 2,048 MB memory. The computations are performed transiently using OpenFOAM solver until quasi-steady-state condition is reached, which is judged based on a steady temperature values at various positions of the wall in x-direction. The time averaged film thickness, velocity profile and wall temperatures are calculated for a quantitative analysis.

2.4.5 Numerical Results and Discussion

The validation of the numerical method has been performed by comparison of the measured liquid film thickness, the wall temperature distribution and the velocity profile in the gas with the numerically computed values. In all the experiments, the inlet gas and liquid temperatures are set to be 24°C. The temperatures are computed at positions $x = 35$, 45, 55 and 65 mm from the liquid inlet. Average film thickness is calculated at a position $x = 35$ mm.

2.4.5.1 Validation by Comparison with Experimental Data

The average film thickness has been computed at different liquid Reynolds numbers ($Re_L = 375$, 525 and 650) and at a fixed gas Reynolds number ($Re_G = 10^4$). Figure 2.16 shows the comparison between average liquid film thickness at $x = 35$ mm from liquid inlet for three different liquid Reynolds numbers by

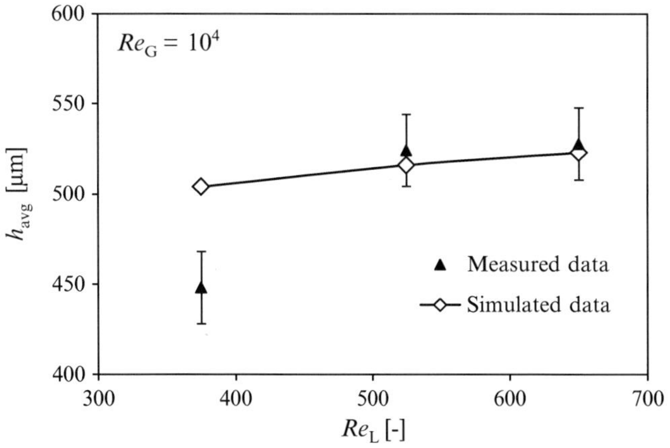

Fig. 2.16 Average film thickness values at various liquid Reynolds numbers

maintaining constant gas Reynolds number. In Fig. 2.16, experimental values (solid triangles) measured using the shadowgraphy method are compared with the results of 3D simulations (diamonds) showing a good agreement.

Liquid film flow is dynamic in nature. The film thickness h is determined by time averaging of the liquid film height at a particular axial position. It is noticed that the film thickness increases with increase in liquid Reynolds number. This is due to increase in liquid mass flow which leads to a thicker film. The slight disagreement has been observed for $Re_L = 375$. The reason for this can be the inability of the shadowgraphy method to capture the transient asymmetric wavy film thickness. Moreover, in the VOF simulations no distinct interface can be determined. Therefore, the presented numerical results are based on criterion $\alpha = 0.5$.

The measured distribution of the air velocity and computed velocity distribution are plotted in Fig. 2.17 for $Re_L = 525$ and $Re_G = 10^4$. Figure 2.17 compares the predicted (curve) velocity profile and the measured experimental values (solid stars) at a position $x = 30$ mm from the liquid inlet. It is notable that the model can predict the position of peak velocity. The calculated gas velocity exceeds the experimental data by 10–15% for the transverse positions (y-direction) between 3 and 10 mm. This discrepancy can be attributed to the limited accuracy of the measurements in this region using a Pitot tube. In experiments, in the regions of gas phase where the fluctuation of velocity is high, the error is higher.

Comparison between the measured and computed wall temperature has been performed for liquid and gas Reynolds numbers 525 and 10^4 respectively at constant wall heat flux of 20 W/cm^2. The simulated wall temperature distribution is plotted against the axial distance in Fig. 2.18. In this figure the black color filled triangles represent the measured temperature values and the squares represent the simulated temperature distribution. It is noted that the measured wall temperature decreases with increasing distance.

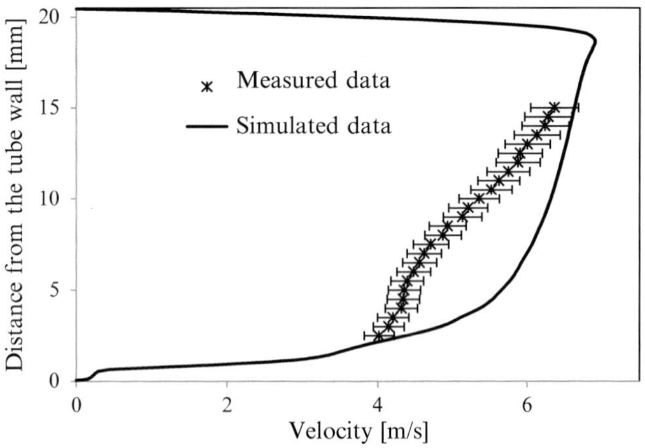

Fig. 2.17 Measured and simulated velocity profiles at $Re_L = 525$ and $Re_G = 4 \cdot 10^4$ at a position $x = 30$ mm

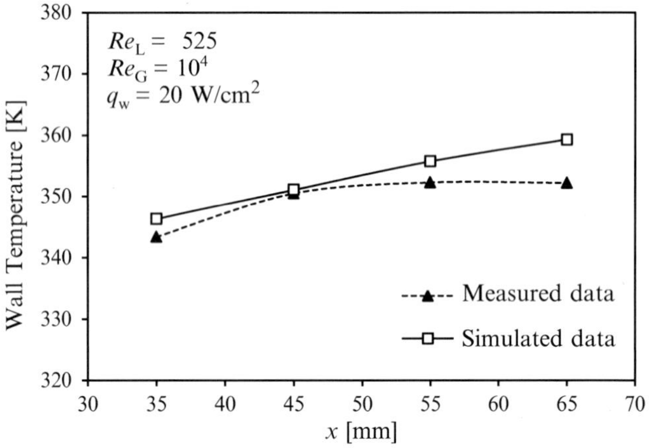

Fig. 2.18 Comparison of computed wall temperature distribution with measurements

The numerical simulations overpredict the experimental data on wall temperatures after a certain axial position. The decrease of measured wall temperature in the downstream direction can be attributed to the heat losses which have not been taken into account in numerical simulations. The evaporation could also result in reduction of the liquid film temperature, and, as a result, the reduction of the wall temperature. The evaporation model has not been included in these computations.

2.4.5.2 Parametric Study

A parametric study has been performed in a two-dimensional domain (see Fig. 2.15c) in order to shorten the computational time. The influence of gas

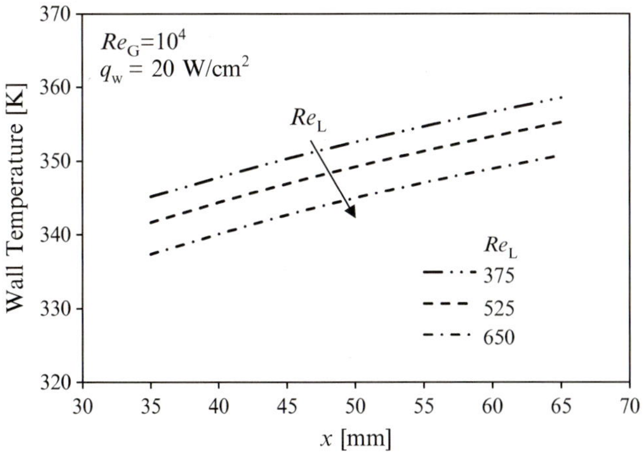

Fig. 2.19 Axial wall temperature distribution at fixed Re_G

and liquid Reynolds numbers on the heat transfer in shear-driven liquid films has been evaluated. Figure 2.19 depicts the wall temperature distribution for a fixed gas Reynolds number ($Re_G = 10^4$) and three different liquid Reynolds numbers ($Re_L = 375$, 525 and 650) at uniform wall heat flux ($\dot{q}_W = 20$ W/cm^2). Strong influence of inlet liquid Reynolds number on temperature profiles can be seen in Fig. 2.19. Temperatures on the wall are lower for high liquid Reynolds numbers. This heat transfer enhancement phenomenon is related to intensification of heat convection. For a fixed Re_G and increasing Re_L the liquid velocity increase, which causes a intensification of heat transfer as well. In order to study the effect of Re_G on heat transfer, simulations are performed for fixed liquid flow rate ($Re_L = 525$) and different air flow rates, corresponding to $Re_G = 10^4$, $4 \cdot 10^4$ and $7 \cdot 10^4$. Figure 2.20 shows the wall temperature profiles obtained at uniform wall heat flux of 20 W/cm^2. For high gas Reynolds number, low wall temperature is predicted, which indicates the heat transfer enhancement. For a particular liquid Reynolds number, an increase of gas flow rate results in an increase of the shear at the interface, in the acceleration of the flow within the liquid film and in intensification of convective heat transport in the system.

It can be seen from Fig. 2.20 that the axial temperature gradient becomes less steep with increasing gas Reynolds number. This behavior suggests that the effect of the gas flow rate on heat transfer is stronger than the effect of the liquid flow rate.

Figure 2.21a shows the wavy film contour (dark colour) at $Re_L = 525$ and $Re_G = 7 \cdot 10^4$. Irregular wavy pattern can be observed in the figure. The instabilities are generated due to high interfacial shear force caused by the gas flow. These instabilities lead to development of wavy film flow. The waves are amplified in the streamwise direction. It can be observed that the large amplitude interfacial waves induce a reverse flow (see Fig. 2.21b). The flow reversal phenomenon plays an important role in heat and mass transfer enhancement [10].

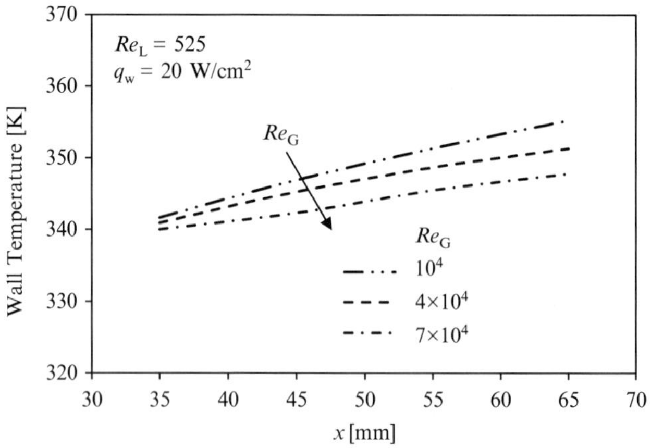

Fig. 2.20 Axial wall temperature distribution at fixed Re_L

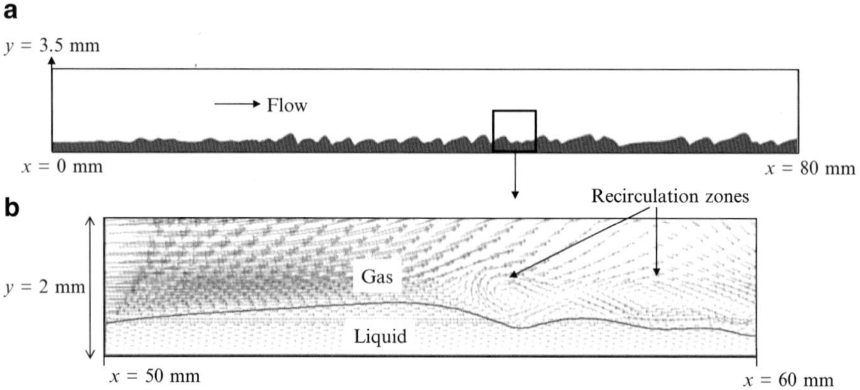

Fig. 2.21 (**a**) Volume fraction contours; (**b**) velocity vectors

2.5 Conclusions

Film thickness evolution and wall temperature distribution in shear-driven film flows on heated surfaces have been studied experimentally. The liquid film thickness strongly decreases with increasing gas velocities and increases with increasing liquid mass flow rates. An increase of liquid film velocity enhances the convective heat transfer rate at the wall. The infrared thermography records of the surface of a film flowing along a heated micro-grooved tube revealed a strongly wavy film dynamics and partially wetted regimes at high values of Re_G. Further experiments will be performed to study the thermo-hydrodynamic mechanisms of a shear-driven liquid film flow on a pyramided structured surface. Air-water two-phase flow with heat transfer has been simulated using a CFD model. The extended OpenFOAM code based on VOF method is a suitable tool for the detailed study of hydrodynamics and heat transfer in shear-driven film flows. The computed gas velocity profiles agree qualitatively with experimental results. The obtained wall temperature profiles have been compared with measured temperature data showing a good agreement. The effects of the inlet conditions on the hydrodynamics, heat transfer and wave behavior have been quantified. Further simulations will include the effect of liquid evaporation on heat and mass transfer in a two-phase flow.

Acknowledgements The authors acknowledge the financial support of the German Science Foundation (Deutsche Forschungsgemeinschaft) in the framework of the Collaborative Research Center 568 (Subproject A2).

References

Project-Related Publications

1. Helbig, K.: Messung zur Hydrodynamik und zum Wärmetransport bei der Filmverdamfung. EPDA Elektronische Publikationen Darmstadt. http://elib.tu-darmstadt.de/diss/000867. Accessed 07 Apr 2011
2. Gambaryan-Roisman, T., Stephan, P.: Analysis of falling film evaporation on grooved surfaces. J. Enhanced Heat Trans. **10**(4), 445–457 (2003)
3. Gambaryan-Roisman, T., Alexeev, A., Stephan, P.: Effect of the microscale wall topography on the thermocapillary convection within a heated liquid film. Exp. Therm. Fluid Sci. **29**, 765–772 (2005)
4. Kabova, Y., Alexeev, A., Gambaryan-Roisman, T., Stephan, P.: Marangoni-induced deformation and rupture of a liquid film on a heated microstructured wall. Phys. Fluids **18**, 012104 (2006)
5. Löffler, K., Yu, H., Gambaryan-Roisman, T., Stephan, P.: Hydrodynamics and heat transfer of thin films flowing down inclined smooth and structured plates. In: Proceedings of the 4th International Berlin Workshop – IBW4 on Transport Phenomena with Moving Boundaries, Berlin (2007)
6. Kunkelmann, C., Ibrahem, K., Schweizer, N., Herbert, S., Stephan, P., Gambaryan-Roisman, T.: The effect of three-phase contact line speed on local evaporative heat transfer: experimental and numerical investigations. Int. J. Heat Mass Trans. **55**(7–8), 1896–1904 (2012)
7. Helbig, K., Alexeev, A., Gambaryan-Roisman, T., Stephan, P.: Evaporation of falling and shear-driven films on smooth and grooved surfaces. Flow Turbul. Combust. **75**, 85–104 (2005)
8. Helbig, K., Nasarek, R., Gambaryan-Roisman, T., Stephan, P.: Effect of longitudinal minigrooves on flow stability and wave characteristics of falling liquid films. ASME J. Heat Trans. **131**, 011601 (2009)
9. Gambaryan-Roisman, T., Stephan, P.: Flow and stability of rivulets on heated surfaces with topography. ASME J. Heat Trans. **131**(3), 033101 (2009)
10. Marati, J., Budakli, M., Gambaryan-Roisman, T., Stephan, P.: Heat transfer in shear-driven thin liquid film flows, MP14. In: Proceedings of the ICHMT International Symposium on Advances in Computational Heat Transfer, July 1–6, Bath, England (2012)

Other Publications

11. Pommersberger, K.: Untersuchung der Gemischaufbereitung in Gasturbinenbrennkammern unter Berücksichtigung der Eigenschaften kommerzieller Flüssigbrennstoffe. PhD Dissertation, TH Karlsruhe, Institut für Thermische Strömungsmaschinen (2003)
12. Arbeiter, F.: Experimentelle Untersuchung zum Emissionsverhalten einer LPP Brennkammer mit Film-verdunstung. Diploma Thesis, TH Karlsruhe, Institut für Thermische Strömungsmaschinen (2002)
13. Batarseh, F.Z., Roisman, I.V., Tropea, C.: Spray generated by an airblast atomizer at high-pressure conditions. In: Proceedings of ASME Turbo Expo, Fostering Gas Turbine Education, Technology & Networking, Montreal (2007)
14. Schober, P., Ebner, J., Schäfer, O., Wittig, S., Experimental study on the effect of a strong negative pressure gradient on a shear-driven liquid fuel film. In: Proceedings of 9th ICLASS, Sorrento, Italy (2003)
15. Lan, H., Friedrich, M., Armaly, B.F., Drallmeier, J.A.: Simulation and measurement of 3D shear-driven thin liquid film flow in a duct. Int. J. Heat Fluid Flow **29**, 449–459 (2008)
16. Elsäßer, A., Samenfink, W., Ebner, J., Dullenkopf, K., Wittig, S.: Dynamics of shear driven liquid films. In: Proceedings of 7th International Conference on Laser Anemometry, Karlsruhe (1997)
17. Asali, J.C., Hanratty, T.J.: Ripples generated on a liquid film at high gas velocities. Int. J. Multiphase Flow **19**, 229–243 (1993)
18. Bar-Cohen, A., Sherwood, G., Hodes, M., Solbreken, G.L.: Gas-assisted evaporative cooling of high density electronic modules. IEEE Trans. CPMT A **18**(3), 502–509 (1995)
19. Kabov, O.A., Zaitsev, D.V., Cheverda, V.V., Bar-Cohen, A.: Evaporation and flow dynamics of thin, shear-driven liquid films in microgap channels. Exp. Therm. Fluid Sci. **35**, 825–831 (2011)
20. Fujita, Y.: Boiling and evaporation of falling film on horizontal tubes and its enhancement on grooved tubes. In: Kakaç, S., Bergles, A.E., Mayinger, F., Yuncu, H. (eds.) Heat Transfer Enhancement of Heat Exchangers, pp. 325–346. Kluwer Academic Publishers, Dordrecht (1998)
21. Wierschem, A., Aksel, N.: Influence of inertia on eddies created in films creeping over strongly undulated surfaces. Phys. Fluids **16**(12), 4566–4574 (2004)
22. Stephan, P., Busse, C.A.: Analysis of heat transfer coefficient of grooved heat pipe evaporator walls. Int. J. Heat Mass Trans. **35**(2), 383–391 (1992)
23. Hirt, C.W., Nichols, B.D.: Volume of fluid (VOF) method for the dynamics of free boundaries. J. Comput. Phys. **39**, 201–225 (1981)
24. Brackbill, J.U., Kothe, D.B., Zemach, C.: A continuum method for modeling surface tension. J. Comput. Phys. **100**, 335–354 (1992)
25. Lam, C.K.G., Bremhorst, K.: A modified form of the k-ε model for predicting wall turbulence. J. Fluids Eng. **103**, 456–460 (1981)
26. OpenFOAM documentation: Programmer's and User's Guide, OpenCFD Limited (2007)
27. Rup, K., Soczowka, M.: An improved low Reynolds number k-ε model for heat transfer calculations. Forschung im Ingenieurwesen **65**, 225–235 (2000)

Chapter 3
Thermodynamically Consistent Modelling of Gas Turbine Combustion Sprays

A. Sadiki, M. Chrigui, and A. Dreizler

Abstract In order to support the design procedure and increase the reliability and safety of combustion engines fired with liquid fuel at a reasonable cost, numerical prediction tools well validated by comprehensive experimental data are needed. As there is today enough evidence that Large Eddy Simulation (LES) is able to well capture intrinsically time and space dependent phenomena, LES will be employed. However, in most LES based spray modules for predicting spray combustion the interactions between both phases and between evaporating droplets and combustion are either not adequately considered or not incorporated at all. The objective of this work is to develop and validate a thermodynamically consistent spray module for Large Eddy Simulation that allows describing accurately the essential processes featuring spray combustion in gas turbine combustion chambers. These include besides the injection of liquid fuel, the turbulent droplet dispersion, the vaporization of the droplets and mixture formation and the subsequent spray combustion. In particular, (1) a physically consistent SGS-model describing the influence of droplet diameter and interface transport on the gas phase turbulence as well as the effect of the droplet evaporation on the mass and scalar transport processes (turbulence modulation) has been adapted for LES into an Eulerian–Lagrangian framework. (2) Apart from classical evaporation models valid in atmospheric

A. Sadiki (✉) • M. Chrigui • A. Dreizler
Institute for Energy and Powerplant Technology, Mechanical Engineering, Technische Universität Darmstadt, Petersenstr. 30, 64287 Darmstadt, Germany

Institute of Reactive Flow and Diagnostics, Mechanical Engineering, Technische Universität Darmstadt, Petersenstr. 30, 64287 Darmstadt, Germany
e-mail: sadiki@ekt.tu-darmstadt.de; mchrigui@ekt.tu-darmstadt.de; dreizler@csi.tu-darmstadt.de

A. Sadiki • A. Dreizler
Center of Smart Interface, Mechanical Engineering, Technische Universität Darmstadt, Petersenstr. 32, 64287 Darmstadt, Germany

J. Janicka et al., *Flow and Combustion in Advanced Gas Turbine Combustors*, Fluid Mechanics and Its Applications 102, DOI 10.1007/978-94-007-5320-4_3,

conditions, an advanced evaporation model, the so called non-equilibrium model, appropriate for gas turbine conditions have been integrated and validated. (3) The chemistry-turbulence interaction under droplet evaporating conditions has been considered according to a presumed (filtered) probability density function while the combustion process itself is described following a tabulated detailed chemistry based on FGM (Flamelet Generated Manifold). (4) All the developed sub-models along with the complete model have been implemented in the working package FASTEST/LAG3D and validated in non-reacting and reacting configurations with available experimental data. Comparisons include exhaust gas temperature, droplet velocities and corresponding fluctuations, droplet mean diameters and spray volume flux at different distances from the exit planes. An overall good agreement with experimental data has been achieved. Parts of this contribution has been already reported as mentioned throughout the paper.

Keywords Thermodynamically consistent modelling • Acetone spray • Evaporation • Partially premixed combustion • Eulerian–Lagrangian method • LES

3.1 Introduction

The necessity for transportation and power generation industries to meet stringent regulations concerning pollutants emissions led to the development of advanced low emission combustor strategies, such as lean-burn technologies, among others. Thereby spray combustion is used extensively, where it follows three preliminary steps: injection of liquid fuel, vaporization of the droplets and mixing of the oxidant and the fuel. Understanding of these fuel preparation processes and their interactions with the combustion process is a milestone for the improvement of combustion efficiency and reduction of pollutant emission. To meet requirements for optimization and design of flow systems in technical and industrial applications of spray flows a comprehensive knowledge about the governing phenomena and micro-processes on-going in these systems is highly needed [21–28, 30].

Three main approaches for simulating turbulent combustion sprays are especially emerging: the Eulerian–Lagrangian method, the Eulerian-Eulerian strategy and the filtered (probability) density function (FDF) [39]. While the flow field of the continuous phase is mostly described by an Eulerian description, the dispersed phase can either be considered in a Lagrangian way as consisting of tracking a large number of particles or discrete entities (Lagrangian approach) or in an Eulerian framework where it is treated like a continuum as a second (semi-) continuous phase or pseudo-fluid (Eulerian approach). In the FDF approach, a transport equation of a multi-dimensional filtered density function is solved that involves the spray source term in closed form [49, 59, 63].

Considerable and important work has been achieved in the investigation of reacting sprays. Bini et al. [28] and Jones et al. [83] used LES to study acetone evaporating spray and kerosene evaporation and combustion in two different configurations, respectively. In [83] the effect of evaporating droplets on the diffusion flame temperature and concentrations using LES-PDF method and Euler-Lagrange procedure was highlighted. In both contributions [28, 83] the authors noticed the importance of subgrid scale fluctuations. Lederlin et al. [26] addressed sub-filter scalar mixing and pollutant formation in an LES reacting two phase flow using unsteady flamelet and 2 steps chemistry, respectively, whilst Senoner et al. [23] and Sanjosé et al. [27] compared Euler-Euler versus Euler-Lagrange method in a monodisperse configuration without chemical reaction. A clear advantage of accuracy could not be evidenced for the polydisperse Lagrangian simulations compared to the monodisperse Eulerian computations. This might be originated from the simplified injection model used. Bellan and Selle [22] derived three subgrid terms in the equations describing the evolution of evaporating two phase flows. They investigated the importance of the terms based on Direct Numerical Simulation (DNS) database and showed that non-equivalent forms in the LES equations were obtained. They also suggested further consideration for reactive spray flows. Patel and Menon [30, 51] carried out LES of liquid fueled lean direct injection combustor by mean of the linear eddy model. Droplets breakup in the vicinity of the fuel injector was found to play a major impact on fuel evaporation. Flame regime was shown to change from diffusion to premixed. Apte et al. [24] developed a stochastic model of atomizing spray in which droplet breakup is validated.

Extensive fundamental and applied researches were dedicated to describe questions that govern the interaction between the reactive multiphase flows. Yet studies have proved a strong dependence of the combustion and disperse phase properties from one another and the difficulty of isolating coupled physical effects. Nevertheless in most of LES based spray modules, the interactions between both phases and between evaporating droplets, turbulence and combustion including finite rate chemistry effects are either partially considered or not incorporated at all.

In this work focus is put on the Eulerian–Lagrangian approach combined with LES. Lagrangian models may provide detailed information on the droplet-droplet interactions, phase changes and droplet combustion and wall-droplet interaction. In addition, application of boundary conditions for the droplets is straightforward. But Lagrangian models are limited by the number of particles that can be tracked in the system, as the computational cost rises steeply with the number of droplets. In order to restrict the associated computational costs, some assumptions have to be made in general. Numerical issues related to unsteady computations along with massively parallel computing of dispersed multiphase flows were addressed in [44, 77, 83], among others.

It is important to note that diffusion, mixed modes or partially premixed combustion regimes generally occur in spray combustion. The mixture fraction formalism is followed in this work. Computationally, spray combustion is thereby captured by tracking the dispersed phase and solving the balance equations for mass, momentum, mixture fraction and species mass fractions for the carrier gas phase. These equations involve unclosed terms that need to be modeled.

The objective of this work is to develop and validate a thermodynamically consistent spray module for Large Eddy Simulation that allows describing accurately the essential processes featuring spray combustion in gas turbine combustion chambers. These include besides the injection of liquid fuel, the micro-processes like, turbulent droplet dispersion, the vaporization of the droplets and mixture formation together with turbulence-droplet interaction, spray combustion along with turbulence-chemistry-interaction, etc.

In the following section the governing equations and modelling used are briefly presented in the LES framework for the carrier phase. The next section (Sect. 3.3) specifies the mathematical formulation and modelling of the dispersed phase. The modeling of dominant micro-processes will be presented. In particular an own turbulence modulation modeling strategy will be highlighted. This will be followed in Sect. 3.4 by the numerical procedure. A review of some achievements dealing with sprays featuring gas turbine combustor flow properties as recently reported in different publications by the authors will be provided in Sect. 3.5. Perspectives of LES methods for simulations of complex sprays configurations will be finally outlined in conclusions.

3.2 Governing Equations and Modelling of Carrier Phase Flow

Within the Eulerian–Lagrangian method the turbulent fluid phase is described following a classical LES approach. The governing equations (continuity equation, Navier–Stokes equations, energy equation etc.) are filtered to separate the large-scale and small-scale turbulence. The large-scale turbulence is thus resolved and solved for by the discretized equations whereas the unresolved small-scale turbulence has to be modelled. Different aspects related to the filtering techniques are treated in a comprehensive manner by Sagaut [71].

3.2.1 Governing Equations

According to the flamelet generated manifold (FGM) based tabulated detailed chemistry applied to model the combustion [49, 66], the filtered transport equations for control variables, namely the mixture fraction and one reaction progress variable (RPV), are solved together with the filtered transport equations for mass conservation and momentum. The unsteady, general form of the filtered transport equations needed emerges as:

$$\frac{\partial\left(\bar{\rho}\,\tilde{\phi}\right)}{\partial t}+\frac{\partial\left(\bar{\rho}\tilde{u}_i\,\tilde{\phi}\right)}{\partial x_i}-\frac{\partial}{\partial x_i}\left(\bar{\Gamma}\,\frac{\partial\tilde{\phi}}{\partial x_i}-\Lambda_i^{sgs}\right)=\bar{S}_\phi+\bar{S}_{\phi,p}+\bar{S}_{\phi,p,v}+\bar{S}_c+\bar{S}_{c,v} \tag{3.1}$$

Table 3.1 Expression of terms in Eq. (3.1)

ϕ	Λ_i^{sgs}	$\bar{S}_{\phi,p} + \bar{S}_{\phi,p,v}$	$\bar{S}_c + \bar{S}_{c,v}$
1	0	$\sum\limits_p \frac{\dot{m}_{p,v} N_p}{V_{i,j,k}}$	
$\tilde{u}_i$	τ_{ij}^{SGS}	$-\sum\limits_p \frac{\dot{m}_p N_p}{V_{i,j,k}} \left[\left(u_{p,i}^{t_n+\Delta t} - u_{p,i}^{t_n} \right) - g_i \, \Delta t \right] + \sum\limits_p \frac{\dot{m}_{p,v} N_p}{V_{i,j,k}} u_{p,i}^{t_n}$	
$\tilde{z}$	J_{zi}^{sgs}	$\sum\limits_p \frac{\dot{m}_{p,v} N_p}{V_{i,j,k}}$	
$\tilde{Y}$	J_{Yi}^{sgs}	$\sum\limits_p \frac{\dot{m}_{p,v} N_p}{V_{i,j,k}}$	$\tilde{S}_c + \tilde{S}_{c,v}$

in which ϕ may represent the mass density, velocity components u_i, mixture fraction z, species mass fraction or RPV Y, as well as subgrid turbulent kinetic energy or/and turbulent dissipation rate, etc., respectively. The quantity Γstands for an effective diffusion coefficient, while $\bar{S}_\phi$ expresses turbulence source terms well known in single phase flows [40, 75, 76].

The additional source terms $\bar{S}_{\phi,p}$ and $\bar{S}_{\phi,p,v}$ in (3.1) represent the volume averaged rate of exchange of mass, momentum, mixture fraction, RPV and of SGS turbulent transported quantities between the gas and the particulate phase, respectively. They express the interaction between the two phases and account for the two-way coupling of phases. $\bar{S}_{\phi,p}$ is due to the inter-phase transport without phase transition and $\bar{S}_{\phi,p,v}$ takes into consideration the transfer caused by the droplet evaporation. The source terms, $\bar{S}_c$and $\bar{S}_{c,v}$, are the filtered chemical reaction rate that account for combustion contributions without evaporation and under interaction with evaporating droplets, respectively. In (3.1) and throughout the paper, quantities denoted as $(\bar{\cdot})$and $(\tilde{\cdot})$express mean filtered and filtered quantities, respectively.

Due to the truncation of scales by filtering, subgrid scale (SGS) unknown quantities, Λ_i^{sgs}, appear that incorporate the effect of the unresolved scales on the resolved scale motion in form of SGS scalar flux. Table 3.1 summarizes all the terms in Eq. (3.1). Thereby u_p, v_p and w_p represent the three velocity components of the parcel. N_p is the number of real droplets represented by one numerical droplet. $V_{ijk}$is the cell volume. The quantity g represents the gravitation and $\dot{m}_p$ the droplet mass flow rate across a control volume (CV) per second. t stands for the Lagrangian integration time step.

3.2.2 *LES Based Modeling of the Turbulent Carrier Phase Flow*

With regard to Eq. (3.1), two main problems have to be faced. The first problem is how the SGS models for the momentum, scalar flux transport and the SGS kinetic energy can be formulated in compatibility with the physics to be described. The second is a precise selection of initial and boundary conditions along with numerical requirements to provide reliable and accurate LES results. The second issue has been extensively addressed in the first workshop on quality assessment of unsteady methods for turbulent combustion prediction and validation [52]. Recent overviews can also be found in [75]. We focus in this paper on the first problem.

Equation (3.1) govern the evolution of the large, energy-carrying, scales of flow and scalar field. In flow and scalar field, the effect of the small scales appears through the SGS stress tensor and the SGS scalar flux vector

$$\tau_{ij}^{SGS} = \widetilde{u_i u_j} - \widetilde{u}_i \widetilde{u}_j \tag{3.2}$$

$$J_{\xi i}^{sgs} = \widetilde{u_i \xi_\alpha} - \widetilde{u}_i \widetilde{\xi}_\alpha,\ \xi \equiv (z,\ Y_\alpha) \tag{3.3}$$

respectively. The two last terms in Eq. (3.1) together with the quantities (3.2) and (3.3) must be modeled in order to obtain a closed set of Eq. (3.1). In this contribution, the linear eddy viscosity based models is used to determine the subgrid scale stresses:

$$T_{ij}^{sgs} - \frac{1}{3} T_{kk}^{sgs} \delta_{ij} \approx \nu_t \left[2\tilde{S}_{ij} - \frac{2}{3} \tilde{S}_{kk} \delta_{ij} \right], \quad \tilde{S}_{ij} = \frac{1}{2} \left(\frac{\partial \tilde{u}_j}{\partial x_i} + \frac{\partial \tilde{u}_i}{\partial x_j} \right) \tag{3.4}$$

Two closure approximations for the sgs turbulent viscosity were applied. For the first closure, we follow the formulation given by Smagorinsky as

$$\nu_t = \left(C_s \bar{\Delta} \right)^2 \sqrt{2 \tilde{S}_{lk} \tilde{S}_{lk}} \tag{3.5}$$

In order to stabilize the model, the modification proposed by Sagaut [71] is applied. In addition a clipping approach will reset negative Germano coefficient C_s to zero to avoid destabilizing values of the model coefficient. No special wall treatment is included in the subgrid-scale model [71, 84]. We rather rely on the ability of the dynamic procedure to capture the correct asymptotic behavior of the turbulent flow when approaching the wall (see e.g. Wegner et al. [72]).

For the second closure we apply the Yoshizawa et al. [86] approach along with an advanced one-equation model. The formulation of Yoshizawa yields

$$\nu_t = \left(C_k \bar{\Delta} \right) \sqrt{\tilde{k}^{sgs}} \tag{3.6}$$

where the sgs turbulent kinetic energy is given by

$$\frac{\partial}{\partial t} (\bar{\rho} \tilde{k}^{sgs}) = -\frac{\partial}{\partial x_i} (\bar{\rho} \tilde{u} \tilde{k}^{sgs}) + \frac{\partial}{\partial x_j} \left(\bar{\rho} \nu_t \frac{\partial \tilde{k}^{sgs}}{\partial x_j} \right)$$
$$+ 2 \bar{\rho} \nu_{sgs} \tilde{S}_{ij} \tilde{S}_{ij} - C_\varepsilon \bar{\rho} \frac{\left(\tilde{k}^{sgs} \right)^{\frac{3}{2}}}{\bar{\Delta}} + \bar{F}_{S_{k^{sgs}}} \tag{3.7}$$

for two-phase flows. All the model parameters in this equation are determined dynamically according to [51]. The last term represents the work done due to the two-phase coupling force in the momentum equation and provides a coupling between the turbulent motion of the droplets and the evolution of the SGS kinetic energy. Consequently, the SGS turbulent kinetic energy is indirectly modified due to particle drag and phase change since the force term will change the resolved velocity field, which in turn, will change the SGS kinetic energy. This term expresses the so-called turbulence modulation and will be addressed in detail below.

To represent the sub-grid scale scalar flux in the mixture fraction and in the RPV equations advanced SGS scalar flux models have been recently suggested as reviewed by Sadiki et al. [85]. Nevertheless the simple gradient approach is used here with a constant turbulent Schmidt number, σ_t:

$$J_i^{sgs} = -\frac{\nu_t}{\sigma_t}\frac{\partial \tilde{z}}{\partial x_i}; \quad \sigma_t = 0.7 \tag{3.8}$$

The remaining terms to be closed, the chemical reaction terms, are modeled following the FGM method as will be explained in the next subsection.

3.2.3 Combustion Modeling

In most LES studies the flamelet approach [31, 33, 34, 47] has been widely used in non-premixed combustion systems including spray combustion devices. Thereby the interactions between evaporating droplets, turbulence and combustion including finite rate chemistry effects are either partially considered or not incorporated at all. Relying on engineering applications, such a reduction technique is pursued here. Especially the so-called Flame Prolongated ILDM (FPI) approach [66], or the Flamelet Generated Manifold (FGM) strategy [49] is used. It has already been proven to be accurate for modeling (partially-) premixed flames in direct numerical simulations (DNS), LES and Reynolds Averaged Navier Stokes (RANS) settings in single phase flows [74].

One of the most common approaches to account for spray in the flamelet approach is to implicitly consider the effect of evaporation on combustion through the changes in local mixture fraction field already modified by droplet evaporation, while model parameters for spray models are consequently calibrated. This means that the classical flamelet approach is used in its original form. According to Domingo et al. [58] this is the case when all droplets are supposed to have evaporated before combustion. Regardless of the droplet evaporation in the physical space, the reaction space is kept the same as for non-vaporizing case [36, 67].

Based on RANS Eguz et al. [87] developed the FGM approach to model the PCCI (Premixed Charge Compression Ignition) in which the fuel was a spray, see also [46]. Similar work was done by Hu et al. [88]. The sprays are injected into a pressurized constant volume chamber that mimics a diesel engine condition at combustion. N-heptane and iso-octane were the two fuels selected for the combustion. These investigations were focused on auto ignition for which the FGM table was pre-integrated using a β pdf yielding a 4D manifold. In the present work the method used in [87, 88] is adapted and applied to model and capture the combustion in a partially premixed configuration. Differently to [87, 88], in which the FGM was built upon a premixed combustion, in this study the FGM table is based on a diffusion regime as it deals with spray combustion. To our knowledge, very rare applications of FGM have been done for spray combustion coupled to LES. In particular, a proper account for the effect of droplet vaporization in the reaction space is still an open issue [56]. A first attempt, in which an appropriate vaporization-reaction equation is being used, is work in progress.

In the framework of this work all droplets are supposed to have evaporated before combustion. So, only the change in local mixture fraction field by droplet evaporation is considered while the source term responsible for evaporation contribution in the transport equation of the RPV is negligible [17–19, 58]. Within the FGM approach, the RPV used in the present work is defined with different species as

$$Y_{RPV} = \frac{Y_{CO_2}}{M_{CO_2}} + \frac{Y_{H_2O}}{M_{H_2O}} + \frac{Y_{H_2}}{M_{H_2}} \tag{3.9}$$

where Y_{CO_2}, Y_{H_2O} and Y_{H_2} are the mass fraction of CO_2, H_2O and H_2, respectively. M represents the molar mass, so its reciprocal is used as a weighting factor for each species. The choice of the species defining the RPV depends on the problem being solved. In case of partially premixed spray combustion, the three major species retained above are considered to properly capture the reaction zone. The two-dimensional manifold $\psi\,(z, Y_{RPV})$ is then generated by simply simulating a set of 1D diffusion flamelets with increasing scalar dissipation rate and thereafter switch to unsteady flamelets when reaching the critical scalar dissipation rate.

The filtered combustion variables are retrieved by integrating over the joint PDF of the mixture fraction and the defined RPV:

$$\tilde{\phi} = \int_0^\infty \int_0^\infty \phi\,(z, y) P\,(z, y)\, dz dy; \;\; \mathrm{y} \equiv \mathrm{Y}_{RPV} \tag{3.10}$$

where, $P\,(z, y)$ is the joint PDF. It is practical to carry out the pre-integration upon the normalized values, i.e., the RPV is normalized by its maximum value restricting the integration domain to lie in [0, 1]. Assuming a statistically independence between the RPV and the mixture fraction yields:

$$\tilde{\phi} = \int_0^1 \int_0^1 \phi\,(z, y) \tilde{P}(z) \tilde{P}\left(y^*\right) dz d y^* \tag{3.11}$$

where the Favre-weighted PDF is derived from a standard one, $\widetilde{P}\,(y^*) = \rho\,(y^*)\, P\,(y^*)\,/\bar{\rho}$, with y* as normalized RPV. A discussion of this issue is reported in [69]. Note that the spray source terms in the transport equation of mixture fraction make this quantity no more conservative and may influence the PDF distributions [48]. Gutheil [59] showed that a comparison of Monte-Carlo PDF with standard *beta*-PDF demonstrates that *beta*-function describes the shape of the PDF differently. Nevertheless a presumed *beta*-PDF distribution is chosen here as first approximation. Thus, the mixture fraction depends on its first and second moments. Effects of this assumption on predicted RANS results were reported in [69]. Advanced approaches were suggested in [50]. On LES results these are not studied in the literature yet [22].

The next question is how to model the PDF of the RPV. As a first-order approach, the δ-function is used. So, the combustion variables are function of the RPV mean values only. This assumption implies that the fluctuations of the RPV are enough resolved or they could be omitted. This is realistic for spray flames under study, since they tend to exhibit diffusion flame behavior in which the RPV fluctuations are not large compared to premixed cases. For partially premixed combustion that occurs in the stabilization of lifted diffusion flames, the suitability of this approach cannot be judged [74]. The joint PDF in (3.11) yields

$$P(z, y*) = P(z) \bullet P(y*) = \beta(z; \tilde{z}, \widetilde{z''^2}) \bullet \delta(\widetilde{y*}) = P(z; \tilde{z}, \widetilde{z'',^2}\widetilde{y*}) \qquad (3.12)$$

Thermo-chemical quantities can then be parameterized and tabulated in so-called pre-integrated tables (tabulated SGS chemical parameters) as function of filtered mixture fraction, its variance and normalized filtered RPV:

$$\tilde{\phi} = f(\tilde{z}, \widetilde{z'',^2}\widetilde{y*}) \qquad (3.13)$$

The reliability of spray combustion models lies primarily on how the fuel-air mixture preparation is accurately described. It is then of interest to better understand the behavior of mixture fraction variance when spray evaporation is present. For this purpose, a transport equation of mixture fraction variance is suitable [29, 54, 59]. Sadiki et al. [17] investigated the impact of the modeling of the evaporation source term in the transport equation of mixture fraction variance on the prediction of combustion properties in RANS context. Thereby the outcomes of two models have been compared. It turned out that the model by Reveillon et al. [57] could deliver more accurate prediction in comparison to the formulation by Hollman et al. (see in [59]). Because the proper contribution of evaporation source term in the equation of the RPV (see Eq. (3.1)) has been neglected as complete evaporation has been assumed before combustion, the mixture fraction variance is obtained simply following the algebraic gradient formulation

$$\widetilde{z''^2} \approx C_{eq}\Delta^2\left(\frac{\partial \tilde{z}}{\partial x_i}\right)^2 \qquad (3.14)$$

which assumes an equilibrium between production and dissipation [73]. The model coefficient, C_{eq}, lies into [0.1, 0.2] and is set to 0.15 in the present work.

The combustion of ethanol is modeled by means of a detailed chemical reaction mechanism as developed and validated by Marinov [89]. It consists of 56 species and 351 reactions.

3.3 Governing Equations and Modeling of the Disperse Phase

3.3.1 Governing Equations

The simulation of the evolution of the dispersed phase by the Lagrangian approach requires, first, the solution of equations of the position x_{pi}, the velocity u_{pi} and temperature T_{pi} along the trajectory of each computational particle in the carrier flow field. Assuming a spherical, single parcel these equations are:

$$\frac{dx_{pi}}{dt} = u_{pi}, \quad m_p \frac{du_{pi}}{dt} = F_i \tag{3.15}$$

$$m_p C_p \frac{dT_{pi}}{dt} = Q \quad \text{and} \quad \frac{dm_p}{dt} = -\dot{m}_{vap} \tag{3.16}$$

where $m_p$is the mass of parcel, C_p the heat coefficient, F_i denotes the summation of all the forces acting on the particle and Q the net rate heat transfer to the particle while $\dot{m}_{vap}$ expresses the droplet vaporization rate, respectively. The particle dynamics is described according to the Basset-Boussinesq-Oseen equation (BBO-equation) with all force contributions (drag, mass added force, Basset force, pressure gradient, buoyancy and gravity effects) in (3.15). While writing (3.16b) temperature variation inside the particle is neglected and thus particle temperature is considered uniform. These equations for a single parcel in a turbulent flow have been already discussed in a vast number of works. A recent review is provided by Mashayek [90]. Since the density ratio between the disperse phase and the carrier phase is approximately three orders of magnitude greater than unity, the forces that contribute to the droplet motion are drag, gravitation and buoyancy forces [19, 21–28, 64]. Equation (3.15b) then reduces to:

$$\frac{du_{p,i}}{dt} = \frac{3}{4}\frac{C_W}{D_p}\frac{\rho}{\rho_p}\left|\vec{u} - \vec{u_p}\right|\left(u_i - u_{p,i}\right) + \frac{\left(\rho_p - \rho\right)}{\rho_p} g_i = \frac{1}{m_p}\sum_j F_j \tag{3.17}$$

The drag coefficient C_W used to model the complex dependency between the droplet and the flow is not constant but depends on relative velocities, viscosity of the disperse phase and carrier phase, the particle rand shape and the roughness of the particle's surface. The drag coefficient used within this work is determined for a spherical, not deformable, droplet according to (see in [19, 21–28]):

$$C_W = \frac{24}{\mathrm{Re}_p}\left(1 + \frac{1}{6}\mathrm{Re}_p^{2/3}\right) \quad \mathrm{Re}_p \leqslant 1{,}000 \tag{3.18}$$

$$C_W = 0.44 \quad \mathrm{Re}_p \geqslant 1{,}000 \tag{3.19}$$

where Re_p denotes the particle Reynolds number calculated using

$$\mathrm{Re}_p = \frac{D_p \left|u - u_p\right|}{\nu} \tag{3.20}$$

In (3.17) the first term includes the particle-relaxation time, τ_d, that is given by

$$\tau_d = \frac{4 D_d \rho_d}{3 C_W \, \bar{\rho} \, |u_i - v_i|} \tag{3.21}$$

Thereby D_d is the particle diameter, ρ_d the density of a particle group. The quantity ν is the kinematic viscosity of the fluid.

3.3.2 *Modeling of Dominant Micro-processes*

3.3.2.1 Turbulence Modulation Modeling and Two-Way Coupling

The interactions between the carrier gas phase and droplets depend on many factors (e.g. gas and particle/droplet physical properties, phase velocities, concentrations, temperature and turbulent fluctuations, swirl etc.). According to [78] they can be classified in three main categories: (a) Turbulent dispersion of the dispersed phase; (b) Modification of the turbulence characteristics of the continuous phase by the dispersed phase or by inter-phase transport, or so-called turbulence modulation; (c) Modification of the inter-phase transport by turbulence. The two first aspects will be addressed here, while the third results basically from the coupling between the phase change process and turbulence or other internal or external effects, like swirl, temperature, etc. With respect to numerical characterization of the interaction between turbulence and vaporization processes in the frame of RANS the reader may refer to Sadiki et al. [78].

In the LES the effects of the small scales on particle dispersion, motion or deposition have been either modelled separately or neglected. One of the early contributions is due to Yeh and Lei [91]. They investigated dispersion of spherical solid particles in a confined isotropic turbulence. The effect of the particles on turbulence was neglected while their motion was described by accounting for the inertia, modified Stokes drag, and gravity forces. In this equation, only the resolved velocity was used, thus neglecting the effects of the velocity fluctuations in the SGS. Thereby the SGS stress tensor was modeled by means of the eddy viscosity model in which the length scale was a constant times the filter size.

Turbulent Dispersion

To quantify the instantaneous fluid velocity, seen in the droplets in Eq. (3.15) and its effect on the droplets distribution, the SGS values of the fluid parcel velocity at the droplet location should be modeled. This can be done using a stochastic Lagrangian process using the fluid turbulent variables [42]. It is the common practice in the Eulerian–Lagrangian LES studies of dispersed flows to neglect the SGS flow scales [21–27]. It has been argued that the long-time particle dispersion is governed by the resolved, larger-scale fluid eddies. As it is known from recent studies by Pozorski et al. [63, 65] the impact of SGS dispersion can vary depending on the particle inertia parameter. In particular, for evaporating spray flow, the droplets become increasingly smaller and their inertia parameter changes, hence sooner or later the droplets unavoidably enter the size range where there is an impact from the flow SGS. For forced isotropic turbulence the direct effect of SGS velocity on the droplet motion has been found by Pozorski et al. [65] to be small in well resolved regions of the flow where the amount of energy in SGS is small (less than 30%). Bini and Jones [28] mentioned that the SGS dispersion plays a crucial role in determining the spreading and properties of the disperse phase. Dianat et al. [35] showed that the incorporation of a SGS dispersion model tends to improve the predictions of concentration. Shotorban et al. [41] demonstrated through an asymptotic approach when the SGS effects are neglected on particles in an isotropic turbulent flow, LES may not be able to predict the particle concentration.

Modulation Modeling

In addition to the turbulent dispersion, the two-way coupling in two-phase flow investigations is fully completed by introducing additional source terms $\bar{S}_{\phi,p}$ and $\bar{S}_{\phi,p,v}$ arising from the dispersed phase description into the equations of the continuous phase (1). These contributions characterise direct influences of the dispersed phase on the gas phase in mass, momentum, mixture fraction, scalar species as well as in turbulent quantities. For LES, details about these terms can be found in [19, 21, 24, 51, 68]. The calculation of these contributions is mostly based on the particle source in cell methodology or a modified PSI-CELL approach and applied at respective cell centroids in a manner consistent with the LES scheme used.

In fact the presence of particles can have twofold effect. The first is directly on the LES resolved transported variables (momentum, mixture fraction, scalar concentration) due to the coupling terms. This is called a simple two-way coupling. In the case of the momentum, the source term will change the resolved velocity field, which in turn will change the resolved kinetic energy. The second (indirect) effect is the modification to the SGS kinetic energy due to the additional source term, which accounts for the work done due to interaction between particles and the small-scale unresolved turbulent field. In addition to the turbulent dispersion, if both effects are accounted for, then a fully two-ways coupling is achieved. In the following, turbulence modulation without phase change is first addressed and then completed by a formulation under phase change conditions.

Only very recently, the two-way coupling was treated by taking into account the effects of particles on the SGS viscosity (see in [68]). Based on the Smagorinsky ansatz and following some RANS approaches the idea consists in heuristically adding a new term in the expression of the SGS viscosity. In this regard Lei et al. [92] presented two interesting works based on the ansatz by Yuu et al. in which the Smagorinsky model with the Germano-procedure (see in [71]) is considered. In first assessment and validation of this model, Lei et al. compared their numerical results with those obtained by using the Smagorinsky model with wall Van-Dries consideration applied to a particle-laden, vertical channel flow.

To well account for the second (indirect) effect in RANS, besides the standard approach, several models exist in the literature. Crowe [93] applied the first law of thermodynamics to well account for the weak turbulence along with the augmentation of the turbulent kinetic energy due the presence of particles. The experiment by Tsuji et al. [94] was used for model validation as Lain and Sommerfeld [95] did for the comparison between the standard approach and the model by Crowe. To account for the modification to the SGS kinetic energy due to the additional source term in the LES context, it is now recognized that a suitable approach is to follow the RANS approaches by modifying the generation and dissipation terms in the SGS turbulent kinetic energy equations, SGS dissipation rate equation as well as Reynolds stresses transport equations (if considered) by means of appropriate source terms.

Patel et al. [30, 51] account for the two-way coupling in their LES of spray combustion in swirling flows. Thereby, the effects of swirl intensity and heat release have been investigated. In the SGS turbulent energy, they considered the standard RANS-formulation by approximating the turbulent energy source term as

$$\bar{S}_p^k = \overline{u'_i S'_{pi}}^m \approx \overline{u_i S_{pi}^m} - \bar{u}_i \bar{S}_{pi}^m \tag{3.22}$$

and adapted it to LES where the averages are carried out for all particle trajectories crossing the cell and cell volume. The index i indicates that the calculation of the particle source term is performed by a summation over all velocity components. This approximation seems to capture well the attenuation phenomena, as reported later on. No special treatment of the SGS dissipation rate was considered.

To consistently describe both the attenuation and the augmentation of the gas phase turbulence due to the presence of particle, and the possible balance of both, Sadiki and Ahmadi (see in [78, 82]) pursue another way. Because all the phenomena involved are thermodynamic processes, they included and exploited the second law of thermodynamics into the modelling. This yields a model compatible with the irreversibility requirements of the second law of thermodynamics which governs every physical process [80]. This model has been expressed in a Eulerian-Lagragian form in [82] and accounts in a physically consistent way for a fully two-way coupling. Recently, this model has being incorporated in a LES code by Groh [68] to account for these effects through the SGS kinetic energy. This relies on the one-equation SGS-model according to the formulation by Yoshizawa (see Eq. 3.7).

To validate and assess the new model, the results from three models representing particle source terms in the SGS kinetic energy equation will be compared to each other and to experimental data, respectively. These are

1. The standard model, expressed in the well-known way as

$$F_{S_{k^{sgs}}} \equiv F_{k,p} = \overline{u_i S_{u_i,p}} - \overline{u_i}\,\overline{S_{u_i,p}} \tag{3.23}$$

2. The Crowe model based on a consideration of the energy balance equation expressing the first law of thermodynamics and reformulated in a Lagrangian framework by Lain and Sommerfeld [95]

$$F_{k,p} = \left(\overline{u_{pi} S_{u_i,p}} - \overline{u_i}\,\overline{S_{u_i,p}}\right) \tag{3.24}$$

3. The so-called thermodynamically consistent model that results from the exploitation of the entropy inequality expressing the second law of thermodynamics as suggested by Sadiki et al. [78, 82]

$$F_{k,p} = \beta\left(\overline{u_{pi} S_{u_i,p}} - \overline{u_i}\,\overline{S_{u_i,p}}\right) + \left(\overline{u_i S_{u_i,p}} - \overline{u_i}\,\overline{S_{u_i,p}}\right) \tag{3.25}$$

with the model parameter

$$\beta = \alpha' + \frac{(1-\alpha')\left(\overline{u_{pi} S_{u_i,p}} - \overline{u_{pi}}\,\overline{S_{u_i,p}}\right)}{\left(\overline{u_{pi} S_{u_i,p}} - \overline{u_i}\,\overline{S_{u_i,p}}\right)} \tag{3.26}$$

The second term in the model (3.25) represents the usual (dissipative) standard contribution, while the first term accounts for the production of the turbulent kinetic energy. This term corresponds to the model by Crowe [25] once $\beta = 1$. The parameter α' depends on the particle properties, in particular on the particle diameter. So, this thermodynamically consistent model is designed to well capture both the enhancement and the diminution of the turbulence of the gas phase due to the presence of both big and small particles/droplets, as it is the case in sprays in internal combustion engines. More details related to the thermodynamically consistent modeling strategy and to the exploitation of the entropy inequality are reported in [78, 82, 101]. In this work, a value of $\beta = 0.1$ has been used.

To account for mass and heat transfer process (interface transport), the new turbulence modulation model needs to be extended by including evaporation process contributions. The resulting expression is given as

$$F_{k,p} = \beta\left(\overline{u_{pi} S_{u_i,p}} - \overline{u_i}\,\overline{S_{u_i,p}}\right) + \left(\overline{u_i S_{u_i,p}} - \overline{u_i}\,\overline{S_{u_i,p}}\right) + \frac{1}{2}\left[\beta\left(\overline{u_i}^2\,\overline{S_{u_i,p,v}} - \overline{u_{pi}^2 S_{u_i,p,v}}\right) + \left(\overline{u_i}^2\,\overline{S_{u_i,p,v}} - \overline{u_i^2 S_{u_i,p,v}}\right)\right] \tag{3.27}$$

The reader must recall that the quantity $\bar{S}_{u_i,p}$ expresses the particle/droplet source term in the Navier–Stokes equation, while $\bar{S}_{u_i,p,v}$ is the droplet source term in the continuity equation of the gas phase under consideration of evaporation process (see Eq. (3.1)).

To account for the dispersion of particles, the stochastic motion of particles due to turbulent motion is described by representing the instantaneous gas phase velocity at the particle position as the sum of the local gas phase filtered velocity and the subgrid-root-mean-square velocity. In the expression of the latter contribution a random number sampled from Gaussian distribution (with zero mean) has been introduced.

Evaporation Process

To account for the phase change (evaporation) a three dimensional vaporization model that integrates interactions with other micro-processes needs to be considered. Its influence on the carrier gas phase is already described by the term,$\bar{S}_{\phi,p,v}$, in Eq. (3.1). With respect to interaction with combustion, an artificially thickened flame model without including interactions of evaporating droplets and combustion has been used in [26, 27]. In [24, 26] a flamelet model with a reactive progress variable was used while in [30, 51] the linear eddy model was applied. In general little attention has been put on evaporating droplet-turbulence-chemistry interactions. Jones et al. [83] developed an LES for spray based on a Eulerian transported probability density function (PDF) method to investigate a high-swirl reacting kerosene flow. A detailed description of the interaction between the subgrid turbulence, evaporation and the chemistry was accounted for.

In existing LES spray investigations the uniform temperature model (UTM) [19, 21–28, 30, 37] for evaporation was often applied based on thermodynamic equilibrium conditions.

A non-equilibrium evaporation model as suggested in [43] (see also Sadiki et al. [78, 82]) applied in this work as an extension of the Abramson and Sirignano based equilibrium model [37, 43, 60]. Non equilibrium effects were captured by determining the departure from thermodynamic equilibrium by adding a deviation term to the molar fraction of the vapor at the droplet surfaces [43]. The model describes the evolution of the droplet's temperature and diameter, i.e. evaporation rate and energy flux through the liquid/gas interface. It is based on the film thickness theory and does not consider any temperature variation in the interior of the droplets (homogenous temperature). However, the temperature variation has an unsteady nature and is accompanied by an unsteady mass transition. As this model does not account for the gradient at the interior of the droplet, the latter ones are not discretized. Therefore, this model does not require long computing times. The assumption of uniform temperature is reasonable since dragged droplets have diameters in the range of 30 μm.

In particular, break-up and coalescence are neglected to ensure that the evolution of the droplet diameter is only due to the evaporation processes. The Weber number near the nozzle (x/D = 0.3), which is used as an indicator for the break-up phenomenon, is less than 0.3 in the configuration. The critical value, however, is about 40 orders of magnitude larger, i.e. $We_{cri} = 12.07$. The Ohnesorge Number $Oh < 0.006$ for all cases at the exit plane, therefore no further drop deformation and break up are possible downstream of the exit plane and thus the changes in the droplet size distribution are due to evaporation only. Recent review of evaporation models has been reported in [38, 43, 45, 60], among others.

The physical properties of the air vapor mixture (in the gaseous film around the droplet) are determined using the reference temperature and mass fraction. They are calculated using the Simpson or Sparrow & Gregg "1/3 rule" [19, 21–28, 43].

3.4 Numerical Procedure

The governing equations of the gas phase are discretized in the 3D low-Mach number LES code FASTEST code featuring flexible, block structured and boundary fitted grids [70]. The code uses the finite volume method in which a co-located grid with a cell-centered variable arrangement is applied. For spatial discretization, specialized central differencing schemes that hold the second order for arbitrary grid cells are used. The convective term in the scalar transport is discretized using non-oscillatory, bounded TVD schemes. For the time stepping multiple stage Runge–Kutta schemes with second order accuracy are used. A fractional step formulation is applied and at each stage a momentum correction is carried out in order to satisfy continuity. Hence a Poisson equation is derived and solved iteratively with multi-grid and SOR relaxation. The parcels are tracked using LAG3D code [78, 82] in which the equation of motion, the temperature evolution and the evaporation rate are discretized using Euler first/second order schemes and solved explicitly. The computational time needed to produce statistically reliable results is 4 weeks on 8 cores parallelized workstation.

The transfer of the carrier phase properties and the disperse phase source terms during the coupling process is set to 0.1% of the turbulent time scale. Simulating two-phase flow, by mean of unsteady coupling in the E-L approach, yields a major difficulty in resolving Partial Differential Equations (PDE) with different time scales for the fluid flow and droplet motion. The droplet time step could reach values below 10^{-7} s in case of droplets having diameter in the range of micrometer intervals, as it is the case during the evaporation process. Moreover every droplet in turbulent flow features its own time scale. The coupling between both phases within every time step borders toward the direct numerical simulation. It certainly reveals the exact dynamic of the dispersed phase; however the computational costs are extensive, particularly for high number of numerically tracked droplets. In order to reduce the computational costs, the coupling between both phases will be controlled not by the droplet time scale but using the time marching algorithm of the continuous phase, i.e. 10^{-4} s.

3.5 Some Achievements and Model Validation

Different configurations have been used for the model validation. The first class of configurations includes a vertical particle laden channel flow [81] and a swirled annular gap two-phase flow with solid particle [96] for validation of the modulation model. The main results have been reported by Groh et al. [68]. The second class consists of a swirled annular gap two-phase flow with evaporating spray [97], an evaporating n-heptane spray in a single combustor [11] and the spray jet series from [53] to validate the droplet evaporation process. The third class is constituted by a series of turbulent spray jets with various liquid fuels such as acetone and ethanol [53] and a reacting n-heptane spray in a single combustor [11] for validation of combustion spray modeling [19]. The experimental validation data were made available in the framework of the international workshop on turbulent spray combustion co-initiated and co-organized by the SFB568 [53]. Thereby valuable and comprehensive validation data have been gathered for a series of spray jets established for different ranges of jet velocity and liquid mass flow rates [53, 61].

In the present work, some achievements are especially selected to illustrate how the models for turbulence modulation and spray combustion have been validated and evaluated. More details about all the achievements accomplished by applying the methodology suggested in this paper contribution are reported in [1–20], among others.

3.5.1 LES of a Swirled Annular Gap Two-Phase Flow

The new model has been first used in predicting the particles properties and the flow field in a well documented, turbulent two-phase flow configuration. The particle-laden swirling, coannular jet, experimentally investigated by Sommerfeld and Qui [96] has been chosen. This configuration with highly controlled conditions and well defined boundary conditions allows to systematically validate LES of dilute particle dynamics, comparable to those in combustion chambers, avoiding additional complexity due to e.g. phase exchange or atomization. In particular, this test facility is chosen to produce recirculating flow conditions as they are typical for gas turbine combustors. The test section consists of a downward facing round pipe. Two separated flows, one primary particle laden jet flow surrounded by a secondary annular swirling flow, are streaming into the combustor. The flow rates of both ducts are measured separately and provided. The swirl of the annular flow is generated by a radial swirl generator. A detailed description of the experiments along with geometry and dimensions of the test section of the investigated case can be found in [96] while the specific flow conditions under investigation here are summarized in Table 3.2.

The 1.5 m long tube surrounding the test section is out of Plexiglas for complete optical access. The tube length ensures insensitiveness to downstream flow phenomena. The rigs is set up vertically with gravity acting in downstream

Table 3.2 Flow conditions under study [96]

Carrier phase	
Primary mass flux	0.0099 kg/s
Secondary mass flux	0.0383 kg/s
Reynolds number	52,400
Swirl number	0.47
Particle phase	
Mean particle diameter	45.5×10^{-6}m
Particle mass density	2,500 kg/m^3
Particle mass flux	0.00034 kg/s
Loading	0.034

direction. PDA measurements are reported at seven cross sections, face normal to the axial direction. At these experimental planes gas phase and particle phase measurements of mean and rms velocity are available, as well as particle size and mass flux distributions.

The computational domain is represented by approximately 1.000 000 control volumes. For the inlet mean velocities are superposed with turbulent fluctuations. These fluctuations are generated using the digital filters method by Klein et al. [98].

While for the mean velocity of both jets and bulk velocity was set according to experimental data, the swirl velocity component of the annular flow was modeled as fixed body rotation.

This study applies an Eulerian–Lagrangian approach as described above. Particles are injected in equilibrium with the gas phase velocity. Particle sizes are generated with a random process to fit the experimental data. Similarly the particle position is chosen, with constant probability in a plane normal to the axis. At the wall, particles are reflected by fully conserving their kinetic energy.

From the simulation of this flow laden by such very small particles under this loading condition it is to expect that the modulation process will not play a significant role in this configuration. Two different model levels have therefore been considered: (a) A one-equation based model as described above that well accounts for the modulation process by means of the thermodynamically consistent model and (b) A zero-equation based formulation following Smagorinsky/Germano model in which account is not made for specific influences of the dispersed phase.

A complete dataset containing statistical data for the continuous and the dispersed phase is provided. In particular the unsteady processes have been captured and reported in [15]. Focused on the turbulent quantities, some LES results are presented in comparison with experimental data in Figs. 3.1 and 3.2 for the calculated mean radial velocity of the carrier and dispersed phases as well as corresponding rms of the carrier phase. No particular effect of the turbulence modulation modeling could be observed.

In general, it is not only possible to capture mean flow phenomena, but also the level of mean velocities and fluctuations are reproduced relatively good. Even the radial velocity component which normally is the easiest to fail due to its smallness is satisfying. The location of the recirculation zone is predicted satisfactorily as

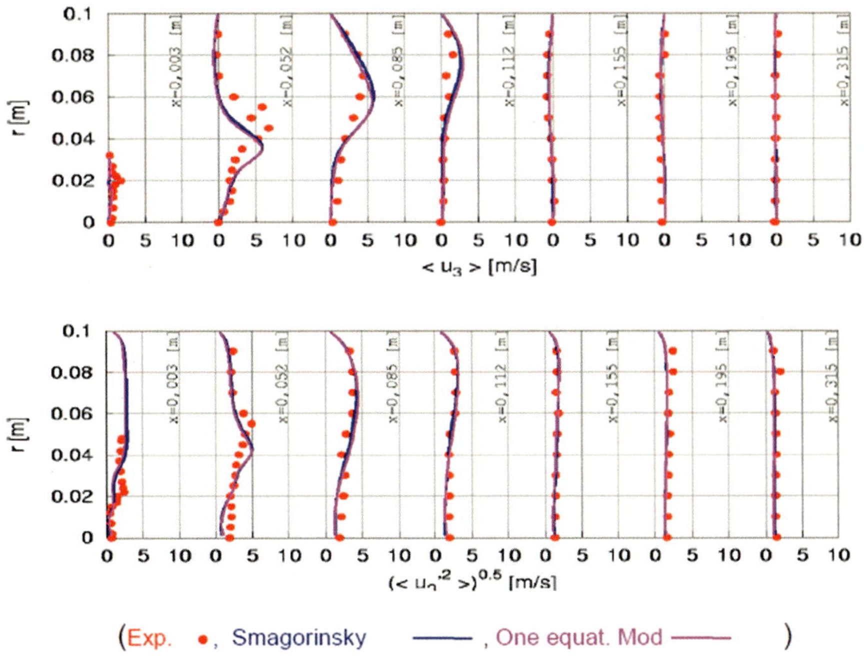

Fig. 3.1 Comparison of calculated mean radial velocity (*top*) and corresponding rms (*bottom*) of the carrier phase

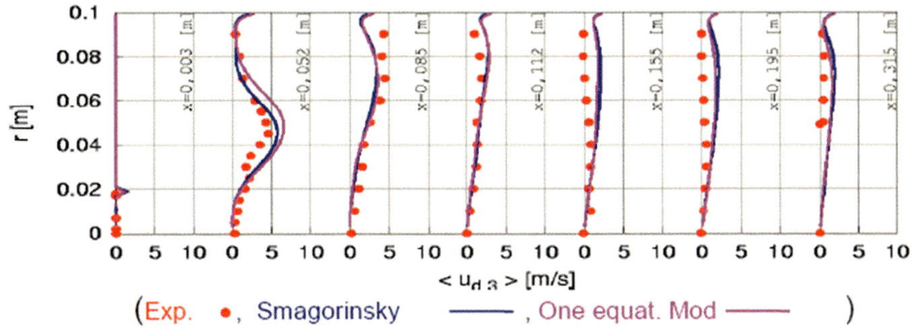

Fig. 3.2 Comparison of calculated mean radial velocity components of the dispersed phase

demonstrated by the velocity axial components (not shown). This influences the dispersed phase prediction as obvious in Fig. 3.2. Both simulations achieve almost the same predictions supporting the fact that modulation processes are negligible in this configuration.

To conclude, let us observe that the chosen grid resolution seemed to be near the lower limit that is necessary for capturing all physics of the continuous phase with such a simple SGS model we used. As long as one relies on simple models, the next

Table 3.3 Particle properties of the considered flow configurations [81]

Case	D_d [m]	ρ_d [m^3/kg]	ϕ_d [−]
A	–	–	0.0
B	1.22×10^{-4}	2, 440	2.3×10^{-5}
C	1.22×10^{-4}	2, 440	4.5×10^{-5}
D	1.22×10^{-4}	2, 440	8.5×10^{-5}
E	4.80×10^{-4}	2, 440	4.1×10^{-4}

step is to use the finest grid possible. In fact, the finer the grid size, the better is the resolution of turbulence scales and the less important are the modeling assumptions. Otherwise it is recommendable to employ the one-equation SGS models as it will be demonstrated in the next validation case. Thereby the volumetric loading ratio is high and particle diameters are not very small. This is the situation in dense spray regions encountered in the near of injector nozzles in gas turbine combustors.

3.5.2 LES of a Particle-Laden Vertical Channel Flow

The modulation model has been especially validated by using data from comprehensive experiments performed at the institute of Energy and Power Plant Technology at the Technical University Darmstadt (TUD) in a special designed vertical closed-circuit wind-tunnel [81]. The considered basic flow is a grid-generated isotropic homogeneous turbulent flow. The hole-diameter M of the grid is 0.012 m and the bulk velocity is up to 12 m/s, corresponding to a Reynolds-number of 64, 000, relying on the channel half-width. The test-section is designed as a vertical square tunnel with inner dimensions of 0.2 m by 0.2 m. Air is used as the continuous phase and solid non-reacting particles with various diameters and different volumetric loading ratios act as the dispersed phase. All properties of the dispersed phase are listed in Table 3.3.

A full data set for this configuration is available from experiments using 2-D phase-Doppler anemometry (PDA) to obtain velocities and higher moments for both phases. Particle concentration measurements are performed by conventional probe (patternator) techniques. All data and a detailed description of the experimental setup can be found in Geiss et al. [38].

We rather focus on comparisons between the experimental data and the numerical simulations. After validating the single phase calculations three simulations were performed. They were achieved by taking into account the two-way coupling effect in the transport equation for momentum and SGS turbulent kinetic energy, in which the different modeled terms (called in the legend as standard (Eq. 3.23), Crowe (Eq. 3.24) and Sadiki and Ahmadi (Eq. 3.25)) have been considered, respectively.

For the LES computation, computational grids ranging from $128 \times 32 \times 32$ to $128 \times 64 \times 64$ cells were used. To track the parcels, 300,000 numerical particles were sufficient, each numerical particle representing up to 50 real particles in experiment.

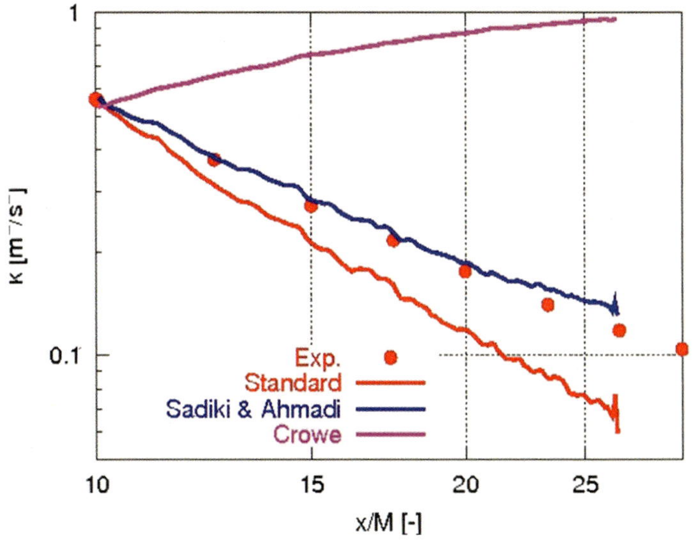

Fig. 3.3 Comparison of calculated decay of turbulent kinetic energy along the centreline for case D (small particles) in logarithmic plotting

The numerical results for the decay of the turbulent kinetic energy produced by the presence of small and big particles, respectively, along the centreline will be discussed. The results in Fig. 3.3 for a particle diameter of 120 μm show that the standard formulation is far too dissipative and cannot predict the expected attenuation of the turbulence induced by the dispersed phase. On the other hand, the turbulence modulation model derived by Crowe predicts even turbulence augmentation, when it is expected, that the turbulence is attenuated by the presence of small particles. The turbulence model formulated by Ahmadi, Sadiki et al. provides results in good agreement with experimental data.

In predicting the turbulence modulation induced by large particles (480 μm), the formulation of Crowe shows tremendous over-prediction for the induced turbulence, while the standard model is still dissipative. Again, nice results are obtained by the thermodynamically consistent model with the same model coefficient of 0.1 applied for the prediction of laden flow with small particles (Fig. 3.4).

In order to examine the sensitivity of the new model in capturing the behaviour of the turbulent kinetic energy as function of the volumetric loading ratios, the new model is used to reproduce the energy spectra characteristics modified by the presence of particles. We display in Fig. 3.5 LES calculated spatial energy spectrum for case A (single flow, here as reference) and for cases B and C (laden particle flow, see Table 3.3). As well known from DNS and experiments [99] one can see the redistribution of turbulence energy due to the presence of particles is not uniform. Furthermore, the energy content of small scales is increased in relation to a flow without particles (case A) when the mass loading is high (case C). Particles transfer

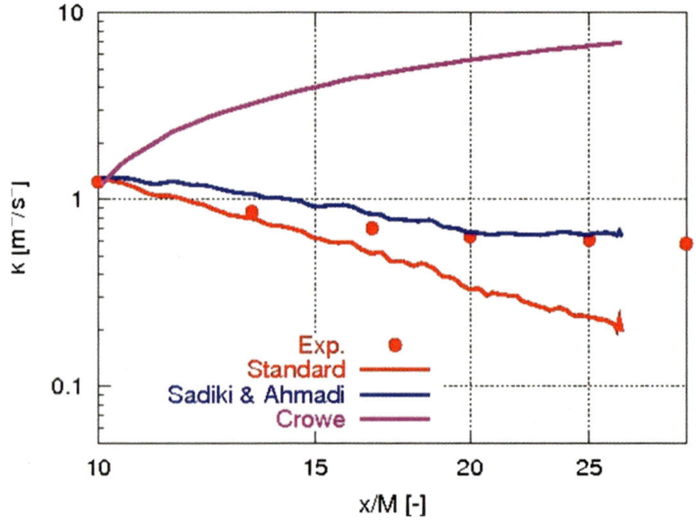

Fig. 3.4 Comparison of calculated decay of turbulent kinetic energy along the centreline for case E (large particles) in logarithmic plotting

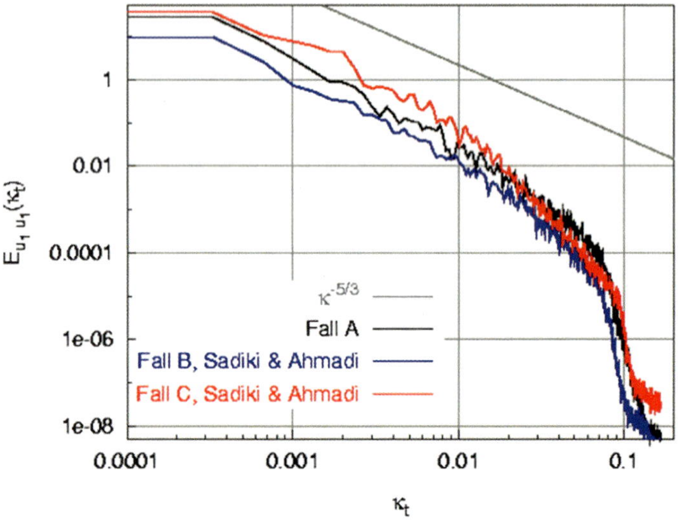

Fig. 3.5 Comparison of LES calculated spatial energy spectrum for single phase (case A) and flow with particles

their momentum to the high wave number motion of the carrier fluid. In contrast, the energy content of small scales decreases when the particle mass loading is two times small (case B). Regarding the region of low wave number motion, the presence of particles attenuate the fluid turbulence level according to the findings in [99].

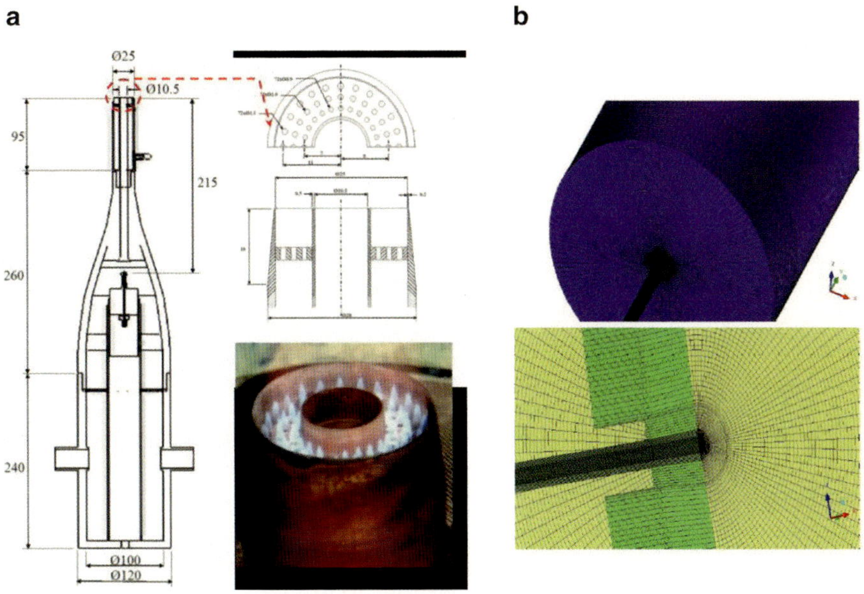

Fig. 3.6 (**a**) Schematic of the spray burner set up, (**b**) Computational domain

3.5.3 LES of Evaporating and Combusting Sprays

The aim of this section is to demonstrate the predictive capability of combustion LES coupled to Lagrangian droplet dynamic model together with a tabulated detailed chemistry based on FGM approach to retrieve the turbulent droplet dispersion, size distribution, evaporating spray evolution and combustion properties for ethanol spray flames.

A detailed description of the experimental setup and apparatus used for the generation of the experimental data is provided by Masri and Gounder [61] (see also [53]). The geometry of the configuration used to study the ethanol spray combustion is shown in Fig. 3.6a. The burner is mounted vertically in a wind tunnel that supplies a co-flowing air stream of 4.5 m/s. The co-flow of diameter that surrounds the burner and the coflow/burner assembly is 104 mm. The contraction of the carrier phase topology has a ratio of 10:1. The outer diameter of the annulus is 25.0 mm whereas the lip thickness is 0.2 mm. The pilot flame that is set to a stoichiometric mixture of hydrogen, acetylene and air has an unburnt bulk velocity of 1.5 m/s. Its border is mounted 7 mm upstream of the nozzle exit and contains 72 holes. The co-flow and nozzle exit plane is 59.0 mm downstream of the tunnel exit plane. This provides an unconfined working section. The spray is initialized 215 mm upstream of the nozzle exit plane and demonstrates a polydisperse behavior after traveling a prevaporization zone in which small classes evaporate before reaching the exit of the nozzle. The resulting ethanol flame features a partially premixed character.

Table 3.4 Different cases investigated and operating conditions [53]

Reacting ethanol spray cases	EtF 3	EtF 8
Bulk jet velocity U_{jet} (m/s)	24	48
Carrier mass flow rate (g/min)	150	301
Liquid fuel injection rate (g/min)	45	45
Measured liquid flow at exit (g/min)	30.7	36.6
Vapor fuel flow at jet exit (g/min)	14.3	8.4
Overall equivalence ratio, $\Phi_{overall}$	2.9	1.4
Equivalence ratio at jet exit, Φ_{exit}	0.85	0.25
Spray jet density (kg/m^3)	1.47	1.35
Spray jet vapour phase viscosity (10^{-5}) (kg/m-s)	1.88	1.96
Jet Reynolds number	19,678	34,848

While in [79] the non-reacting and reacting cases SP1, SP2 and EtF1, EtF4 and EtF7 have been investigated using RANS, the investigated configurations in this work are EtF3 and EtF8 (see Table 3.4). Keeping the same liquid fuel injection rate, the jet Reynolds number is varied as well as the carrier mass flow rate along with the spray jet density. All the boundary conditions are provided in Table 3.4. A decreasing mass loading in the inner jet from 30 to 15% could be calculated. The velocity components of the carrier phase are given as block profile at the inlets and the Reynolds number varies between 19,678 and 34,848 attesting a highly turbulent two phase flow. As the carrier phase travels a distance 20D to reach the nozzle exit plane, the flow develops turbulent structures, even with block velocity profiles.

The computational domain consists of 17 blocks that count 1.1×10^6 control volumes (cv), Fig. 3.6b. The number of parcels injected within one coupling time step is 2,500. The number of time steps achieved between both phases, that represent the fluid data and/or source term transfer, exceeds 320,000 couplings. Thus the averaging of the spray properties is performed over more than 750×10^6 parcels. The disperse phase properties are statistically independent and not conditioned on the number of parcels tracked or coupling time steps. The TVD scheme is applied for the velocity exit boundary with a condition with 6 m/s. The boundary condition for the RPV is set to zero in the entire domain except at the pilot flame inlet, which is set to the maximum absolute value that equals 0.0101.

The spray is injected by an ultrasonic nebulizer. The droplet size distribution produced by the ultrasonic nebulizer is known to be approximately lognormal. Close to the jet wall the measured droplet size PDF shows a bias towards small droplets. The droplets deformation regime map as a function of Weber and Ohnesorge (Oh) numbers, provided by Faeth [100], was used for determining if further droplet deformation or break up occurred in the spray jets investigated here. The Weber number was found to be less than 0.3 and Oh was less than 0.006 for droplets in the sizes range $40 < D_p < 50$ μ indicating that the droplets are spherical and will not undergo any further deformation due to droplet break up, downstream of the nozzle exit plane. The simulations are performed using 12 different classes of droplets.

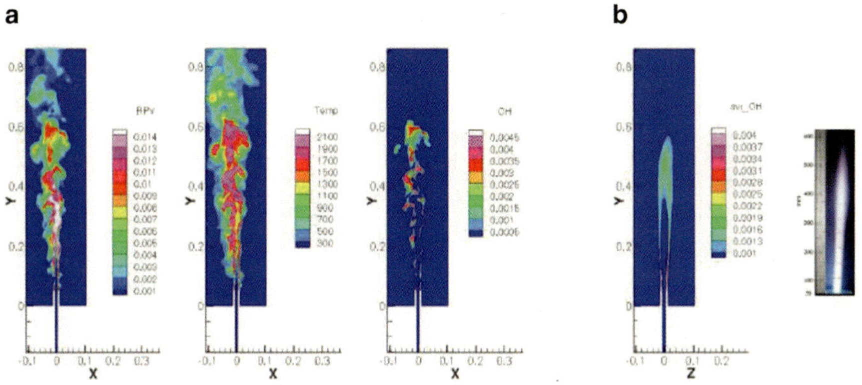

Fig. 3.7 (**a**) Instantaneous plots of the RPV, temperature and OH mass fraction of the ethanol spray flame, EtF3, at the axis-symmetrical cross section; (**b**) Numerical averaged OH mass fraction of the Ethanol spray flame EtF3 (*left*). Experimental real photos of the Ethanol spray flame EtF3 (*right*)

Figure 3.7a shows instantaneous contour plots of the RPV, temperature, CO2 and OH in a cross section along the symmetry axis of the configuration for the test case EtF3. The RPV varies between "0" and 0.14. The maximum value is obviously registered at the reaction zone where mixture fraction is close to stoichiometry. Downstream of the nozzle exit and at the centerline of the configuration, no flame is observed, probably caused by the high strain rate at the nozzle exit plane or by a lack of sufficient heat to maintain the combustion at the centerline. Further downstream, at $y = 0.3$, due to momentum transfer from axial to radial direction the RPV reaches the maximum value. The temperature plot is a mimic of the RPV and the mixture fraction. At the centerline and downstream the nozzle the temperature should reduce its value below the inlet boundary condition because of spray evaporation. This heat loss effect is not considered in the modeling yet. Depending on the evaporation rate the temperature plot is changing. If droplets evaporate within the pre-vaporization zone completely, the combustion regime will likely to demonstrate a premixed nature.

For very slow vaporization, i.e. most droplets exit the nozzle without evaporation, the spray flame is likely to have a diffusion flame behavior. Worth mentioning that, numerically, the combustion cannot be sustained without setting the RPV boundary condition at the pilot flame location to its maximum value. An initialization of the RPV at the nozzle exit plane is not enough to make the mixture burn, because the RPV is transported downstream and the mixture blows out. In order to stabilize the flame an initialization requires a recirculation zone which is not present in this configuration. The CO_2 contour (not shown) replicates the temperature distribution pattern since it is the main species describing the RPV. It varies between the maximum values of ca. 0.15 and its minimum mass fraction in the air which is $0.359 \cdot 10^{-3}$. The OH mass fraction, which is an indicator of the position of the flame varies between "0" and $4.5 \cdot 10^{-3}$. The averaged value of the OH mass fraction

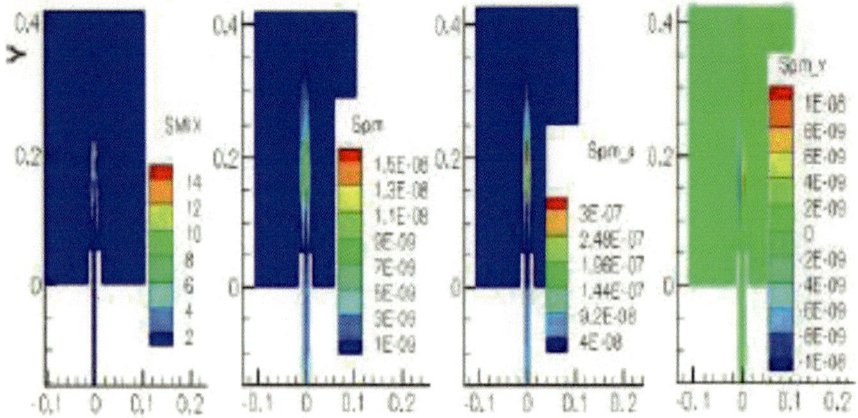

Fig. 3.8 Source terms of the RPV, mass conservation/mixture fraction, axial and radial momentums for the test case EtF3

which is shown in Fig. 3.7b and compared to the experimental measurement gives the height of the ethanol spray flame EtF3 (h = 0.56 m) and EtF8 (h = 0.52 m, not shown). A good agreement with the experiment is observed.

Figure 3.8 shows the contour plots of the source terms for the RPV, mass conservation and momentums in axial and radial direction. The first source term is provided by the FGM table whereas the three others are computed by the Lagrangian code and represent the absolute value needed during the coupling between both codes. The disperse phase source terms depend on the dimension of the numerical cv, the time needed to cross a cell and on the number of droplets that crossed the cv. A smooth distribution of the source terms has the advantage of accelerating convergence of the carrier phase simulation as well as the convergence of the coupling within one time step. Droplets start to evaporate immediately after injection.

When reaching the nozzle exit their surface temperature is reduced because of loss of the latent heat, thus the evaporation decreases. By reaching the combustion region, the carrier phase temperature increases and provides the necessary heat to make the spray evaporates. This explain the large value of the mass source term at region of combustion, i.e. y = 0.2 m. The axial momentum source terms (Spm_u) are important at the inlet and at the combustion region. This is caused by the high relative velocities in these regions. At the inlet, droplets are generated by an ultrasonic nebulizer, which necessitate a certain relative velocity to drag the spray. The radial source (Spm_v) terms are symmetric to the centerline featuring a negative sign because of radial velocity profile. The axial source terms are 30 times order of magnitude larger than the radial ones.

Figure 3.9a, b show the axial droplet velocities and corresponding fluctuations of both test cases EtF3 and EtF8. Reasonable agreement is observed in the first two cross sections. At x/D = 20 and x/D = 30, small discrepancies are observed in the

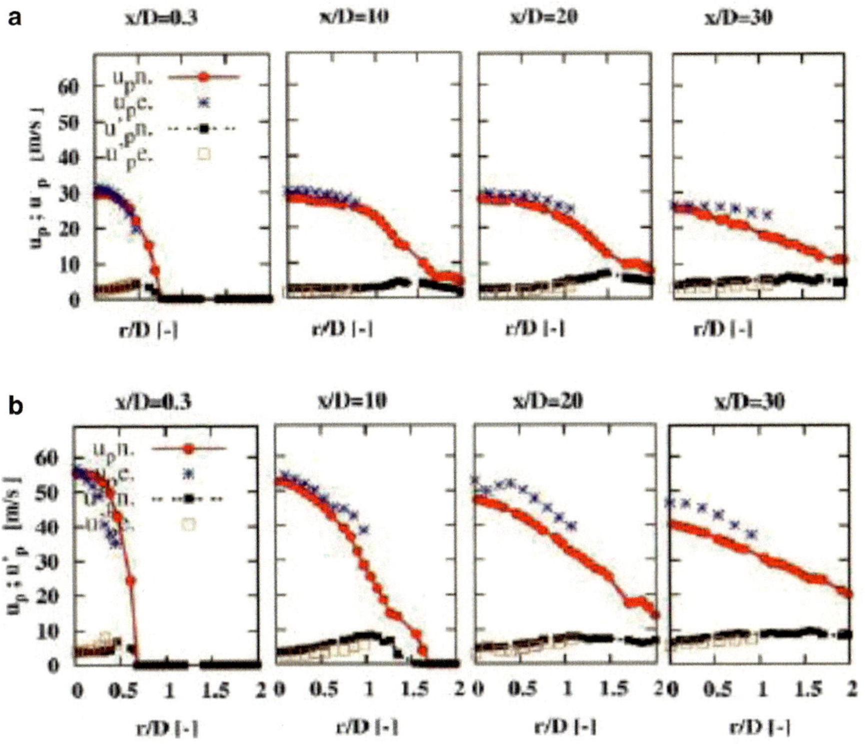

Fig. 3.9 (**a**) Radial profiles of the droplet averaged axial velocity and its fluctuations: EtF3, (**b**) Radial profiles of the droplet averaged axial velocity and its fluctuations: EtF8

averaged droplet velocity. Unfortunately a comparison between simulated gas phase velocity (that may help to clarify this discrepancies) and experimental data of the carrier phase (that are not available) is not possible.

Disagreement at the last cross sections may be originated from the presence of big droplets which are not following the carrier phase. In the EtF8 case droplets are dragged faster than in EtF3, they have less time for evaporation. At the last cross section larger droplets are therefore more important. They tend to impose their own dynamics and do not follow the carrier phase.

Droplet radial velocity profiles and corresponding fluctuations are depicted in Fig. 3.10a, b. Good agreement between numerical and experimental results is observed. The averaged droplet radial velocity increases with increasing inlet carrier phase velocity as the momentum transfer between axial and radial direction becomes more important. The droplet radial velocity fluctuation has the same order of magnitude as the velocity itself.

Capturing the evaporation correctly is extremely essential for a correct prediction of the spray combustion. Errors in the evaporation will be transmitted to the determination of the mixture fraction that leads to a disagreement in the carrier phase properties, which in turn influence the spray variables.

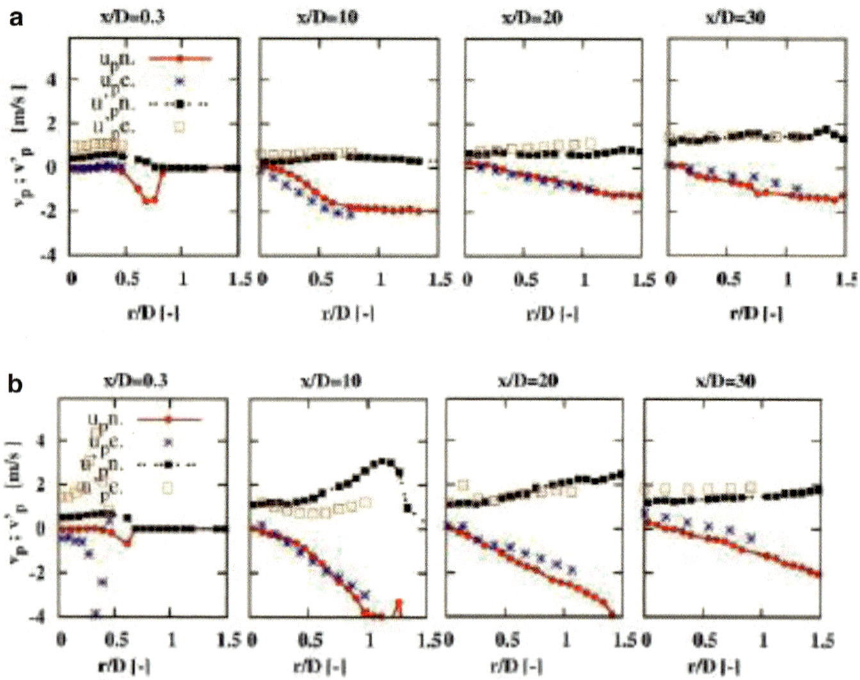

Fig. 3.10 (**a**) Radial profiles of the droplet averaged radial velocity and its fluctuation: EtF3, (**b**): Radial profiles of the droplet averaged radial velocity and its fluctuation: EtF8

Figure 3.11a, b shows the spray volume flux, that is an indicator of the phase transition, for the test cases EtF3 and EtF8. Comparisons demonstrate a good agreement between numerical and measured data. The numerical results slightly over-predict the volume flux. Experimentally it is not possible to capture all droplets that cross a considered cv. Part of the spray is crossing the measuring section without being detected. Thus, the experimental profiles lie under the numerical curves. Further, the profiles of the droplet volume flux give an indication about the spray opening angle which is well captured for the two test cases. Deviations are observed close to the nozzle exit. The experimental measurements have recorded a small film that has developed at the edge of the nozzle. The film formation is caused by the spray wall interaction which is treated in this work as a simple reflection. Numerically, effects like collision, coalescence or turbulent dispersion at the wall edge should be accounted for along with the four-way coupling in order to improve the prediction. Note that considerations of group combustion phenomena [55, 62] as observed in experiments have not been accounted for.

Figure 3.12a, b display the excess temperature profile for EtF3 and EtF8 at different cross section. Reasonable agreement is observed for EtF3. The under-prediction of the excess temperature in EtF8 at $x/D = 10$ could be originated from the high carrier phase velocity. As droplets travel with double axial velocity they

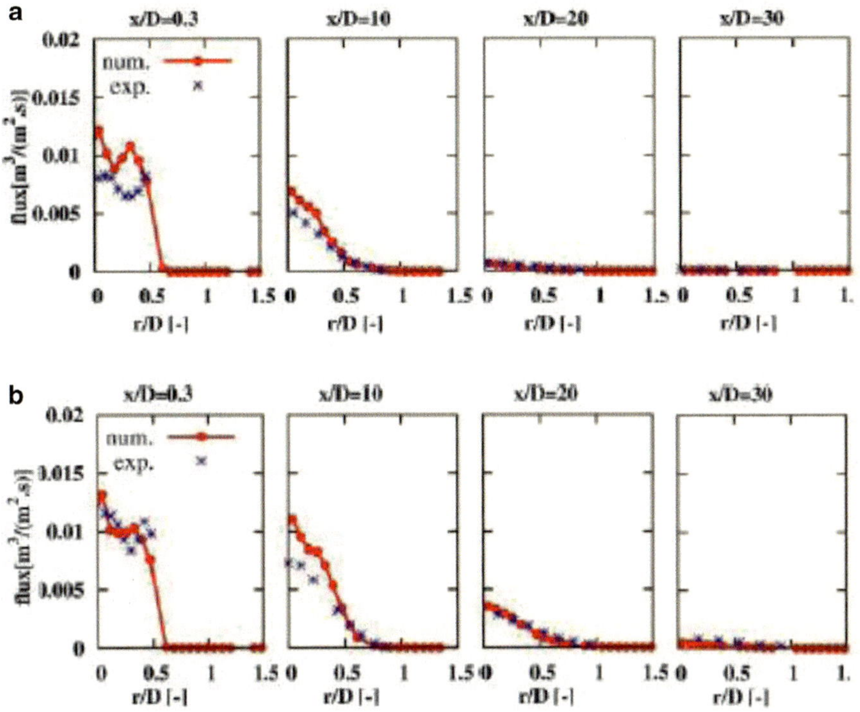

Fig. 3.11 (**a**) Radial profiles of the droplet volume flux: EtF3, (**b**) Radial profiles of the droplet volume flux: EtF8

experience only half residence time in the pre-vaporization region, i.e. before reaching the combustion region. Thus, the mixture fraction close to the nozzle exit plane is lean and the combustion in this region is weak. The flame breadth is good predicted, revealing that the opening angle of the spray is predicted correctly. The flame becomes thinner and shorter when increasing the carrier phase inlet velocity. The maximum value of the temperature is found in the interface between the pure oxidizer and the released vapor representing approximately the opening angle of the spray. The stoichiometry of the mixture fraction is registered in this location. Further improvements of the temperature prediction along the center line, i.e. rich region, could be achieved by including CO in the definition of the RPV.

3.6 Conclusions

To illustrate different aspects of the SGS turbulence modulation modelling two configurations have been presented. The first configuration dealt with the way how to validate the Lagrangian particle tracking scheme as well as the accuracy of the

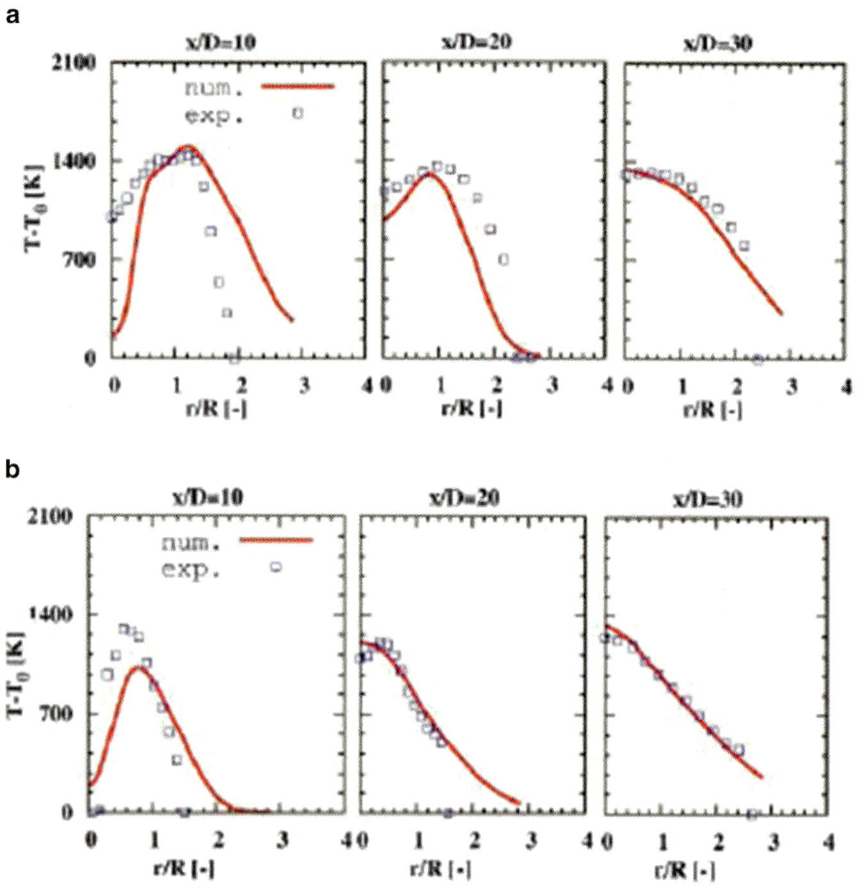

Fig. 3.12 (**a**) Radial profiles of the carrier phase excess temperature at different distances from the nozzle exit plane: EtF3, (**b**) Radial profiles of the carrier phase excess temperature at different distances from the nozzle exit plane: EtF8

numerical method for LES of complex flow configuration. Thereby, the need of incorporating the SGS velocity fluctuation for the description of the dispersion was noticeable. An advanced dispersion model is recommendable. A compromise between suitable grid resolution, related computational costs and SGS models used has to be found.

In the second configuration current models used for the turbulence modulation description along with the full two-way coupling have been evaluated in the context of LES. The results pointed out that the new proposed model, the thermodynamically consistent model, is able to well account for both the augmentation and the attenuation of turbulence due to the presence of large particles (i.e. large particle Reynolds number) and/or small particles (i.e. small particle Reynolds number), respectively. Even though the effect of particle diameter on the turbulence properties of the continuous phase is more complex, a subtle compensation between particle-induced dissipation and production of turbulence could also be well captured. This model is recommendable for LES simulation of sprays, in which such subtle compensations between particle-induced dissipation and production of turbulence due to polydispersity occur in a natural way.

Ethanol spray flames (cases EtF3 and EtF8) were investigated using LES coupled to Lagrangian droplet tracking together with a FGM tabulated chemistry. The coupling between the numerical codes, which are implemented in a modular manner, was assured via transfer of source terms and flow field properties. Adopting a coupling time step, set to 10% of the turbulent time scale, and neglecting the effects of the SGS velocity of the carrier phase on the drag of droplets, this investigation delivers reasonable results for the disperse phase properties and achieves overall plausible agreement with experimental data. Especially the combustion properties, like flame height, which is determined by the contour plot of the OH mass fraction and temperature of the gas phase, are well predicted confirming the predictive capability of the LES modeling suggested. However, further improvements are needed: (a) The influence of the four ways coupling near the nozzle exit region along with the impact of the dispersion modeling on the droplet diameter distribution should be investigated. (b) In particular, the heat losses due to evaporation which would be important for the droplet phase transition close to the nozzle exit need also to be integrated into the FGM approach. (c) The consideration of the evaporation source term in the RPV equation must be carefully addressed along with (d) a proper account of the effect of droplet vaporization in the reaction space while generating the flamelet tables.

Acknowledgment The authors are grateful to the financial support by the German Research Council (DFG) through the SFB568.

References

Project-Related Publications

1. Sadiki, A., Goryntsev, D., Wegner, B., Janicka, J.: Design of technical combustion systems using LES: state of the art and perspectives. In: 7th International ERCOFTAC Symposium on Engineering Turbulence Modelling and Measurements, (ETMM7), Limassol, Cyprus, 4–6 June 2008
2. Olbricht, C., Hahn, F., Sadiki, A., Janicka, J.: Analysis of subgrid scale mixing using a hybrid LES-Monte-Carlo PDF method. Int. J. Heat Fluid Flow **28**(6), 1215–1226 (2007)
3. Chrigui, M., Sadiki, A., Janicka, J., Zgahl, A.: Study of n-heptane spray evaporation and dispersion within premixed combustion in complex geometry configuration. In: Accepted to the 32th International Symposium on Combustion, McGill University, Montreal, Canada (2008)

4. Chrigui, M., Sadiki, A., Janicka, J.: Numerical analysis of spray dispersion, evaporation and combustion in a single gas turbine combustor. In: ASME TURBO-EXPO, GT2008-51253, Berlin, Germany (2008)
5. Chrigui, M., Batarseh, F.Z., Sadiki, A., Roisman, I., Tropea, C.: Numerical and experimental study of spray produced by an airbalst atomizer under elevated pressure conditions. In: ASME TURBO-EXPO, GT2008-51305, Berlin, Germany (2008)
6. Ahmad, W., Chrigui, M., Sadiki, A., Ngoma, G.D.: Effect of evaporation on the combustion behaviour of kerosene spray flame. In: ASME Turbo Expo 2010 (GT2010-22641), Glasgow, Scotland, UK, 14–18 June 2010
7. Chrigui, M., Sadiki, M., Ngoma, G.D.: Unsteady, turbulent, two-phase flow using an Euler/Lagrange approach devoted to two-way coupling conditions. In: International Conference on Multiphase Flow 2010 (ICMF-2010), Florida, USA, 30 May–4 June 2010
8. Chrigui, M., Hage, M., Sadiki, A., Janicka, J., Dreizler, A.: Experimental and numerical analysis of spray dispersion and evaporation in a combustion chamber. At. Spray **19**, 929–955 (2009)
9. Chrigui, M., Roisman, I., Batarseh, F., Sadiki, A., Tropea, C.: Spray generated by an airblast atomizer under elevated ambient pressures. J. Propuls. Power AIAA **26**(6), 1170–1183 (2010)
10. Chrigui, M.: N-Hpetane spray evaporation and dispersion in turbulent flow within a complex geometry configuration. J. Comput. Therm. Sci. **2**(1), 55–78 (2010)
11. Chrigui, M., Sadiki, A., Janicka, J.: Evaporation and dispersion of N heptane droplets within premixed flame. J. Heat Mass Trans. **46**(8–9), 869–880 (2010)
12. Chrigui, M., Schneider, L., Zghal, A., Sadiki, A., Janicka, J.: Droplet behavior within a LPP ambiance. J. Fluid Dyn. Mater. Process. **6**(4), 399–408 (2010)
13. Pantangi, P., Sadiki, A., Janicka, J., Hage, M., Dreizler, A., van Oijen, J.A., Hassa, C., Heinze, J., Meier, U.: LES of pre-vaporized kerosene combustion at high pressures in a single sector combustor taking advantage of the flamelet generated manifolds method. In: Proceedings of ASME Turbo Expo 2011 (GT2011-45819), Vancouver, Canada, 6–10 June 2011
14. Chrigui, M., Moesl, M.K., Ahmadi, W., Sadiki, A., Janicka, J.: Partially premixed prevaporized kerozene spray combustion in turbulent flow. Exp. Therm. Fluid Sci. **34**(1), 308–315 (2010)
15. Hahn, F., Sadiki, A., Janicka, J.: Large eddy simulation of a particle laden swirling flow based on an Euler-Lagragian approach. In: 6th International Conference on Multiphase Flow (ICMF2007), Leipzig, Germany (2007)
16. Chrigui, M., Zghal, A., Sadiki, A., Janicka, J.: Spray evaporation and dispersion of n-heptane droplets within premixed flame. Heat Mass Trans. **46**, 869–880 (2010)
17. Sadiki, A., Ahmadi, W., Chrigui, M., Janicka, J.: Towards the impact of fuel evaporation-combustion interaction on spray combustion in gas turbine combustion chambers. Part I: effect of partial fuel vaporization on spray combustion. In: Merci, B., Roeckaerts, D., Sadiki, A. (eds.) Experiments and Numerical Simulations of Diluted Spray Turbulent Combustion, Proceedings of the 1st International Workshop on Turbulent Spray Combustion, Chap. 3, pp. 69–110. Springer, Dordrecht/Heidelberg/London/New York (2011)
18. Sadiki, A., Ahmadi, W., Chrigui, M., Janicka, J.: Towards the impact of fuel evaporation-combustion interaction on spray combustion in gas turbine combustion chambers. Part II: influence of high combustion temperature on spray droplet evaporation. In: Merci, B., Roeckaerts, D., Sadiki, A. (eds.) Proceedings of the 1st International Workshop on Turbulent Spray Combustion, Chap. 3, pp. 111–132. Springer, Dordrecht/Heidelberg/London/New York (2011)
19. Chrigui, M., Gounder, J., Sadiki, A., Masri, A.R., Janicka, J.: Partially premixed reacting acetone spray using LES and FGM tabulated chemistry combustion and flame http://dx.doi.org/10.1016/j.combustflame.2012.03.009 (2012)
20. Chrigui, M.: N-Heptane spray evaporation and dispersion in turbulent flow within a complex geometry configuration. J. Comput. Therm. Sci. **2**(1), 55–78 (2010)

Other Publications

21. Oefelein, J.C.: Large eddy simulation of turbulent combustion processes in propulsion and power systems. Prog. Aerosp. Sci. **42**, 2–37 (2006)
22. Bellan, J., Selle, L.C.: Large eddy simulation composition equations for single-phase and two-phase fully multicomponent flows. Proc. Combust. Inst. **32**(2), 2239–2246 (2009)
23. Senoner, J.M., Sanjosé, M., Lederlin, T., Jaegle, F., García, M., Riber, E., Cuenot, B., Gicquel, L., Pitsch, H., Poinsot, T.: Eulerian and Lagrangian large-eddy simulations of an evaporating two-phase flow. Comptes Rendus Mécanique **337**(6–7), 458–468 (2009)
24. Apte, S.V., Mahesh, K., Moin, P.: Large-eddy simulation of evaporating spray in a coaxial combustor. Proc. Combust. Inst. **32**(2), 2247–2256 (2009)
25. Pitsch, H., Desjardins, O., Balarac, G., Ihme, M.: Large-eddy simulation of turbulent reacting flows. Prog. Aerosp. Sci. **44**(6), 466–478 (2008)
26. Lederlin, T., Pitsch, H.: Large-eddy simulation of an evaporating and reacting spray. In Center for Turbulence Research, Annual Research Briefs, pp. 479–490. Stanford University (2008)
27. Sanjosé, M., Lederlin, T., Gicquel, L., Cuenot, B., Pitsch, H., García-Rosa, N., Lecourt, R., Poinsot, T.: LES of two-phase reacting flows. In: Center for Turbulence Research Proceedings of the Summer Program, pp. 251–263. Stanford University (2008)
28. Bini, M., Jones, W.P.: Large eddy simulation of an evaporating acetone spray. Int. J. Heat Fluid Flow **30**(3), 471–480 (2009)
29. Pera, C., Réveillon, J., Vervisch, L., Domingo, P.: Modeling subgrid scale mixture fraction variance in LES of evaporating spray. Combust. Flame **146**(4), 635–648 (2006)
30. Patel, N., Menon, S.: Simulation of spray–turbulence–flame interactions in a lean direct injection combustor. Combust. Flame **153**(1–2), 228–257 (2008)
31. Bray, K.N.C., Peters, N.: Laminar flamelets in turbulent reacting flows. In: Libby, P.A., Williams, F.A. (eds.) Turbulent Reacting Flows, pp. 63–113. Academic, London (1994)
32. Hanjalic, K.: Will RANS survive LES: a view of perspectives. ASME J. Fluids Eng. **127**, 831–839 (2005)
33. Carbonell, D., Perez-Segarra, C.D., Coelho, P.J., Oliva, A.: Flamelet mathematical models for non-premixed laminar combustion. Combust. Flame **156**(2), 334–347 (2009)
34. Mortensen, M., Bilger, R.W.: Derivation of the conditional moment closure equations for spray combustion. Combust. Flame **156**(1), 62–72 (2009)
35. Dianat, M., Yang, Z., McGuirk, J.J.: Large-Eddy Simulation of a Two-Phase Plane Mixing-Layer. Advances in Turbulence XII, Springer Proceedings in Physics, Part 11, vol. 132, pp. 775–778. (2009)
36. Yuichi, I., Nobuyukiles, T.: LES of spray combustion flows. J. Jpn. Sci. Technol. (J-EAST) **7**, 27–28 (2005)
37. Abramzon, B., Sirignano, W.A.: Droplet vaporization model for spray combustion calculations. Int. J. Heat Mass Trans. **32**, 1605–1618 (1989)
38. Apte, S.V., Gorokhovski, M., Moin, P.: LES of atomizing spray with stochastic modelling of secondary breakup. Int. J. Multiphase Flow **29**, 1503–1522 (2003)
39. Riber, E., Moureau, V., García, M., Poinsot, T., Simonin, O.: Evaluation of numerical strategies for large eddy simulation of particulate two-phase recirculating flows. J. Comput. Phys. **228**(2), 539–564 (2008)
40. Pitsch, H.: Large-eddy simulation of turbulent combustion. Annu. Rev. Fluid Mech. **38**, 453–482 (2006)
41. Shotorban, B., Zhang, K.K.Q., Mashayek, F.: Improvement of particle concentration prediction in large-eddy simulation by defiltering. Int. J. Heat Mass Trans. **50**(19–20), 3728–3739 (2007)
42. Fede, P., Simonin, O.: Numerical study of the subgrid fluid turbulence effects on the statistics of heavy colliding particles. Phys. Fluids **18**, 045103 (2006)

43. Miller, R.S., Harstad, K., Bellan, J.: Evaluation of equilibrium and non-equilibrium evaporation models for many gas–liquid flow simulations. Int. J. Multiphase Flow **24**, 1026–1055 (1998)
44. Garcia, M., Sommerer, Y., Schönfeld, T., Poisot, T.: Assessment of Euler-Euler and Euler-Lagrange strategies for reactive large-eddy simulation. In: CERFACS, IMFT – Toulouse, France, pp. 1–10 (2004)
45. Harstad, K., Bellan, J.: Modeling evaporation of Jet A, JP-7, and RP-1 drops at 1 to 15 bars. Combust. Flame **137**, 163–177 (2004)
46. Bekdemir, C., Somers, L.M.T., de Goey, L.P.H.: First application of the flamelet generated manifold (FGM) approach to the simulation of an igniting diesel spray. In: 19th International Multidimensional Engine Modeling User's Group Meeting at the SAE Congress, Detroit, Michigan, April 2009
47. Peters, N.: Fifteen lectures on laminar and turbulent combustion, ERCOFTAC Summer School, p. 174 (1992)
48. Ge, H.-W., Gutheil, E.: Simulation of a turbulent spray flame using coupled PDF gas phase and spray flamelet modeling. Combust. Flame **153**, 173–185 (2008)
49. Bastiaans, R.J.M., van Oijen, J.A., de Goey, L.P.H.: Application of flamelet generated manifolds and flamelet analysis of turbulent combustion. Int. J. Multiscale Comput. Eng. **4**(3), 307–317 (2006)
50. Baurle, R.A., Girimaji, S.S.: Assumed PDF turbulence-chemistry closure with temperature-composition correlations. Combust. Flame **134**, 131–148 (2003)
51. Patel, N., Kırtaş, M., Sankaran, V., Menon, S.: Simulation of spray combustion in a lean-direct injection combustor. Proc. Combust. Inst. **31**(2), 2327–2334 (2007)
52. Workshop on Quality Assessment of Unsteady Methods for Turbulent Combustion Predictions and Validation, Seeheim-Jugenheim, Germany, 16.–17.06.2005, see www.sfb568.de/workshops (2005)
53. Merci, B., Rockaerst, D., Sadiki, A. (eds.): Experiments and Numerical Simulations of Diluted Spray Turbulent Combustion, Proceedings of the 1st International Workshop on Turbulent Spray Combustion. Springer, Dordrecht (2011)
54. Meftah, H., Reveillon, J., Mir, A., Demoulin, F.X.: SGS analysis of the evolution equations of the mixture fraction and the progress variable variances in the presence of spray combustion. Int. J. Spray Combust. Dyn. **2**(1), 21–48 (2010)
55. Chiguier, N.: Group combustion and laser diagnostic methods in sprays: a review. Combust. Flame **51**, 127–139 (1983)
56. Luo, K., Pitsch, H., Pai, M.G.: DNS of three-dimensional swirling n-heptane spray flames, In Center for Turbulence Research, Annual Research Briefs, pp. 171–183. Stanford University (2009)
57. Reveillon, J., Vervisch, L.: Analysis of weakly turbulent diluted-spray flames and spray combustion regimes. J. Fluid Mech. **537**, 317–347 (2005)
58. Domingo, P., Vervisch, L., Reveillon, J.: DNS analysis of partially premixed combustion in spray and gaseous turbulent flame-bases stabilized in hot air. Combust. Flame **140**, 172–195 (2005)
59. Gutheil, E.: Modeling and Simulation of Droplet and Spray Combustion, Handbook of Combustion. Wiley-VCH Verlag GmbH & Co. KGaA, Weinheim (2010)
60. Sazhin, S.S.: Advanced models for fuel droplet heating and evaporation. Prog. Energy Combust. Sci. **32**, 162–214 (2006)
61. Masri, A.R., Gounder, J.D.: Turbulent spray flames of acetone and ethanol approaching extinction. Combust. Sci. Technol. **182**(4–6), 702–715 (2010)
62. Chiu, H.H., Kim, H.Y., Croke, E.J.: Internal group combustion of liquid droplets. In: The Combustion Institute (ed.) Proceedings of 19th Symposium on Combustion, Pittsburgh, Pennsylvania (1982)
63. Pozorski, J., Apte, S.A.: Filtered particle tracking in isotropic turbulence and stochastic modeling of subgrid-scale dispersion. Int. J. Multiphase Flow **35**(2), 118–128 (2009)

64. Apte, S.V., Mahesh, K., Lundgren, T.: Accounting for finite-size effects in disperse particle-laden flows. Int. J. Multiphase Flow **34**, 260–271 (2008)
65. Pozorski, J., Luniewski, M.: Analysis of SGS effects on dispersed particles in LES of heated channel flows. In Quality and Reliability of LES II, ERCOFTAC Series, pp. 171–180. Springer, ISBN 978-1-4020-8577-2, e-ISBN: 978-1-4020-8578-9 (2010)
66. Fiorina, B., Gicquel, O., Vervisch, L., Carpentier, S., Darabiha, N.: Premixed turbulent combustion modeling using tabulated detailed chemistry and PDF. Proc. Combust. Inst. **30**, 867–874 (2005)
67. Weber, J., Peters, N.: Calibration of spray model constants for CFD-simulations of DI diesel engines using the representative interactive flamelet (RIF) model. Int. J. Veh. Des. **41**(1–4), 143–164 (2006)
68. Groh, B.: Grobstruktursimulation turbulenter Mehrphasenströmungen mit und ohne Phasenübergang. In: Dissertation, Fortschritt-Berichte VDI, Reihe 7 Strömungstechnik, Nr. 467, pp. 112–119. VDI Verlag GmbH, Düsseldorf (2005)
69. Landenfeld, T., Sadiki, A., Janicka, J.: A turbulence-chemistry interaction model based on a multivariate presumed Beta-PDF method for turbulent flames. Flow Turbul. Combust. **68**, 111–135 (2002)
70. Lehnhäuser, T., Schäfer, M.: Improved linear interpolation practice for finitevolume schemes on complex grids. Int. J. Numer. Meth. Fluids **38**(7), 625–645 (2002)
71. Sagaut, P.: Large Eddy Simulation for Incompressible Flows. Springer, Berlin (2001)
72. Wegner, B., Maltsev, A., Schneider, C., Sadiki, A., Dreizler, A., Janicka, J.: Assessment of unsteady RANS in predicting swirl flow instability based on LES and experiments. Int. J. Heat Fluid Flow **25**, 528–536 (2004)
73. Pierce, C.D., Moin, P.: Large eddy simulation of a confined coaxial jet with swirl and heat release, AIAA paper 98–2892 (1998)
74. Vreman, A.W., Albrecht, B.A., van Oijen, J.A., de Goey, L.P.H., Bastiaans, R.J.M.: Premixed and non-premixed generated manifolds in large-eddy simulation of Sandia flame D and F. Combust. Flame **153**, 394–416 (2008)
75. Poinsot, T., Veynante, D.: Theoretical and Numerical Combustion, 3rd edn. CNRS, Toulouse (2011)
76. Janicka, J., Sadiki, A.: Large eddy simulation of turbulent combustion systems. Proc. Combust. Inst. **30**, 537–547 (2005)
77. Balachandar, S., Eaton, J.K.: Turbulent dispersed multiphase flow. Annu. Rev. Fluid Mech. **42**, 111–133 (2010)
78. Sadiki, A., Chrigui, M., Janicka, J., Maneshkarimi, M.R.: Modeling and simulation of effects of turbulence on vaporization, mixing and combustion of liquid fuel sprays. Flow Turbul. Combust. **75**, 105–130 (2005)
79. De, S., Lakshmisha, K.N., Bilger, R.W.: Modeling of nonreacting and reacting turbulent spray jets using a fully stochastic separated flow approach. Combust. Flame **158**(10), 1992–2008 (2011)
80. Sadiki, A.: Extended thermodynamics as modeling tool of turbulence in fluid flows. In: Trends in Applications of Mathematics to Mechanics, pp. 451–462. Shaker Verlag, Aachen (2005)
81. Geiss, S., Dreizler, A., Stojanovic, Z., Chrigui, M., Sadiki, A., Janicka, J.: Investigation of turbulence modification in a non-reactive two-phase flow. Exp. Fluids **36**, 344–354 (2004)
82. Chrigui, M., Ahmadi, G., Sadiki, A.: Study on interaction in spray between evaporating droplets and turbulence using second-order turbulence RANS models and a Lagrangian approach. Prog. Comput. Fluid Dyn. **4**(3–5), 162–174 (2004)
83. Garcia, M.: Développement et validation du formalisme Euler-Lagrange dans un solveur parallèle et non-structuré pour la simulation aux grandes INP Toulouse, France Thèse (2009)
84. Lee, J., Choi, H., Park, N.: Dynamic global model for large eddy simulation of transient flow. Phys. Fluids **22**(7). dx.doi.org/10.1063/1.3459156 (2010)
85. Löffler, M., Pfadler, S., Beyrau, F., Leipertz, A., Dinkelacker, F., Huai, Y., Sadiki, A.: Experimental determination of the sub-grid scale scalar flux in a non-reacting jet flow. Flow Turbul. Combust. 1386–6184, 1573–1987 (Online) (2007)

86. Yoshizawa, A., Horiuti, K.: A statistically-derived subgrid-scale kinetic energy model for large eddy simulation of turbulent flow. J. Phys. Soc. Jpn. **54**(8), 2834–2839 (1985)
87. Eguz, U., Somers, L.M.T., de Goey, L.P.H.: Modeling of PCCI combustion with the FGM approach. In: 13th International Conference on Numerical Combustion, Corfu, Greece, 27–29 April 2011
88. Hu, Z., Cracknel, R., Somers, L.M.T.: Computational study of fuel effects in premixed charge compression ignition (PCCI) engine combustion. In: 8th International Symposium Towards Clean Diesel Engines TCDE 2011, Chester, U.K., 8–9 June 2011
89. Marinov, N.M.: A detailed chemical kinetic model for high temperature ethanol oxidation. Int. J. Chem. Kinet. **31**, 183–220 (1999)
90. Mashayek, F., Taulbee, D.B., Givi, P.: Modeling and simulation of two phase turbulent flow. In: Roy, D.G. (ed.) Propulsion Combustions: Fuels to Emissions, Kapitel 8, pp. 241–280. Taylor & Francis, Washington, D.C. (1998)
91. Yeh, F., Lei, U.: On the motion of small particles in a homogeneous turbulent shear flow. Phys. Fluids **3**(11), 2758–2776 (1999)
92. Lei, K., Taniguchi, N., Kobayashi, T.: A new dynamic SGS-Model for large eddy simulation of particle-laden flows. In: Third AFOSR International Conference on DNS/LES (TAICDL) (2001)
93. Crowe, C.T.: A review of carrier-phase turbulence in dispersed flows. In: Proceedings of the 4th International Conference on Multiphase Flow, Paper-No. 604 (2001)
94. Tsuji, Y.: Activities in discrete particle simulation in Japan. Powder Technol. **113**, 278–286 (2000)
95. Lain, S., Sommerfeld, M.: Turbulence modulation in dispersed two-phase flow laden with solids from a Lagrangian perspective. Int. J. Heat Fluid Flow **24**, 616–625 (2003)
96. Sommerfeld, M., Qui, H.H.: Detailed measurements in a swirling particulate two-phase flow by a phase-Doppler-anemometer. Int. J. Heat Fluid Flow **12**, 20–28 (1991)
97. Sommerfeld, M., Qui, H.H.: Experimental studies of spray evaporation in turbulent flow. Int. J. Heat Fluid Flow **19**, 10–22 (1998)
98. Klein, M., Sadiki, A., Janicka, J.: A digital filter based generation of inflow data for spatially developing direct numerical or large eddy simulation. J. Comput. Phys. **186**, 652–665 (2003)
99. Elghobashi, S., Truesdell, G.C.: On the two-way interaction between homogenous turbulence and dispersed solid particles I: turbulence modification. Phys. Fluids A **5**, 1790–1796 (1993)
100. Faeth, G.M.: Mixing, transport and combustion in sprays. Prog. Energy Combust. Sci. **13**, 293–345 (1987)
101. Ahmadi, G., Cao, J., Schneider, L., Sadiki, A.: A thermodynamically formulation for chemically active multiphase turbulent flows. Int. J. Eng. Sci. **44**, 699–720 (2006)

Part II
Combustion

This part of the book dwells on experimental, theoretical and numerical description of gaseous combustion in gas turbine combustors. It includes the development of methods involving reduced reaction mechanisms for complex fuels, soot modeling and the formulation of appropriate turbulence-chemistry interaction models for different combustion regimes. In particular, innovative measurement techniques allow time-resolved characterization of flame stability and extinction behavior and generation of comprehensive experimental data.

In this context Chap. 4 presents the designed experiments and measurement techniques for improving the understanding in turbulent combustion with a focus on turbulent flow and scalar fields as well as on their mutual interactions. The results are restricted to generic gaseous turbulent flames that feature different characteristics important to practical applications. They include combined scalar and flow measurements that can significantly improve the understanding of turbulence-chemistry interactions. Further achievements based on recently developed high-speed laser imaging are presented that complement methods at low repetition rate, commonly used to generate statistical moments and probability density functions.

Despite permanently growing computer power, the simulation of a reaction–diffusion-convection system involving a large scale chemical combustion mechanism is still far from reach. However, the prediction of minor species, e.g. soot, NO_x, and other pollutants, needs detailed mechanisms. Chapter 5 focusses on an effective dimension reduction of chemical combustion mechanisms. This is firstly based on an efficient use of the intrinsic low dimensional manifold method, and finally on optimization methods.

In Chap. 6 three different aspects of Large Eddy Simulation (LES) of combustion processes are covered using tabulated chemistry models in the line of the Flamelet-Generated Manifold approach (FGM) to describe the chemical reactions occurring in gas turbines. These are the extension of the FGM to describe pollutants such as NO_x and CO, the "Transported Filtered Density Function" method (Transported FDF), and the "Artificially Thickened Flame" (ATF) model. They have been analyzed and validated in various generic combustion systems.

Soot resulting from combustion processes is known to have a negative impact on health and environment. It also may lead to material damages especially in gas turbines. Therefore a deeper insight in the processes leading to the formation and consumption of the soot precursors and soot particles is needed. Chapter 7 dwells on describing these processes in the gas and the particle phase and provides mathematical models for simulating soot formation in complex technical systems such as gas turbines using LES.

Focusing on premixed combustion, Chap. 8 uses a description following a two-phase flow where the flame is modeled as a non-material interface. This induces jumps in velocity and pressure at the interface, which depend on the density ratio. In classical numerical methods for single-phase flows, such as finite volume, it is difficult to represent jumps accurately and stable, i.e. without oscillations, at the same time. In this chapter an extension to the Discontinuous Galerkin method is presented which is able to exactly represent jumps, with sub-cell accuracy.

Chapter 4
Advanced Laser Diagnostics for Understanding Turbulent Combustion and Model Validation

B. Böhm, D. Geyer, M.A. Gregor, C. Heeger, A. Nauert, C. Schneider, and A. Dreizler

Abstract This contribution is not an original publication but a report of cumulative work that was carried out within the framework of SFB 568. The work was published in different archival journals and figures and text passages have been taken from different journal articles as indicated by the references. The aim of this report is to present experiments in projects B1 and B3 for improving our understanding in turbulent combustion with a focus of turbulent flow and scalar fields as well as their mutual interactions. The report is restricted to generic gaseous turbulent flames that feature different characteristics important to practical applications. The methods presented here are feasible to study boundary conditions, flow and scalar fields and are based all on interactions between laser light and matter. Following a brief introduction, two target flames are discussed in Sect. 4.2. Sections 4.3 and 4.4 exemplify flow and scalar measurements. Section 4.5 discusses combined scalar/flow measurements that can significantly improve our understanding of turbulence-chemistry interactions. In Sect. 4.6 new developments based on high-repetition-rate imaging are highlighted. These diagnostics complement methods at low repetition rate commonly used to generate an understanding by statistical moments and probability density functions. High repetition rate imaging techniques presently are an emerging field. Although the most recent developments achieved in the funding period of the Collaborative Research Center are included to this report, near-future progress in this field will lead to even more interesting insights into combustion phenomena.

Keywords Combustion laser diagnostics • Raman/Rayleigh • PIV • OH-PLIF • LDA

B. Böhm • D. Geyer • M.A. Gregor • C. Heeger • A. Nauert • C. Schneider • A. Dreizler (✉)
Reactive Flows and Diagnostics, Center of Smart Interfaces, Mechanical Engineering, Technische Universität Darmstadt, Petersenstr. 30, D-64287 Darmstadt, Germany
e-mail: dreizler@csi.tu-darmstadt.de

J. Janicka et al., *Flow and Combustion in Advanced Gas Turbine Combustors*, Fluid Mechanics and Its Applications 102, DOI 10.1007/978-94-007-5320-4_4,

4.1 Introduction to Experimental Combustion Research

Turbulent combustion is the backbone of primary energy conversion. Although it is desirable that regenerative energy conversion processes such as solar or wind energy gain in their weight, combustion will keep its dominant role in the foreseeable future. This fact and recent public disputes on global change enforce that turbulent combustion processes in their various applications such as electrical power and heat generation, propulsion and mobility must be further improved in terms of efficiency and pollutant emissions.

Different pathways exist to advance combustion technologies. Figure 4.1 highlights the role of experimental methodologies. Experiments may serve either in a **direct** manner to measure key-quantities of a practical combustion process for subsequent improvements or experimental studies serve in an **indirect** way to support a mathematical/numerical model of the combustion process to be designed in future. Figure 4.1 additionally shows that experimental studies can gain from numerical modelling and numerical recipes.

For an experimental investigation of a combustion process two ingredients are necessary: (1) a test rig where the combustion process of interest can be operated and (2) suitable methods for its experimental characterisation. Both aspects will be discussed here to some extent but the discussion is restricted only to **generic, gaseous and continuous combustion processes** that feature some characteristics of general interest and to **laser-optical measuring techniques**. Topics such as coal or spray combustion are apparently excluded here.

Optical- and especially laser-based methods are commonly used to study combustion processes in detail. This is attributed to the following features: (1) Optical methods are non- or minimal invasive. (2) Optical techniques can be applied in-situ. The only prerequisite is a suitable optical access. (3) The temporal resolution in comparison to probe-sampling-techniques (gas chromatography, mass

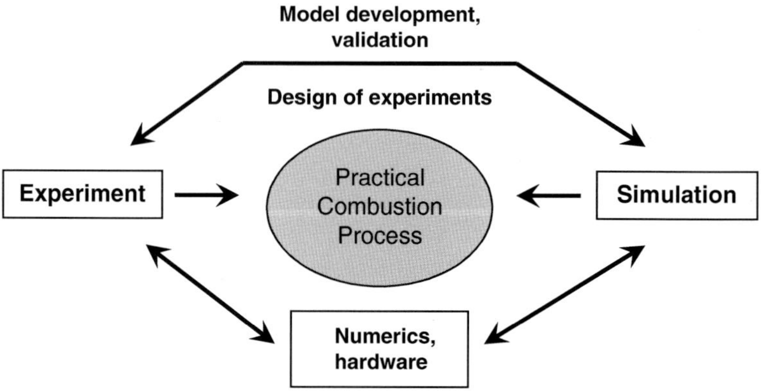

Fig. 4.1 Mutual interconnections between experiments, numerical simulations/modelling and numerical recipes/hardware developments to improve current combustion technology

spectrometry, etc.) is extremely high. Typical time scales of a practical turbulent combustion process can be resolved. (4) The spatial resolution of optical combustion diagnostics is reasonable. Although it is in general better than intrusive probe-techniques (thermocouple, hot-wire anemometer, etc.) smallest length scales of highly turbulent combustion processes often cannot be resolved. Therefore the sensitivity of the probe-volume size on the result must be examined for each measurand.

To investigate a turbulent combustion process in detail many different quantities are of interest. These quantities of interest can be divided into flow field quantities, scalar field quantities and inflow/boundary conditions.

A turbulent flow field is characterised by its mean and rms-values (root mean square = second statistical moment) of all three velocity components. Higher moments than the second may be in special cases of some interest but impact of experimental noise and other experimental limitations on precision and accuracy are not well studied. In addition to this pointwise information on mean and rms-values some information on length scales such as integral or Kolmogorov length scales or elements of the rate-of-strain and rate-of-vorticity tensor is important.

The scalar field of a combustion process contains the distribution of the temperature, chemical species concentrations and quantities derived from temperatures and/or concentrations. Prominent examples for the latter one are the mixture fraction or scalar dissipation rate in non- and partially-premixed flames or the reaction progress in premixed flames. Linear as well as non-linear spectroscopic methods are widely used to precisely and accurately measure these various key-quantities of the scalar field. However, as quantification in many circumstances is difficult some of these spectroscopic techniques are often used in a qualitative manner only. For example relative distributions of flame-generated radicals such as hydroxyl (OH) serve to divide unburned from burnt regions.

Turbulent combustion is complicated due to the fact that chemical kinetics and turbulent fluid flow mutually interact with each other. More insights into this mutual interplay can therefore be gained from simultaneous flow and scalar field measurements. In Sect. 4.5 combinations of PIV/PLIF and LDV/PLIF show that these simultaneously applied diagnostics allow to switch the point of view from a laboratory to a flame-fixed coordinate system. Averaging and wash-out due to intermittency inherent to turbulent flames is thereby avoided. Furthermore, statistics conditioned on the intermittent flame front becomes feasible.

Boundary and inlet conditions determine many features of turbulent flames. Therefore they need to be recorded and controlled thoroughly. Dependent on the type of process examined, not only inlet profiles at the nozzle exit but profiles inside the nozzle must be known. In case of enclosed combustion, transfer processes (heat, mass, momentum) to the wall should be studied as well.

Laser diagnostics are operated in general at rather low repetition rates. In consequence the samples are statistically uncorrelated. To derive reliable statistics and build up probability density functions (PDF) statistically uncorrelated sampling is a prerequisite. However, transient processes may occur in turbulent combustion processes that are not matched appropriately by PDF. For example, cyclic variations in IC engines, ignition, extinction or flash back phenomena may occur. For an improved understanding these transients must be studied additionally from a different viewpoint. For this purpose Sect. 4.6 discusses high-speed diagnostics with repetition rates in the kilo-Hertz (kHz) regime recording statistically correlated information. These newly developed diagnostics allow temporally tracking the evolution of flow and scalar field during individual transient events. This cinema-like information complements the traditional way of understanding reactive fluid flows by statistical moments and PDF.

4.2 Generic Turbulent Flames with Optical Access

As outlined above this report is restricted to continuous, gaseous, and turbulent flames. These flames mimic certain aspects in gas turbines and central heating applications. However, methods discussed in subsequent chapters are applicable to spray flames or internal combustion (IC) engines as well.

To examine a turbulent flame by optical methods an appropriate optical access is mandatory. Whereas for unconfined, atmospheric conditions an optical access entails no principal difficulty, pressurized conditions increase the effort dramatically. In any case an optical access to the interior of a nozzle is hindered by its rather small dimensions causing interfering reflections.

In the following, two examples of generic turbulent flames are presented. The first example is an atmospheric turbulent opposed jet burner that was used to study partially-premixed flames. The second example is an unconfined premixed swirl-burner. Both turbulent flames feature certain aspects but none the entire complexity of a real-world application. However, by joining these different pieces of information and by parametric variation of inflow/boundary conditions, understanding of turbulent flames is improved.

4.2.1 *Turbulent Opposed Jet Flame and in-Nozzle Measurements*

The turbulent opposed flow configuration shown in Fig. 4.2 is well suited to investigate the transition from stable to extinguishing flames. The burner as well as various optical diagnostics used for characterization of flow and scalar fields is described in more detail in [1–4, 14–17]. It consists of two identical vertically-opposed contoured nozzles. A 9:1 contraction ends in a 10 mm long straight section with a diameter of $D = 30$ mm. Perforated plates with a blockage of $b = 45\%$ and hexagonally arranged holes with $d = 4$ mm diameter are placed at the end of the contoured nozzle, followed by a 50 mm long tube.

The perforated plates were selected from a series of plates of different blockage ratios and hole diameters to produce strong fluctuations and almost isotropic

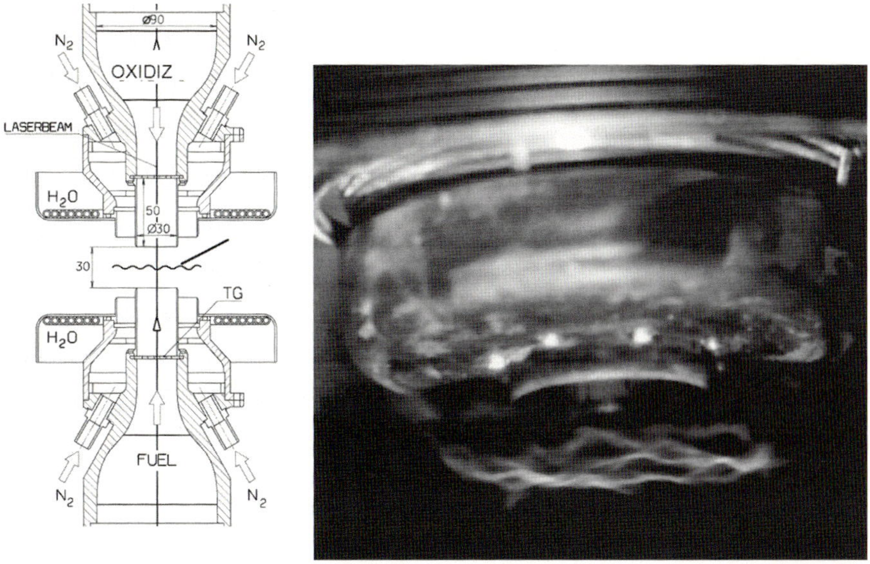

Fig. 4.2 Schematic of the turbulent opposed jet burner (*left*) and snapshot of the luminous flame using a CMOS-camera operated with an exposure time of 2 ms (Reprinted from Geyer et al. and [1, 14])

Table 4.1 Variation of Re-number and equivalence ratio of the fuel-feeding nozzle. This parametric variation allows to study the transition from stable to extinguishing flames

Re_{air}	a_m (1/s)	$\Phi = 3.18$	$\Phi = 2.0$	$\Phi = 1.6$	$\Phi = 1.2$
3,300	115	TOJ1A			
4,500	158	TOJ1B	TOJ2B	TOJ3B	TOJ4B
5,000	175	TOJ1C	TOJ2C		
6,650	235	TOJ1D	TOJ2D		
7,200	255		TOJ2E		

turbulence at the nozzle exit. In the nozzle contraction, two annular meshes (5 mm wide) were installed to avoid flow separation from the walls. Setting the distance between the nozzles (H = 30 mm) equal to the nozzle diameter (H/D = 1.0) allowed to keep the axial dimension of the flow field compact. The half nozzle distance is used as the origin of the axial coordinate z. Consequently, the lower and upper nozzle are located at z = −15 mm and z = +15 mm respectively. An annular coflow of nitrogen with a diameter of 60 mm and a bulk velocity of one third the fuel/air mixture bulk velocity shielded the inner jets from the surrounding air. Optical access is obtained without additional measures. Notice that the surrounding nozzles for the concentric coflow are shortened to maximize the angle of observation.

Table 4.1 summarizes various flow conditions for stable and extinguishing flames.

Table 4.2 Flow field quantities for flame TOJ2D

Bulk velocity W_b	3.4 m/s
Turbulent Re-number Re_t	90
Bulk strain rate $a_b = (-W_{b,O} + W_{b,F})/H$	231 s^{-1}
Residence time in mixing layers $t_{res} = a_b{}^{-1}$	4.3 ms
Large-eddy turnover time $t_{ov} = l_0 \big/ (2k)^{1/2}$	16.2 ms
Integral time scale T at nozzle exit	1.6 ms
Integral length scale l_0 at nozzle exit	4.7 mm
Kolmogorov length scale η_K at nozzle exit	0.16 mm
Obhukov-Corrsin scale η_C at nozzle exit	0.18 mm

A dry and dust-free air stream emanated from the upper nozzle with a bulk velocity W_b and a bulk Reynolds number Re_b based on D and W_b. The air stream impinged on a partially-premixed methane/air stream issuing from the lower nozzle. In order to establish the stagnation plane at the half nozzle distance $H/2$ the momenta of these two opposing jets were set equal. Accordingly, the bulk strain rate is $a_b = (-W_{b,O} + W_{b,F})/H$, $W_{b,O}$ and $W_{b,F}$ are the bulk velocities at oxidizer and fuel sides, respectively. A mean residence time t_{res} in the stagnating flow region was estimated by $t_{res} = a_b^{-1}$. The integral time scale t_0 of the turbulent flow field was computed from auto-correlation functions of hot-wire anemometry data 0.5 mm downstream of the oxidizer nozzle exit. Using t_0 and the axial velocity at the centerline, the integral length scale $l_0 = 4.7mm)$ at the nozzle exit was calculated utilizing Taylor's hypothesis. Thus, the integral length scale was slightly larger than the hole diameter d of the perforated plates.

Based on l_0 and the turbulent kinetic energy $\left(k = 1/2\left(w^2 + 2 \times u^2\right)\right)$ a local turbulent Reynolds number, Re_t, was calculated at the centerline ($z = 14.5$ mm). Here, w and u denote axial and radial velocity fluctuations, employing the experimentally verified radial symmetry for u and v. In case of TOJ2D, detailed in Table 4.2, the velocity fluctuations in both directions increased from ~0.3 ms^{-1} at the nozzle to ~0.42 ms^{-1} in the stagnation plane. Since the large-eddy turnover time, $t_{ov} = l_0/(2/3k)^{1/2} = 16.2$ ms, was approximately four times the residence time in the mixing layers t_{res}, a 'young-turbulence' emerged. This means, that the residence time was too short to allow the internal dynamics of turbulence to transfer a significant part of the kinetic energy contained in large eddies to the smallest length scales. Consequently, mixing processes at smaller length scales were less intense than in fully developed turbulent flows. The Kolmogorov length scale η_K as smallest flow scale was estimated using the turbulent Reynolds number and the integral length scale as ~0.16 mm. Since the Prandtl number Pr, the ratio of kinematic viscosity ν and thermal diffusivity D_T, in this flow is less than one (~0.65 to ~0.80), the smallest scalar length scale to be resolved is larger than η_K. For $Pr < 1$ the smallest spatial extension of scalar gradients to be resolved and hence the scalar dissipation rate is active on the Obukhov-Corrsin scale ($\eta_C = \left(D_T{}^3/\varepsilon\right)^{1/4}$, ε is the dissipation) [15]. Hot-wire anemometry has been used to estimate the dissipation ε at the oxidizer nozzle exit. From this information the value for η_c at the nozzle exit was determined as 0.18 mm.

Flame extinction is the ultimate consequence of intense turbulence – chemistry interaction. Therefore, extinction limits were experimentally determined with a precision better than 0.7% using calibrated mass-flow controllers. A flame was defined as 'extinguishing' if the flame could not be stabilized at least twice for a time period of 30 s in 50 consecutive attempts. The extinction limit of the flow for the current fuel composition was reached at $Re_b = 7{,}200$, which corresponds to a bulk strain rate of 255 s^{-1}.

The design of the perforated plates to enhance turbulence shows significant impact on the flame. In accordance to this experimentally observed sensitivity, large-eddy simulations [2, 16] (LES) critically depended on the initial choice of the flow properties through the perforated plates. Therefore an experimental analysis of the developing turbulence directly downstream of the perforated plates inside the nozzles is desirable. This is achieved by replacing the 50 mm straight metal tube that is attached to the contoured nozzles by a quartz tube. Of course the concentric nozzle for the N_2-coflow was dismounted. Using particle imaging velocimetry (PIV) detailed in Sect. 4.3 the flow field emanating from the individual 4 mm-dia. holes was investigated [3, 17] quantifying the initial turbulence level as well as the break-up of the individual jets.

4.2.2 Premixed Swirl Flame

The swirl burner introduced in this section [4, 5, 18, 19] is designed to study premixed flames that feature characteristics already very similar to practical designs. Parametric variation of *Re*-number/thermal power is easily feasible by increasing mass flow rates. In addition, by variation of the swirl intensity, the transition from a statistically stable to a processing flame that finally flashes back is possible [6, 20, 21].

A schematic of the unconfined premixed swirl burner is shown in Fig. 4.3a. The burner consisted of a 15 mm wide annular slit surrounding a central bluff body with $d = 30$ mm. Upstream of the nozzle, swirl was generated by a moveable block geometry. Theoretical swirl numbers $S_{0,\mathrm{th}}$ could be adjusted in the range from 0 to 2.0. Values of $S_{0,\mathrm{th}}$ exceeding 0.8 resulted in flash back. 70 mm upstream of the moveable block, natural gas was injected into the combustion-air flow at 300 K using a perforated ring line. Homogeneous mixing was confirmed experimentally during the development phase of the burner by seeding the fuel with submicron particles and performing planar Mie scattering in a 40x40mm^2 plane monitoring the flow downstream the nozzle exit.

To achieve defined boundary conditions, the burner was placed into a coaxial air flow, emanating from an annular orifice with $D = 220$ mm in diameter.

The mean velocity of this co-flowing air was set to 0.5 m/s. All flow rates were controlled electronically by high precision mass flow meters (Bronckhorst). The flow field was investigated both for combusting and non-reacting (isothermal) conditions (pure air, same momentum at the nozzle exit as the respective combusting

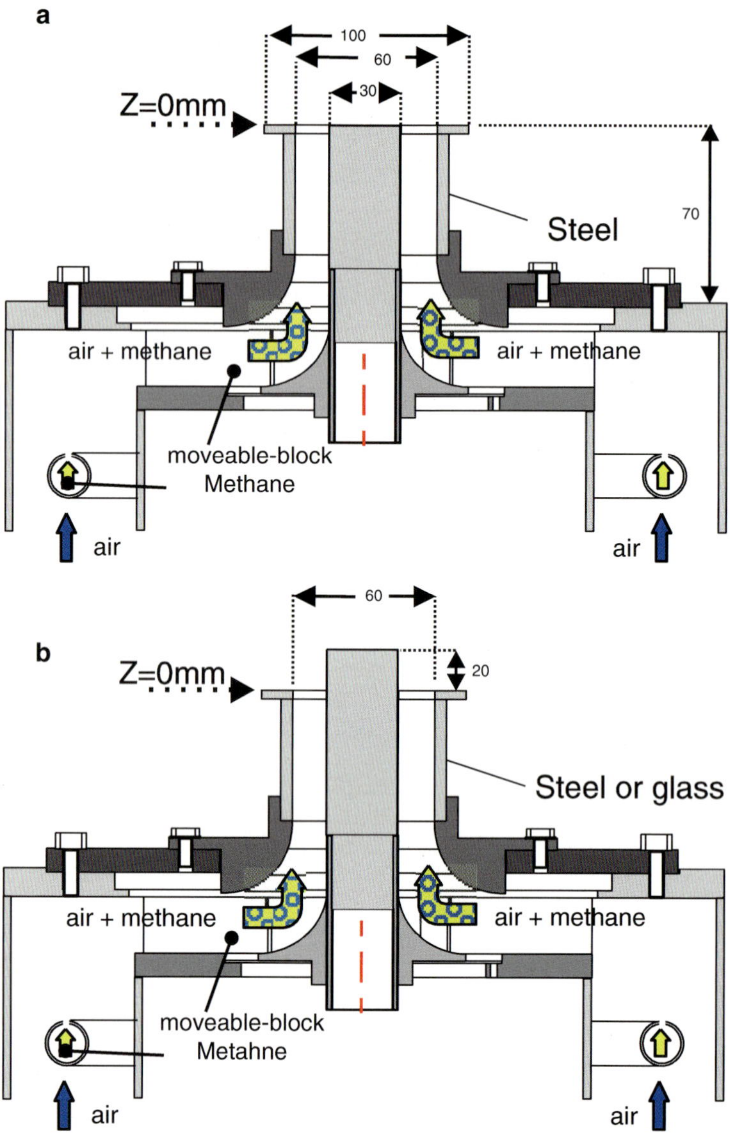

Fig. 4.3 Top: Cross-sectional view of the nozzle used for stable flame operation. *Bottom*: Adapted nozzle by extended central bluff-body and optical access to the annular slot. This configuration is designed for studying processing flames and flashes back into the nozzle (Reprinted from Schneider et al. [4, 18])

case). Flow conditions are listed in Table 4.3. The number within the acronym denotes the thermal power in kW.

In order to classify turbulent premixed flames a regime diagram initially proposed by Borghi [22], revised by Peters [23] and others, is commonly used where the oncoming turbulence intensity u' normalised by the laminar burning velocity s_L is

Table 4.3 Different flow configurations investigated on the premixed swirl burner. The *Re*-number is calculated for the nozzle exit, bulk velocity and exit area. In the corresponding isothermal cases fuel was substituted by air keeping the momentum at the nozzle exit

		PSF-30	PSF-90	PSF-150
$S_{0,th}$	[−]	0.75	0.75	0.75
P	[kW]	30	90	150
ϕ	[−]	0.833	0.833	1.0
Q_{gas}	$[m_n^3/h]$	3.02	9.06	15.1
Q_{air}	$[m_n^3/h]$	34.91	104.33	145.45
$Re_{tot.}$	[−]	10,000	29,900	42,300
s_L	[m/s]	0.36	0.36	0.42
l_F	[m]	$0.26.10^{-3}$	$0.26.10^{-3}$	$0.18.10^{-3}$

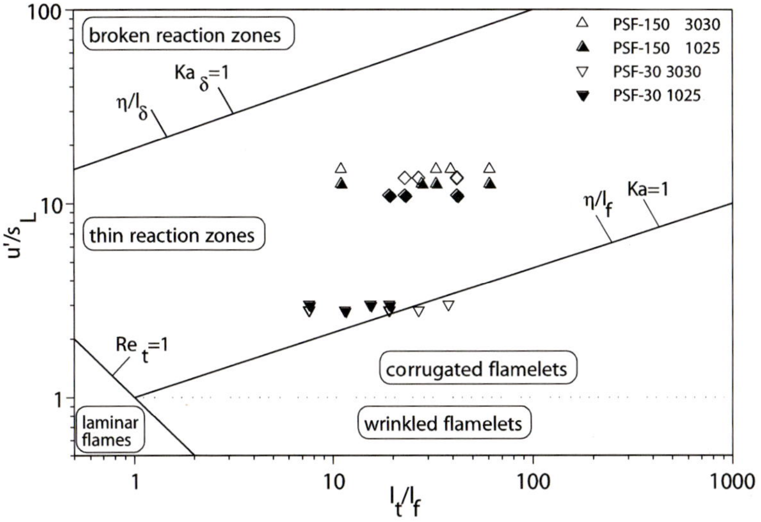

Fig. 4.4 Phase diagram to classify turbulent premixed combustion. The flames PSF-30 and -150 are charted by symbols (Reprinted from Schneider et al. [4, 18])

plotted versus the turbulent length scale l_0 normalised by the laminar flame thickness l_F. Notice, that for this classification in general equal diffusivities and a Schmidt number of $Sc = 1$ are assumed. Turbulence intensities and turbulent length scales were measured by single and two point LDV, respectively, as discussed in Sect. 4.3. Laminar burning velocities of different equivalence ratios are taken from [23], and the laminar flame thickness is deduced from $l_F = D/s_L$. The molecular diffusion coefficient $D = \left(\lambda/c_p\right)_0/\rho_u$ is calculated from the equivalence ratio $\varphi = 1/\lambda$, specific heat capacity c_p and density of unburned premixed gases ρ_u taking values from [23] and assuming a mean temperature of 1,500 K within the internal reaction zone. Resulting values for s_L and l_F are included to Table 4.3. Using spatial positions of ($x = 10$ mm, $r = 25$ mm) and ($x = 30$ mm, $r = 30$ mm) as representative positions within highly turbulent shear layers, the respective classification of the investigated flames within the regime diagram is shown in Fig. 4.4. The configuration PSF-30 is located in the cross-over of corrugated flamelets to thin reaction zones, whereas PSF-90 and PSF-150 are clearly located in the regime of thin reaction zones.

Flashback was observed at swirl numbers exceeding a critical value $S_{th,crit}$. It could be associated with the onset of a precessing vortex core (PVC), analogous to the non-reacting case causing a coherent precession of the flame and eventually reaching back into the annular slot.

4.2.3 The Role of Flame Sequences

In terms of a better understanding of turbulent combustion and in terms of providing experimental data for validation of numerical simulations it is crucial to vary parametrically *Re*-number, swirl intensity, pressure, equivalence ratio, and fuel composition. Parametric variations provide insight into mutual dependencies of boundary conditions and the turbulent flame. Of course not all of these variations are feasible using a single test rig. Therefore different generic flames, as exemplified above, need to be investigated using various diagnostic methods. This makes such fundamental experiments studying turbulent combustion so cumbersome and expensive. This fact calls for international cooperation of academia and industry. The International Workshop on Measurement and Computation of Turbulent Nonpremixed Flames (TNF) [24] may serve as an excellent example of a fruitful international cooperation.

4.3 Flow Field Measurements

This section highlights the application of 1- and 2-point LDV in the swirling premixed unconfined flame detailed in Sect. 4.2.2 [4, 18]. Application of PIV will be presented in Sect. 4.5.2 in measurements simultaneously with planar laser-induced fluorescence.

4.3.1 Flow Properties at the Nozzle Exit – Combusting Conditions

To allow for axial velocity measurements close to the nozzle exit the LDV sending/receiving optics were inclined by 6°. The resulting bias was ∼0.5% off the absolute value and could be therefore neglected. Figure 4.5 shows first and second moments of all three velocity components as well as two Reynolds-stress components. To facilitate a direct comparison of the three *Re*-number cases (compare Table 4.3), data are normalized to the respective maximum axial velocities specified in the figure caption. Axial and tangential velocity components are of the same order of magnitude whereas the radial component catches less than 20% of

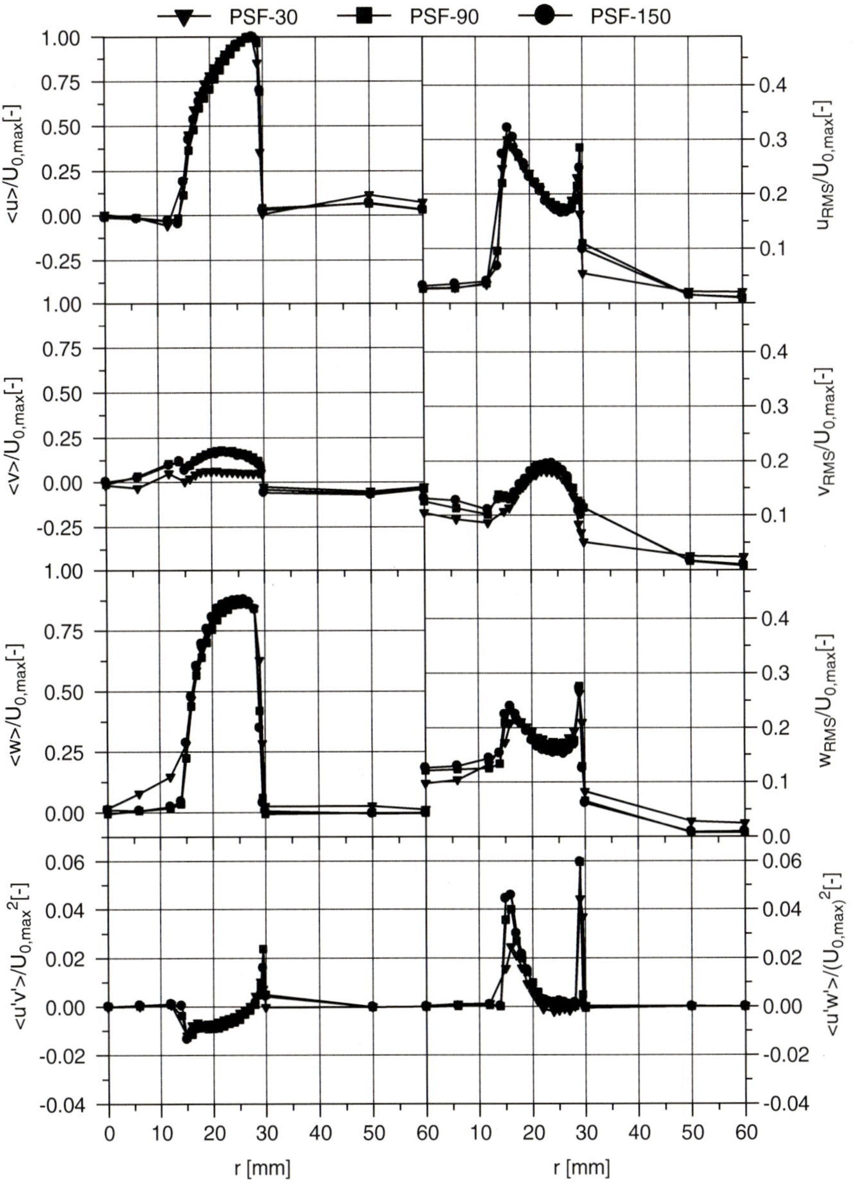

Fig. 4.5 Comparison of radial profiles of first and second statistical moments of all velocity components and two Reynolds-stresses at $x = 1$ mm. Data are normalized on the respective maximum nozzle exit velocity. PSF-30: $U_{0,max} = 6.48\ ms^{-1}$, PSF-90: $U_{0,max} = 20.2\ ms^{-1}$, PSF-150: $U_{0,max} = 28.0\ ms^{-1}$ (Reprinted from Schneider et al. [4, 18])

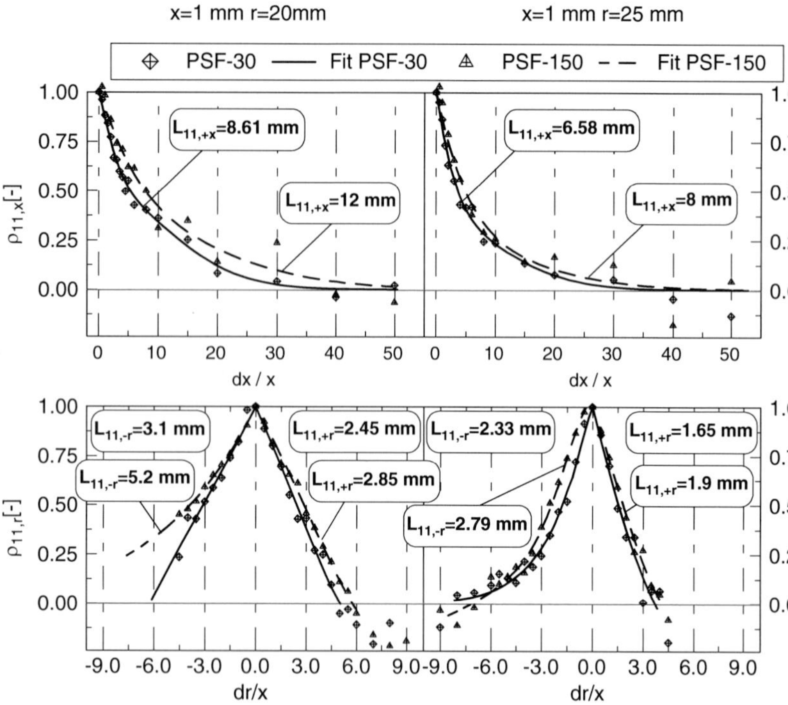

Fig. 4.6 Normalised spatial cross-correlations of PSF-30 and -150 measured at two different radii at $x = 1$ mm. Longitudinal ($L_{11,+x}$) and lateral length scales ($L_{11,\pm r}$) deduced from integration of the fitted polynomials are inserted. Notice that axial (*dx*) and radial shifts (*dr*) of the moved LDV-system are normalised by the axial height (Reprinted from Schneider et al. [4, 18])

the maximum axial velocity. Using the most common definition of the swirl number by Gupta et al. [25] and neglecting the pressure term, the actual swirl number at this location was 0.64 – a value reduced by 15% with respect to the theoretical swirl number of 0.75. Two shear layers are observed, located at $r = 15$ and 30 mm corresponding to the bounds of the annular slit. In the shear layers, fluctuations and Reynolds stresses reaches their maximum values.

Figure 4.6 shows spatial cross-correlations $\rho_{11,+x;\pm r}$ as measured by two-point-LDV at $r = 20$ and 25 mm ($\rho_{11,-x}$ cannot be measured at this axial height). For the sake of clarity PSF-90 is not included to the figure. Straight and dashed lines denote results from the polynomial fit. In general, length scales of the high Re-number case are larger. For both cases, length scales at $r = 25$ mm located closer to the outer shear layer are clearly reduced by ~30%, while the ratios of longitudinal to transversal length scales $L_{11,+x}/L_{11,\pm r}$ of 3.1 and 3.3, respectively, are changing by less than 10% between these two locations.

In Fig. 4.7 temporal auto-correlations and respective E_{11}-components deduced from the auto-covariance are presented for the same locations. For a direct comparison, the abscissa is normalised by the respective turbulent time scales. The temporal

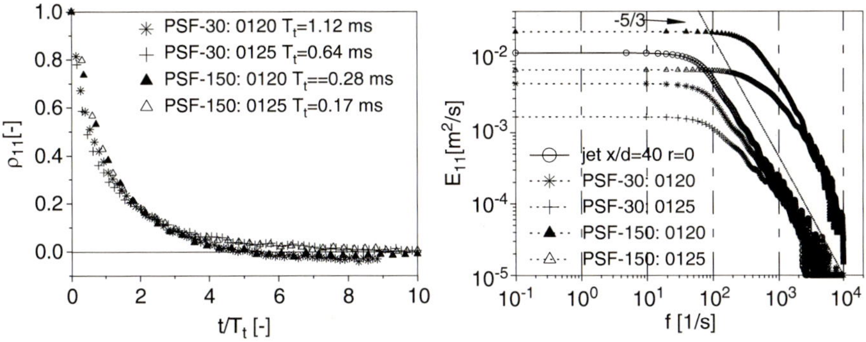

Fig. 4.7 *Left*: Temporal auto-correlations measured at the same locations as in Fig. 4.6 (denotation: 0120 denotes $x = 1$ mm, $r = 20$ mm). To enable a direct comparison of the different Reynolds-number cases, time is normalised by corresponding integral time scales specified in the insert. *Right*: PSD deduced from respective auto-covariance. For comparison results obtained from an isothermal jet are included (Reprinted from Schneider et al. [4, 18])

decays are clearly similar. In accordance to the length scales, time scales T_{11} closer to the outer shear layer are remarkably reduced relative to their values at $r = 20$ mm. For both *Re*-numbers the PSDs show a distinctive inertial subrange.

The viscous subrange could not be resolved with the current data rate. Due to the higher fluctuation levels in the axial and tangential velocity component for a given *Re*-number, at $r = 20$ mm the amplitudes are higher but the characteristic frequency where the spectrum fades into the inertial subrange is independent of the position. Therefore enhanced turbulence levels at $r = 20$ mm can be attributed solely to large scale motions. Changing the *Re*-number from ~10,000 to ~42,000 shifts this characteristic frequency from ~100 to ~300 Hz.

4.3.2 *Flow Properties Downstream the Nozzle – Combusting Conditions*

Exemplary for PSF-150, Fig. 4.8 presents a vector plot of the flow field including mean axial and radial velocities. Typical for swirling conditions the flow expands when leaving the nozzle. As a consequence by vortex breakdown [26] an internal recirculation zone (IRZ) is generated due to positive pressure gradients along the symmetry axis of the central bluff body. For the case PSF-150 the IRZ axially extends approximately 120 mm. At $x = 60$ mm the width increases to its maximum of ~5 mm. The axial velocity component approximately is maintained, but the radial velocity component up to $x = 30$ mm increases on the cost of tangential momentum.

Figure 4.9 presents radial profiles at $x = 30$ and 90 mm of the same quantities as shown in Fig. 4.5. The cases PSF-90 and -150 show a Reynolds-similarity whereas PSF-30 exhibits a reduced spreading. Fluctuation levels and the Reynolds-stress

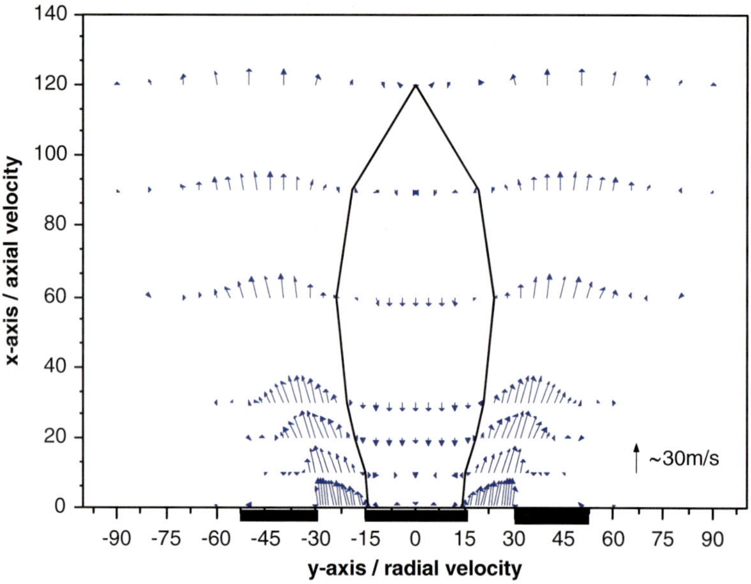

Fig. 4.8 Vector plot for the case PSF-150 showing means of axial and radial velocity components in a plane cutting the symmetry axis (Reprinted from Schneider et al. [4, 18])

component $\overline{u'v'}$ peak at radial locations where gradients of mean axial and tangential velocity components are at maximum.

Figure 4.10 compares spatial cross-correlations for positions at $x = 10$ mm, $r = 25$ mm and $x = 90$ mm, $r = 30$ mm. For clearness, only PSF-30 and -150 are shown. As observed for the nozzle exit flow, for these specific locations the higher *Re*-number case PSF-150 is characterised by significantly increased length scales except of $L_{11,\pm x}$ at $x = 90$ mm. At $x = 10$ mm the asymmetries in $L_{11,\pm x;\pm r}$ can be attributed to the influence of combustion. By traversing the moveable LDV-measurement volume in positive axial or negative radial directions the flame zone is entered which is characterised by reduced density. Here, $L_{11,+x}$ and $L_{11,-r}$ are clearly enlarged.

The ratio of longitudinal to transversal length scales is dependent on the *Re*-number and the location within the turbulent flow field. As shown in Table 4.4 there is a general trend of decreasing $L_{11,x}/L_{11,r}$ ratios in downstream direction (exception is for PSF-30, $x = 90$ mm). Compared to the ratio at the nozzle exit the respective values already at $x = 10$ mm are reduced remarkably. For isotropic conditions a ratio of 2.0 would be expected. The clear deviation from this value shows the strong anisotropic character of this flow. This implies that for numerical simulation of this type of flow simple turbulence models assuming local isotropy are physically incorrect (Fig. 4.11).

Temporal correlations and respective PSDs are presented in Fig. 4.12. As already discussed for Fig. 4.9, the case PSF-30 does not show Reynolds-similarity. This

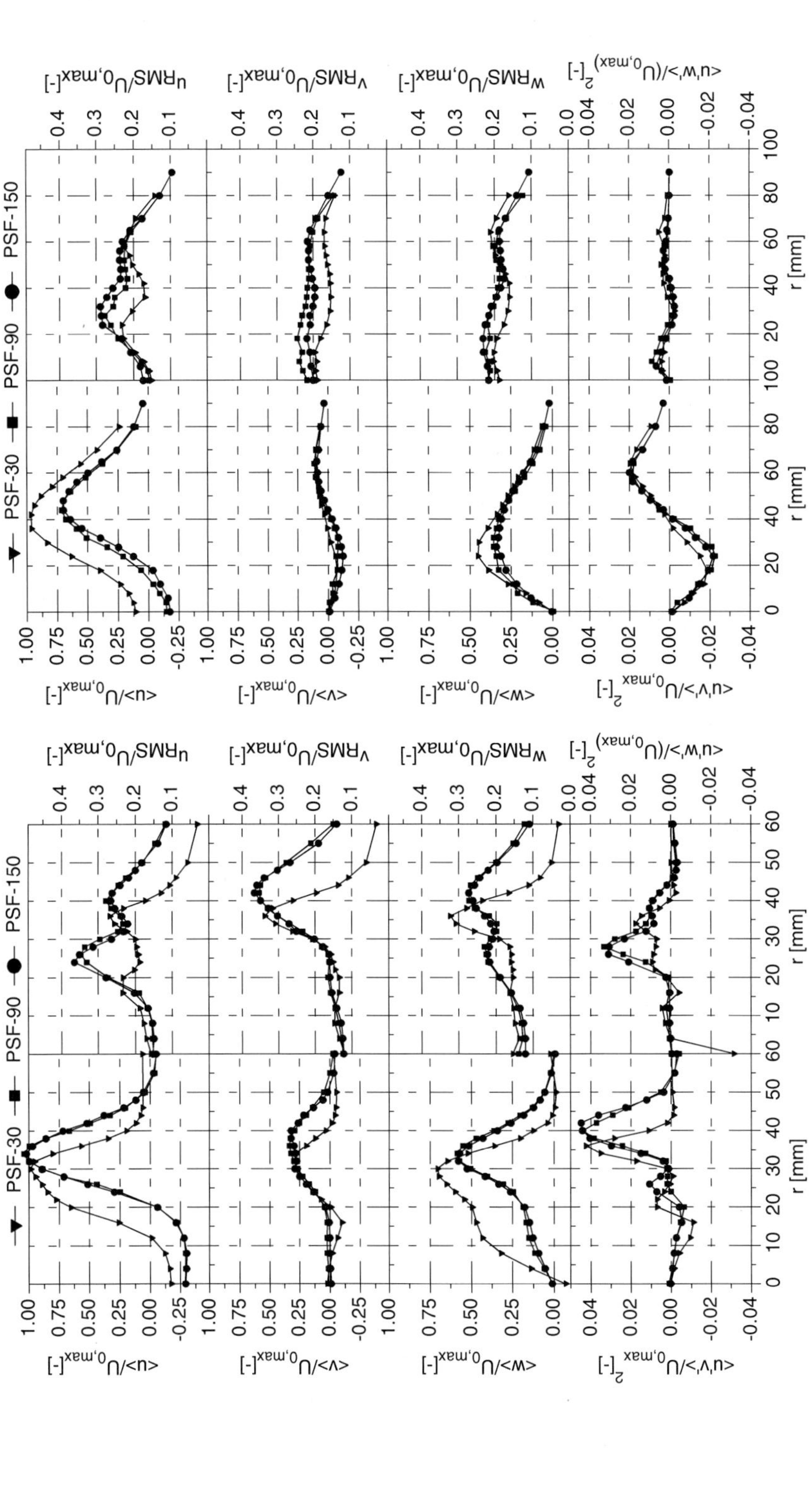

Fig. 4.9 Radial profiles of the same quantities as shown in Fig. 4.5 at an axial height of $x = 30$ mm (*left*) and 90 mm (*right*) (Reprinted from Schneider et al. [4, 181])

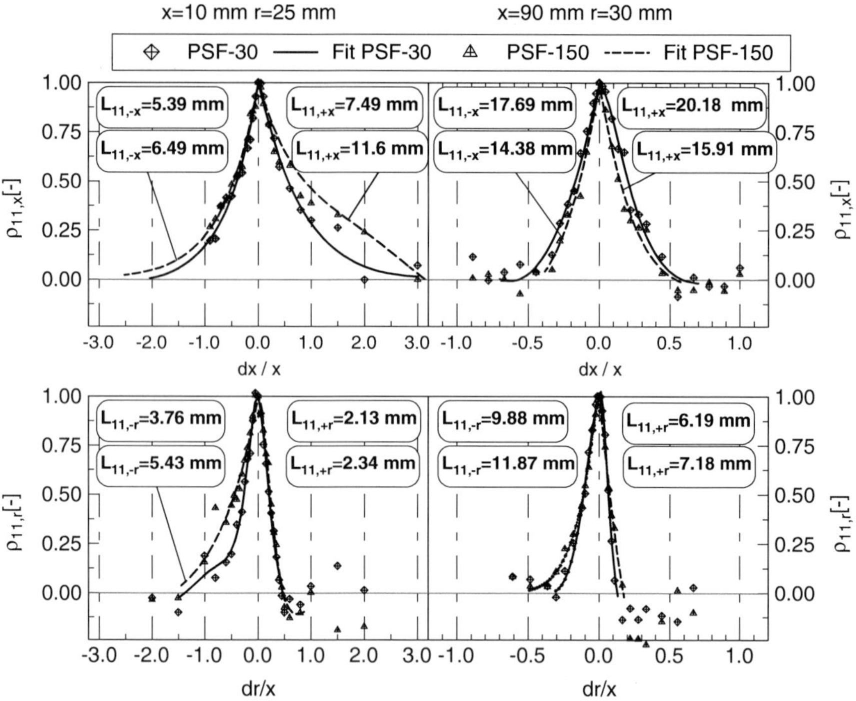

Fig. 4.10 Normalised spatial cross-correlations of PSF-30 and -150 measured at two different radii at $x = 10$ and 90 mm. Longitudinal ($L_{11,\pm x}$) and lateral length scales ($L_{11,\pm r}$) deduced from integration of the fitted polynomials are inserted. Notice that axial (*dx*) and radial shifts (*dr*) of the moved LDV-system are normalised by the axial height (Reprinted from Schneider et al. [4, 18])

Table 4.4 Comparison of longitudinal to lateral length scale $L_{11,x}/L_{11,r}$

	$x = 10$ mm $r = 25$ mm	$x = 30$ mm $r = 20$ mm	$x = 30$ mm $r = 30$ mm	$x = 90$ mm $r = 30$ mm
PSF-30	2.85	1.85	2.05	2.36
PSF-150	2.33	2.06	1.77	1.59

indicates that for PSF-30 turbulence is not yet fully developed. Due to increased velocities, the higher *Re*-number case shows a factor 4–5 shorter time scales. For both cases, from $x = 10$ to 90 mm the time scales increase by a factor of ~2.5 to 3 indicating a relaminarization due to viscous forces. Comparing both axial positions for different flames, distinct differences between the temporal decay of ρ_{11} is observed within the first few time scales. This appears more clearly by different slopes exceeding 40 Hz (PSF-30) and 300 Hz (PSF-150), respectively, in the PSDs.

At upstream locations such as $x = 10$ mm, $r = 25$ mm the slope is approximately −5/3 indicating an inertial subrange similar to isothermal jets. Further downstream

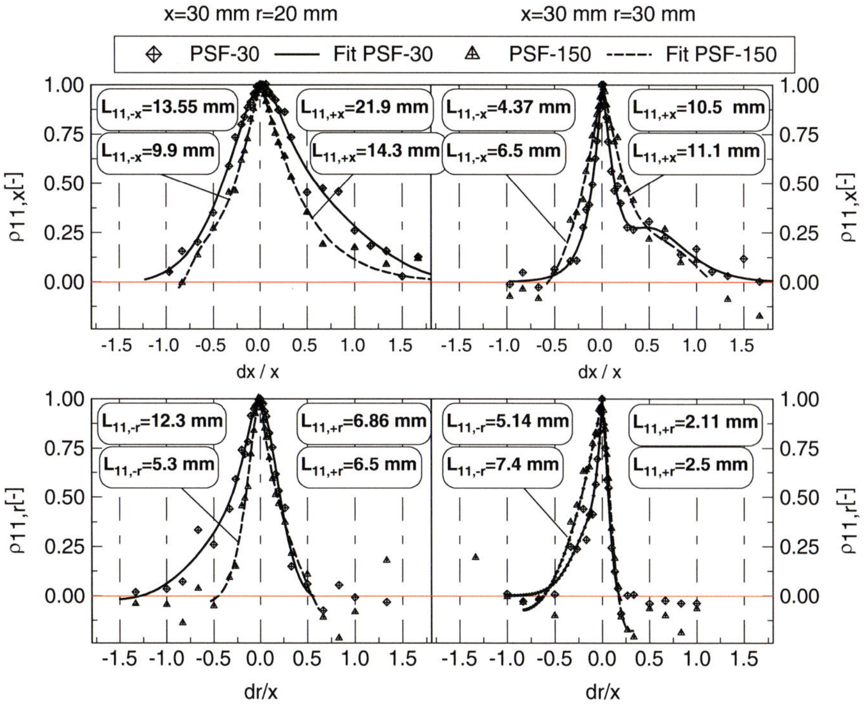

Fig. 4.11 Same quantities as in Fig. 4.10 but at axial location of $x = 30$ mm for two different radii (Reprinted from Schneider et al. [4, 18])

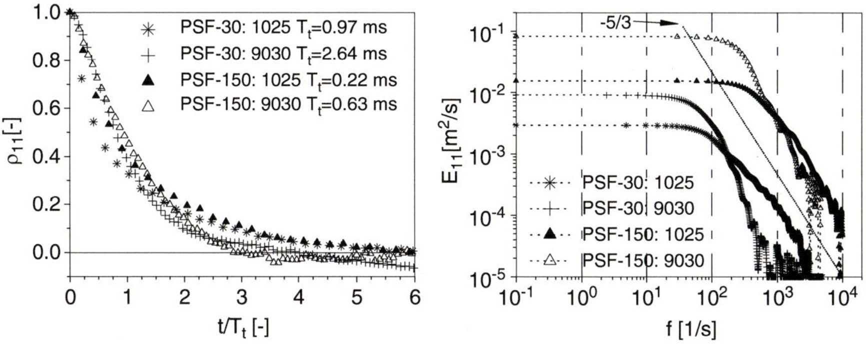

Fig. 4.12 Left: Temporal auto-correlations measured at the same locations as in Fig. 4.10. Time is normalised by corresponding integral time scales specified in the insert. Right: PSD deduced from respective auto-covariance (Reprinted from Schneider et al. [4, 18])

at $x = 90$ mm dissipation is severely enlarged resulting in a much steeper decay. Therefore turbulence structure further downstream is characterised by higher energy content in low frequent eddies but much reduced energy in the upper frequency range. From this observation one can state that turbulence structure defined by its

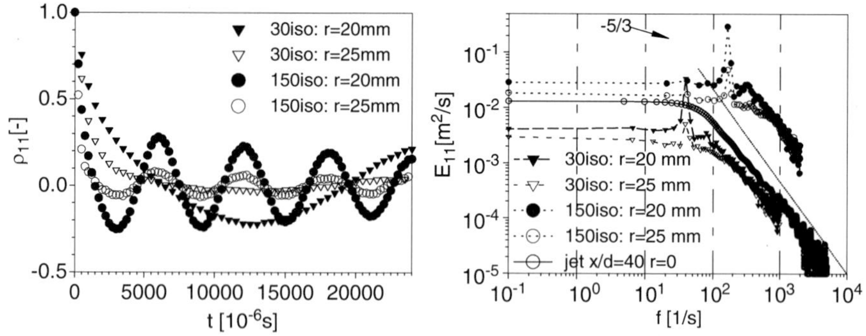

Fig. 4.13 *Left*: Temporal auto-correlations for isothermal conditions. The amplitude of the coherent motion is strongly dependent on the measurement location. *Right*: Corresponding PSD. Characteristic frequencies are independent on measurement location (Reprinted from Schneider et al. [4, 18])

energy cascade and relative energy distribution in eddies of different frequency is space dependent as expected but very different from locally homogeneous isotropic turbulence.

4.3.3 Isothermal Case

To serve a data base for validation of numerical modelling it proved to be of crucial importance to start with a non-combusting case in order to check whether the chosen turbulence model predicts important features of the flow field correctly. In addition to the flow field in the present case mixing with ambient air is of interest as well [27]. As for the reacting case, results of the isothermal case are divided into nozzle exit flow measured at $x = 1$ mm and radial profiles at downstream positions spanning from $x = 10$ to 120 mm. Just a selection of data is shown.

4.3.3.1 Flow Properties at the Nozzle Exit – Isothermal Conditions

First and second statistical moments at the nozzle exit plane at x = 1 mm show very similar profiles as the respective combusting cases and are therefore not shown. Only in the range spanning from r = 0 to 15 mm second moments of all three velocity components exhibit a significant increase. This region corresponds to the IRZ which in contrast to the combusting cases shows a coherent precession as revealed in Fig. 4.13.

After an initial decay temporal correlations show strong oscillations. The initial decay is attributed to turbulence, while the oscillations indicate a precession vortex core (PVC) around the symmetry axis. On the one hand, dependent on the measurement location, correlation amplitudes of the oscillations vary from 0 to 0.5. Therefore, the coherent structure at some locations contributes significantly to overall fluctuation levels. For example, at x = 1 mm, r = 20 mm approximately 28% of the fluctuations are due to the PVC. Amplitudes correlate well with the level of axial velocity fluctuations (not shown). Characteristic frequencies on the other hand are independent on the location but strongly dependent on the Re-number. As obvious from the corresponding PSDs characteristic frequencies peak at 38 and 195 Hz for PSF-30 and -150, respectively. Using the Strouhal-number $St = L \cdot f / U$ as dimensionless parameter, similarity of the different cases can be evaluated. With f as the lowest frequency peak in the PSD, L the hydraulic diameter of the nozzle and U the mean axial exit velocity, independent of the Re-number a value of $St \approx 0.25$ is observed. Therefore the precession frequency linearly scales with the flow rate indicating that from this view the asymptotic limit already is reached by the present configurations.

4.3.3.2 Flow Properties Downstream the Nozzle – Isothermal Conditions

In Fig. 4.14 at $x = 30$ mm radial profiles of statistical moments and Reynolds-stresses are compared to the combusting case PSF-150. Spreading of the flow is reduced in the isothermal case as can be seen from mean axial velocity profiles. Inner and outer shear layers characterised by high fluctuation levels are consequently shifted to smaller radii as well. The redistribution from tangential into radial momentum is retarded. For radial positions up to ~35 mm particularly second moments are enlarged. This observation can be explained similar to the nozzle exit flow by precession of the IRZ.

The region between the inner shear layers is characterised by positive gradients of the circulation

$$\Omega = \frac{\partial}{\partial r}\rho(\bar{w} \cdot r)^2.$$

According to the Rayleigh-criterion this indicates stable flow conditions and can be viewed as a solid body rotation. In the vicinity of the outer shear layer Ω is negative. This corresponds to unstable flow conditions because for increasing radii the tangential momentum decreases. This instable region may bear up coherent structures. However, this type of analysis is limited to isothermal flows as long as no simultaneous velocity-density measurements are performed.

To highlight the character of the PVC more clearly Fig. 4.15 shows temporal correlations measured by using both 1D-LDV-systems simultaneously. Stars denote the temporal auto-correlation at a spatially fixed position (x = 30 mm, r = 20 mm) where the amplitudes of the oscillation and the axial velocity fluctuations are at their peak values. Triangles represent temporal correlations at different radial locations measured by the second LDV-system, ranging from r = 20 mm to -20 mm. At identical measurement locations auto-correlations of both channels

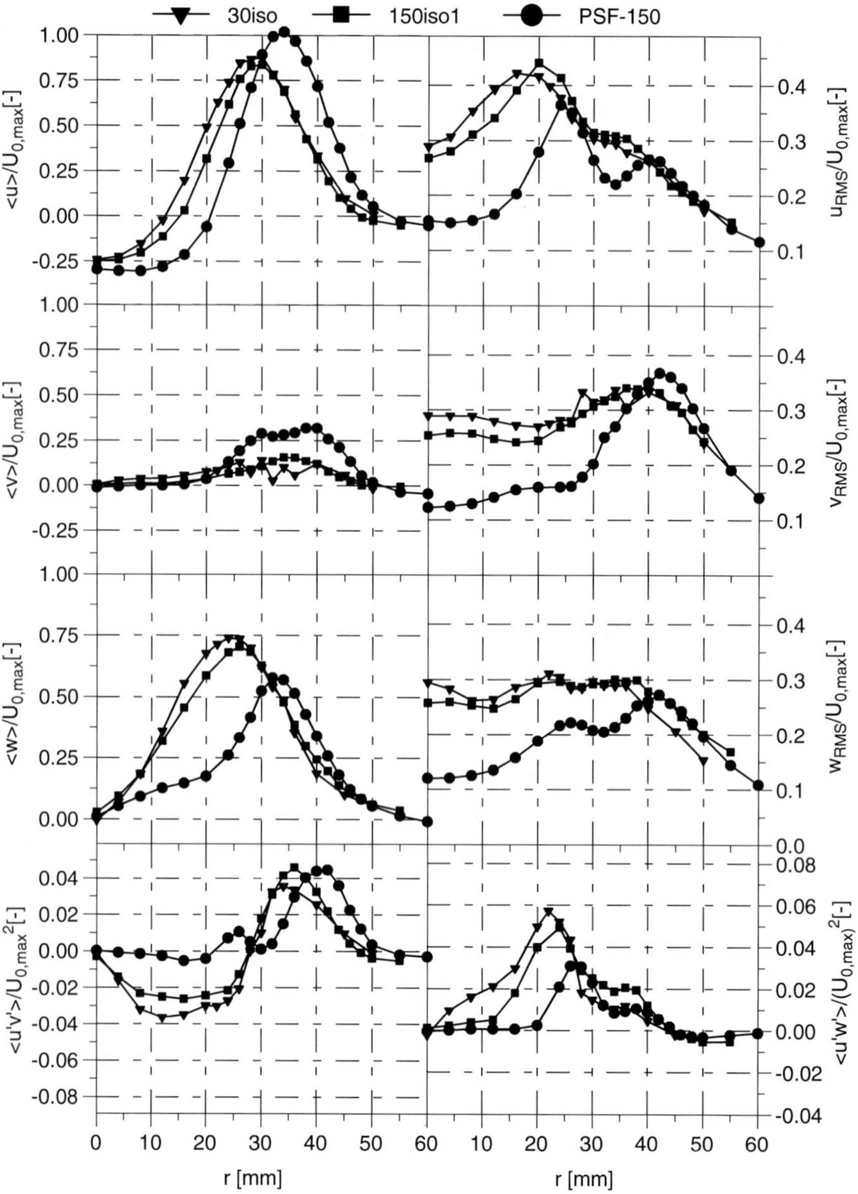

Fig. 4.14 Isothermal conditions: Comparison of radial profiles of first and second statistical moments of all velocity components and two Reynolds-stresses at x = 30 mm. Data are normalized on the respective maximum nozzle exit velocity. PSF-30: U0,max = 6.48 ms^{-1}, PSF-150: U0,max = 28.0 ms^{-1} (Reprinted from Schneider et al. [4, 18])

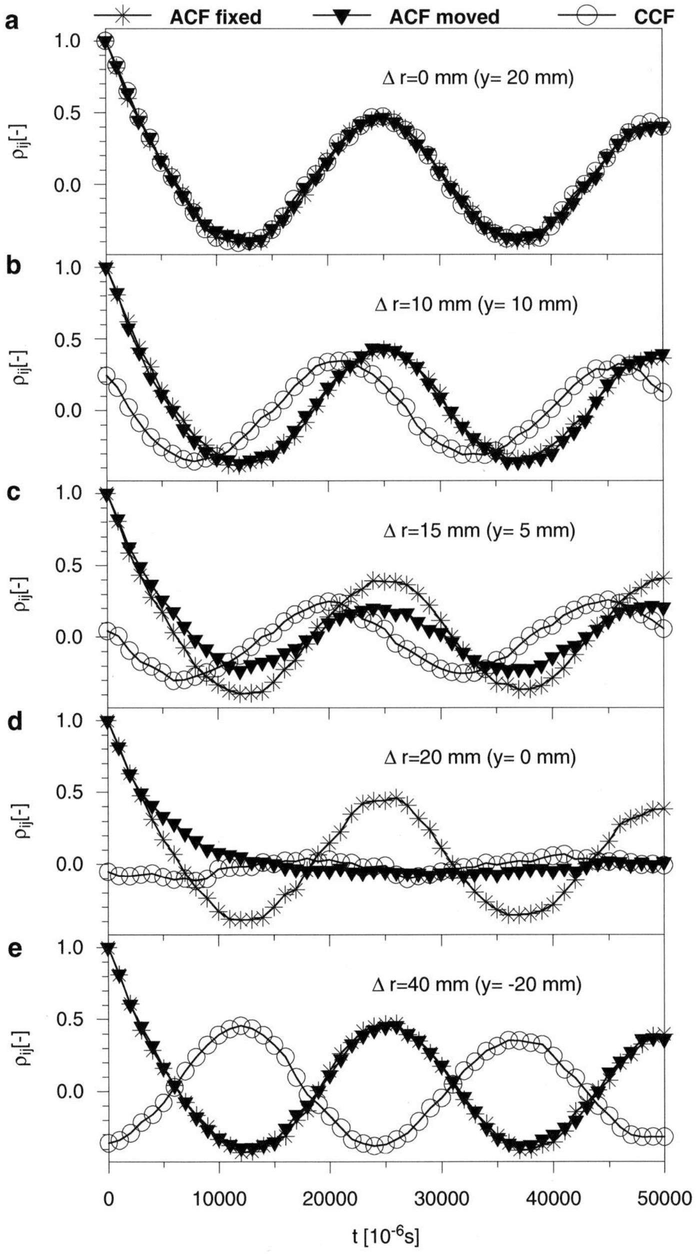

Fig. 4.15 Temporal auto-correlations measured by two independent LDV-systems at an axial height of $x = 30$ mm. The first measurement volume was fixed to radial positions (denoted here by y) of 20 mm (**a**), the second was shifted radially to 10 mm (**b**); 5 mm (**c**); 0 mm (**d**); −20 mm (**e**). Temporal dependency of the cross-correlation (*open circles*) shows variation of amplitude and phase shifts in the oscillations dependent on the spatial separation of the measurement loci (Reprinted from Schneider et al. [4, 18])

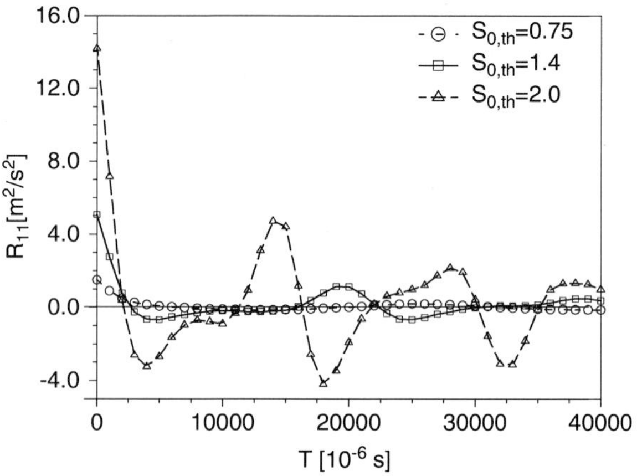

Fig. 4.16 Temporal auto-covariance for varying theoretical swirl numbers at x = 10 mm, r = 25 mm (Reprinted from Schneider et al. [4, 18])

coincide (Fig. 4.15a) as expected. A negative radial shift of the second LDV-measurement location results in a decreasing amplitude of the oscillation as obvious from Figs. 4.15b, c, d. Due to rotational symmetry of the setup, at r = −20 mm the oscillations amplitude is fully recovered (Fig. 4.15e). For cross-correlations (open circles) two observations can be made: Firstly, a phase shift appears increasing with displacement of the LDV-measurement locations. At r = − 20 mm a phase shift of π is reached indicating an almost perfect axial symmetry of the PVC. Secondly, the cross-correlations amplitudes decrease until Δr = 20 mm is reached, but increase again for increasing spatial separation of the two LDV-measurement locations and fully recover for Δr = 40 mm. The amplitude variation can be easily explained by the space dependent auto-correlation amplitude of the moved LDV-measurement location.

More difficult is a sound interpretation of the phase shift. On average the phase shift implies a bended geometry of the PVC because for increasing radii parts of the PVC lag behind. This hypothesis is sustained by negative circulation observed in the outer shear layer. However, because a phase shift allows not to deduce a direction unambiguously, the PVC at larger radii could – from observations presented in Fig. 4.15 – hurry ahead as well. A visualization of transient eddies by particle imaging velocimetry promises here a clarification.

Due to the almost perfect symmetry obvious from Fig. 4.15e and slow decay of auto-correlation amplitudes at large time intervals, it appears that velocities within the PVC sustain for hundreds of milliseconds. This is worth mentioning as overall fluctuation levels reach up to 40% especially in the shear layers where amplitudes of the coherent motion are at maximum.

To investigate the influence of tangential momentum on the PVC the theoretical swirl number was varied. Figure 4.16 shows temporal auto-covariances for

S0,th = 0.75; 1.4; 2.0 at x = 10 mm and r = 25 mm. For S0,th = 0 no oscillations appeared as expected. With increasing swirl, amplitudes and characteristic frequencies both increase. As obvious from the corresponding PSDs higher swirl cases are characterised by more distinct overtones.

Summarizing, the most distinct differences between isothermal and combusting cases are dynamical properties such as the PVC. It is speculated that the PVC in the combusting case was not observed because of a reduced effective swirl number. A reduction of the swirl number is simply achieved by decreased density in the burnt gases. From this perspective as well, combined simultaneous velocity/density measurements similar exemplified in [28] are required.

4.4 Scalar Field Measurements

In addition to instantaneous flow field measurements by LDV and/or PIV, spectroscopic methods are exploited to measure instantaneous temperature and/or concentration fields. Although various methods exist this report focuses only on laser-induced fluorescence (LIF) and Raman/Rayleigh scattering.

Laser-induced fluorescence with excitation in the UV is frequently used for thermometry and the detection of minor combustion-related molecules such as NO, OH, CH, C_2, HCHO and others. Fluorescing organic molecules have found an interest as tracers that allow quantitatively observing fluid mixing processes as well as fuel concentration mapping. Raman scattering is used to measure major species concentrations such as H_2, O_2, N_2, CO, H_2O, CO_2, and CH_4 precisely. Often it is applied simultaneously with Rayleigh scattering which is used to measure temperatures assuming the validity of the ideal gas law. Rayleigh-based thermometry requires information on the effective cross-section depending on the mole fractions of the main species. Moreover its signal wavelength occurs at the same wavelength as the excitation. Therefore it is prone to spurious light scattered from surfaces. For this reason CARS-thermometry is better suited for studying confined combustion processes.

4.4.1 1D-Raman/Rayleigh Applied to Turbulent Opposed Jet Flames

Turbulent, non-premixed flames that are stabilized in stagnation point geometries, have been studied to gain insights into flame stabilization and turbulent mixing. Partially-premixed and non-premixed turbulent opposed jet configurations have been investigated experimentally and theoretically as summarized in [1, 7, 14, 29]. Detailed experimental information on these flames allows a critical assessment of models for turbulent combustion, as for example mixing models.

From a computational point of view, the opposed jet configuration offers significant simplifications since the flow field can be represented reasonably well by a one-dimensional (1D) formulation [30]. Especially for the probability density function (PDF) methods, the 1D-simplification reduces the stochastic errors by running the simulations with large numbers of Monte Carlo (MC) particles, as demonstrated in a previous study of isothermal turbulent opposed flows [31]. However, model development was limited due to the lack of an appropriate experimental data set containing spatially and temporally resolved results on scalars as well as on velocities. To close this gap and to provide data suitable for model validation, temperatures, major species concentrations, scalar dissipation rates [7, 29], spatial OH-distributions [8, 32], axial and radial velocity components between the nozzles [9, 33] and even inside the nozzles [3, 17] have been measured using various diagnostics such as single-shot 1D-Raman/Rayleigh scattering, laser-induced fluorescence, PIV/PTV/OH-PLIF, high-speed PIV and 2C-LDV.

The MC-PDF-model used in [31] has been developed in several areas including the incorporation of a modelled equation for the scalar dissipation rate and an adaptive particle distribution scheme. Mixing models have been investigated in detail in partially-premixed jet-flames [24] but these models were hardly validated in other flames with simple flow fields. The partially-premixed turbulent opposed jet flame is distinctly different from commonly studied jet flames and exhibits a relatively simple flow field. However, a comparison of MC-PDF-simulations with spatially and temporally resolved experimental data, particular on the scalars, have been not conducted so far. This section examines the performance of mixing models in a turbulent opposed jet. Thus, an assessment of the models for a simple flow different from jet flames is provided. Note that for validation purposes the focus here is on temperature, mole fractions of main species and the flow field, whereas in [16] the focus is on scalar dissipation rates.

4.4.2 Model Formulation and Numerical Method

As a good approximation the opposed jet flame is described by the 1D-formulation [30] using the Reynolds stress model by Launder et al. [34] for modelling the turbulent flow field. The scalar dissipation rate is computed directly by its modelled equation [35]. With the mixture fraction variance, the mixing frequency is determined without making any assumption about the time scale ratio between turbulence and mixing. Monte Carlo simulation is conducted for solving the joint scalar PDF with a non-uniform distribution of particles per grid point. This distribution is adapted according to the mean temperature profile so that more particles are placed in the reaction zone [36]. The resulting non-uniform distribution contains a maximum of 1,600 particles per grid point near the stagnation plane and 400 particles near the jet nozzle exits. As such, the stochastic errors can be reduced without a significant increase in computing time.

The 12-step CH_4-air reduced mechanism with 16 species developed by Sung et al. [37] is used with ISAT [38] to further reduce computing time. Simulations are carried out with several mixing models including modified Curl, IEM, EMST [39], and a 1D-flamelet like mixing model [36]. The 1D-flamelet like mixing model considers the particles to undergo a 1D-diffusion process resembling the transient flamelet model. Monte Carlo particles in each grid are arranged in a 1D-domain according to their respective mixture fraction values. In the 1D-domain, the particles are placed with equal distance and the following diffusion equation is solved for each scalar

$$\frac{\partial \varphi_i}{\partial t} = D_M \frac{\partial^2 \varphi_i}{\partial x^2}, i = 1, K$$

where D_M is the model diffusion coefficient and K is the total number of scalars. During the simulation, D_M/L^2 is set proportional to the mixing frequency, where L is the size of the 1D-domain. The proportionality is selected to yield the correct level of mixture fraction fluctuations.

4.4.3 Results and Discussion

Simulations with the modified Curl and the IEM-mixing models lead to flame extinction even after the bulk strain rate is reduced to one fifth. Similar findings for the extinction behaviour of the modified Curl and the IEM-model are reported by Subramaniam and Pope [39]. Hence, results will be presented only from two mixing models, namely the 1D-flamelet like and the EMST-model. An analysis of simulation results indicates that in the flame zone, the present turbulent opposed jet flame falls into the flamelet regime based on the criterion by Bilger [40]. Mixing models, such as the modified Curl and IEM-models, do not have the localness property and fail to predict burning flames. In contrast, the EMST-model was developed particularly to include the localness leading to results consistent with those from a flamelet like model.

4.4.3.1 Axial Profiles

Figure 4.17 presents a comparison of experimental and numerical results of velocities. The top of Fig. 4.17 shows radial profiles of mean and normalized *rms* of axial velocity as well as the turbulent kinetic energy. The velocity exit profile has characteristics in between the plug flow and potential flow conditions. Plug flow characteristics since mean radial velocities (not shown) are close to zero at $z = -13$ mm, and potential flow characteristics because the mean axial velocity profile is not constant but has a minimum at the centerline. While a large constant

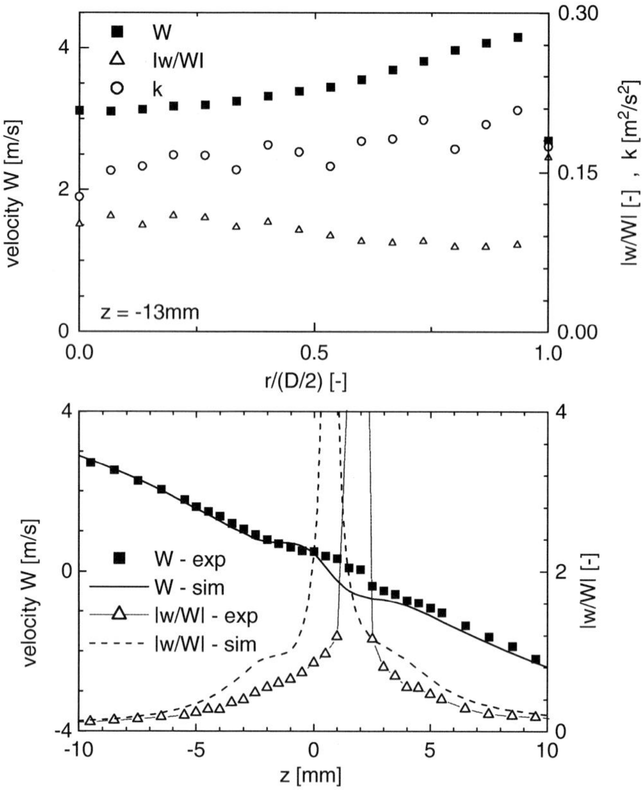

Fig. 4.17 *Top*: Radial profiles of mean axial velocity W, normalized rms of axial velocity w/W and turbulent kinetic energy k measured by 2D-LDV at $z = -13$ mm (fuel jet). *Bottom*: Comparison of profiles for mean axial velocity and normalized axial velocity fluctuations from experiment and simulation. Position $z = 0$ represents the half distance between the nozzles. Notice, that both mixing models resulted in almost identical axial velocity profiles (Reprinted from Geyer et al. [1, 14])

value of approximately 0.1 was found for the normalized fluctuation of the axial velocity around the centerline, a slow increase of turbulent kinetic energy can be observed towards the nozzle edge.

The bottom of Fig. 4.17 shows comparisons of axial profiles for mean and normalized *rms*-values of the axial velocity component.

Except in the region around the stagnation point ($W = 0$ m/s), numerical results are in reasonable agreement with the measured velocity component. Deviations in the vicinity of the stagnation point are at least in part due to the LDV-measurements featuring a bias, as seeding densities in the fuel and oxidizer flows are not perfectly matched. From a statistical point of view, volume elements from the flow with slightly higher seeding density are over represented, resulting in a bias. Since the skewness of the axial velocity is largest around the stagnation plane, an

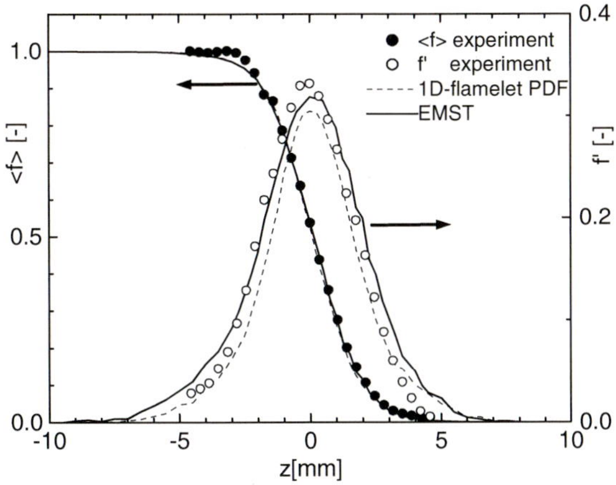

Fig. 4.18 Axial profiles of mean and rms of mixture fraction (Reprinted from Geyer et al. [1, 14])

over-seeding of the fuel stream is likely. However, quantifying this bias as discussed by Korusoy and Whitelaw [41] was not attempted so far. Therefore, for model validation purposes the biased stagnation point should not be overvalued. The normalized fluctuations approaches infinity where the mean velocity tend to zero. As the predicted and measured stagnation points are at slightly different locations, the experimental and numerical results peak at different positions. Viewing the neighbouring regions, one can see that the measured and predicted values at the tails agree within a factor of approximately two. Taking into account the uncertainties in both model and measurements, this agreement in the second moments is reasonable.

Figure 4.18 shows axial profiles of the mean and variance of mixture fraction. The comparisons focus on a region of z = ±10 mm since the whole mixing layer remains within this range. The location of the stoichiometric mixture fraction ($f_{stoich} = 0.53$) and stagnation plane ($f_{stag} = 0.50$) are almost identical, the first is found at z = −0.1 mm while the latter is located at z = +0.1 mm. Simulations show a good agreement with experimental findings for both the width of the mixing layer and the level of fluctuations. This agreement is independent of the mixing model. The peak value of mixture fraction fluctuation ($f'_{max} = 0.33$) is well matched by the simulations, although the predicted maxima are slightly shifted towards the oxidizer nozzle.

A comparison of temperature profiles is presented in Fig. 4.19. The peak value in the mean temperatures is lower more than 600 K than the adiabatic flame temperature. This results from flame stretch and the pronounced movement of the turbulent flame, which dithers around the stagnation point. Two-dimensional OH-radical distributions [8, 32] measured by planar laser-induced fluorescence have revealed that the mean reaction zone moves within $z = \pm 1.1$ mm. Subsequently, the temperature fluctuations corresponding to this flame motion are as high as 600 K. The axial profile, as seen in Fig. 4.19, shows two almost symmetric maxima at the axial positions $z = \pm 2$ mm. The simulations reproduce the mean temperature profile

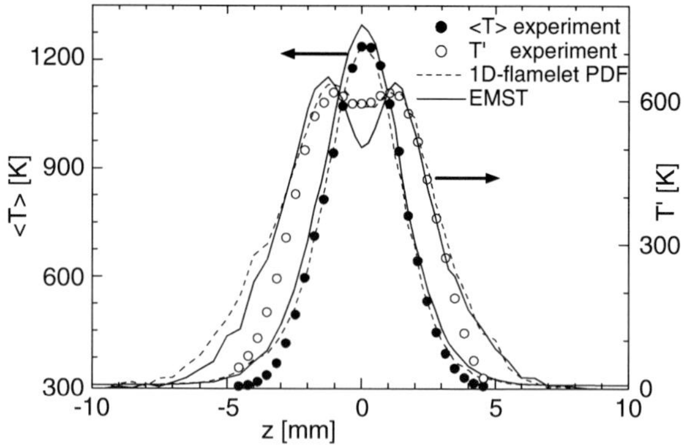

Fig. 4.19 Axial profile of mean and rms of temperature (Reprinted from Geyer et al. [1, 14])

and the temperature fluctuations well. The peak values of the mean temperature predicted by both EMST and 1D-flamelet like model coincide within 4.5% with experimental data. At the tails of the fluctuation profile simulations indicate a mixing zone 1–2 mm wider. The EMST-model predicts the local minimum of temperature fluctuations in the vicinity of the stagnation point particularly well.

4.4.3.2 Scatter Plots

For a more detailed analysis of model performance, Fig. 4.20 shows scatter plots of H_2O- and CH_4-mass fraction. As both mixing models behave similar, only the scatter plots for the EMST-model are shown. On the lean side ($f \in [0; 0.53]$), a satisfying agreement of experimental and numerical results is found. In the experiment more scatter is observed. This in parts results from experimental error that has been estimated to be 5%, based on calibration measurements in laminar opposed jet flames. At this stage already, turbulent mixing seems to be underestimated in the PDF-simulation independent of the mixing model. At the rich side, however, especially in a mixture fraction range spanning from 0.6 to 0.85, both mixing models overestimate the H_2O-concentration. These deviations clearly are outside the experimental range of confidence and can be explained by inspecting the CH_4-scatter plots. In contrast to experimental results, the turning point in the simulated scatter plots is shifted towards the rich side by approximately $\Delta f \approx 0.15$. As seen in Fig. 4.20, the simulation results are close to laminar flamelets for low strain ($10\ \text{s}^{-1} < a < 400\ \text{s}^{-1}$). In contrast, the experimental scatter is enclosed by flamelets with strain rates ranging from $a = 200$ to $750\ \text{s}^{-1}$. The experimental conditional averages (top row of Fig. 4.20) are fairly well represented by a laminar flamelet with $a = 400\ \text{s}^{-1}$ over a wide range of mixture fractions. Evidently, the turbulent flame is influenced by much higher strain than predicted by PDF-modelling. Correspondingly, at the rich side, simulated H_2O-levels exceed experimental values by far.

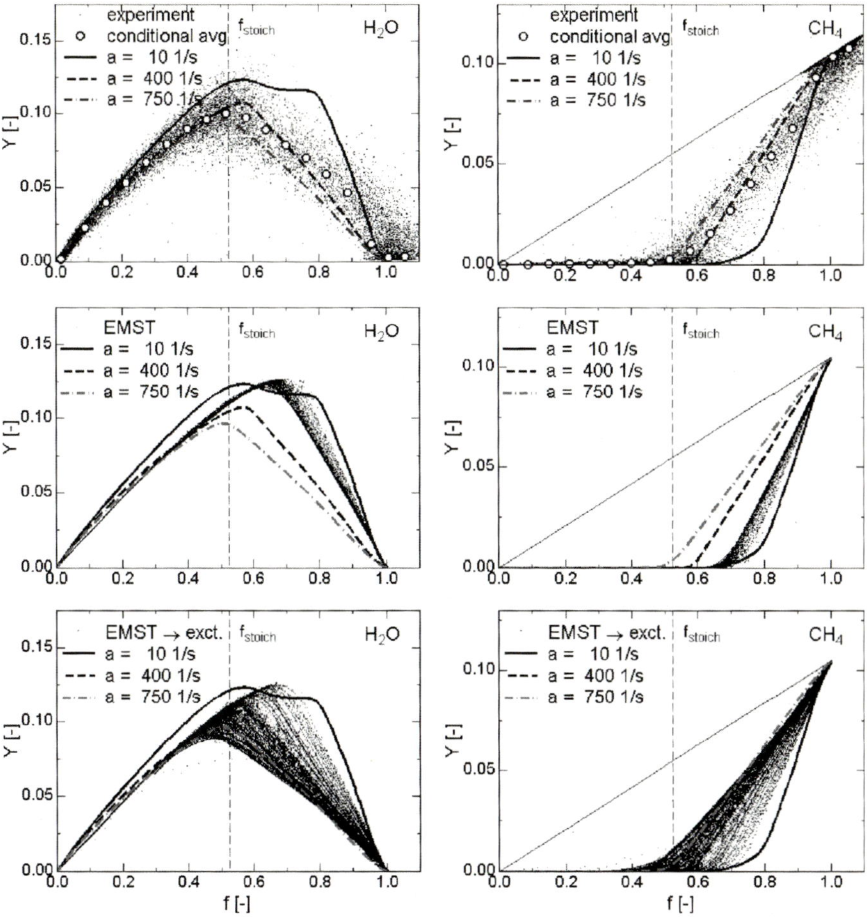

Fig. 4.20 Scatter plots of H_2O- (*left*) and CH_4-mass fraction (*right*). *Top*: experiment, *middle*: EMST-mixing model, *bottom*: results from EMST-model close to global extinction. Using experimental scatter plots at $f = 1$, measurement accuracy can easily be illustrated. Laminar flamelet calculations of varying strain rate with full transport are included for comparison (Reprinted from Geyer et al. [1, 14])

For the EMST-model the bottom row of Fig. 4.20 shows scatter plots for conditions close to global extinction. Obviously, a much wider range of conditions, especially in the rich part, is accessed in comparison to the experiment. The plot reveals that some low strain rate events might sustain a flame in simulations, while it is already extinguished at that mean bulk strain rate in the experiment. However, the majority of the samples converges against results from laminar flamelet calculations with $a = 750\ \mathrm{s}^{-1}$, the extinction limit for the laminar flame. Contrary to turbulent jet flames [24], local extinction and re-ignition events prior to global extinction

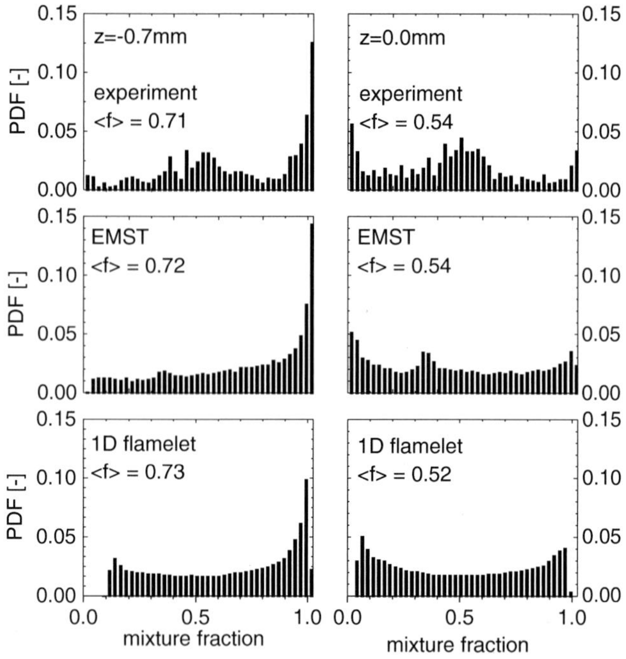

Fig. 4.21 Mixture fraction PDFs at a location z = −0.7 mm and at z = 0.0 mm. *Top*: experiment, *middle*: EMST-mixing model, *bottom*: 1D-flamelet like model (Reprinted from Geyer et al. [1, 14])

were found neither for experiment nor for simulations. This is a strong indication that turbulence–chemistry interaction in turbulent opposed jet flames is distinctly different from jet flames.

4.4.3.3 Mixture Fraction PDF-Distribution

Detailed PDF-distributions of mixture fraction are compared in Fig. 4.21 at two locations, namely z = −0.7mmand z = 0.0 mm. These two locations correspond to a rich flame ($f \sim 0.71$) and a stoichiometric flame ($f \sim 0.53$), respectively. For the rich flame, the predicted shapes by both models are similarly exhibiting a gradual decrease in the PDF from the fuel side and an almost uniform distribution once the mixture fraction drops below 0.8. The measurements reveal a similar initial decrease in the PDF from the fuel side but have more variations for mixture fractions below $f = 0.9$, resembling a bell shape with its peak at $f \sim 0.53$. For the stoichiometric flame, model predictions show an inversed bell shape while experimental data resemble again a bell shape around $f \sim 0.5$ with two distinct peaks for pure fuel and oxidizer. These experimental data reveal the strong intermittent nature between fuel and oxidizer and the rapid mixing between them. Compared to experimental data, simulation results show a much slower mixing process with less intermittency.

4.4.3.4 Extinction Limit

Global flame extinction can be viewed as an ultimate test of intense turbulence–chemistry interaction. As mentioned above, global extinction was observed at a bulk strain rate of $a = 255\ s^{-1}$. Numerically, the extinction limit was evaluated by varying parametrically the bulk strain rate via axial inflow velocities in a series of different simulations. In contrast to the experimental data, extinction in the simulation occurred around $a = 560\ s^{-1}$, close to the extinction limit of $a = 750\ s^{-1}$ from laminar flame calculations. The much higher resistance to flame extinction is consistent with the high flame temperatures predicted at the rich side.

4.4.3.5 Assessment of Numerical Models

The above comparisons between experimental and numerical results provide a basis for the discussion of the performance of various submodels. As the 12-step chemistry has been used previously in modelling partially-premixed jet flames with the same fuel mixture ($\Phi = 2$), showing good results in the rich parts of flame [24], the reduced chemistry mechanism is believed to be adequate. The comparison of mixture fraction PDF in Sect. 4.4.3.3 suggests that the numerical models predict a less intermittent flow with a slower mixing process compared to that implied by experimental data. The predicted mixing process is influenced by both the flow field model and the mixing model used in the PDF-approach as well as by their coupling. The observed strong intermittency between fuel and oxidizer is induced by the large flow motions around the stagnation plane. A strong intermittency is likely to associate with high turbulence fluctuations. Quantities, such as turbulent kinetic energy and axial fluctuation components, are essential to provide a critical assessment of model performance. A closer examination of Fig. 4.17 suggests that the noted discrepancies between experimental and numerical results near the stagnation point could be significant. Due to the uncertainties of LDV near the stagnation point, measurements of the velocity field need to be improved to yield more accurate second moment statistics to allow for a definite conclusion on the flow field prediction.

The EMST-model provided good results in modelling local extinction and re-ignition [24] and hence it is capable of modelling strong turbulence-chemistry interactions. Yet, both EMST and the flamelet like mixing models yield similar results in the rich parts of the flame. Both mixing models require the input of a mixing frequency that is determined only by the mean scalar dissipation in the present simulations. At the rich part of the flame, the predicted mean scalar dissipation rates are lower than the measured values by a factor of two [2, 16]. In addition, large variations in the scalar dissipation rate were measured peaking near $f \sim 0.75$. From previous large-eddy simulations with the transient flamelet model by Pitsch [42], the variation of scalar dissipation rate was found to be the key in predicting the proper CO-levels in rich parts of flame. If variation was not properly accounted for, the predicted CO-levels were too high in comparison to data [42].

Consequently, in addition to the uncertainty in the flow field prediction, future model improvements need to include proper treatment of the scalar dissipation rate and its fluctuations.

4.4.4 Conclusions of Measurements in Turbulent Opposed Jet Flames

Experimental results obtained from scalar and velocity field measurements using 1D-Raman/Rayleigh scattering and two-dimensional LDV were compared to numerical results from one-dimensional Monte Carlo PDF-simulations with a 12-step reduced chemistry and different mixing models. These comparisons revealed that the property of localness as in the EMST- and flamelet like model is essential for describing the mixing processes physically correctly whereas the IEM- and modified Curl model failed because of extinction at much too low bulk strain rates. Although extinction limits computed with these two mixing models were overestimated by a factor of more than two in terms of bulk strain rates, simulations in general revealed a very good agreement for scalar and flow field properties in the physical space. The average location and the width of the reaction zone were reproduced remarkably well, whereas deviations of the flow properties observed in the vicinity of the stagnation plane may partly be due to uncertainties in the LDV-measurements. Detailed comparisons in the mixture fraction coordinates were carried out, using scatter plots and PDF-distributions. At the lean side experimental and numerical results agreed well, while at the rich side, large discrepancies were found, indicating that the nature of both mixing and intense intermittency of the turbulent opposed jet flame are underestimated by the numerical model. These findings were supported by mixture fraction PDFs. Future model improvements are needed to properly account for the experimentally observed strong intermittency and mixing. Finally, the necessity to validate mixing models in simple flow fields that differ from jet flames, in order to devise generally applicable models, was demonstrated.

4.5 Simultaneous Flow and Scalar Field Measurements

4.5.1 Combined LDV- and OH-PLIF-Measurements in Swirling Premixed Flames

Premixed combustion is of increasing technological relevance due to lower NOx-emissions at lean equivalence ratios. A proper design of future low-NOx-combustors depends on more reliable numerical simulations of turbulent premixed combustion processes. Although a variety of models such as the Bray-Moss-Libby (BML) model [43, 44], the thickened flame model [45], the linear-eddy model (LEM) [46, 47], or the G-equation approach [48, 49] have been applied in a RANS (Reynolds averaged Navier–Stokes)- or LES (large-eddy simulation)-context during the last two decades [50], there is still an ongoing scientific debate how to model turbulent premixed flames. To check the validity of single unclosed terms in the transport equations or to validate integral models determining entire turbulent flames, there is a need for experimentally well investigated target flames.

Whereas much progress has been attained already for non-premixed flames [24], much less comprehensive experimental data suitable for model validation of turbulent premixed target flames exist. Comprehensive studies including flow and scalar field information measured by different laser diagnostics in a highly stretched, piloted Bunsen flame are documented in [51] offering insights into turbulence–chemistry interaction and turbulence structures. A variety of other experimental studies in different lab-type flames have been conducted, for example in turbulent opposed flows [52–54], low swirl flames [55–57], and wire-stabilised flames [58, 59]. The emphasis was often on single quantities, such as turbulent fluxes that have been measured in jet flames [60, 61], wire stabilized flames [62], and bluff-body stabilized flames [28, 62]. For more complex nozzle geometries and strong swirling flames, that are important for industrial applications, much fewer detailed experimental investigations have been performed.

In modelling premixed combustion the Favre-averaged progress variable $\tilde{c} = \overline{\rho c}/\bar{\rho}$ has been used successfully [63]. The reaction progress c is a normalized temperature given by $c = (T - T_u)/(T_b - T_u)$with T being a temperature and the subscripts "u" and "b" denoting unburned and burnt gases, respectively. Flamelet-based models relying on the progress variable assume an infinite thin flame structure. This directly implies c to be a step function between unburned and burnt gases with a bimodal probability density function (PDF) [64]. Different planar laser diagnostic methods such as filtered Rayleigh scattering [62] or planar laser induced fluorescence (PLIF) for detection of spatial hydroxyl radical (OH) distributions [60, 61] have been used to measure mean spatial distributions of reaction progress directly.

Alongside the mean reaction term, the transport equation of $\tilde{c}$ exhibits the turbulent scalar flux $\overline{\rho u'' c''}$that needs to be modelled [65]. Within the formalism of the BML-model the axial turbulent flux of $\tilde{c}$ is expressed by [63]

$$\overline{\rho u'' c''} = \bar{\rho}\tilde{c}\,(1 - \tilde{c})\,(\overline{u_b} - \overline{u_u}) \tag{4.1}$$

The quantities $\overline{u_b}$ and $\overline{u_u}$ denote conditional mean axial velocity components within the burnt and unburned gases, respectively. Equivalent equations apply for the two other directions. Counter-gradient diffusion, associated with thermal expansion and acceleration of fluid passing the interface dividing unburned and burnt gases exists when $\overline{u_b} > \overline{u_u}$leading to $\overline{\rho u'' c''} > 0$. Gradient diffusion occurs when $\overline{u_b} < \overline{u_u}$ and the flux is assumed to be down the gradient of $\tilde{c}$.

For rather weak turbulent premixed flames counter-gradient diffusion has been observed experimentally as summarized by Veynante et al. [65]. Gradient diffusion is expected to take place when the flame is dominated by turbulent motions rather than thermal dilatation due to chemical reactions. Based on direct numerical simulations of flames exposed to isotropic turbulence, these two regimes of gradient and counter-gradient fluxes have been predicted [65].

With the turbulence intensity u', the laminar flame speed s_L, the heat release factor $\tau = (T_b - T_u)\,1!/1!T_u$, and the efficiency function α both regimes of turbulent transport can be distinguished by

$$\begin{aligned} N_b &\equiv \frac{\tau s_L}{2\alpha u'} \geqslant 1 \text{ counter - gradient diffusion} \\ N_b &\equiv \frac{\tau s_L}{2\alpha u'} < 1 \text{ gradient diffusion} \end{aligned} \tag{4.2}$$

According to this criterion high values of the ratio u'/s_L and low heat release factors favour gradient diffusion. The efficiency function α is dependent on the length scale ratio L_t/l_F (with L_t the integral length scale and l_F the laminar flame thickness). Decreasing L_t/l_F promote an increase of N_b and thereby counter-gradient diffusion. A crossover from a counter-gradient regime to a gradient regime has been proven experimentally [61] for a piloted Bunsen-type natural-gas/air flames with various equivalence ratios and turbulence intensities. In a different study investigating wire and bluff-body stabilized CH_4/air flames [62], regimes of gradient and counter-gradient diffusion have been observed depending on local flame conditions.

The overall motivation for this chapter is to measure the turbulent flux for the premixed swirl flames presented in Section 4.2.2. Strong swirl and high ratios of u'/s_L are used. By varying the *Re*-number between 10,000 and 29,900 the regime of thin reaction zones and corrugated flamelets is accessed. Turbulent flux is deduced from combined laser Doppler velocimetry (LDV, Sect. 4.5.1) and OH-planar laser-induced fluorescence (PLIF, Sect. 4.5.2). Hereby, OH-contours are interpreted as interfaces between burnt and unburned gases.

4.5.1.1 Optical Setup and Synchronisation of Simultaneous LDV- and OH-PLIF-Measurements

According to Eq. (4.1) the measurement of the turbulent flux rests upon **conditional** velocities. In the present study LDV and OH-PLIF have been therefore applied simultaneously. The transient OH-contour deduced from single-shot PLIF-images was used to determine the radial displacement Δr of the actual velocity measurement from the turbulent flame front allowing designation of conditional velocities. These simultaneous measurements were restricted to a few selected locations within the flame brush as the repetition rate of the optical method was terminated by the read-out times of the ICCD used within the PLIF-setup.

Close to a flame front LDV-seeding particles can be exposed to large temperature gradients. Thermophoretic effects might therefore influence the measured gas velocities.

Thermophoretic velocities u_{TP}normal to the flame front and down the temperature gradient can be estimated by [66]

$$u_{TP} \approx 0.5\nu \frac{\nabla T}{T}$$

with the kinematic viscosity ν. In [61] u_{TP} was deduced from strained laminar flame calculations. For stoichiometric conditions thermophoretic velocities are below 0.09 ms^{-1} and can be therefore neglected in comparison to convection velocities.

For OH-PLIF a tuneable KrF-Excimer laser (EMG 150 Lambda Physik) was used to excite the $P_1(8)$-line within the $A^2\Sigma \leftarrow X^2\Pi$ (3, 0)-band. By a $f = 1{,}000$ mm cylindrical lens a 17 mm high and 0.5 mm thick light sheet was formed in the measurement plane intersecting the flame at its vertical symmetry axis. Using a 105 mm f/4.5 UV-lens, non-resonant OH-fluorescence within the (3, 2)-band around 295 nm was imaged perpendicularly to the laser pointing onto a 1,300 × 1,300 pixel, thermoelectrically cooled, and intensified charge-coupled device (ICCD, PiMax, Roper Scientific). In front of the ICCD-camera a band pass filter (central wavelength 297 nm, FWHM ~45 nm, arrangement build out of four dichroic mirrors) was placed. Gating the intensifier with 200 ns long exposure times prevented any spurious radiation from the laboratory environment and flame chemiluminescence. By limited interline transfer rates, the camera read-out times were restricted to two samples/sec determining the overall repetition rate of simultaneous LDV/PLIF-measurements.

LDV is random and intermittent in nature. For this reason validated Doppler bursts have been used as master trigger in the experiment. To avoid any feedback on the LDV-signal processing by electrical interferences from the thyratron, the excimer laser and the ICCD-camera were triggered 5 μs after a validated LDV-burst. According to [18] integral time scales were measured to be of the order of 0.17 to 1.12 ms, depending on the case and the location within the flow field. This is sufficiently larger than the 5 μs time lag between the LDV- and OH-PLIF-measurement. From this point of view the LDV- and PLIF-measurements can be taken as simultaneously. Allocation of PLIF-shots and validated LDV-signals was achieved by synchronized master clocks within the PC recording the PLIF-data and the LDV-controller.

Due to the presence of seeding particles the intense excimer laser pulses excited laser-induced incandescence (LII) in addition to the OH-fluorescence. As LII exhibits a wide spectral distribution of thermal radiation, the ICCD, although band pass filtered, detected significant amounts of LII-signals. For this reason seeding density was reduced furthermost to keep cross-talk and spectral interferences at minimum. However, data post-processing described below was still crucial to discriminate OH-PLIF- from LII-signals.

For LDV a conventional fiber-optic system was used in backscatter mode (Dantec) recording two velocity components employing a multi-line Ar^+-laser (Coherent). The measurement volume extended approximately 1 mm in length and 80 μm in cross direction, respectively. Doppler bursts were evaluated electronically

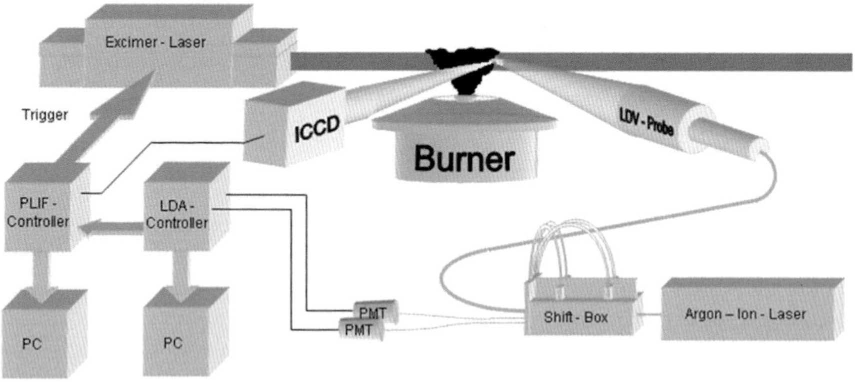

Fig. 4.22 Experimental setup for combined LDV- and OH-PLIF-measurements based on conventional equipment (Reprinted from Nauert and Dreizler [10, 67])

(controller TSI IFA 750). Magnesium oxide (MgO) served as seeding material. This choice was motivated by its thermal stability and small slip. Point-wise axial and radial velocity components were measured within selected locations in the flame brush. The overall error of the mean conditional velocity components shown in the results section was $\sim$3%.

Figure 4.22 shows the experimental setup for the combined OH-PLIF- and LDV-measurements. For each axial height, the LDV-measurement position was fixed at a radial position R_{LDV} corresponding to a location of high probability to find the OH-contour. The angle between pointing of the LDV-probe and the PLIF-laser sheet was 81°. So far the resulting bias for the radial velocity component at order of 4% was not accounted for in the following treatment. Typically, up to 25,000 statistically uncorrelated samples suitable for the determination of conditional velocities were recorded. This large number of data is crucial to allow for detection of small changes in mean conditional velocities. Notice, that the actual number of recorded samples was much higher as approximately 20% of the samples were rejected due to too strong interference from LII (compare Fig. 4.23).

4.5.1.2 Data Post-processing

Electronic excitation of OH was based on the rather weak (3–0)-band within the A-X-electronic system. This necessitated intense UV-radiation around 248 nm that led to strong LII-signals in the particle-laden flow. In many cases LII was not observed at the LDV-measurement location, as one would expect, but randomly distributed in the imaged area. Presumably measurable LII stemmed from large (clustered) seeding particles that exceeded the LDV-fringe separation ($<$4 μm). Particles substantially larger than the fringe separation decrease the contrast in the LDV-burst and are not validated.

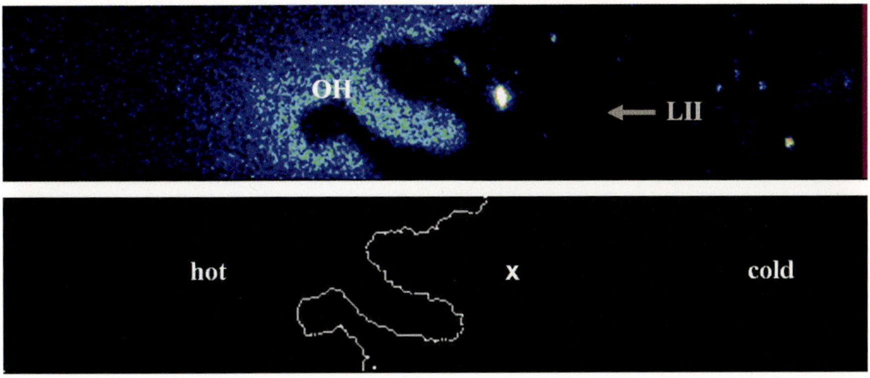

Fig. 4.23 *Top*: Original intensity image showing OH-PLIF- and LII-signals (marked by an *arrow*) at much higher signal intensities. The image corresponds to 50.8 × 10.2 mm. *Bottom*: Interface dividing hot and cold gases reconstructed from the OH-intensity image resulting in an OH-contour. The spatially fixed LDV-measurement location is marked by the *white cross* (Reprinted from Nauert and Dreizler [10, 67])

Generally, a distinction between OH-PLIF- and strong LII-signals was possible because of the high LII-signal levels. This enabled the usage of an intensity-threshold to distinguish LII from OH-PLIF. After subtraction of the dark current and normalization on the laser intensity profile, the areas showing LII-signals were set to zero in the original image. The image was converted subsequently into a binary image by using a threshold 50% of the maximum observed OH-PLIF-intensity. By using a-priori-knowledge regarding the rough flame location, the OH-interface dividing hot products and unburned gases was reconstructed by finding the edge in the binary image. Remaining noise was smoothed by a 3x3 pixel median filter. Finally, a spatial resolution of 300 μm was achieved. Figure 4.23 summarizes the steps from the intensity image to a contour plot. The intensity-image shows a typical occurrence of LII that was successfully removed by image post-processing.

4.5.1.3 Results and Discussions

Planar distributions of OH-radicals are used to locate the flame brush and the mean reaction progress. For this purpose instantaneous two-dimensional OH-contours (as presented in Fig. 4.23) are ensemble averaged resulting in a two-dimensional histogram. This histogram provides the Reynolds mean progress variable $\bar{c}$ [60] from which the extension of the flame brush and the spreading rate of the flow can be derived.

For the 30 kW-case, the left hand side of Fig. 4.24 presents a radial cut through the two-dimensional histogram at an axial height of $x = 30$ mm. Obviously, the distribution is almost symmetric (notice, further downstream slight asymmetries occur towards the unburned gases that are due to imperfections in the data post-processing

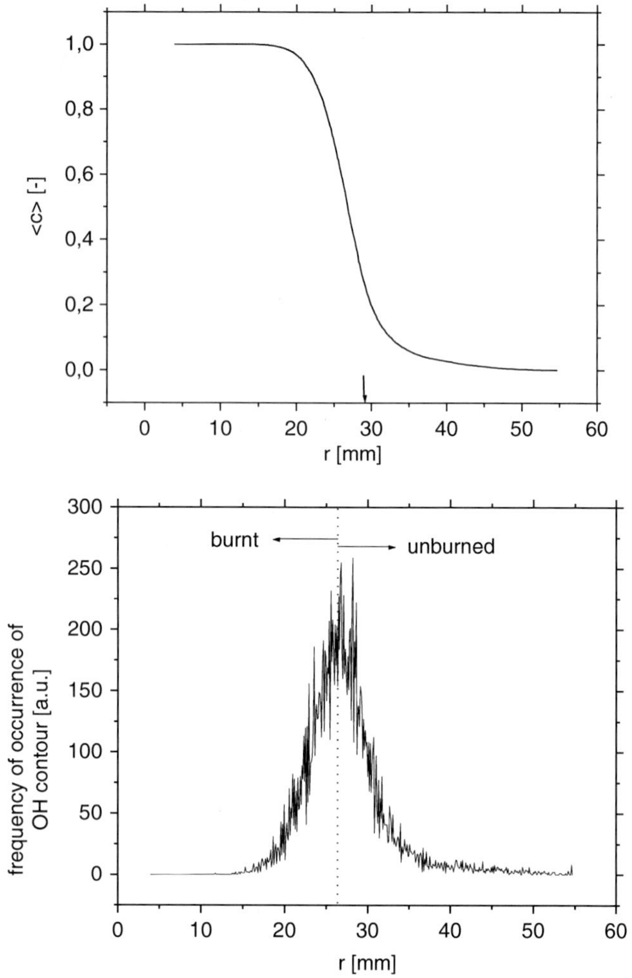

Fig. 4.24 *Bottom*: One-dimensional histogram obtained by a radial cut through the ensemble averaged 2D-OH-contours at an axial height of $x = 30$ mm. The vertical dashed line marks the average position of the cold–hot interface. For this distribution 16,000 OH-PLIF-single-shots have been processed. *Top*: Mean reaction progress resulting from integration of the figure at the l.h.s. The arrow marks the spatially fixed position of LDV-measurements (R_{LDV}) taken at this respective height (Reprinted from Nauert and Dreizler [10, 67])

removing parasitic effects from LII). Using results from a Gaussian fit (not shown), the probability to measure the OH-contour tops at $r = 26.5$ mm. According to the thin flame assumption this histogram can be identified as the probability to find burnt or unburned gas mixtures, respectively. For radii smaller than 26.5 mm hot products occur in average whereas further outside cold gases are present more often. Within this concept any finite rate chemistry effects are neglected and the temperature of hot products is given by the adiabatic flame temperature T_b of the gas mixture. It is worth mentioning, that local equivalence ratios can differ significantly from the gas mixture emana-ting from the nozzle caused by secondary air entrainment. In an ongoing effort secondary air entrainment is measured by one-dimensional Raman/Rayleigh spectroscopy but is not considered any further here.

By integrating the mean reaction progress, $\bar{c}(r)$ can be obtained from the normalised histogram. The result is shown at the right hand side of Fig. 4.24 and can be interpreted as a non-linear transformation of laboratory into flame coordinates. A value of $\bar{c} = 0.5$ corresponds to the average location of the cold – hot interface ($r = 26.5$ mm) at this axial height. The separation between $\bar{c} = 0.1$ and $\bar{c} = 0.9$ can be taken as the radial extension of the flame brush. For the 30 kW-case the width is increasing from 10.5 mm at $x = 30$ mm to approximately 28 mm at $x = 50$ mm. The spreading rate of the flame can be quantified by the radial position of the $\bar{c} = 0.5$-values ($r_{\bar{c}=0.5}$). As stated above, at $x = 30$ mm in average the flame is positioned at $r_{\bar{c}=0.5} = 26.5$ mm. Downstream the corresponding positions are $r_{\bar{c}=0.5} = 35$ mm at $x = 40$ mm and $r_{\bar{c}=0.5} = 42$ mm at $x = 50$ mm.

As described in the previous sections, axial (u) and radial (v) velocity components were measured simultaneously with the spatial OH-distribution tagging the interface. For instantaneous realisations, the radial displacement $\pm\Delta r$ of the actual LDV-measurement location from the cold – hot interface was determined. Results of these measurements can be accordingly termed conditional velocities $\overline{u_u}$, $\overline{u_b}$, $\overline{v_u}$, and $\overline{v_b}$. For PSF-30 at various axial heights conditional mean velocities are shown in Fig. 4.25. Data are binned in 4 mm wide intervals (symbols). Additionally a floating average is presented (lines). Varying radial LDV-measurement locations for each height are specified in Table 4.5. Obviously, for the axial velocity component at low axial heights $\overline{u_b} < \overline{u_u}$. Further downstream this turns around and for x = 50 mm $\overline{u_b} > \overline{u_u}$. This crossover is not observed for the mean conditional radial velocity component. For all locations investigated, radial velocities for the cold gases exceed the hot gas velocities.

For a more detailed discussion, conditional velocities in the vicinity of the cold–hot interface are considered now. For this purpose conditional velocities are ensemble averaged within $\Delta r = \pm 2.5mm$mm wide stripes. For the lowest axial position at $x = 20$ mm, stripe-conditioned velocities in the burnt region are lower than in the region of unburned gases. The axial component decreases from $\overline{u_u} = 6.0$ ms^{-1} to $\overline{u_b} = 5.5$ ms^{-1}. This corresponds to a 9%-change of the conditional axial velocity across the interface. The corresponding values for the radial velocity component decrease from 0.6 to 0.3 ms^{-1} (50%). At this location turbulent velocity fluctuations exceed variations of the velocity induced by gas expansion effects. At $x = 30$ mm the differences between $\overline{u_u}$ and $\overline{u_b}$ decreased already to 0.2 ms^{-1} or 3% which is at the same order of magnitude as the experimental error. However, the relative changes in the conditional radial velocity did not change significantly. Further downstream, at $x = 40$ mm, the mean conditional axial velocity component is nearly independent on Δr, whereas the radial component still exhibits higher velocities at the unburned side of the interface. For the same width of the stripes, the radial velocity component decreases already by nearly 15%. At $x = 50$ mm

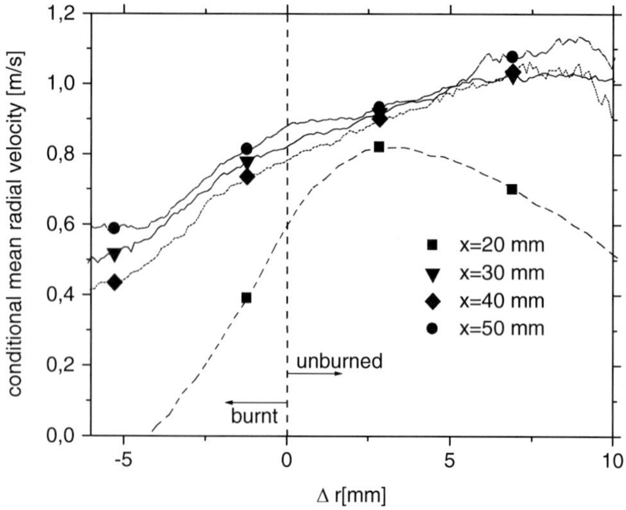

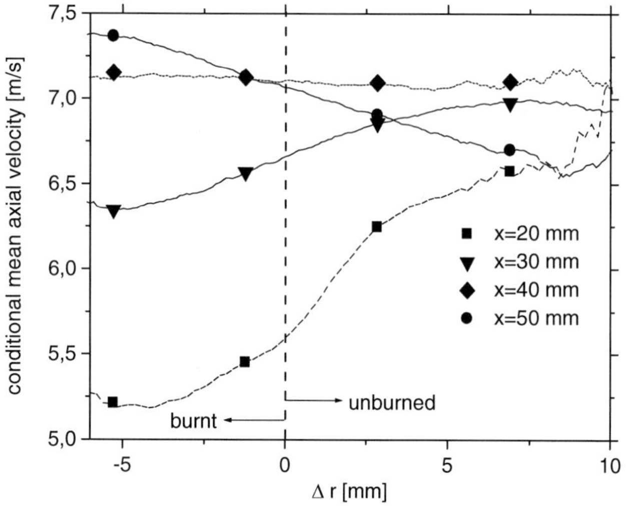

Fig. 4.25 Conditional velocities measured at various axial heights for the 30 kW-case. Different lines represent results from a floating average, symbols represent results from fixed 4 mm wide bin intervals. Measurement locations R_{LDV} are listed in Table 4.5. *Left*: Mean axial conditional velocity component. *Right*: Mean radial conditional velocity component (Reprinted from Nauert and Dreizler [10, 67])

the value of $\overline{u_u}$ equals 7.0 ms^{-1}, slightly below $\overline{u_b}$ =7.15 ms^{-1}. The trend from $\overline{u_b} < \overline{u_u}$ at $x = 20$ mm to $\overline{u_b} > \overline{u_u}$ at $x = 50$ mm is in contrast to the progression observed for the radial component where $\left(\overline{V}_b - \overline{V}_u\right) < 0$ remains, independent on the axial position. The different behaviour for axial and radial velocity components at $x = 50$ mm indicates that gas expansion effects can dominate axially whereas in radial direction turbulent fluctuations determine the direction of turbulent transport.

Table 4.5 Favre-mean fluxes. LDV-measurement locations at different heights correspond to various values of the mean reaction progress. Counter-gradient diffusion is apparent for positive fluxes

		20 mm	30 mm	40 mm	50 mm
R_{LDV}	[mm]	24	29	34	38.5
$\bar{c}$	[–]	0.34	0.22	0.57	0.66
$\overline{u''c''}Z$	[ms^{-1}]	−0.033	−0.008	0.004	0.026
$\overline{v''c''}Z$	[ms^{-1}]	−0.020	−0.004	−0.016	−0.017

The experimental data can be used to determine two components of the turbulent flux presented in Eq. (4.1). For this purpose the corresponding stripe-conditioned axial and radial velocity components are used.

In conjunction with Favre-averaged reaction progress [61]

$$\tilde{c} = \frac{\bar{c}}{\bar{c} + \rho_b/\rho_u\,(1-\bar{c})} \tag{4.3}$$

Favre-mean fluxes can be calculated and are listed in Table 4.5. For each axial location the fixed LDV-measurement location R_{LDV} (compare arrow at r.h.s. of Fig. 4.24) corresponds to a certain value of the mean reaction progress $\bar{c}$.

Whereas for the Favre-mean radial flux in all cases gradient diffusion was observed, for the axial component a cross-over from gradient diffusion at $x = 20$ mm to counter-gradient diffusion at $x = 50$ mm was apparent.

These experimental results can be used to check the validity of the criterion introduced in [65] and presented in Eq. (4.1). The efficiency function α was introduced to take into account the variable ability of turbulent eddies to act on the flame front and is dependent on the length scale ratio L_t/l_F. For the configuration under investigation, longitudinal and transversal length scales have been measured by two-point LDV [4, 18]. Longitudinal length scales at the nozzle exit plane $L_{11,x=1\,\mathrm{mm}}$ were between 6 and 8 mm, depending on the radial position. Due to viscous forces this length scale in general increases axially. Therefore length scale ratios L_t/l_F, using the laminar flame thickness l_F from Table 4.3, exceeded clearly values of 20. According to Figure 15 in [65] for values $L_t/l_F > 20$ the efficiency function α is approximately unity. Within the flame brush the ratio between turbulence intensity and laminar flame speed, u'/s_L, does not vary significantly between $x = 20$ and 50 mm ($u'/s_L \approx 5.5$). With the heat release factor τ (neglecting any effects of secondary air entrainment on adiabatic flame temperature) one finally finds that $N_b \equiv \frac{\tau s_L}{2\alpha u'} \approx 0.5$, approximately independent of the axial heights considered. This value implies higher velocities in the unburned regions ($\overline{u_u}, \overline{v_u} > \overline{u_b}, \overline{v_b}$) and thus gradient diffusion. This provision is in agreement with conditional mean radial velocity measurements, but for the axial component the trend from $\overline{u_u} < \overline{u_b}$ to $\overline{u_u} > \overline{u_b}$ is not in accordance with the DNS-based predictions. Notice, that isotropic turbulence was assumed for the DNS-calculations but the present experimental configuration shows clearly anisotropic features. However, due to the small relative

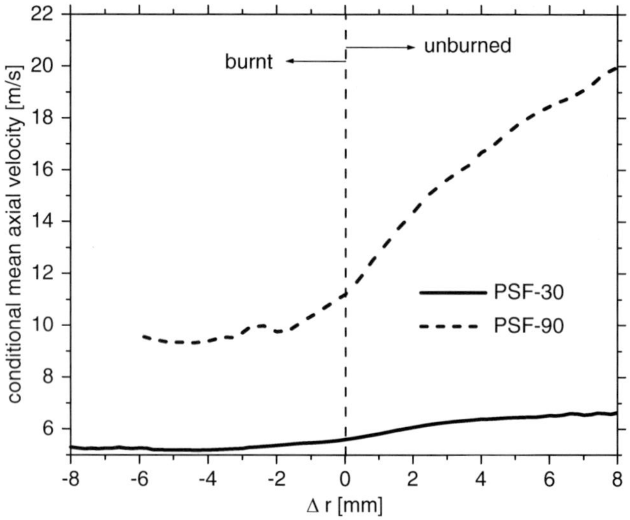

Fig. 4.26 Comparison of conditional mean axial velocities for different Reynolds-numbers. The Re-numbers for PSF-30 and PSF-90 were 10,000 and 29,900, respectively (Reprinted from Nauert and Dreizler [10, 67])

changes observed for the conditional axial velocities that are at the same order of magnitude as experimental uncertainties, for swirling flames a final conclusion on the validity of the criterion proposed in [65] cannot be drawn. More work and especially velocity measurements with even less experimental uncertainties are needed to fully understand the transition from gradient to counter-gradient diffusion. Additionally, an investigation of the tangential velocity component is desirable and is planned in a future experiment.

The influence of increasing *Re*-number on conditional axial velocities is shown in Fig. 4.26 for a selected height at $x = 20$ mm. Except of the *Re*-number, that has been increased from 10,000 (*PSF-30*) to 29,900 (*PSF-90*), all other parameters have been kept constant. The much higher absolute turbulence levels clearly promote gradient diffusion. The relative change of the stripe-conditioned axial velocity component across the cold – hot interface increased from 9% for *PSF-30* to 28% for *PSF-90*. This strongly supports the findings from previous studies that gradient diffusion is promoted by increasing ratios of u'/s_L. Consequently, flames classified in the regime of thin reaction zones show more likely gradient diffusion than corrugated flames.

4.5.1.4 Conclusions Conditional LDV Measurements

Joint velocity/scalar measurements have been performed in swirling premixed natural gas/air flames with two Reynolds-numbers using combined laser Doppler velocimetry and planar laser-induced fluorescence. For various locations in the flame brush conditional mean axial and radial velocities have been measured. The focus of the discussion was on stripe-conditioned velocities in the vicinity of the cold – hot interface that was tagged by two-dimensional OH-distributions. In general and in accordance with previous investigations, high levels of turbulence promote gradient diffusion. For this reason turbulent fluxes of flames within the regime of thin reaction zones are more likely dominated by gradient diffusion. However, mean conditional axial and radial velocity components can exhibit different schemes of turbulent transport. It was shown that with increasing distances from the burner for the axial velocity component a crossover from gradient to counter-gradient diffusion takes place whereas the radial component constantly shows gradient diffusion. The cross-over from gradient to counter-gradient diffusion is shown to be in disagreement to a criterion deduced from previous DNS-calculations. It should be considered that the DNS-data were obtained for homogeneous isotropic turbulence, not matching the conditions of the present turbulent flow. However, a final conclusion on the validity of the DNS-based criterion should not be drawn from the current data because of these different flow characteristics. Furthermore, the behaviour of the conditional tangential velocity needs to be studied in future experiments as well.

The present study is part of an ongoing effort. Flow and scalar field are characterised in detail using various laser diagnostics such as Raman/Rayleigh spectroscopy, laser-induced fluorescence, laser Doppler velocimetry, and particle imaging velocimetry. This data base already serves for validating numerical simulations such as LES [68].

4.5.2 Combined PIV- and OH-PLIF-Measurements in Partially-Premixed Turbulent Opposed Jet Flames

Recent improvements in combustion technology are built on an advanced understanding of physical and chemical processes underlying turbulent flames. As outlined in the introduction this knowledge is usually gained from generic configurations such as turbulent jet [69], opposed jet [7, 29] or bluff-body stabilized [4, 18, 70] flames. While stable non-premixed flames are well studied and are most accessible for deterministic numerical simulations [50], transient phenomena such as flame extinction, founded on the complex interaction between turbulent mixing and chemical reactions, are much less understood [24].

In non- and partially-premixed flames, the scalar dissipation rate was identified as a suitable measure to quantify finite rate chemistry effects [23]. Using 1D-Raman/Rayleigh scattering [7, 29, 71] or combined 2D-Rayleigh/planar laser-induced fluorescence (PLIF) [72], scalar dissipation rates were measured in hydrocarbon-fuelled piloted jet and opposed jet flames. As scalar dissipation rates in reaction zones are determined by local properties of the turbulence field such as strain or vorticity in the region of the reactive mixing layer, additional conditional flow field measurements relying on combined flow/scalar measurements are imperative.

Although simultaneous Raman/Rayleigh/laser Doppler velocimetry (LDV) approaches were shown to be feasible [73, 74], the data rate precludes building up a statistically reliable data base. Instead, a simultaneous application of particle imaging velocimetry (PIV) and a visualization of the reaction zone by PLIF are preferable because the presence of seed particles impacts much less on the fluorescence signals. Furthermore, the two-dimensional nature of PIV provides estimates of vorticity and dilatation. Transient locations of the reaction zones can be identified by radical distributions such as OH or CH measurable by PLIF and serve to deduce local flow field properties conditioned by the reaction zone [10, 67, 75, 76].

Turbulent opposed jet flames are prominent configurations for studies of flame extinction [7, 29, 77–79] as summarized recently in [7, 29]. Mean scalar dissipation rates can be easily increased by increasing the momentum of the impinging gas flows. Hence, these flames can be operated very close to extinction. However, conditional flow field properties, as demonstrated in [80] for premixed stagnation-type flames, in the vicinity of the reaction zone influencing local scalar dissipation rates were not measured for non- or partially-premixed opposed jets.

The inhomogeneous density fields associated with combustion pose a challenge for PIV-application. In the exhaust gas regions seeding particle densities are reduced approximately by $\rho_{burnt}/\rho_{unburnt}$ with ρ being the density. For example, in [81] this was compensated by extremely high seeding densities that inhibited PIV-data processing in unburned regions. Consequently it is impossible to achieve velocity vectors with the same resolution in the burnt and unburned regions with standard PIV-algorithms. For the opposed jet configuration flame characteristics like the extinction limits, in addition, are very sensitive to disturbances such as radiative heat losses by seed particles. This requires low densities of seeding particles, reducing the achievable spatial resolution of PIV, particularly in the exhaust regions. To avoid a significant alteration of the extinction limits and to provide a sufficient spatial resolution, velocities in the burnt gas regions are deduced in the present study from a particle tracking algorithm (PTV), while upstream of the flame PIV is used. A single-shot illustration of this procedure for an extinguishing flame is discussed subsequently (Fig. 4.27).

To allow comparison to 1D-Raman/Rayleigh measurements and to complement the data base of [7, 8, 29, 32] suitable for combustion LES-validation, the identical burner shown in Fig. 4.2 and operational points including those from [7, 29] are used. Since conditional scalar dissipation rates were previously measured, it is crucial to relate the PIV/PTV-measurements to the transiently intermitting reaction zone of the turbulent flame. This is accomplished by recording OH-distributions by PLIF simultaneously with PIV/PTV. With the help of flamelet models for instance the Raman/Raleigh results can be conditioned on the OH-zone as well. Conditional flow field statistics are reported for varied equivalence ratios of the fuel jet and two different Reynolds-numbers, including on and off-axis data evaluation of the large data base.

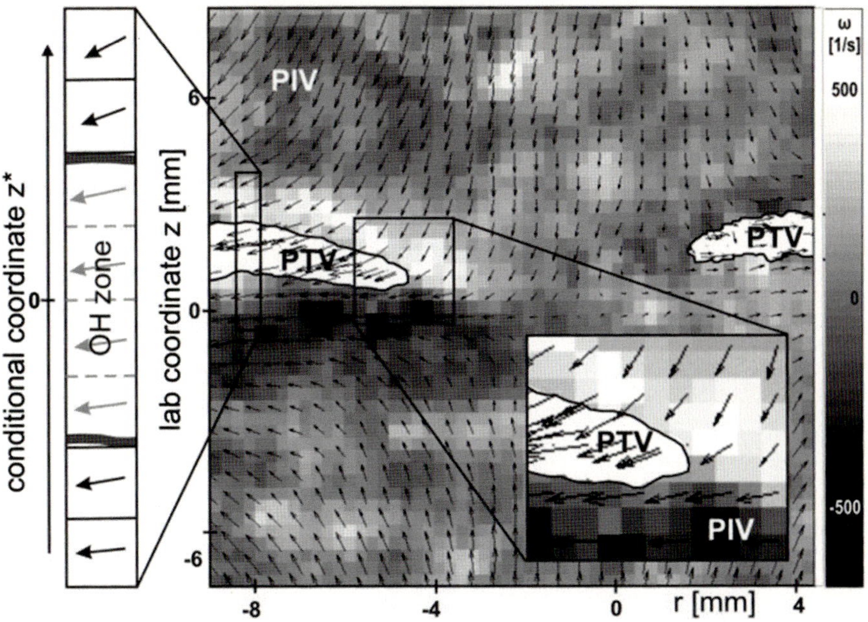

Fig. 4.27 Single-shot of OH-PLIF and PIV/PTV during flame extinction showing the transient OH-distribution (*white areas*) and velocity vectors resulting from the PIV-algorithm outside the OH-layer and the PTV-algorithm inside the OH-layer. Outside the OH-layer the vorticity field is shown in shades of *grey*. The edge of the extinguishing flame including the local flow and vorticity field is inset. At the *left* the definition of the conditional coordinate z^* is highlighted. As the PIV-interrogation areas are fixed to the upper and lower OH-contour and of a certain size, the PTV-area is divided into four subregions while the origin of z* is located at the middle of the OH-zone (Reprinted from Böhm et al. [9, 33])

4.5.2.1 Optical Setup

The combined PIV/PTV/OH-PLIF-setup is shown in Fig. 4.28. Two separate laser systems were used. The laser beams were formed into sheets and overlapped to illuminate a planar region of the flame cutting the vertical centerline of the burner. For OH-PLIF, a frequency doubled Nd:YAG-laser (Pro270, Quanta Ray) pumped a dye-laser (Sirah Precision Scan). After frequency doubling of the tunable radiation, the UV-laser beam was formed into a sheet with a height of 16 mm and a thickness of ~250 μm in the probe volume region. OH was excited via the $R_1(8)$-transition of the (0,1)-band in the A-X-system at 281.3 nm with a temperature-insensitive population distribution. Fluorescence from the A-X (1,1)- and (0,0)-bands was detected using a UV-lens (Nikkor, f = 105 mm, f/# = 4,5), a dichroic long pass filter and two *WG305*-filter (Schott) in front of an intensified CCD-camera (Roper Scientific, PiMax, 1,300 × 1,340 pixels). The field of view was 20 × 20 mm^2 and the

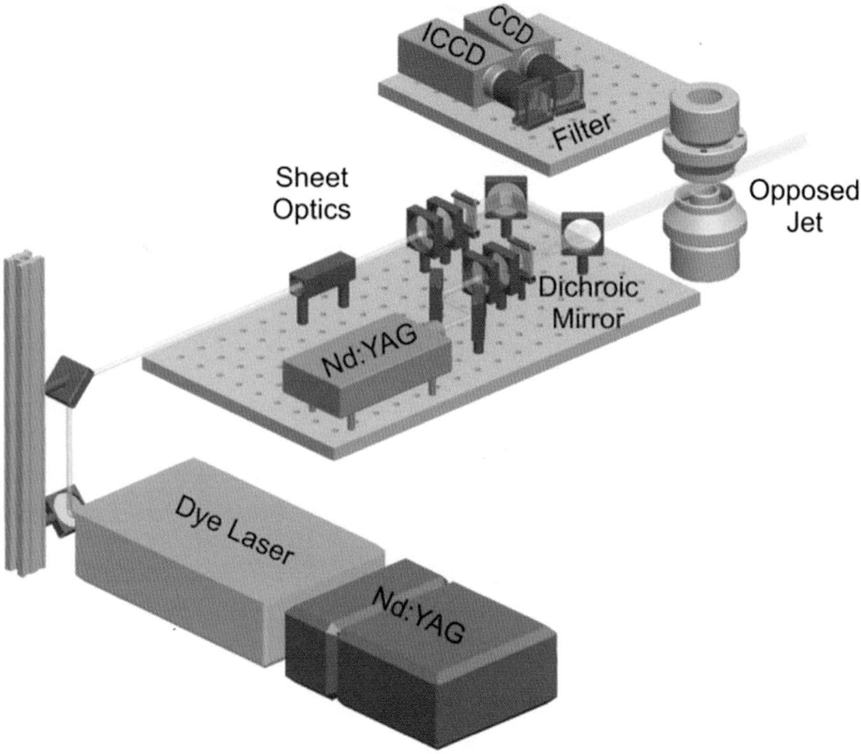

Fig. 4.28 Optical layout for simultaneous PIV/PTV/OH-PLIF-measurements (Reprinted from Böhm et al. [9, 33])

measured resolution 70 μm. After background subtraction and correction for laser profile variations, OH-contours were deduced using a threshold-method. Variations in the contour detection were insensitive to the findings discussed below.

Due to the slow three-body recombination rates of OH, the use of OH-PLIF as a reaction zone marker has been questioned [82]. It has been shown additionally that OH-mole fractions show a similar but retarded dependence as heat release on increasing strain [83]. However, laminar opposed-jet flamelet calculations [8, 32] yielded that the lower OH-boundary can be considered as an approximate reaction zone marker in the opposed jet configuration.

The beams of two frequency doubled Nd:YAG-lasers (New Wave) were overlapped to provide successive illumination for PIV/PTV. The energy in each pulse was set to 40 mJ and the pulse separation was $\Delta t = 50$ μs. In the probe volume the light sheet had a height of 24 mm and a thickness of ~600 μm. A frame-transfer-CCD (PCO, $1{,}376 \times 1{,}040$ pixels) was used with an imaged area of 20x24mm^2. Cameras for PIV/PTV and PLIF were aligned using a backside-illuminated mask consisting of a 2D-array of 90 holes each with a diameter of 150 μm. De-warping and identification of corresponding pixels on both CCDs was accomplished based on this mask in a post-processing step.

For the combined PIV/PTV-approach dry MgO with a mean diameter of about 1 μm was used as seed material. Approaching the reaction zone the particle temperature increased causing additional radiative heat losses. This disturbance impacted the flame as seeding was increased and shifted the extinction limit to lower Reynolds-numbers relative to un-seeded flames. To assure that changes in extinction limits were less than 5%, the seeding density of approximated 8 g/Nm^3 gas was controlled carefully. Two seeders were installed to seed both jets individually. To assure stable and reproducible seeding densities, a combination of static meshes and rotating brushes were installed at the bottom of a MgO-storage tank. The speed of the motor driving the rotating brushes determined the seeding density. Mie scattering intensities 3 mm downstream of the nozzle exits were monitored to achieve as much as possible constant and equal seeding in both jets.

4.5.2.2 PIV/PTV-Data Post-processing

The PIV-data processing is performed on a regular grid consisting of 32 × 32 pixels (or 16 × 16 pixel) with a resolution of 600 μm (300 μm) and fixed to the OH-front. In a first step a standard PIV-algorithm using multi-pass interrogation with window-shifting and three-point Gaussian subpixel interpolation is used [84]. Starting with interrogation areas of 64 × 64 pixels, the initially determined displacement is used as an estimate for successively smaller subregions of size $S = 32$ and 16 pixels each side. This improves the search for spatially refined correlations.

To relate the PIV/PTV-measurements to the transient reaction zone of the turbulent flame and to avoid a bias due to different seeding densities in cold and hot gases, simultaneous contributions from the reactant and product regions to the velocity vector resulting from a single PIV-interrogation area must be avoided. For each image, a coordinate system is introduced that is fixed to the reaction zone at the radial position of current interest as deduced from the OH-image. Starting from the upper OH-contour, flame-fixed axial coordinates z_u* are positive, and starting from the lower OH-contour, flame-fixed coordinates z_l* are defined as negative. Thus, z* denotes the flame-fixed axial coordinate and displacements are determined on a shot-by-shot basis relative to the reaction zone. The definition of the flame-fixed coordinate system is visualized by the insert left of Fig. 4.27. The mean of the instantaneous upper and lower OH-contours in the laboratory system are located at $z_{\text{upper OH-contour}} = 1.4$, 1.9, 2.7 and $z_{\text{lower OH-contour}} = 0.2$, 0.5, 1.6 mm for the flames TOJ2D, TOJ2C, and TOJ1C, respectively, with z being the laboratory coordinate system. These positions are assigned by vertical lines in Figs. 4.29, 4.30 and 4.32 where the focus is on a region $z^* = \pm 6$ mm.

Seeding densities are reduced significantly in the burnt gas regions by approximately 85%. Since successful PIV-post-processing relies on the statistical average of about ten particles per subregion [85], the PIV-algorithm outlined above would be accessible only on a grid consisting of at least 48 × 48 pixels. Higher seeding densities are, as discussed, prohibitive because of their impact on flame characteristics. Therefore particle tracking velocimetry (PTV) is used in regions with high temperatures where the seeding density is low. An initial guess for

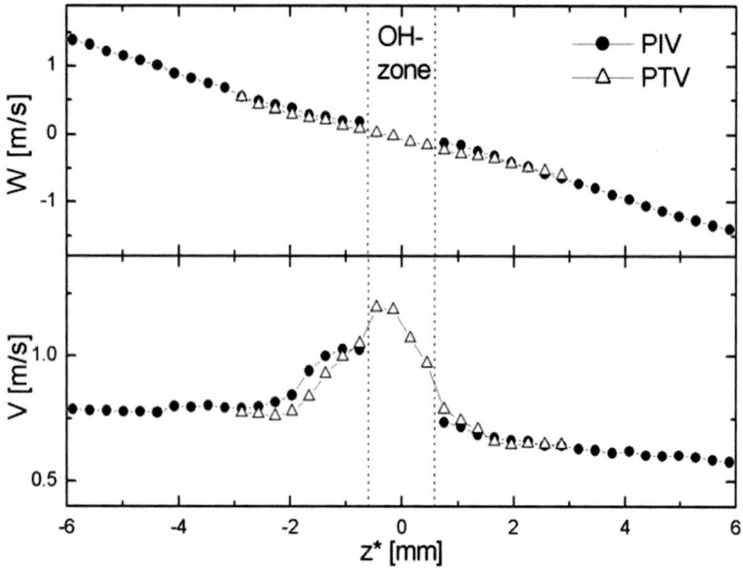

Fig. 4.29 Mean axial (W) and radial (V) velocities of flame TOJ2D at $r = 4$ mm conditioned on the transient OH-contours. The abscissae, z^*, is a spatial coordinate fixed to the transient OH-contours. The *vertical lines* denote the location of the upper and lower mean OH-contours and their separation is the mean thickness of the OH-layer. Axial velocities are positive when pointing in $+z^*$-direction. Radial velocities are positive when pointing away from the centerline. PTV-results are extended into the region of unburned gases for comparison to PIV (Reprinted from Böhm et al. [9, 33])

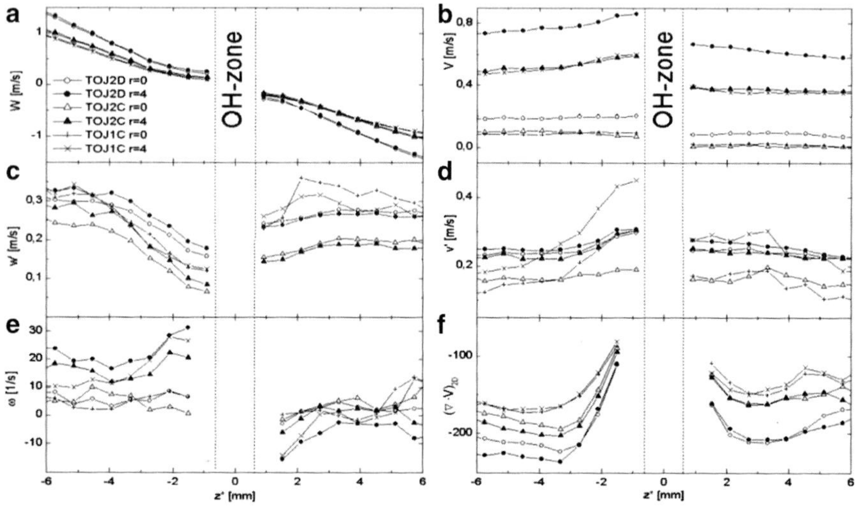

Fig. 4.30 Axial profiles of conditional mean velocities (**a**) and (**b**), velocity fluctuations (**c**) and (**d**), vorticity (**e**) and dilatation (**f**). *Vertical lines* denote mean upper and lower OH-contours (see text). All three flames and two radial locations ($r = 0$ and 4 mm) are included (Reprinted from Böhm et al. [9, 33])

velocity in the PTV-regions is obtained from the PIV-algorithm using a 32×32 pixels subregion after spurious vectors are removed and replaced by interpolation. Identification of single particles is done in the first PIV-image. Subsequently, subregions of 5×5 pixels are centered around the currently considered particle in the first image and the respective location in the second image deduced from the initial guess of the velocity vector. A cross-correlation is applied resulting in an improved estimation of the single velocity vector. Finally, the OH-layer at the radius of current interest is subdivided into four squared subregions and PTV-results in the corresponding subregions are statistically averaged. For the case of TOJ2D, these resulting subregions are $S = 16 \pm 4$ pixels each side (this corresponds to an average thickness of the OH-layer of 1.2 ± 0.3 mm). Obviously, PTV and PIV exhibit very similar spatial resolution. However, with a reduced PTV-subregion of 3×3 pixels, PTV-results proved to be consistent with PIV-data prior to the reaction zone as demonstrated in Fig. 4.29.

PTV is more accurate and precise than correlation-based PIV-algorithms, comparably insensitive to out-of-plane motion and accepting of larger velocity gradients [86]. However, thermophoretic effects are inherent to any velocimetry relying on seed particles [87]. Thermophoretic velocities u_{TP} normal to the reaction zone and down the temperature gradient can be estimated by $u_{TP} \approx 0.5\nu\nabla T/T$ [66] with the kinematic viscosity ν. For stagnation point flows such as turbulent opposed jets, thermophoretic effects are considerable. According to estimations by Sung et al. [88] in a comparable opposed jet configuration u_{TP} is 0.15 ms^{-1} at worst for the flame TOJ2D.

In summary, the flame-fixed coordinate system and the improved PIV-algorithm for un-reacted gases in connection with PTV in the vicinity of the reaction zone is a suitable way to deduce conditional flow field properties in flames. This holds even more in flames highly sensitive to external disturbances by seeding. The following results comprise for each flame up to 2,000 PIV/PTV/PLIF-single-shots.

4.5.2.3 Results and Discussions

In this section, conditional flow field properties are shown along the centerline ($r = 0$ mm) and one off-centerline location at $r = 4$ mm. All velocities, vorticity, dilatation or strain rates are shown as values conditioned on the instantaneous flame location.

Mean and fluctuations of the axial and radial velocity components as well as mean vorticity and dilatation are presented in Fig. 4.30 for flames TOJ1C, TOJ2C and TOJ2D listed in Tables 4.1 and 4.2. This figure exhibits results only from PIV-data post-processing using a 32×32 pixel resolution. The absolute value of the mean axial velocity component $|W|$ decreases almost constantly approaching the respective OH-contour. As expected for higher bulk exit velocities (TOJ2D) the profiles show a steeper slope which is independent on the radial position. For the same Re (TOJ1C and TOJ2C) but different fuel mixtures ($\phi = 3.18$ and 2.0) the profiles coincide. The mean radial velocity (V) centerline profile shows constant values independent of the axial displacement from the respective OH-contour. On the air-side V is very close to V = 0 ms^{-1} as expected but on the fuel side a slight deviation from zero is observed (TOJ2D: 0.2 ms^{-1}; TOJ1C, TOJ2C: 0.1 ms^{-1}). This indicates small imperfections of the symmetry of the flow system that are increased slightly for higher Re. Consistently higher mean radial velocities are observed for the off-axis profiles at $r = 4$ mm. The enhanced mean radial velocities on the fuel side are due to the higher exit velocities of the methane/air mixture which are required to balance the higher density of the air in order to achieve an equal momentum for both gas streams. The fluctuation levels of axial and radial velocity components show an opposite trend when they approach the respective OH-contour. Whereas the axial fluctuations decrease, the radial fluctuations slightly increase. This behavior is more prominent on the fuel side. The turbulent kinetic energy (TKE) $k = 1/2\left(w'^2 + 2v'^2\right)^{1/2}$ (compare Fig. 4.32 for TOJ2D) exhibits an increase particularly at the lower OH-contour indicating the presence of enhanced axial strain.

At off-axis locations, the modulus of the out-of-plane vorticity-component $\omega = 1/2\,|\partial w/\partial r - \partial v/\partial z*|$ grows as the OH-contour is approached. Notice that the algebraic signs of vorticity are opposite for fuel and air flow, respectively, following the definition of z^* and the velocity directionality. The rise of $1/2\,|\partial w/\partial r - \partial v/\partial z*|$ is already seen close to the OH-distribution on the left hand side of Fig. 4.27. In the ensemble-averaged data presented in Fig. 4.30e this trend is observable particularly for the off-axis profiles at $r = 4$ mm and the fuel side where the stoichiometric contour is located close to the OH-contour. The correlation of vorticity with the onset of the reaction zone observed here is similar to findings in [89] and might be due to inhibited momentum mixing across the reaction zone.

At the centreline, vorticity is approximately independent on the displacement from the respective OH-contours and close to zero as expected. This indicates that eddies are generated preferably close to the reaction zone particularly at off-centerline positions. This conclusion is consistent with the observation of larger radial velocity fluctuations at off-axis locations.

Visual inspection of high speed movies by operating a CMOS-camera (Photron) at 1 kHz and exciting Mie scattering at 10 kHz using a diode pumped intra-cavity frequency doubled Nd:YLF-laser (Coherent) at 527 nm additionally confirm the formation of eddies in a zone of high strain near the reaction zone. Only a fraction of these time-sequences is shown in Fig. 4.31. Light-grey areas show flame luminosity, the 10-dot-arrays (partly blurred and appearing as lines for low velocities) are due to Mie scattering of a single seed particle illuminated ten times per camera exposure. Consequently, particle trajectories can be tracked in time. In this sample an eddy is formed in the close vicinity of the lower contour of the luminous flame area.

The 2D-dilatation $(\nabla V)_{2D} = (\partial w/\partial z * + \partial v/\partial r)$ presented in Fig. 4.30f is negative in the whole region. It features a weak minimum around $z^* = \pm 3$ mm. For smaller values of z^*, the modulus $|\partial w/\partial z * + \partial v/\partial r|$ is decreasing with a rather steep slope, whereas for $z^* > 3$ mm this slope is much less.

As $\partial v/\partial r$ is always almost positive, axial strain (acting compressive) exceeds radial strain (acting excessive). Thus, $|\partial w/\partial z*| > |\partial v/\partial r|$ holds. This can be

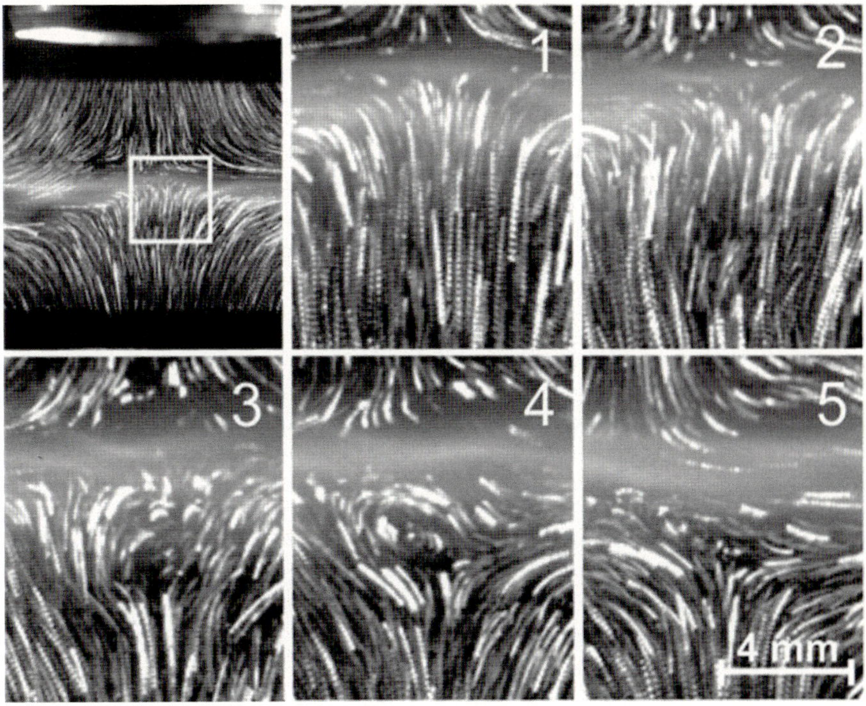

Fig. 4.31 Simultaneously measured flame luminosity (*grey areas*) and Mie scattering from ten laser shots during each 1 kHz camera exposure. Temporal sequence (images 1–5) is from the $11.5 \times 11.5\ \text{mm}^2$ *white box* in the enlarged *left top* image

explained by the fact that the time-averaged flame is aligned horizontally posing density gradients which cause again axial velocity gradients - mainly in the axial direction. However, when the respective OH-contour is approached, the difference ($|\partial w/\partial z * -\partial v/\partial r|$) diminishes since gas expansion by heat release predominantly accelerates radial velocities in the vicinity of the stagnation plane (compare Fig. 4.29).

Figure 4.32a, b shows axial profiles of conditional vorticity, dilatation, TKE, axial and radial strain resulting from the combined PIV/PTV-evaluation for the case TOJ2D at the off-axis position $r = 4$ mm. Here, the enhanced spatial resolution based on the 16×16 pixels subregions is used. The PTV-region is extended by $z^* \approx \pm 2$ mm outside the OH-layer. This is motivated by the fact that in the vicinity of the OH-layer, the seed particle density declined significantly and PTV is believed to be more reliable in this region. The profiles from the PIV- and PTV-parts lead very smoothly into each other and supplement the data presented in Fig. 4.30. The vorticity profile shows a change of sign within the OH-layer, crossing zero shortly after the lower OH-contour. At the same position, the dilatation exhibits a local minimum. Peak values of all profiles presented in Fig. 4.32a, b coincide to each other fairly well at $z^* \approx -1.4$ and $z^* \approx +0.6$ mm. As stated earlier, axial and radial

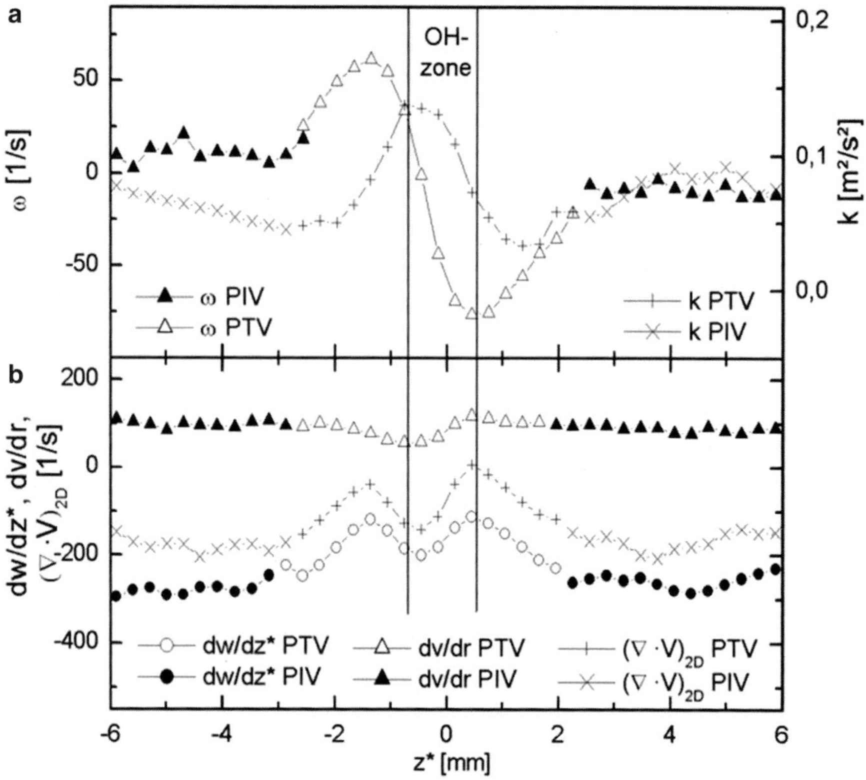

Fig. 4.32 Axial profiles of conditional vorticity, TKE (**a**), dilatation, axial and radial strain (**b**) at the off-axis position $r = 4$ mm for the case TOJ2D. *Vertical lines* denote the mean locations of upper and lower OH-contours (Reprinted from Böhm et al. [9, 33])

strain possess opposite signs. Minima in radial strain coincide with maxima in axial strain and vice versa. The off-axis axial strain profile here is in close agreement to laminar flamelet calculations. This indicates a relaminarisation of flow and scalar fields close to the heat release zones in agreement to [1, 16, 89].

With respect to the lower OH-contours associated with the reaction zone, the profiles are approximately symmetric. Contrary, profiles are asymmetric relative to the upper OH-contour. This might be explained by the fact that the upper OH-contour is extended at the lean side of the reaction zone by slow three-body reactions.

4.5.2.4 Conclusions of Simultaneous PIV/OH PLIF Measurements

Flames burning very close to extinction are very sensitive to external disturbances. Particle seeding required for velocimetry can cause significant shifts of the extinction limits. Thus seeding densities must be reduced and this entails poor spatial resolution for PIV. In the present study a combined PIV/PTV-post-processing was developed and applied to turbulent opposed jet flames. Cold gas regions were processed by an improved PIV-algorithm and hot gas regions by PTV. Overlapping PTV- and PIV-profiles showed that both methods are in agreement in overlapped regions. Despite a rather low seeding density, an in-plane resolution of 300 μm was achieved.

This method was applied with simultaneous OH-PLIF-measurements to enable the evaluation of conditional flow field statistics. The results show that vorticity is generated particularly close to the reaction zone where axial strain and dilatation exhibit local minima. Thus, the emergence of flow disturbances directly in the reaction zone could be observed with the presented method. Sequences of simultaneous Mie scattering/chemiluminescence imaging at high repetition rate were additionally used to temporally track the generation of eddies in the vicinity of the reaction zone verifying qualitatively the results from quantitative conditional PIV/PTV. Measurements for different equivalence ratios and Re showed that approaching the extinguishing flame conditions only gradually changes the conditional flow field statistics.

4.6 High Repetition-Rate Diagnostics, Application to Turbulent Opposed Jet Flames

In the previous section the focus was on laser-based measurement of velocity components, gas temperatures and species concentrations. Point-wise as well as planar techniques were discussed that allow determining mean values and fluctuations. Common to all these approaches is the fact that the repetition rate was fairly low. For many combustion processes it is desirable to complement this information by additional insights. In the following, examples are highlighted, namely gas-phase diagnostics at high repetition rates ("high-speed diagnostics").

In case of low repetition rates (typically below 150 Hz) and high levels of turbulence experimental samples are statistically uncorrelated. To calculate time averaged mean quantities or *rms*-values, statistically uncorrelated sampling is a prerequisite. However, the view on specific processes occurring in turbulent flames, such as island formation in lifted jet flames, flame extinction, ignition or flash back, using first and second statistical moments of different physical and chemical quantities only is not sufficient. This can be related to the fact that the sampling rate of conventional combustion diagnostics (>5 ms) is slow compared to the time scales typical for flame transients such as extinction, ignition or flash back (time scales typically below 5 ms). Therefore tracking the temporal evolution of such an event and building up statistics of individual realizations of those processes urgently requires data rates in the kHz-regime. Ideally, repetition rates of the laser diagnostics are adaptable to the time scales of the transient process under investigation.

New developments in the fields of diode-pumped solid-state laser and CMOS-camera technology opened up new and exciting perspectives for turbulent flame diagnostics. Compared to previous approaches [90], much longer sequences can be monitored consisting of thousands of frames recorded during one individual process. Furthermore, due to the cyclic data storage architecture of current CMOS-cameras, the experiment can be run until the transient event occurred. Then the experimentalist releases the trigger and actually stops further data acquisition. Those frames taken before the trigger signal was released (during the transient process of interest) are stored. Thereby the "success rate" of capturing the transient event, such as turbulent flame extinction or flash back, poses no longer any limitation to the statistics.

Laser diagnostics at high sampling rates can be based on different methods. A major difficulty is that solid-state laser pulse energies at several kHz repetition rate are still limited to few milli-Joules (unless one goes to pulse-burst laser systems [91]). This low energy favours the application of linear methods with high quantum yields. Techniques proved to be applicable at several kHz repetition rates are Rayleigh scattering [11, 92, 93], Mie scattering for PIV [94] and PLIF [95–97]. Of course, interesting information on transients in turbulent flames can be drawn from chemiluminescence even without a laser [98].

The present section presents some more information on the instrumentation of present high speed technology for planar laser-induced fluorescence (PLIF) applications and some recent advances of diagnostics at repetition rates up to 30 kHz. Examples of applications are turbulent mixing, flame extinction in a partially-premixed turbulent opposed jet burner (Fig. 4.2), flashback in a premixed swirl-flame (Fig 4.3b) and cyclic variations in a direct-injection spark-ignition (DISI) Otto engine. OH PLIF measurements are presented to show the feasibility of planar flame-radical detection at 5 kHz. A recent review with the focus on multi-parameter high-speed imaging in combustion applications can be found in [12, 99].

4.6.1 Instrumentation and Setup

Different types of lasers were used in the present study. For Mie scattering and PIV a dual-cavity laser was used. Pulse energies were up to 1 mJ/pulse and maximal rates of double-pulses up to30 kHz (Nd:YVO_4 IS4II-DE, EdgeWave).

For planar laser-induced fluorescence (PLIF) of radicals relevant to combustion processes such as hydroxyl (OH), tuneable radiation in the ultraviolet spectral range is required. For this purpose an extra-cavity frequency-doubled diode-pumped solid-state Nd:YLF-laser (Edge Wave) was operated at 5 kHz repetition rate to pump a tuneable dye laser. The average power of the pump laser peaked up to 17 W. This corresponded to single-pulse energies of 3.4 mJ. The pulse length was below 10 ns. The dye laser was based on a conventional system designed originally for 10 Hz operation (Radiant Dyes). To adapt the dye laser to the needs of high repetition rates, several modifications of the laser were essential. The low pump energies allowed pumping only the oscillator and the preamplifier. Consequently no main amplifier was operated. The dye cuvette was connected to a dye circulator equipped with a stronger pump than in 10 Hz-applications and a larger dye reservoir. Flow rates of the dye were 12 l/min. The delay-line of the beam pumping the preamplifier was shortened. A general issue was the long-term stability. To reduce thermal drifts, the rear panel of the cuvette mount was water-cooled. The output coupling mirror was shielded by masks from both sides to prevent heating of the mount by radiation other than the laser beam. The maximum average output power achieved at 283 nm was 150 mW. This corresponds to pulse energies of 30 μJ. The spectral width was approximately 0.04 cm^{-1} based on etalon measurements.

Signals from chemiluminescence (CL), Mie scattering and fluorescence, respectively, were monitored by a state-of-the art CMOS-camera (HSS5 or HSS6, LaVision). At maximal repetition rates of 30 kHz the exposure time was 10 μs. At this high repetition rates 256×256 pixels of the CMOS (HSS5) were active. At 10 kHz the usable area of the array increased to 512×512 pixels. The A/D-conversion had a dynamic range of 10 bit. If not stated otherwise the CMOS-camera was equipped with a lens-coupled two-stage image intensifier (High Speed IRO, LaVision). In case of 5 kHz-PLIF the gate width of the IRO was set to 100 ns. For Mie and CL-measurements the gate width was in the order of few μs.

The turbulent opposed jet flame was investigated (Fig. 4.2). In case of reacting conditions CL and Mie were measured simultaneously during extinction of a partially-premixed flame. In this case both flows were seeded. The fuel was composed of 17.4 vol% CH_4 in air. The Reynolds-numbers were 7,200 corresponding to flames at the extinction limit (compare Sect. 4.2.2).

With the OH-PLIF-experiments at a repetition rate of 5 kHz new ground was broken. So far OH-PLIF-measurements were restricted to 1 kHz [100] to the best of the authors´ knowledge. The dye laser was tuned to the $Q_1(5)$-line in the $A^2\Sigma X^2\Pi$ (1–0)-band. The laser sheet height was approximately 15 mm. OH-radicals were monitored in the flame shown in Fig. 4.2. The scope was to measure the temporal variation of the OH-radical distribution in combination with 2-component-PIV. During this study the signals were not corrected for pulse-to-pulse fluctuations of the total energy or variations in the laser profile.

4.6.2 Results and Discussion

4.6.2.1 Simultaneous Mie Scattering and Chemiluminescence

Under chemically reacting conditions Mie scattering can be combined with chemiluminescence. Whereas the particles observed by Mie scattering represent the turbulent fluid motion in a two-dimensional plane, chemiluminescence gives an estimation of the approximate flame position. In the example shown in Fig. 4.31 the CMOS-camera (without IRO) was operated at 1 kHz while Mie scattering of 0.5 μm-MgO-particles was laser-excited at 10 kHz. Each exposure shows particle trajectories of up to 10 subsequent particle positions in top of the chemiluminescence image. In the selected example the formation of an eddy is highlighted. The formation of such eddies generate vorticity when the fluid approaches the turbulent flame located in the vicinity of the mean stagnation plane.

4.6.2.2 High Speed PIV Conditioned on Extinction

Replacing the chemically inert MgO-particles by oil droplets and instead of using chemiluminescence, the transient location of the flame front can be monitored in a two-dimensional plane due to droplet evaporation in areas of elevated temperature.

The Mie scattering from the cold gas can be used simultaneously for processing instantaneously two velocity components. In the following example, double pulses from the Nd:YVO4-laser with a pulse-to-pulse separation of 50 μs were generated at an overall repetition rate of 2.5 kHz. The CMOS-camera was operated without IRO. The data post-processing was based on a PIV-algorithm presented in Böhm et al. [9, 33]. Interrogation areas of 16×16 pixels corresponded to 400x400μm^2 in the measurement plane. The top of Fig. 4.33 shows one individual velocity map out of a sequence consisting of up to 4,500 exposures.

The aim of this experiment was to track the temporal development of eddies approaching the flame front and causing flame extinction. For this purpose a proper orthogonal decomposition (POD) of the considered data set was conducted [101]. POD provides an optimal set of basic functions for the selected ensemble of data. For each velocity map time-dependent coefficients were extracted resulting in an amplitude spectrum. This spectrum was band-pass filtered. A band-pass filter was chosen such that mean velocity and large-scale shear was subtracted. Finally, in a back-transformation a velocity map was reconstructed that carried mainly the information of the vorticity.

In Fig. 4.33 the instantaneous flame position in the measurement plane is shown by white areas. The selected sequence shows the onset of flame extinction in a time-resolved manner. It can be observed that the flame is thinned over a period of few milliseconds. This thinning is a consequence of a counter-rotating pair of vortices located on the fuel side. During the extinction process the location of the counter-rotating vortices is moving only marginally. In between the vortices a high local fluid flow is observed penetrating into the flame.

Inspection of 14 individual data sequences revealed two classes of extinction events. For eight cases we observe two counter-rotating vortices from the fuel side convecting towards the lower OH contour (denoted as Class I, shown in Fig. 4.34a). These eddies align below the flame. By expansion due to heat release at the flame front the vortices distorted radially. They lingered in approximately the same spatial location as their vorticity rapidly increased. The centers of rotation of each eddy stabilized approximately 2–3 mm below the lower OH contour and were radially separated from each other by 5–6 mm.

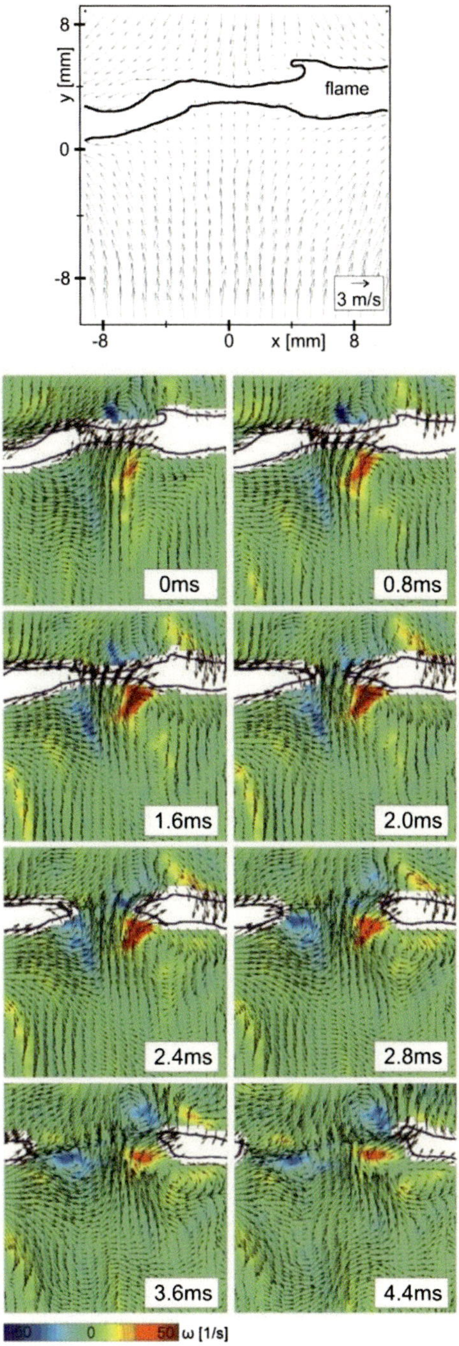

Fig. 4.33 The *top* of the figure shows the transient flow field prior to extinction. The location of the flame is represented by the white areas. PIV-evaluation was possible only in the cold gases where the aerosol droplets have not yet evaporated. The *bottom* of the figure shows the filtered velocity maps during extinction. Filtering was based on a POD-decomposition

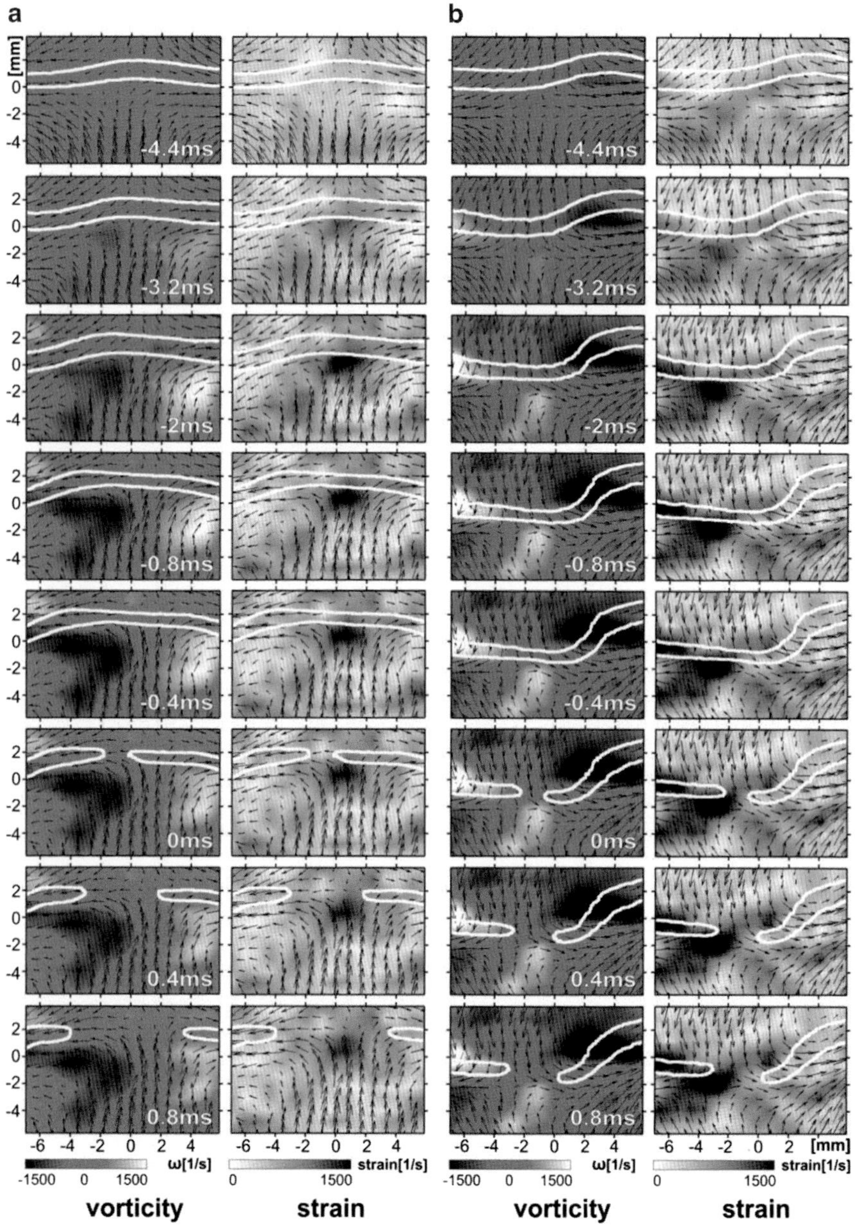

Fig. 4.34 Two individual sequences of extinction. Vorticity and strain are superimposed by false colors in *top* of the in-plane velocity vectors. Look-up tables in all cases are identical. Instantaneous locations of the OH contours are presented by *black lines* (Reprinted from Böhm et al. [13, 102])

The magnitude of the radial separation was likely driven by the length scale of the turbulence enhancement plates (interhole spacing: 5 mm). The counter-rotating vortices transported fuel that impinged perpendicularly (in a jet-like manner) upon the flame front to produce greater compressive and extensive strains. Strain increased with vorticity until t* = –0.8 ms. By that time the OH-contour began to contract and was finally breached at t* = –0 ms. Convection pushed the flame tips in opposing radial directions. After the breach occurs, convection appears to dominate the flame's radial propagation. In rare cases (not shown here), flame breaches were observed to occur in the vicinity of the mean stagnation point and then collapse and convect outwards to allow the reestablishment of a self-sustaining turbulent flame.

The remaining six extinction events were classified as Class II events. In general they showed less regularity than Class I extinctions. Figure 4.34b shows one of six events. A counter-rotating vortex pair originating from the air side and a single vortex from the fuel side surround the flame starting from t* = –5 to –4 ms. These vortices were smaller and more distorted than in Fig. 4.34a. During the vortex-flame interaction the flame becomes significantly distorted. In a manner reminiscent of Class I events however, a region of high strain developed just below the lower OH contour. For both Class I and II type events, peak strain values are observed to be of the same order of magnitude. In Class II events flame tips resisted the radial direction convection for a longer period. This could be caused either by different local mixture fraction distributions or by lower instantaneous radial velocity at the location of the flame tip. No experimental evidence of the former is present, however evidence of the latter was observed at the left side of the flame between t* = 0 and 0.8 ms, as shown in Fig. 4.34b.

These single temporal sequences show that vorticity does play a vital, but not exclusive, role in understanding extinction of turbulent opposed jet flames. A few ms prior to extinction, multiple vortices acting coherently generate strain over extended regions close by the flame. According to Lemaire et al. [103] peak strain was identified to be a significant controlling parameter. However, in Katta et al. [104], based on numerical simulations, unsteady flames were observed to resist strain twice the extinction level of laminar flames. Single sequences shown in Fig. 4.34a, b exhibit strain increased locally up to 1,500 s^{-1}. To draw further conclusions, extinction progress must be averaged appropriately. This is accomplished by multidimensional conditioning in space and time, as detailed in the previous section.

Figure 4.35 shows the conditional average of POD-filtered out-of-plane vorticity and two-component strain for all 14 experiment runs (I + II, left), as well as for Class I (middle) and Class II cases (right) separately. The black lines denote the mean OH contours. Vorticity fields are presented for t* = –4.4, –2.4 and –0.4 ms whereas strain is shown only for t* = –0.4 ms. Conditional averages of the temporal evolution of both vorticity and strain for Class I as well as for Class II support the conclusions drawn already from the single events. Class I showed primarily strong counter-rotating vortex pairs from the fuel side, causing a distinct region of peak strain values. Class II showed less regular vortices predominantly from

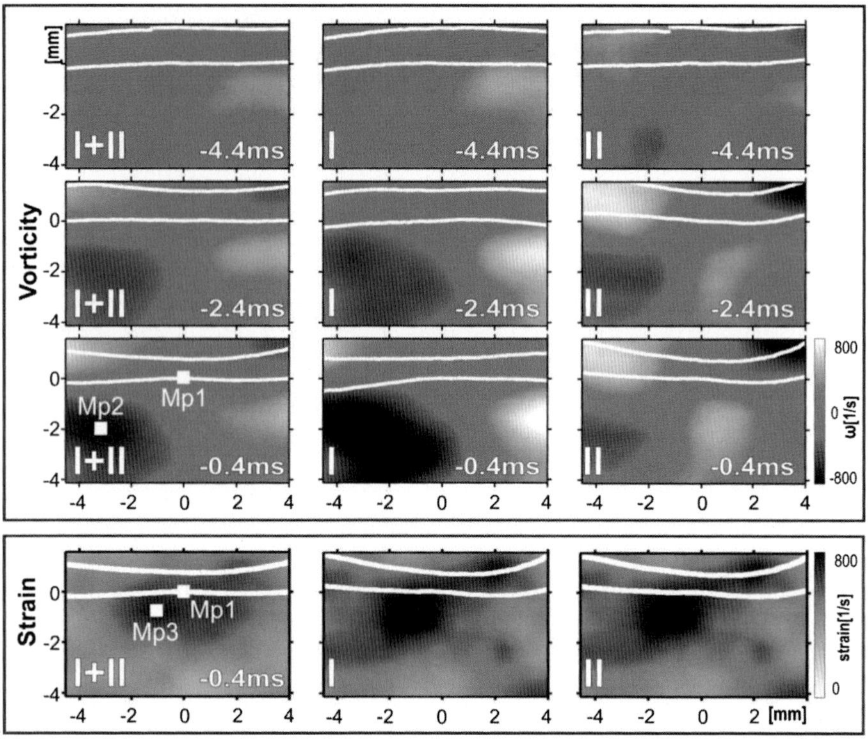

Fig. 4.35 Conditional averages of vorticity and strain for all events (I + II), class I and class II events at distinct times prior to extinction (Reprinted from Böhm et al. [13, 102])

the air side. The shape of individual vortices and regions of high strain differed for individual extinction sequences. Consequently, peak values in the conditional mean data were reduced by a factor of approximately two compared to individual progressions. Despite the differences in the conditionally averaged vorticity fields areas of maximum strain were observed to align in all cases (I + II) slightly below the lower OH contour.

The measured strain components can only give an approximation of the principal strain rates because the flow measurements are limited to two components of the flow field. The absolute strain given in this work includes only two components of strain while the flame experiences an additionally extensive component out of the measurement plane. Because the flame is approximately horizontal with slight inclinations caused by eddies it can be assumed to be normal to the measurement plane and thereby the out of plane component is approximately a principal strain rate [71]. The conditional average of the axial strain component acting compressive (negative) and radial component extensive (positive) are shown in Fig. 4.36 for all 14 runs (I + II). Two local maxima of extensive strain begin to form at $t = -4.4$ ms in the region of the lower OH-contour ~1–2 mm radially to the right and left of the

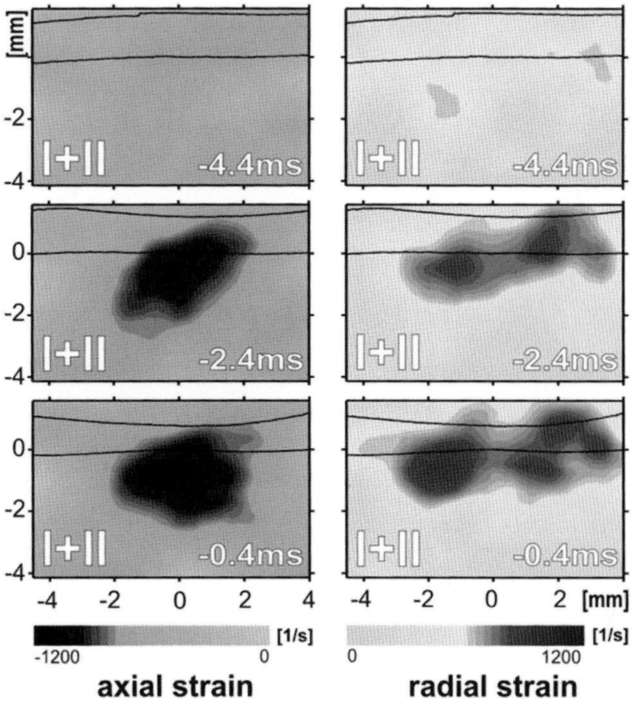

Fig. 4.36 Conditional averages of axial strain (acting compressive) and radial strain (extensive) for all events (I + II) at distinct times prior to extinction (Reprinted from Böhm et al. [13, 102])

point where extinction is going to occur ($z^* = 0$, $r^* = 0$ mm) while a maximum of compressive strain is formed in-between these peaks at the lower OH-contour just below the region where the OH layer is breached first.

For a more quantitative representation of the conditional averages, Fig. 4.37 shows temporal evolutions of strain and vorticity recorded at various monitor points. The spatial location of each monitor point is shown in Fig. 4.35. The temporal traces proceed only up to $t^* = -0.2$ ms as the definition of z^* is based on continuous OH contours.

MP1 corresponds to the origin of the newly introduced conditional co-ordinate system. MP2 is at the maximum of the conditionally averaged vorticity and MP3 is at maximum strain (compare Fig. 4.35, left column). Independent of classification, at MP1 and MP3 the conditional average of the strain was constantly 400 s^{-1} up to 5 ms prior to extinction. These early times correspond to statistically stationary turbulence. Strain increased from a conditional average of 400 s^{-1} at $t^* = -4.4$ ms to 700 s^{-1} (MP1) and 800 s^{-1} (MP2) at $t^* = -1.4$ ms. The conditional average subsequently decreased slightly until the flame breached. Minimum and maximum values of conditional strain are shown by bars in Fig 4.37d. These bars do not indicate error ranges but rather the variability due to randomness of the turbulent

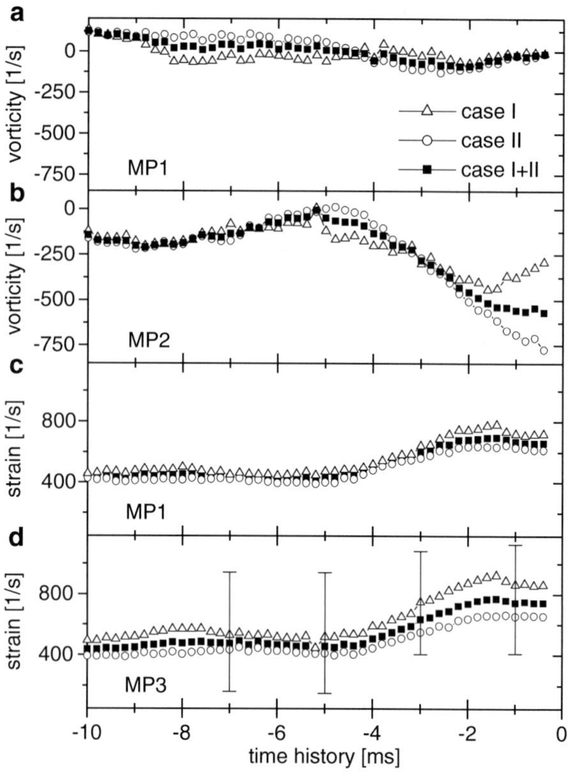

Fig. 4.37 Temporal traces of conditional averaged vorticity and strain at monitor points (MP) assigned in Fig. 4.35 (Reprinted from Böhm et al. [13, 102])

flow field. In contrast to strain at MP1, conditional vorticity remained close to zero at all times. However, at MP2 the absolute value of conditional vorticity increased in a similar manner as the strain until it reached a plateau at $t^* = -1.4$ ms (I + II). Differences in vorticity between Classes I and II were observed. For Class II, absolute values still increased, whereas in Class I, vorticity decreased.

MP1 is the location where the flame is first breached. Only conditionally averaged strain increased at this location. This indicates that for this type of turbulent flame, while vorticity acts to generate strain at distinct locations close to the flame, the magnitude of the vorticity is not as important as that of the strain for determining extinction. Traces of conditionally averaged strain pass a weak maximum at $t^* = -1.4$ ms. This is a clear indication that peak strain values alone are not an appropriate measure to determine extinction for these unsteady flame conditions.

This finding is in agreement with Katta et al. [104]: Strain describes the flow structure. If strain is generated faster than diffusion time-scales, diffusion is delayed with respect to velocity imposition. Consequently non-equilibrium flow may occur, as described in Katta et al. [104]. During non-equilibrium situations, either of the two following mechanisms might be responsible for final extinction:

1. An increased flux of reactants into the reaction zone cools the flame, reduces reaction rates and promotes extinction.
2. The local Damköhler number will be decreased, also promoting extinction.

Increasing strain creates large scalar gradients. These gradients should reveal themselves by high values of scalar dissipation rate (SDR). Unfortunately, the temporal evolution of SDR conditioned to regions of high strain was not measurable in this experiment. Therefore, the hypothesis from Katta et al. [104] that even SDR is an insufficient measure of extinction in unsteady flames and for chemical non-equilibrium conditions (chemistry- rather than diffusion-limited conditions) could not be evaluated experimentally.

4.6.2.3 Conclusions on Turbulent Opposed Jet Investigations at High Repetition Rate

Previous experimental studies using simultaneous PIV/OH-PLIF on a turbulent, opposed jet flame were extended from 1 Hz to 5 kHz. This enabled the acquisition of statistically dependent datasets and allowed for tracking of the evolution of vortex-flame interactions in real time. Previous studies of PDFs of scalar dissipation rate (SDR) have displayed statistical errors at the peripheries, which are exactly the regions of the PDF most closely associated with extinction (as high SDRs are believed to cause extinction). The technique employed in this study allows one to focus specifically on the small fraction of measurement time directly relevant to extinction and thus vastly increases the accuracy, resolution and physical insight of experiments relating to this important phenomenon.

For the flame studied, two classes of flow structure leading to global extinction were identified. The newly introduced technique of multidimensional conditioning used new criteria to compare individual extinction events in a meaningful, statistical manner and avoid smearing important spatiotemporal information. It was found that turbulent vortices generate strain close to the flame and that extinction occurred when the resulting increase in strain exceeded critical values for sufficient time. Although strain was a more important measure for extinction than vorticity, peak strain alone was not a sufficient measure to predict extinction.

4.7 Conclusions

The overall aims of projects B1 and partly B3 were to (1) enhance our understanding of gaseous turbulent flames with a focus on turbulence-chemistry interaction as one of the key issues in practical combustion processes and (2) to provide comprehensive and reliable data of well-documented target flames necessary for validation of numerical simulation methods such as RANS- or LES-approaches. To achieve these objectives two turbulent flames were designed and build. The first flame was a turbulent opposed jet flames operated as partially-premixed flame, the second flame was a swirling lean premixed flame. For various operation conditions flow and scalar fields were extensively studied. For this purpose laser-diagnostics have been adapted and developed such as planar imaging at high repetition rates and combined flow and scalar field methods. These investigations significantly improved the understanding of underlying physical and chemical processes. The data bases are now well-accepted in the community and are indeed used nationally and internationally for validation of numerical simulations.

Acknowledgements Financial support of Deutsche Forschungsgemeinschaft through SFB 568 Projects B1 and B3 and TU Darmstadt is kindly acknowledged.

References

Project-Related Publications

1. Geyer, D., et al.: Finite rate chemistry effects in turbulent opposed flows: comparison of Raman/Rayleigh measurements and Monte Carlo PDF simulation. Proc. Combust. Inst. **30**, 711–718 (2005)
2. Geyer, D., et al.: Scalar dissipation rates in isothermal and reactive turbulent opposed-jets: 1D-Raman/Rayleigh experiments supported by LES. Proc. Combust. Inst. **30**, 681–689 (2005)
3. Böhm, B., et al.: In-Nozzle measurements of a turbulent opposed jet using PIV. Flow Turbul. Combust. **85**, 73–93 (2010)
4. Schneider, C., Dreizler, A., Janicka, J.: Fluid dynamical analysis of atmospheric reacting and isothermal swirling flows. Flow Turbul. Combust. **74**, 103–127 (2005)
5. Gregor, M.A., et al.: Multi-scalar measurements in a premixed swirl burner using 1D Raman/Rayleigh scattering. Proc. Combust. Inst. **32**, 1739–1746 (2009)
6. Nauert, A., et al.: Experimental analysis of flash back in lean premixed swirling flames: conditions close to flash back. Exp. Fluids **43**, 89–100 (2007)
7. Geyer, D., et al.: Turbulent opposed-jet flames: a critical benchmark experiment for combustion LES. Combust. Flame **143**, 524–548 (2005)
8. Omar, S.K., et al.: Investigation of flame structures in turbulent partially premixed counter-flow flames using laser-induced fluorescence. Prog. Comput. Fluid Dyn. **4**, 241–249 (2004)
9. Böhm, B., et al.: Simultaneous PIV/PTV/OH PLIF imaging: conditional flow field statistics in partially-premixed turbulent opposed jet flames. Proc. Combust. Inst. **31**, 709–718 (2007)
10. Nauert, A., Dreizler, A.: Conditional velocity measurements by simultaneously applied laser Doppler velocimetry and planar laser-induced fluorescence in a swirling natural gas/air flame. Z. Phys. Chem. **219**, 635–648 (2005)
11. Bork, B., et al.: 1D high-speed Rayleigh measurements in turbulent flames. Appl. Phys. B **101**, 487–491 (2010)
12. Böhm, B., et al.: New perspectives on turbulent combustion: multi-parameter high-speed laser diagnostics. Flow Turbul. Combust. **86**, 313–341 (2011)
13. Böhm, B., et al.: Time-resolved conditioned flow field statistics in extinguishing turbulent opposed jet flames using simultaneous highspeed PIV/OH-PLIF. Proc. Combust. Inst. **32**, 1647–1654 (2009)

Other Publications

14. Geyer, D., et al.: Finite rate chemistry effects in turbulent opposed flows: comparison of Raman/Rayleigh measurements and Monte Carlo PDF simulation. Proc. Combust. Inst. **30**, 711–718 (2005)
15. Tennekes, H., Lumley, J.L.: A First Course in Turbulence. The MIT Press, Cambridge, MA (1972)
16. Geyer, D., et al.: Scalar dissipation rates in isothermal and reactive turbulent opposed-jets: 1D-Raman/Rayleigh experiments supported by LES. Proc. Combust. Inst. **30**, 681–689 (2005)
17. Böhm, B., et al.: In-Nozzle measurements of a turbulent opposed jet using PIV. Flow Turbul. Combust. **85**, 73–93 (2010)
18. Schneider, C., Dreizler, A., Janicka, J.: Fluid dynamical analysis of atmospheric reacting and isothermal swirling flows. Flow Turbul. Combust. **74**, 103–127 (2005)
19. Gregor, M.A., et al.: Multi-scalar measurements in a premixed swirl burner using 1D Raman/Rayleigh scattering. Proc. Comb. Inst. **32**, 1739–1746 (2009)
20. Heeger, C., et al.: Experimental analysis of flashback in lean premixed swirling flames: upstream flame propagation. Exp. Fluids **49**, 853–864 (2010)
21. Nauert, A., et al.: Experimental analysis of flash back in lean premixed swirling flames: conditions close to flash back. Exp. Fluids **43**, 89–100 (2007)
22. Borghi, R.: On the structure and morphology of turbulent premixed flames. In: Bruno, C., Casci, C. (eds.) Recent Advances in Aeronautical Science. Pergamon Press, London (1984)
23. Peters, N.: Turbulent Combustion, p. 304. Cambridge University Press, Cambridge (2000)
24. Barlow, R.S.: International Workshop on Measurement and Computation of Turbulent Non-premixed Flames (TNF), Sandia National Laboratories (2012)
25. Gupta, A.K., Lilley, D.G., Syred, N.: Swril Flows. Abacus, Cambridge/Tunbridge Wells (1984)
26. Escudier, M.P., Keller, J.J.: Recirculation in swirling flows: a manifestation of voretx breakdown. AIAA J. **23**, 111–116 (1985)
27. Freitag, M., et al.: Mixing analysis of a swirling flow using DNS and experimental data. Int. J. Heat Fluid Flow **27**, 636–643 (2006)
28. Ferrao, P., Heitor, M.V.: Simultaneous velocity and scalar measurements in premixed recirculating flames. Exp. Fluids **24**, 399–407 (1998)
29. Geyer, D., et al.: Turbulent opposed-jet flames: a critical benchmark experiment for combustion LES. Combust. Flame **143**, 524–548 (2005)
30. Maury, F.A., Libby, P.A.: Nonpremixed flames in stagnating turbulence. Part I–The k-e theory with equilibrium chemistry for the methane-air system. Combust. Flame **102**(3), 341–356 (1995)
31. Eckstein, J., et al.: Modeling of turbulent mixing in opposed jet configuration: one-dimensional Monte Carlo probability density function simulation. Proc. Combust. Inst. **28**, 141–148 (2000)
32. Omar, S.K., et al.: Investigation of flame structures in turbulent partially premixed counter-flow flames using laser-induced fluorescence. Prog. Comput. Fluid Dyn. **4**, 241–249 (2004)
33. Böhm, B., et al.: Simultaneous PIV/PTV/OH PLIF imaging: conditional flow field statistics in partially-premixed turbulent opposed jet flames. Proc. Combust. Inst. **31**, 709–718 (2007)
34. Launder, B.E., Reece, G.J., Rodi, W.: Progress in the development of a Reynolds-stress turbulence closure. J. Fluid Mech. **68**(3), 537–566 (1975)
35. Chou, C.P., et al.: Modeling of turbulent opposed-jet mixing flows with k-e model and second-order closure. Int. J. Heat Mass Transf. **47**(5), 1023–1035 (2004)
36. Permana, A.: Turbulent Combustion in Opposed Jet Flows, in Berkeley. University of London, London (2003)

37. Sung, C.J., Law, C.K., Chen, J.-Y.: An argumented reduced mechanism for methane oxidation with comprehensive global parametric validation. Proc. Combust. Inst. **27**, 295–304 (1998)
38. Pope, S.B.: Computationally efficient implementation of combustion chemistry using in situ adaptive tabulation. Combust. Theory Model. **1**, 41–63 (1997)
39. Subramaniam, S., Pope, S.B.: A mixing model for turbulent reactive flows based on Euclidean minimum spanning trees. Combust. Flame **115**(4), 487–514 (1998)
40. Bilger, R.W.: The structure of turbulent nonpremixed flames. Proc. Combust. Inst. **22**, 475–488 (1988)
41. Korusoy, E., Whitelaw, J.H.: Extinction and relight in opposed flames. Exp. Fluids **33**, 75–89 (2002)
42. Pitsch, H.: Improved pollutant predictions in large-eddy simulations of turbulent non-premixed combustion by considering scalar dissipation rate fluctuations. Proc. Combust. Inst. **29**, 1971–1978 (2002)
43. Bray, K.N.C., Libby, P.A., Moss, J.B.: Flamelet crossing frequencies and mean reaction rates in premixed turbulent combustion. Combust. Sci. Technol. **41**, 143–172 (1984)
44. Libby, P.A.: Theory of normal premixed turbulent flames revisited. Prog. Energy Combust. Sci. **11**, 83–96 (1985)
45. Butler, T.D., O'Rourke, P.J.: A numerical method for two-dimensional unsteady reacting flows. Proc. Combust. Inst. **16**, 1503–1515 (1977)
46. Kerstein, A.R.: Linear-eddy modeling of turbulent transport, part 6. Microstructure of diffusive scalar mixing fields. J. Fluid Mech. **231**, 361–394 (1991)
47. Menon, S.: Subgrid combustion modeling for large-eddy simulations. Int. J. Engine Res. **1**, 209–227 (2000)
48. Markstein, G.: Nonsteady Flame Propagation. Pergamon Press, Oxford (1964)
49. Williams, F.A.: Combustion Theory. Addison-Wesley, Reading (1985)
50. Janicka, J., Sadiki, A.: Large eddy simulation of turbulent combustion systems. Proc. Combust. Inst. **30**, 537–547 (2005)
51. Chen, Y.C., et al.: The detailed flame structure of highly stretched turbulent premixed methane-air flames. Combust. Flame **107**, 223–244 (1996)
52. Sardi, E., Taylor, A.M.K.P., Whitelaw, J.H.: Extinction of turbulent counterflow flames under periodic strain. Combust. Flame **120**(3), 265–284 (2000)
53. Bourguignon, E., et al.: Experimentally measured burning rates of premixed turbulent flames. Proc. Combust. Inst. **26**, 447–453 (1996)
54. Li, S.C., Libby, P.A., Williams, F.A.: Experimental investigation of a premixed flame in an impigning turbulent stream. Proc. Combust. Inst. **25**, 1207–1214 (1994)
55. Bédat, B., Cheng, R.K.: Experimental study of premixed flames in intense isotropic turbulence. Combust. Flame **100**, 485–494 (1995)
56. Plessing, T., et al.: Measurements of the turbulent burning velocity and the structure of premixed flames on a low-swirl burner. Proc. Combust. Inst. **28**, 359–366 (2000)
57. Kortschik, C., Plessing, T., Peters, N.: Laser optical investigation of turbulent transport of temperature ahead of the preheat zone in a premixed flame. Combust. Flame **136**, 43–50 (2004)
58. Soika, A., Dinkelacker, F., Leipertz, A.: Measurement of the resolved flame structure of turbulent premixed flames with constant Reynolds number and varied stoichiometry. Proc. Combust. Inst. **27**, 785–792 (1998)
59. Sattler, S., Knaus, D.A., Gouldin, F.C.: Determination of three-dimensional flamelet orientation in turbulent V-flames from two-dimensional image data. Proc. Combust. Inst. **29**, 1785–1792 (2002)
60. Kalt, P.A.M., Frank, J.H., Bilger, R.W.: Laser imaging of conditional velocities in premixed propane-air flames by simultaneous OH PLIF and PIV. Proc. Combust. Inst. **27**, 751–758 (1998)
61. Frank, J.H., Kalt, P.A.M., Bilger, R.W.: Measurement of conditional velocities in turbulent premixed flames by simultaneous OH PLIF and PIV. Combust. Flame **116**(1–2), 220–232 (1999)

62. Most, D., Dinkelacker, F., Leipertz, A.: Direct determination of the turbulent flux by simultaneous application of filtered Rayleigh scattering thermometry and particle imaging velocimetry. Proc. Combust. Inst. **29**, 2669–2678 (2002)
63. Bray, K.N.C.: Turbulent flows with premixed reactants. In: Libby, P.A., Williams, F.A. (eds.) Topics in Applied Physics, vol. 44, pp. 115–183. Springer, Berlin (1980)
64. Bray, K.N.C., Moss, J.B.: A unified statistical model for the premixed turbulent flame. Acta Astonautica **4**, 291–319 (1977)
65. Veynante, D., et al.: Gradient and counter-gradient scalar transport in turbulent premixed flames. J. Fluid Mech. **332**, 263–293 (1997)
66. Gomez, A., Rosner, D.E.: Thermophoretic effects on particles in counterflow laminar diffusion flames. Combust. Sci. Technol. **89**, 335–362 (1993)
67. Nauert, A., Dreizler, A.: Conditional velocity measurements by simultaneously applied laser Doppler velocimetry and planar laser-induced fluorescence in a swirling natural gas/air flame. Z. Phys. Chem. **219**, 635–648 (2005)
68. Wegner, B., et al.: Assessment of unsteady RANS in predicting swirl flow instability based on LES and experiment. Heat Fluid Flow **25**, 528–536 (2004)
69. Barlow, R.S., Karpetis, A.N.: Scalar length scales and spatial averaging effects in turbulent piloted methane/air jet flames. Proc. Combust. Inst. **30**, 673–680 (2004)
70. Dally, B.B., et al.: Instantaneous and mean compositional structure of Bluff-Body stabilized nonpremixed flames. Combust. Flame **114**, 119–148 (1998)
71. Barlow, R.S.: Laser diagnostics and their interplay with computations to understand turbulent combustion. Proc. Combust. Inst. **31**, 49–75 (2007)
72. Frank, J.H., Kaiser, S.A., Long, M.B.: Polarized/depolarized Rayleigh scattering for determining fuel concentrations in flames. Proc. Combust. Inst. **29**, 2687–2694 (2002)
73. Dibble, R.W., et al.: Conditional sampling of velocity and scalars in turbulent flames using simultaneous LDV-Raman scattering. Exp. Fluids **5**, 103–113 (1987)
74. Goss, L.P., Trump, D.D., Roquemore, W.M.: Combined CARS/LDA instrument for simultaneous temperature; velocity measurements. Exp. Fluids **6**, 189–198 (1988)
75. Donbar, J.M., Driscoll, J.F., Carter, C.D.: Reaction zone structure in turbulent nonpremixed jet flames—from CH-OH PLIF images. Combust. Flame **122**(1–2), 1–19 (2000)
76. Han, D., Mungal, M.G.: Stabilization in turbulent lifted deflected-jet flames. Proc. Combust. Inst. **29**, 1889–1895 (2002)
77. Bray, K.N.C., Champion, M., Libby, P.A.: Flames in stagnating turbulence. In: Libby, P.A., Williams, F.A. (eds.) Turbulent Reacting Flows, pp. 573–607. Academic, London (1994)
78. Mastorakos, E., Taylor, A.M.K.P., Whitelaw, J.H.: Scalar dissipation rate at the extinction of turbulent counterflow nonpremixed flames. Combust. Flame **91**, 55–64 (1992)
79. Kitajima, A., et al.: Experimental study of extinction and its quantification in laminar and turbulent counterflow CH4-N2/O2-N2 nonpremixed flames. Combust. Flame **137**(1–2), 93–108 (2004)
80. Kalt, P.A.M., Chen, Y.-C., Bilger, R.W.: Experimental investigation of turbulent scalar flux in premixed stagnation-type flames. Combust. Flame **129**(4), 401–415 (2002)
81. Paone, N.: Velocity measurements in turbulent premixed flames: development of a PIV measurement system and comparison with LDV. In Seventh International Symposium on Application of Laser techniques to Fluid Mechanics, Lisbon, Portugal (1994)
82. Barlow, R.S., et al.: Effect of Damköhler number on super equilibrium OH concentrations in turbulent nonpremixed jet flames. Combust. Flame **82**, 235–251 (1990)
83. Nalm, H.N., et al.: Western States Section/The Combustion Institute, Spring Meeting. Sandia National Laboratories (1997)
84. Stanislas, M., Okamoto, K., Kähler, C.J., Westerweel, J.: Main results of the second international PIV challenge. Exp. Fluids **39**, 170–191 (2005)
85. Raffel, M., Willert, C., Kompenhans, J.: Particle Imaging Velocimetry: A Practical Guide. Springer, Berlin (1998)
86. Cowen, E.A., Monismith, S.G.: A hybrid digital particle tracking velocimetry technique. Exp. Fluids **22**, 199–211 (1997)

87. Muniz, L., Martinez, R.E., Mungal, M.G.: Application of PIV to turbulent reacting flows. In Eighth International Symposium on Application of Laser Techniques to Fluid Mechanics, Lisbon, Portugal (1996)
88. Sung, C.J., Law, C.K., Axelbaum, R.L.: Thermophoretic effects on seeding particles in LDV measurements of flames. Combust. Sci. Technol. **99**, 119–132 (1994)
89. Rehm, J.E., Clemens, N.T.: Local extinction in an unsteady methane-air jet diffusion flame. Proc. Combust. Inst. **27**, 1113–1120 (1998)
90. Kaminski, C.F., Long, M.B.: Multi-dimensional diagnostics in space and time. In: Kohse-Höinghaus, K., Jeffries, J.B. (eds.) Diagnostic Challenges for Gas Turbine Combustor Model Validation, pp. 224–251. Taylor & Francis, New York (2002)
91. Jiang, N.B., Webster, M.C., Lempert, W.R.: Advances in generation of high-repetition-rate burst mode laser output. Appl. Opt. **48**, B23–B31 (2009)
92. Wang, G.H., Clemens, N.T., Varghese, P.L.: Two-point, high-repetition rate Rayleigh thermometry in flames: techniques to correct for apparent dissipation induced by noise. Appl. Opt. **44**, 6741–6751 (2005)
93. Bork, B., et al.: 1D high-speed Rayleigh measurements in turbulent flames. Appl. Phys. B **101**, 487–491 (2010)
94. Upatnieks, A., Driscoll, J.F., Ceccio, S.L.: Cinema particle imaging velocimetry time history of the propagation velocity of the base of a lifted turbulent jet flame. Proc. Combust. Inst. **29**, 1897–1903 (2002)
95. Kaminski, C.F., Hult, J., Aldén, M.: High repetition rate planar laser-induced fluorescence of OH in a turbulent non-premixed flame. Appl. Phys. B **68**, 757–760 (1999)
96. Li, D., et al.: Diode-pumped efficient slab laser with two Nd:YLF crystals and second-harmonic generation by slab LBO. Opt. Lett. **32**, 1272–1274 (2007)
97. Fajardo, C.M., Sick, V.: Sustained simultaneous high-speed imaging of scalar and velocity fields using a single laser. Appl. Phys. B **85**, 25–31 (2006)
98. Ahmed, S.F., Balachandran, R., Mastorakos, E.: Measurement of ignition probability in turbulent non-premixed counterflow flames. Proc. Combust. Inst. **31**, 1507–1513 (2007)
99. Böhm, B., et al.: New perspectives on turbulent combustion: multi-parameter high-speed laser diagnostics. Flow Turbul. Combust. **86**, 313–341 (2011)
100. Winkler, A., Wäsle, J., Sattelmayer, T.: Laserinduzierte Fluoreszenz in Echtzeit zur Bestimmung des Flammenlärms. Fachtagung "Lasermethoden in der Strömungsmesstechnik" **12**, 18-1–18-8 (2004)
101. Adrian, R.J., Christensen, K.T., Liu, Z.-C.: Analysis and interpretation of instantaneous turbulent velocity fields. Exp. Fluids **29**, 275–290 (2000)
102. Böhm, B., et al.: Time-resolved conditioned flow field statistics in extinguishing turbulent opposed jet flames using simultaneous highspeed PIV/OH-PLIF. Proc. Combust. Inst. **32**, 1647–1654 (2009)
103. Lemaire, A., et al.: PIV/PLIF investigation of two-phase vortex-flame interactions: effects of vortex size and strength. Exp. Fluids **36**, 36–42 (2004)
104. Katta, V.R., et al.: Insights into non-adiabatic equilibrium flame temperatures during millimeter-size vortex/flame interactions. Combust. Flame **132**, 639–651 (2003)

Chapter 5
Simplified Reaction Models for Combustion in Gas Turbine Combustion Chambers

Dirk Lebiedz and Jochen Siehr

Abstract The aim of this work is an effective dimension reduction of chemical combustion mechanisms. Despite permanently growing computer power, the simulation of a reaction-diffusion-convection system involving a large scale chemical combustion mechanism is still far from reach. On the other hand, prediction of e.g. soot, NOx, and other pollutants needs detailed mechanisms. Here, model reduction methods can be used for generation of small models. At first, the focus of this work has been on an efficient use of the intrinsic low dimensional manifold method. Later, a new method has been developed based on optimization methods. An efficient tool for solving these optimization problems has been developed and reaction models up to the size of syngas combustion have been reduced.

Keywords Model reduction • Slow invariant manifold • Chemical kinetics • Nonlinear optimization

5.1 Introduction

The interplay between flow (convection), diffusion, and chemical reaction processes in a reactive flow simulation is fairly complex if the model is based on a detailed combustion mechanism involving a large number of chemically reactive species. Here, complexity reduction and model reduction methods can be used effectively.

D. Lebiedz (✉)
Institute for Numerical Mathematics, University of Ulm, Helmholtzstraße 20, 89081 Ulm, Germany
e-mail: dirk.lebiedz@uni-ulm.de

J. Siehr
Interdisciplinary Center for Scientific Computing (IWR), University of Heidelberg, Im Neuenheimer Feld 368, 69120 Heidelberg, Germany
e-mail: jochen.siehr@iwr.uni-heidelberg.de

J. Janicka et al., *Flow and Combustion in Advanced Gas Turbine Combustors*, Fluid Mechanics and Its Applications 102, DOI 10.1007/978-94-007-5320-4_5,

5.1.1 The Aim of Model Reduction

The aim of many model reduction approaches is an identification of so called slow invariant manifolds (SIM). These manifolds are located in the chemical composition space. The general reaction transport equation in the variable ζ, which can be mass fractions, partial density, temperature, or any variable describing the system state depending on time t and space x, can be written as

$$\partial_t \zeta = S(\zeta) + \mathrm{T}(\zeta, \partial\zeta_x, \partial_x^2\zeta),$$

where S is the chemical source term and T the physical transport operator, i.e. convection and diffusion. Many methods for model reduction are applied to the chemical source term of this system of reaction transport partial differential equations (PDE), which describe the reactive flow. This means the model reduction method is only regarding the system of ordinary differential equations (ODE) modeling the kinetic source term

$$\mathrm{d}_t \zeta := \frac{\mathrm{d}}{\mathrm{d}t}\zeta = S(\zeta).$$

Trajectories in the chemical composition space are relaxing to SIMs of low dimension on their way towards equilibrium. In this sense, these SIMs represent the slow chemistry. The existence of a SIM is closely related to multiple time scales and a spectral gap in the eigenvalues of the Jacobian of the chemical source term. The model reduction method can be seen as a method for computation of such a manifold pointwise. In general, some parameterization variables will be chosen, known as reaction progress variables and the method delivers the values of the other variables in sense of species reconstruction. This can be used to simplify the computation of the full system of reaction transport PDE via applying it *in situ* or by evaluation of a tabulated manifold [38] and hence saves a lot of computation time.

5.1.2 Common Model Reduction Methods

Early model reduction approaches like the quasi steady-state (QSSA) and partial equilibrium assumption (PEA) [47] have been performed "by hand". Hence, special expert knowledge about the chemistry model has to be available. In contrast, modern numerical model reduction methods compute automatically a reduced model. Many of these modern techniques are based explicitly or implicitly on a time-scale analysis of the underlying ODE system with the purpose to identify a slow attracting manifold in phase space where – after a short initial relaxation period – the system dynamics evolve. For a detailed overview, see e.g. [24] and references therein.

An important method of this type is the intrinsic low dimensional manifold method (ILDM). Maas and Pope introduced the ILDM method in 1992 [32].

It became very popular and widely used in the reactive flows community. A local time-scale analysis is performed on the Jacobian J_S of the system of ODEs modeling the chemical kinetics S. An upper block triangular matrix is computed with the Schur decomposition.

The fast and slow submatrices can be fully decoupled with the so called Sylvester equation [23]. Fast time scales are assumed to be fully relaxed. Together with the parameter equation the ILDM equations can be stated as

$$\begin{pmatrix} T_{\mathrm{r,\,fast}}^{-1} S(\zeta) \\ P^{\mathrm{T}}\zeta - c \end{pmatrix} = 0,$$

where ${T_{\mathrm{r,fast}}}^{-1}$ is the projection into the fast eigenspace and the second line the parameterization of the SIM. A drawback of this method is that a solution of the ILDM equation does not have to exist. It has been shown that the ILDM method identifies the SIM to first order in the timescale separation [28, 40]. For recent developments and extensions of the ILDM method, see e.g. [14] and references therein.

Another popular technique is the computational singular perturbation (CSP) method proposed by Lam in 1985 [30, 31]. The basic concept of this method is a representation of the dynamical system in a set of "ideal" basis vectors such that fast and slow modes are decoupled. This method improves the SIM approximation with one order per CSP iteration [48]. Also some iterative methods came into application that are based on an evaluation of functional equations suitably describing the central characteristics of a slow attracting manifold, for example invariance and stability. Examples are Fraser's algorithm [18, 21, 34] and the method of invariant grids [15, 16, 24, 25]. Other methods are the invariant constrained equilibrium edge preimage curve (ICE-PIC) method [37, 39], flamelet-generated manifolds [19] and many more.

5.2 Models for Combustion Chemistry

We list formulae necessary for the formulation of a combustion model in this section. For details we refer to text books as e.g. [29, 47].

A model comprises n_{spec} chemical species composed by n_{elem} chemical elements. The chemical source term S obeys the law of elemental mass conservation and an energetic balance has to be fulfilled.

5.2.1 Mass Balances

In the following, the variables z_i are given in terms of specific moles, which are defined as the amount of species i (n_i) divided by the total mass (m) of the system, which is the same as the mass fractions (w_i) of the species i divided by its molar mass (M_i):

$$z_i = \frac{n_i}{m} = \frac{w_i}{M_i}, \quad i = 1, ..., n_{\text{spec}}.$$

The elemental mass conservation of each element in the system is formulated as

$$\bar{z}_i = \sum_{j=1}^{n_{\text{spec}}} \chi_{ij} z_j, \quad i = 1, ..., n_{\text{elem}},$$

where χ_{ij} is the atomic composition coefficient – the number of element i in species j. There is also a restriction to the choice of the elemental specific moles, which requires that the mass fractions sum to one

$$\sum_{i=1}^{n_{\text{elem}}} \bar{M}_i \bar{z}_i = 1,$$

with the molar mass of element i. This ensures the conservation of the total mass of the system.

5.2.2 *Energy Balance*

We consider systems within one of the four standard thermodynamic environments which is either in an isothermal or adiabatic and either in an isochoric or isobaric state. In the isothermal case only the temperature is fixed while in the adiabatic cases the specific enthalpy h or the specific internal energy e respectively, are fixed.

5.2.3 *ODE Models*

The ODE model is considered as a constraint in the optimization problem for the computation of reduced models as discussed in a later section. Therefore, the ODE reaction system that is given by the source term (called S until now) is discussed in the following. This ODE system includes mass action kinetics and obeys the elemental mass conservation laws.

The mass balance in specific moles z_i can be formulated as

$$\mathrm{d}_t z_i = S^{\mathrm{m}} := \frac{\omega_i}{\rho}, \qquad i = 1, ..., n_{\text{spec}}.$$

The symbol ρ refers to the overall mass density in the system in the right hand side of this equation. It is given via the ideal gas law

$$\rho = \frac{m}{V} = \frac{p\bar{M}}{RT}.$$

The total mass and volume V are needed in the isochoric case; the total pressure p, the gas constant R, the temperature T, and the mean molar mass are necessary in the isobaric case for computation of the right hand side. The molar net chemical production rate is denoted by the vector ω. It has to be computed based on a set of chemical elementary reactions and their kinetic parameters as described in Sect. 5.2.4.

In our case, we formulate the energy balance via the right hand side of the temperature equation $d_t T = S^e$. In the isothermal case $S^e = 0$ holds. In the adiabatic cases energy or enthalpy conservation defines the concise form of S^e.

5.2.4 Chemical Kinetics

We subsume the computation of ω in the remainder of this section. A chemical combustion mechanism generally is given as a set of n_{reac} elementary reactions involving n_{spec} species (and eventually a third body M)

$$\sum_{i=1}^{n_{\mathrm{spec}}} \nu'_{ij} \mathrm{X}_i \leftrightarrow \sum_{i=1}^{n_{\mathrm{spec}}} \nu''_{ij} \mathrm{X}_i, \quad j = 1, \ldots, n_{\mathrm{reac}}$$

with the chemical species X_i and the forward and reverse stoichiometric coefficients ν_{ij}' and ν_{ij}''. The forward and reverse rate of reaction j is

$$r_{\mathrm{f},j} = k_{\mathrm{f},j} \prod_{i=1}^{n_{\mathrm{reac}}} c_i^{\nu'_{ij}}$$

$$r_{\mathrm{r},j} = k_{\mathrm{r},j} \prod_{i=1}^{n_{\mathrm{reac}}} c_i^{\nu''_{ij}}$$

with the concentrations c_i of species i and the rate constants $k_{\mathrm{f},j}$ and $k_{\mathrm{r},j}$. With the net stoichiometric coefficient

$$\nu_{ij} = \nu''_{ij} - \nu'_{ij}$$

the net rate of change ω_i of species i is given as

$$\omega_i = \sum_{j=1}^{n_{\text{reac}}} \nu_{ij}(r_{\text{f},j} - r_{\text{r},j}).$$

In case a third body M takes part in reaction j, third body collision efficiencies α_i for all species $i = 1, \ldots, n_{\text{spec}}$ are to be given. The third body concentration

$$c_{\text{M}} = \sum_{i=1}^{n_{\text{spec}}} \alpha_i c_i$$

is then multiplied to the products above.

Formulas for the computation of the rate coefficients are stated in the following. The elementary reactions in the mechanism considered here are given in Arrhenius format and pressure dependent Troe format, resp. The three parameters A, b, and E_{a} are given in the Arrhenius case for each reaction. The forward reaction rate coefficient is computed via the extended Arrhenius formula

$$k_{\text{f},j} = AT^{\frac{b}{1\text{K}}} \text{e}^{\frac{-E_{\text{a}}}{RT}}.$$

A more complicated formula applies in case of pressure dependent reactions. Here, the forward rate constant is computed using Troe fall off curves, see e.g. the original publications [22, 45].

The reverse reaction rate constant of a reaction is computed via evaluation of the equilibrium constant of the reaction. The equilibrium constant of reaction j in terms of concentrations is given as

$$K_{c,j} = \left(\frac{p^\circ}{RT}\right)^{\nu_j} \exp\left(\frac{\Delta S^\circ_{\text{r},j}}{R} - \frac{\Delta H^\circ_{\text{r},j}}{RT}\right),$$

with the standard pressure p° and the net change of the number of species

$$\nu_j = \sum_{i=1}^{n_{\text{spec}}} \nu_{ij}.$$

The change in entropy and enthalpy can be computed by using an evaluation of the NASA polynomials for their molar values of the species involved [13]. The reverse rate of reaction j is

$$k_{\text{r},j} = \frac{k_{\text{f},j}}{K_{c,j}}.$$

5.3 Optimization Problem

Lebiedz introduced a model reduction method based on the minimization of entropy production rate along trajectories in chemical composition space in [1]. The basic idea is that the SIM should be characterized by maximum relaxation of the system dynamics under given constraints of fixed reaction progress variables. This approach has been extended and refined in a number of following publications [2, 4, 6–8].

An optimization problem has to be solved for the approximation of points on the manifold. It can be written in specific moles and temperature as minimization of an objective functional

$$\min_{z,T} \int_{t_0}^{t_\mathrm{f}} \Phi(z(t))\,\mathrm{d}t$$

subject to

$$\begin{aligned}
\mathrm{d}_t z(t) &= S^{\mathrm{m}}(z(t), T(t)) \\
\mathrm{d}_t T(t) &= S^{\mathrm{e}}(z(t), T(t)) \\
0 &= C(z(t_*), T(t_*)) \\
0 &= z_j(t_*) - z_j^{t_*}, \quad j \in I_{\mathrm{rpv}} \\
0 &\le z(t), T(t)
\end{aligned}$$

and

$$\begin{aligned}
t &\in [t_0, t_\mathrm{f}] \\
t_* &\in [t_0, t_\mathrm{f}] \text{ (fixed).}
\end{aligned}$$

The dynamics of the model to be reduced are given via the ODE system in the constraints. This guarantees to identify as solution of the optimization problem a special solution trajectory (piece).

All conservation laws, that the system obeys, are combined in the (nonlinear) function C. The dynamics contain differential forms of the balances of mass and energy. The values of the conserved quantities have to be specified at some (fixed) point in time along a solution which we choose to be t_*. They are also fulfilled along the whole time interval via the dynamics. This is the conservation of each chemical element and a specification of a fixed temperature, enthalpy or energy depending on the assumed thermodynamic environment.

In order to approximate the SIM, the parameterization needs to be specified. The species (specific moles z_i) which serve as reaction progress variables and their number have to be specified in advance. The number of progress variables determines the chosen dimension of the SIM to be approximated. The indices of the reaction progress variables are collected in the index set $I_{\rm rpv}$, a subset of $\{1,\ldots,n_{\rm spec}\}$, and their values are fixed in the optimization problem. Hence, these values are parameters of the optimization problem and change between different calls for the solution of the problem for deriving different points on the manifold.

Only positive values of specific moles and temperature have a physical meaning. This physical inequality constraint is included in the optimization problem. This is also of technical importance in the optimization context with a realistic combustion mechanism as negative values of z_i and T can result in undefined values (logarithm of a negative number) of the right hand sides $S^{\rm m}$ and $S^{\rm e}$ for pressure dependent reactions. The positivity and the linear mass conservation relations define a polytope in the composition space, where the dynamics evolve.

The crucial point in time t_* in $[t_0, t_{\rm f}]$ specifies the position where along the optimal trajectory piece the fixation of the reaction progress variables (and the constraint C) is applied. In first publications, e.g. [1, 3, 4], $t_* = t_0$ has been chosen. The inverse idea is the fixation at the end point $t_* = t_{\rm f}$. The solution point and solution trajectory piece of the optimization problem is supposed to be part of the SIM. Therefore, also in backward direction of time the trajectory is supposed to be relaxed.

To conclude, it can be stated that the essential degrees of freedom of the optimization problem are the effective phase space dimension $n_{\rm spec} - n_{\rm elem}$ minus the number of reaction progress variables $|{\rm I}_{\rm rpv}|$. The goal of solving the optimization problem is the determination (species reconstruction) of the "missing" values as a mapping of the parameters into full composition space.

5.3.1 *The Criterion*

The choice of the criterion Φ is crucial. It should measure the degree of chemical force relaxation along a trajectory, which should be minimal on the SIM. Several criteria have been tested, especially in [4, 5].

5.3.1.1 Entropy Based Criteria

In [1], the author introduced the concept of optimization based model reduction by identification of so called *minimal entropy production trajectories* (MEPT). The concept is based on choosing trajectories along which the entropy production is minimal. The entropy production rate of elementary reaction j (due to irreversible processes) is defined as

$$\frac{\mathrm{d_i}S_j}{\mathrm{d}t} = R(r_{\mathrm{f},j} - r_{\mathrm{r},j}) \ln \left(\frac{r_{\mathrm{f},j}}{r_{\mathrm{r},j}} \right).$$

The total entropy production rate is the sum over all elementary reactions. Hence, the criterion used in the optimization problem (the different suggestions for Φ are numbered in the following) would be

$$\Phi_1(z(t)) = \sum_{j=1}^{n_{\mathrm{reac}}} \frac{\mathrm{d_i}S_j}{\mathrm{d}t}.$$

In [2], this idea is extended and discussed with the definition of entropy in a mathematical sense.

5.3.1.2 Curvature Based Criteria

If we regard the formula for entropy production rate in comparison to the formula for the net chemical production rate, it is obvious that there is a connection between minimizing the entropy production and the chemical production. The only difference is the mass density and a weighting induced by the multiplication with the net stoichiometric coefficients instead of the logarithmic ratio of the rates. Curvature based criteria are an attempt in the same direction which starts from another point of view.

The SIM to be approximated is considered to represent the slow dynamics. This means the rate of change of the variable values is supposed to be small. The rate of change is related to the curvature of the trajectories as geometrical objects in phase space, it is the second derivative

$$\ddot{z} = J_{S^{\mathrm{m}}}(z(t), T(t)) S^{\mathrm{m}}(z(t), T(t)),$$

where we denote the Jacobian of a function S as J_S. This can be interpreted as a directional derivative of the chemical source term with respect to its own direction. In a notation that coincides with the general optimization problem, a choice for the criterion Φ would be

$$\Phi_2(z(t)) = ||J_{S^{\mathrm{m}}}(z(t), T(t)) S^{\mathrm{m}}(z(t), T(t))||_2^2.$$

Interestingly, this can be seen as minimizing the length of a trajectory piece in a suitable Riemannian metric. The length of a continuously differentiable curve z on a Riemannian manifold is defined as the curve integral

$$L(z) = \int_z \sqrt{g_z(\dot{z}(t), \dot{z}(t))}\, \mathrm{d}t$$

with a scalar product g_z on the tangent space of the curve in each point. The choice of $g_z(S,S) = S^T J_S{}^T J_S S$ with the symmetric positive definite (if there are no conservations) matrix $J_S{}^T J_S$ leads to Φ_2. Therefore, the solution curve can be interpreted as a geodesic, i.e. a curve which minimizes the length of path between two points along the manifold. This duality of minimal velocity change and minimal length arises naturally as a minimal curvature coincides with a minimal distance that is covered and vice versa.

5.3.2 *Theoretical Results*

5.3.2.1 Existence of Solutions

In [6] and [8], it has been shown that there always exists a solution of the optimization problem subject to a model which describes a combustion process as introduced before if there exists a feasible solution at all.

The key is the proof of the compactness of the feasible domain of the optimization problem. In case of linear (mass conservation) constraints together with positivity, the chemically realizable space is a closed, bounded polytope in the chemical composition space, which is compact by definition. In case of the nonlinear enthalpy or energy conservation, proper maps ensure compactness. Proper maps are defined as maps where the preimage of a compact set is compact. In this way, the feasible domain is the intersection of compact spaces and, therefore, also compact.

The objective function is a continuous function in the state (z,T) at t_*. It takes its minimum on a compact feasible set. Hence, a solution exists if the feasible set is not empty.

5.3.2.2 Quality of Solutions

In [6], it has been shown that the reverse mode of the method identifies the correct SIM for infinite time horizon in case of a linear test model and the Davis–Skodje test model [18], which have an analytically defined one-dimensional SIM. Hence, we use the reverse mode for all numerical results presented here.

The proof exploits the fact that the general linear system as well as the Davis–Skodje model has an analytical solution. This solution is directly evaluated in the objective function. The solution is identified via the standard sufficient conditions for optimality. A consideration of the limit $t_0 \to -\infty$ finally shows that the deviation between the given exact SIM and the solution of the optimization problem vanishes. The same result holds in case of an infinite spectral gap for a finite time horizon.

5.4 Numerical Methods

The functions in the semi-infinite optimization problem have to be discretized. The resulting nonlinear programming (NLP) problem can be solved either by a sequential quadratic programming (SQP) [36] or an Interior Point (IP) method, see e.g. the review [20].

5.4.1 Discretization

There are mainly two ways for discretization and solution of the semi-infinite optimization problem: the sequential and the simultaneous approach.

In the sequential approach, the solution of the ODE and the optimization are fully decoupled. The initial values $(z(t_0), T(t_0))$ serve as optimization variables. Starting at t_0 the system is integrated with a stiff ODE solver, e.g. via a BDF scheme [17]. The integrand in the objective function is integrated itself, and the end point is evaluated in sense of a Mayer term objective functional. The optimization iteration is performed after that based on the results of the integration and computed derivative information. A single shooting is appropriate in the cases $t_* = t_0$ and $t_* = t_f$ as we deal with a stable ODE system, whereas a double shooting is needed in case of t_* in (t_0, t_f) for the values at t_*.

It might be beneficial in case of unstable or extremely stiff systems to use an all-at-once approach. The solution of the dynamic constraints and the optimization are coupled in this so called simultaneous approach. The interval $[t_0, t_f]$ is divided into several sub-intervals. Polynomials are constructed via, e.g., a collocation method [10] on each sub-interval tangent to the vector field of the dynamics approximating their solution, and the corresponding formulae are treated as constraints in the optimization iteration. We use a Gauss–Radau formula in a collocation approach with linear, quadratic, and cubic polynomials, respectively, because they have stiff decay [10].

5.4.2 Solution of the Finite Dimensional Optimization Problem

In both cases, the result of the discretization is a finite-dimensional NLP problem. It can be solved using an SQP algorithm or an IP method. The SQP algorithm treats the inequality constraints with an active set strategy, see e.g. [35]. Newton's method is applied to the first order optimality conditions of a quadratic approximation of the objective function of the NLP problem and a linear approximation of the constraints. Here, only equality and active inequality constraints are taken into account. The active set in the solution is iteratively identified by activating and deactivating constraints. In contrast, the inequality constraints are coupled to the objective function in an IP method via a barrier term forcing the iterates strictly into the interior of the feasible domain. The resulting equality constrained NLP problem is solved with a homotopy method. Newton's method is applied to the first order optimality conditions. The barrier parameter is driven to zero in progress of optimization. In this way a homotopy path to the solution of the NLP problem is followed.

5.4.3 Algorithms and Software

Application results of the model reduction method are presented in the following section. In general, for the solution of the problem we only use open source software and our own developments. We use the software package IPOPT [46] for numerical optimization.

For the solution of linear equation systems within the optimization algorithm, HSL routines [27] and MUMPS [9], resp., are used. We use a collocation approach based on a Gauss–Radau method [10] for discretization of the optimization problem. Alternatively, we use a shooting approach with a BDF integrator that has been developed in [42, 43]. We use MATLAB for plotting as the only commercial software.

Derivatives needed for the optimization are computed with the automatic differentiation package CppAD [11, 12]. The strategy of CppAD is based on operator overloading. The computer code which is supposed to be differentiated is saved within a CppAD tape. This tape can be "optimized" before evaluation. Furthermore, the criterion $J_S S$ can be seen as a directional derivative where the use of multiple levels of AD (the evaluation of a tape within a tape) can be beneficial. The BDF integrator used in the sequential optimization approach directly uses such a tape as input variable. For generation of sensitivities, it incorporates several strategies of solving sensitivity differential equations and adjoint differential equations [42].

A user usually computes not only a single approximation point on the SIM but a sequence of points needed e.g. in a CFD simulation. Hence, it is necessary to have an effective warm start strategy that has been implemented in our code with a homotopy method. A homotopy parameter drives the iterates from a previous solution to the solution of new problem.

As most numerical codes which simulate combustion processes, the code also reads input files for thermodynamic and kinetic parameters. In our case, it is the standard format for NASA polynomial coefficients [13] and the reaction mechanism with the kinetic parameters for each reaction.

5.5 Results

Results for the application of the optimization based model reduction method are shown in this section.

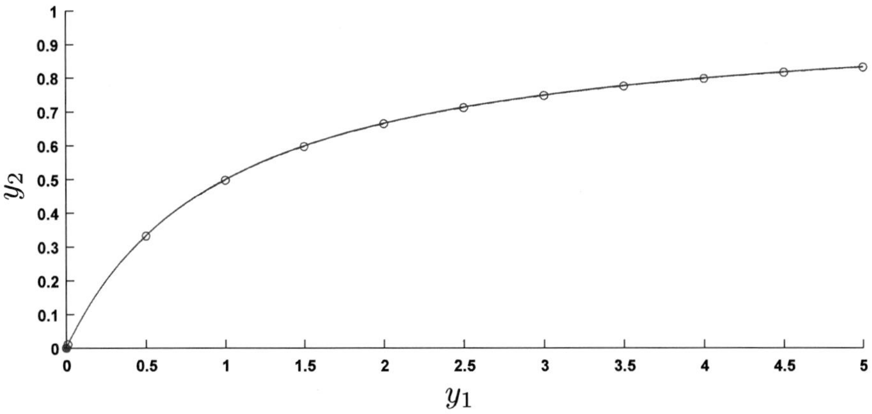

Fig. 5.1 Results for the solution of the optimization problem in case of the Davis–Skodje model with a spectral gap $\gamma = 3$. The rings are the solution points for several values of the reaction progress variable y_1, for each of which the optimization is solved

5.5.1 Davis–Skodje Mechanism

The Davis–Skodje mechanism serves as first test case [18, 41]. It is given via a two-dimensional ODE system

$$\frac{dy_1}{dt} = -y_1$$

$$\frac{dy_2}{dt} = -\gamma y_2 + \frac{(\gamma - 1)y_1 + \gamma y_1^2}{(1 + y_1)^2},$$

where $\gamma > 1$ is a measure for the spectral gap (time scale separation). The Davis–Skodje model is a standard test model for model reduction methods due to the adjustable time scale separation and an analytically given SIM as the graph of the function

$$y_2 = \frac{y_1}{1 + y_1}.$$

Results for the solution of the optimization problem are shown in Figs. 5.1 and 5.2. In both cases, a small value for γ has been chosen, which poses a high challenge for model reduction algorithms. The open rings denote solutions of the optimization problem at the end point $t_* = t_f$. The optimization problem is solved with criterion Φ_2 in the objective function along a time horizon of $t_f - t_0 = 5$, where here y_i play the role of z_i $(i = 1, 2)$ in the problem formulation. Here y_1 is chosen as reaction progress variable and fixed to several different values. Additionally, trajectory pieces are shown which emanate from the solutions points (as solutions

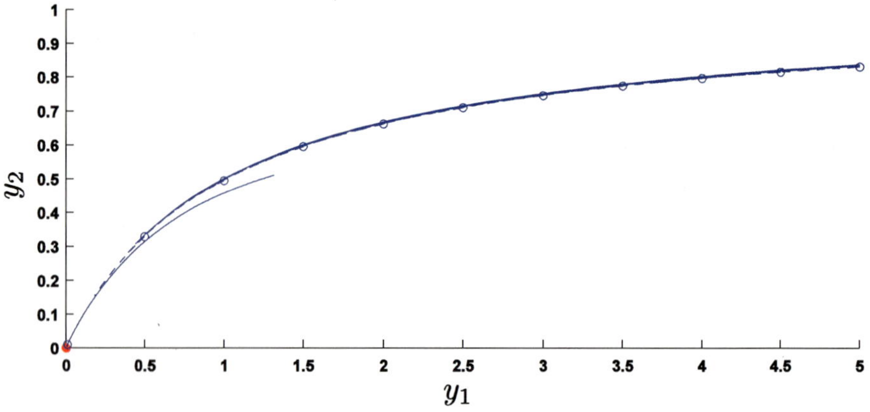

Fig. 5.2 Results for the solution of the optimization problem in case of the Davis–Skodje model as in Fig. 5.1 but with a spectral gap $\gamma = 1.1$

Table 5.1 Simple hydrogen combustion test mechanism from Gorban et al. [26]

			k_f	k_r
H_2	$\leftrightarrow$	2H	2.0	216.0
O_2	$\leftrightarrow$	2O	1.0	337.5
H_2O	$\leftrightarrow$	H + OH	1.0	1,400.0
$H_2 + O$	$\leftrightarrow$	H + OH	1,000.0	10,800.0
$O_2 + H$	$\leftrightarrow$	O + OH	1,000.0	33,750.0
$H_2 + O$	$\leftrightarrow$	H_2O	100.0	0.7714

Forward and backward rate coefficients do not depend on temperature

of the initial value problem) and converge to equilibrium, shown as full dot. The reverse part of the trajectory through a solution point is the optimal trajectory piece computed in the optimization process. It can be seen that the results deliver an invariant one-dimensional manifold. Only for the small value $y_1 = 0.01$ the solution is not invariant in Fig. 5.2. Invariance could be achieved with a larger time horizon as has been proved in [6].

5.5.2 *Six Species Test Mechanism*

Another small test mechanism, which has been introduced for model reduction test purposes in [26] consists of six chemical species involved in six (in each case forward and reverse) elementary reactions, cf. Table 5.1.

This test mechanism is regarded within an isochoric and isothermal thermodynamic environment. Hence, a choice of $\rho = 1$ kg/m^3 leads to the same numeric values of concentrations and specific moles

$$z_i = c_i \,\text{kg/m}^3 \quad \forall i = 1, \ldots, n_{\text{spec}},$$

and we arrive at our standard formulation in specific moles. This mechanism is only an academic example. Units are meaningless here and are skipped in the following.

There are only the chemical elements hydrogen and oxygen involved. The corresponding mass conservation relations are chosen to be

$$2z_{H_2} + 2z_{H_2O} + z_H + z_{OH} = 1$$

$$2z_{O_2} + z_{H_2O} + z_O + z_{OH} = 2.$$

Hence, this mechanism yields a system with effectively four degrees of freedom. We want to compute a two-dimensional manifold. We chose the values of the fuel z_{H2} and the reaction product z_{H2O} to serve as reaction progress variables. Their values are fixed at different values in the area where the value of z_{H2} is higher than the equilibrium value and z_{H2O} is lower than its equilibrium value. Combinations which violate the mass conservation and positivity are identified and ignored automatically by the code.

The results are shown in Fig. 5.3, where again the open rings refer to different solutions of the optimization problem and trajectories emanating from these are converging to equilibrium, shown as a full dot with a large value of H_2O and nearly all H_2 is "burned." An invariant two-dimensional manifold can be seen, where there is no relaxation phase of the trajectories from the solution points onto the manifold.

5.5.3 *Ozone Decomposition*

We consider an ozone decomposition mechanism, which includes atomic oxygen, dioxygen, and ozone, as a first realistic test problem. The mechanism is given in Table 5.2 with the coefficients as in [33] for the forward reactions.

The thermodynamic data is used in form of NASA polynomial coefficients. The model is formulated as in [8] within the framework described in Sect. 5.2. Reverse rate coefficients are derived from equilibrium thermodynamics. For mass conservation the elemental specific mole has to be given. It is

$$\bar{z}_O = (0.015\,999\,\mathrm{kg/mol})^{-1} = 62.5\,\mathrm{mol/kg}.$$

We consider a density of $\rho = 0.2$ kg/m^3 in the isochoric case and a pressure of $p = 10^5$ Pa for isobaric conditions, respectively.

No positivity is demanded in the results presented in Figs. 5.4, 5.5, 5.6, and 5.7 for the species, so negative values can be seen in the plots. The model leaves two degrees of freedom; we choose to compute approximations of a one-dimensional manifold with z_O as reaction progress variable. Again the results are shown as in the examples before. It can be seen that there is always a relaxation phase of the optimized trajectory piece onto that manifold, especially in the adiabatic cases shown in Figs. 5.6 and 5.7 far from equilibrium.

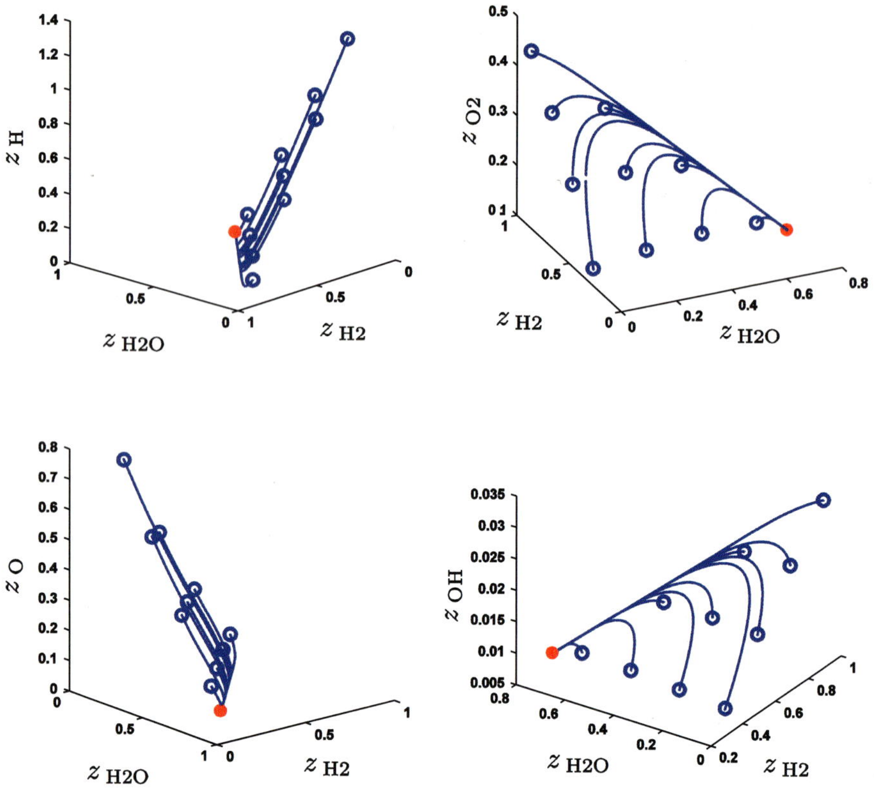

Fig. 5.3 Results (two-dimensional SIM) for the six species test mechanism. The value of z_{H2} and z_{H2O} serve as reaction progress variables. The optimization problem is solved several times for different combinations of these. We use the reverse mode ($t_f = t_*$) with an integration interval of $t_f - t_0 = 10^{-5}$. The free z_i are plotted versus the progress variables

Table 5.2 Ozone decomposition mechanism for the forward rates as in [33]

			A/cm, mol, s	b	E_a/kJ mol
$O + O + M$	$\leftrightarrow$	$O_2 + M$	2.9×10^{17}	-1.0	0.0
$O_3 + M$	$\leftrightarrow$	$O + O_2 + M$	9.5×10^{14}	0.0	95.0
$O + O_3$	$\leftrightarrow$	$O_2 + O_2$	5.2×10^{12}	0.0	17.4

Collision efficiencies in reactions including M: $\alpha_O = 1.14$, $\alpha_{O2} = 0.40$, $\alpha_{O3} = 0.92$

5.5.4 Syngas Combustion Mechanism

We use a syngas combustion which is a part of the GRI 3.0 mechanism [44] as an example of a full detailed chemistry combustion mechanism. It consists of 33 reactions of the GRI 3.0 mechanism which involve no other species than O, O_2, H,

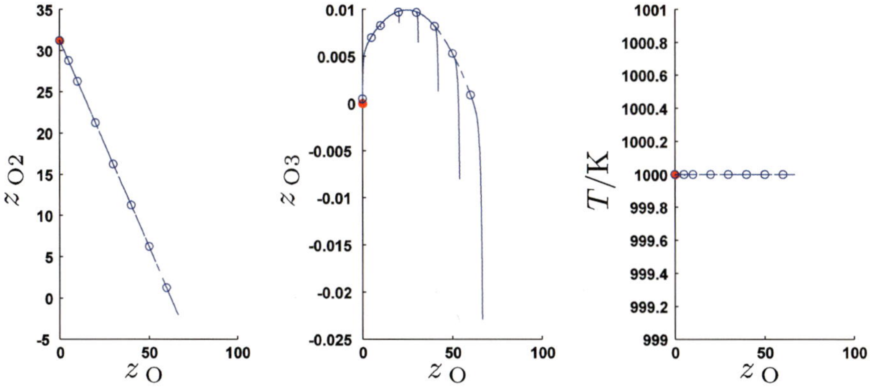

Fig. 5.4 Results of the approximation of a one-dimensional SIM for the ozone decomposition mechanism within an isothermal and isochoric environment. The value of z_{O} serves as reaction progress variable. The optimization problem is solved several times for different values of z_{O} between zero and the largest possible value allowed by elemental mass conservation. The integration interval is $t_{\mathrm{f}} - t_0 = 10^{-6}$ s. The free z_i (in mol/kg) and temperature are plotted versus z_{O}

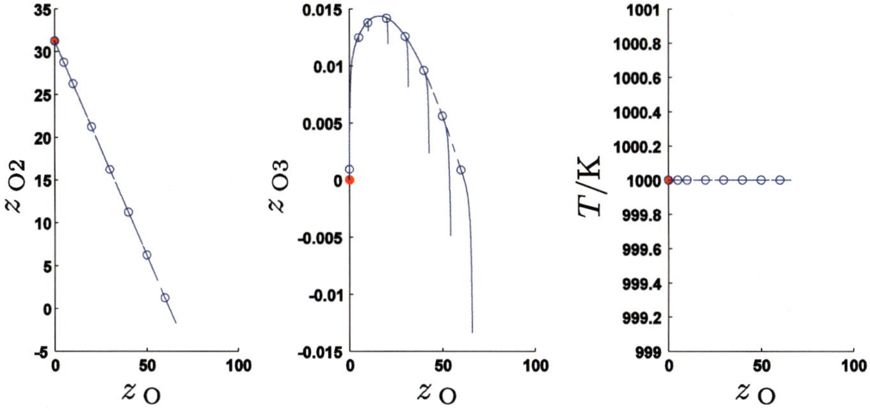

Fig. 5.5 Results of the approximation of a one-dimensional SIM for the ozone decomposition mechanism modeled within an isothermal and isobaric environment. Again the integration interval of $t_{\mathrm{f}} - t_0 = 10^{-6}$ s is used

OH, H_2, HO_2, H_2O_2, H_2O, N_2, CO, and CO_2. Those 33 reactions are 31 Arrhenius-type and two pressure-dependent ones.

The overall syngas combustion reaction is

$$H_2 + CO + O_2 \rightarrow H_2O + CO_2$$

where N_2 only serves as a collision partner. We assume a stoichiometric mixture of syngas with air in an adiabatic and isochoric environment. As fixed mass density we use $\rho = 0.3$ kg/m^3.

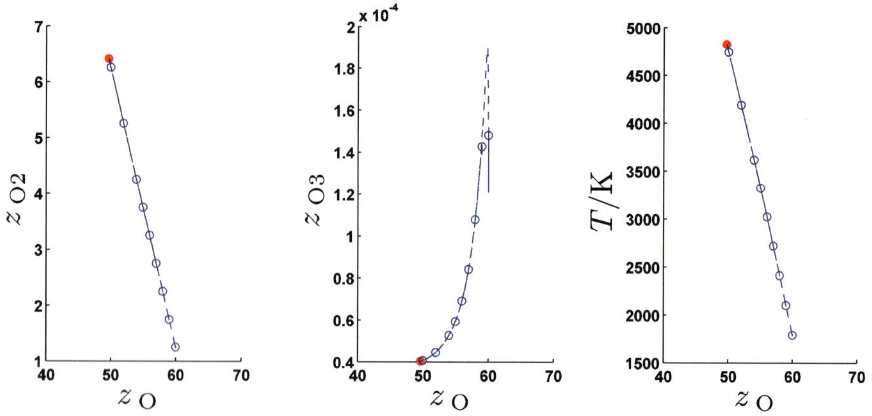

Fig. 5.6 Results (one-dimensional SIM) for the ozone decomposition mechanism modeled within an adiabatic and isochoric environment. As in the adiabatic case the reaction is faster, the integration horizon is reduced to $t_f - t_0 = 10^{-8}$ s

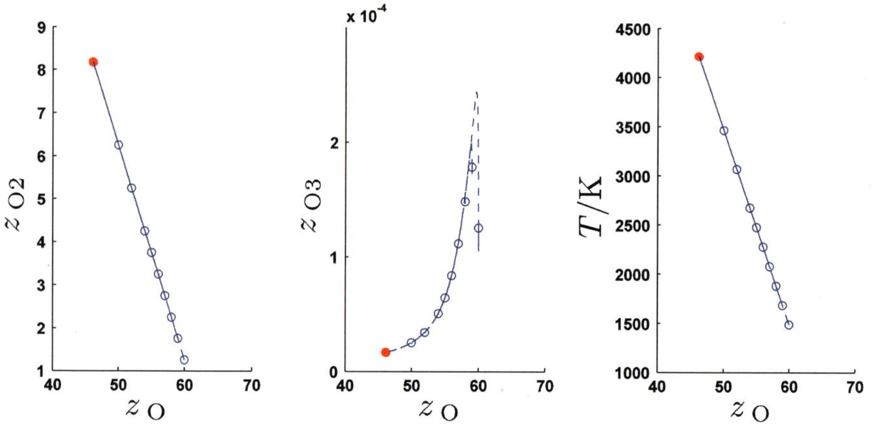

Fig. 5.7 Results (one-dimensional SIM) for the ozone decomposition mechanism modeled within an adiabatic and isobaric environment. Again the integration horizon is $t_f - t_0 = 10^{-8}$ s

We assume a ratio of n_{H2}:$n_{CO} = 1$:1, and a ratio of n_{O2}:$n_{N2} = 1$:3.76 similar to air. This leads to an unburned mixture of $z_{CO} = z_{H2} = 5.973$ mol/kg and $z_{N2} = 22.46$ mol/kg. The specific internal energy of this mixture at a temperature of $T = 1{,}000$ K is used as a fixed specific internal energy for SIM computation.

Results for the solution of the optimization problem are shown in Fig. 5.8. The model has eight degrees of freedom. We aim to compute an approximation of a two-dimensional manifold near equilibrium. We chose the products of the overall reaction as parameterizing variables, namely z_{H2O} and z_{CO2}. The system has much more time scales involved in comparison to the examples before. The computational effort is much higher. In case of an application of the method within a CFD application, a higher dimensional parameterization would certainly be desirable.

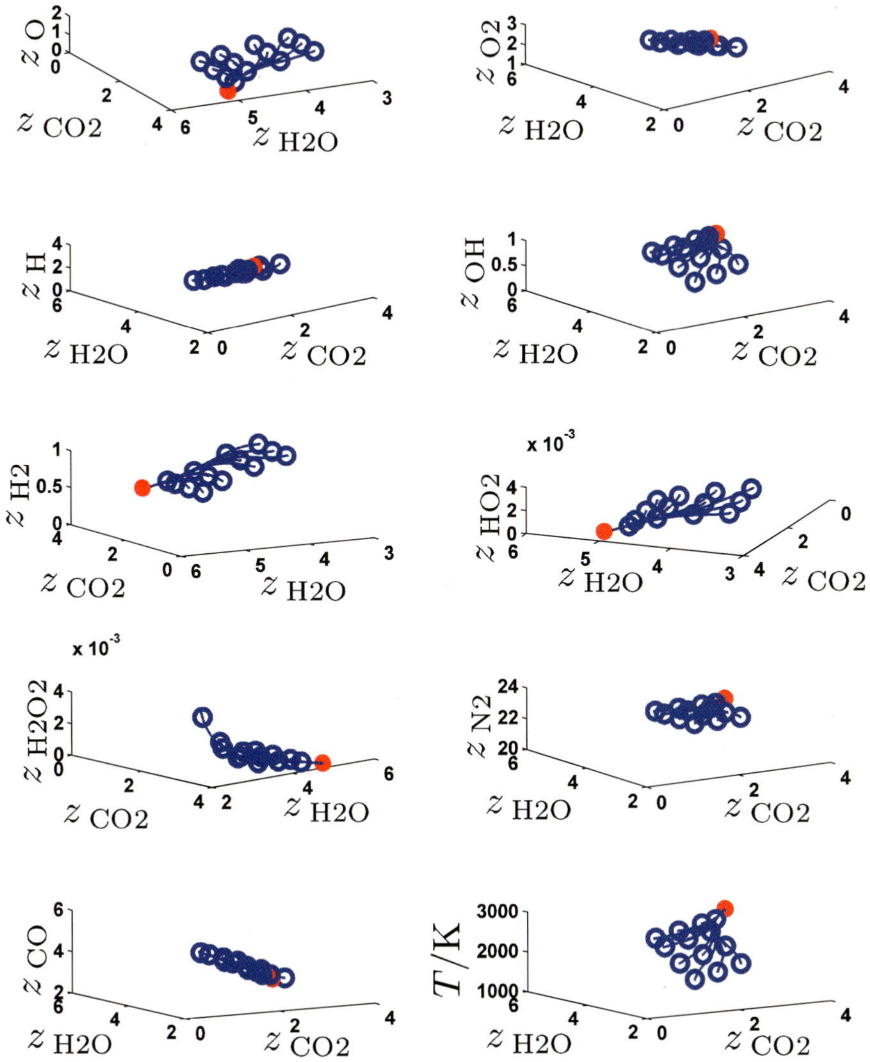

Fig. 5.8 Plot of the results of an approximation of a two-dimensional SIM in the phase space of the syngas combustion model. The values of z_{H2O} and z_{CO2} are used as reaction progress variables. An integration horizon of $t_{\mathrm{f}} - t_0 = 10^{-7}$ s is chosen

5.6 Conclusion

We have implemented a code for model reduction of combustion processes. The model reduction method is based on a suitable criterion for the minimization of a relaxation of chemical forces along reaction trajectories. It can be shown that there always exists a solution of the optimization problem which defines the approximation of the slow manifold. Furthermore, it can be shown that the method identifies the exact slow manifold in case of a linear system and the Davis–Skodje test model. The method has been successfully applied to many example problems and models which arise in combustion chemistry.

Acknowledgements The authors acknowledge the financial support from the German Research Council (DFG) through the project B2 within SFB 568.
The authors wish to thank the late Jürgen Warnatz (IWR, Heidelberg) for providing professional mentoring for combustion research.

References

Project-Related Publications

1. Lebiedz, D.: Computing minimal entropy production trajectories: an approach to model reduction in chemical kinetics. J. Chem. Phys. **120**(15), 6890–6897 (2004)
2. Lebiedz, D.: Entropy-related extremum principles for model reduction of dynamical systems. Entropy **12**(4), 706–719 (2010)
3. Lebiedz, D., Reinhardt, V., Kammerer, J.: Novel trajectory based concepts for model and complexity reduction in (bio)chemical kinetics. In: Gorban, A.N., Kazantzis, N., Kevrekidis, I.G., Theodoropoulos, C. (eds.) Model Reduction and Coarse-Graining Approaches for Multi-scale Phenomena, pp. 343–364. Springer, Berlin (2006)
4. Lebiedz, D., Reinhardt, V., Siehr, J.: Minimal curvature trajectories: Riemannian geometry concepts for slow manifold computation in chemical kinetics. J. Comput. Phys. **229**(18), 6512–6533 (2010)
5. Lebiedz, D., Reinhardt, V., Siehr, J., Unger, J.: Geometric criteria for model reduction in chemical kinetics via optimization of trajectories. In: Gorban, A.N., Roose, D. (eds.) Coping with Complexity: Model Reduction and Data Analysis, pp. 241–252. Springer, Dordrecht (2010)
6. Lebiedz, D., Siehr, J., Unger, J.: A variational principle for computing slow invariant manifolds in dissipative dynamical systems. SIAM J. Sci. Comput. **33**(2), 703–720 (2011)
7. Reinhardt, V., Winckler, M., Lebiedz, D.: Approximation of slow attracting manifolds in chemical kinetics by trajectory-based optimization approaches. J. Phys. Chem. A **112**(8), 1712–1718 (2008)
8. Siehr, J., Lebiedz, D.: An optimization approach to kinetic model reduction for combustion chemistry. Submitted (2012)

Other Publications

9. Amestoy, P.R., Duff, I.S., Koster, J., L'Excellent, J.Y.: A fully asynchronous multifrontal solver using distributed dynamic scheduling. SIAM J. Matrix Anal. A **23**(1), 15–41 (2001)
10. Ascher, U., Petzold, L.: Computer Methods for Ordinary Differential Equations and Differential Algebraic Equations. SIAM, Philadelphia (1998)
11. Bell, B.M.: Automatic differentiation software CppAD (2010). URL http://www.coinor.org/CppAD/

12. Bell, B.M., Burke, J.V.: Algorithmic differentiation of implicit functions and optimal values. In: Bischof, C.H., Bücker, H.M., Hovland, P.D., Naumann, U., Utke, J. (eds.) Advances in Automatic Differentiation, pp. 67–77. Springer, Dordrecht (2008)
13. Burcat, A., Ruscic, B.: Third millennium ideal gas and condensed phase thermochemical database for combustion with updates from active thermochemical tables. Technical report, Argonne National Laboratory (2005)
14. Bykov, V., Maas, U.: Reaction-diffusion manifolds and global quasi-linearization: two complementary methods for mechanism reduction. Open Thermodyn. J. **4**, 92–100 (2010). doi:10.2174/1874396X01004010092
15. Chiavazzo, E., Gorban, A.N., Karlin, I.V.: Comparison of invariant manifolds for model reduction in chemical kinetics. Commun. Comput. Phys. **2**(5), 964–992 (2007)
16. Chiavazzo, E., Karlin, I.V., Gorban, A.N., Boulouchos, K.: Combustion simulation via lattice Boltzmann and reduced chemical kinetics. J. Stat. Mech. Theory Exp. **2009**(6), P06,013 (2009). doi:10.1088/1742-5468/2009/06/P06013
17. Curtiss, C.F., Hirschfelder, J.O.: Integration of stiff equations. Proc. Natl. Acad. Sci. USA **38**, 235–243 (1952)
18. Davis, M.J., Skodje, R.T.: Geometric investigation of low-dimensional manifolds in systems approaching equilibrium. J. Chem. Phys. **111**, 859–874 (1999)
19. Delhaye, S., Somers, L.M.T., van Oijen, J.A., de Goey, L.P.H.: Incorporating unsteady flow-effects in flamelet-generated manifolds. Combust. Flame **155**(1–2), 133–144 (2008)
20. Forsgren, A., Gill, P.E., Wright, M.H.: Interior methods for nonlinear optimization. SIAM Rev. **44**(4), 525–597 (2002)
21. Fraser, S.J.: The steady state and equilibrium approximations: a geometrical picture. J. Chem. Phys. **88**, 4732–4738 (1988)
22. Gilbert, R., Luther, K., Troe, J.: Theory of thermal unimolecular reactions in the fall-off range. II. Weak collision rate constants. Ber. Bunsenges Phys. Chem. **87**, 169–177 (1983)
23. Golub, G.H., Loan, C.F.V.: Matrix Computations. Johns Hopkins University Press, Baltimore (1996)
24. Gorban, A.N., Karlin, I.V.: Invariant Manifolds for Physical and Chemical Kinetics. Lecture Notes in Physics, vol. 660. Springer, Berlin/Heidelberg/New York (2005)
25. Gorban, A.N., Karlin, I.V., Zinovyev, A.: Invariant grids: method of complexity reduction in reaction networks. Complexus **2**, 110–127 (2005). doi:10.1159/000093684
26. Gorban, A.N., Karlin, I.V., Zinovyev, A.Y.: Constructive methods of invariant manifolds for kinetic problems. Phys. Rep. **396**, 197–403 (2004). doi:10.1016/j.physrep. 2004.03.006
27. HSL: A collection of Fortran codes for large-scale scientific computation (2007). URL http://www.hsl.rl.ac.uk
28. Kaper, H.G., Kaper, T.J.: Asymptotic analysis of two reduction methods for systems of chemical reactions. Phys. D **165**, 66–93 (2002)
29. Kee, R.J., Coltrin, M.E., Glarborg, P.: Chemically Reacting Flow: Theory and Practice. Wiley-Interscience, Hoboken (2003)
30. Lam, S.: Recent advances in the aerospace sciences. In: Casci, C. (ed.) Singular Perturbation for Stiff Equations Using Numerical Methods, pp. 3–20. Plenum Press, New York/London (1985)
31. Lam, S.H., Goussis, D.A.: The CSP method for simplifying kinetics. Int. J. Chem. Kinet. **26**, 461–486 (1994)
32. Maas, U., Pope, S.B.: Simplifying chemical kinetics: intrinsic low-dimensional manifolds in composition space. Combust. Flame **88**, 239–264 (1992)
33. Maas, U., Warnatz, J.: Simulation of thermal ignition processes in two-dimensional geometries. Z. Phys. Chem. N. F. **161**, 61–81 (1989)
34. Nguyen, A.H., Fraser, S.J.: Geometrical picture of reaction in enzyme kinetics. J. Chem. Phys. **91**, 186–193 (1989)
35. Nocedal, J., Wright, S.J.: Numerical Optimization. Springer Series in Operations Research and Financial Engineering, 2nd edn. Springer, New York (2006)

36. Powell, M.J.D.: A fast algorithm for nonlinearly constrained optimization calculations. In: Dold, A., Eckmann, B. (eds.) Numerical Analysis. Lecture Notes in Mathematics, vol. 630, pp. 144–157. Springer, Berlin (1978)
37. Ren, Z., Pope, S.: Species reconstruction using pre-image curves. Proc. Combust. Inst. **30**, 1293–1300 (2005). doi:10.1016/j.proci.2004.07.017
38. Ren, Z., Pope, S.B.: The use of slow manifolds in reactive flows. Combust. Flame **147**, 243–261 (2006). doi:10.1016/j.combustflame.2006.09.002
39. Ren, Z., Pope, S.B., Vladimirsky, A., Guckenheimer, J.M.: The invariant constrained equilibrium edge preimage curve method for the dimension reduction of chemical kinetics. J. Chem. Phys. **124**, 114,111 (2006)
40. Rhodes, C., Morari, M., Wiggins, S.: Identification of low order manifolds: validating the algorithm of Maas and Pope. Chaos **9**(1), 108–123 (1999)
41. Singh, S., Powers, J.M., Paolucci, S.: On slow manifolds of chemically reactive systems. J. Chem. Phys. **117**(4), 1482–1496 (2002). doi:10.1063/1.1485959
42. Skanda, D.: Robust optimal experimental design for model descrimination of kinetic ODE systems. PhD thesis, University of Freiburg. Submitted (2012)
43. Skanda, D., Lebiedz, D.: An optimal experimental design approach to model discrimination in dynamic biochemical systems. Bioinformatics **26**(7), 939–945 (2010). doi:10.1093/bioinformatics/btq074
44. Smith, G.P., Golden, D.M., Frenklach, M., Moriarty, N.W., Eiteneer, B., Goldenberg, M., Bowman, C.T., Hanson, R.K., Song, S., Gardiner Jr., W.C., Lissianski, V.V., Qin, Z.: GRIMech 3.0. Published online (1999). URL http://www.me.berkeley.edu/grimech
45. Troe, J.: Theory of thermal unimolecular reactions in the fall-off range. I. Strong collision rate constants. Ber. Bunsenges Phys. Chem. **87**, 161–169 (1983)
46. Wächter, A., Biegler, L.T.: On the implementation of a primal-dual interior point filter line search algorithm for large-scale nonlinear programming. Math. Program. **106**(1), 25–57 (2006)
47. Warnatz, J., Maas, U., Dibble, R.W.: Combustion: Physical and Chemical Fundamentals, Modeling and Simulation, Experiments, Pollutant Formation, 4th edn. Springer, Berlin (2006)
48. Zagaris, A., Kaper, H.G., Kaper, T.J.: Analysis of the computational singular perturbation reduction method for chemical kinetics. J. Nonlinear Sci. **14**(1), 59–91 (2004). doi:10.1007/s00332-003-0582-9

Chapter 6
Large Eddy Simulation of Combustion Systems at Gas Turbine Conditions

J. Janicka, J. Kuehne, G. Kuenne, and A. Ketelheun

Abstract Three different aspects of Large Eddy Simulation (LES) of combustion processes are covered in this chapter. All three are based on using tabulated chemistry models to cover the chemical reactions occurring in gas turbines. The chosen approaches are all based on flamelet models. One part of the work deals with the investigation of subgrid scale models using Transported Eulerian Monte Carlo Probability Density Function (PDF) methods. The chemistry is dealt with using non-premixed flamelets and premixed Flamelet Generated Manifolds (FGM). The second aspect covered is the extension of the FGM method towards premixed flames where unresolved subgrid flame structures need to be handled. Therefore, the FGM approach was coupled with the Artificially Thickened Flames (ATF) model. The third aspect of combustion LES discussed here deals with the inclusion of more detailed reaction kinetics in the FGM approach in order to better predict minor species like nitric oxides or carbon monoxide, which are important development goals in today's gas turbine industry. As all three aspects discussed in this chapter are located on the smallest scales in a combustion system, either the flow or flame subgrid structures, they are closely related to each other.

Keywords Large Eddy Simulation • Turbulent Combustion • Premixed Flames • Monte Carlo PDF • Pollutant Prediction

J. Janicka • J. Kuehne • G. Kuenne • A. Ketelheun (✉)
Department of Mechanical and Processing Engineering, Institute for Energy and Power Plant Technology, Technische Universität Darmstadt, Petersenstr. 30, 64287 Darmstadt, Germany
e-mail: ketelheun@ekt.tu-darmstadt.de

J. Janicka
Center of Smart Interfaces, Technische Universität Darmstadt, Petersenstr. 32, 64287 Darmstadt, Germany

J. Janicka et al., *Flow and Combustion in Advanced Gas Turbine Combustors*, Fluid Mechanics and Its Applications 102, DOI 10.1007/978-94-007-5320-4_6,

6.1 Introduction

In today's gas turbine development two issues are of main importance. First, the reduction of emissions is very important, mostly due to regulations becoming stricter. Second, as fuel prices rise and resources are getting less, fuel efficiency is another aspect requested by the customers. Therefore, a trend towards lean premixed combustion can be observed. This still requires quite some research in the basic modeling and understanding of turbulent combustion processes. This chapter provides an overview over three current issues regarding turbulent combustion. At first, as a step towards the general understanding of combustion principles and models, Filtered Density Functions and Transported Probability Density Function Models are investigated. Second, specialized modeling for premixed flames in the context of Large Eddy Simulation (LES) will be investigated. The model employed uses Artificially Thickened Flames (ATF) to thicken the thin, non-resolved flame front. The third aspect covered in this chapter will be the investigation of advanced models for the prediction of typical gas turbine emissions like nitrogen oxides (NO_x) and carbon monoxide (CO) will be presented. These three subjects will be presented applied to configurations with flow features typical for gas turbines, like swirl and bluff bodies to induce recirculation zones.

The chapter is organized as follows. The subsequent section recalls the basics of tabulated chemistry approaches, especially the Flamelet Generated Manifolds (FGM) approach used in this work. Thereafter, the transported Euler Monte Carlo PDF method will be briefly described and its application to a non-premixed bluff-body stabilized flame will be shown. Section 6.4 presents the ATF modeling in combination with FGM, including the application to a premixed swirl flame. Section 6.5 covers the modeling of pollutant formation in the FGM context, as applied to the flame from Sect. 6.3. In each section, a motivation and background of the respective modeling will be given. Section 6.6 is devoted to conclusions.

6.2 Tabulated Chemistry Approaches

Fuels used in stationary and airborne gas turbines typically include hydrocarbons of different order. These react via a large number of intermediate steps towards carbon dioxide (CO_2) and water (H_2O) as the main reaction products. A number of – mostly unwanted – reaction byproducts may occur. However, even the reaction path of the simplest hydrocarbon, methane (CH_4), includes dozens of species. A typical example for a detailed reaction mechanism used to describe the combustion of methane is the GRI3.0 mechanism [6]. It consists of 53 species interacting in 325 elementary reactions. For all these species a transport equation would have to be solved during the LES. All of them contain a chemical source term and are stiffly coupled with each other. This reduces the possible time step size and leads to very high computational costs.

The other extreme are reduced reaction mechanisms. Here, only a small number of species is considered, typically the main reactants and products and some major by-products. Although this approach strongly reduces the computational cost, it does not give sufficient information on the minor species, such as the pollutants.

In order to combine the advantages of both worlds, the detailed information on the minor species and the low computational cost, the usage of tabulated chemistry is a rather common approach. Several models exist within this framework, such as the flamelet-based models, following an approach by Peters [7], like non-premixed flamelets or Flamelet Generated Manifolds (FGM) [8] or Flame Prolongation of ILDM (FPI) [9]. As this project was based on flamelet models, only these will be described in more detail in the following.

The basic idea of the flamelet models is that a turbulent three-dimensional flame can be described by means of one-dimensional laminar flame segments, the flamelets, whose structure is not affected by the turbulence. The flamelets are parameterized by a small number of control variables (typically one or two) and computed in a pre-processing step. In this pre-processing step, the detailed chemistry is used to compute the composition. Via a coordinate transformation the species, density, temperature, etc. can be stored depending on the control variables. Only for those a transport equation needs to be solved during the LES, all other variables are stored in a table as functions depending on them.

The controlling variables are typically the mixture fraction, often defined as a combination of element mass fractions, describing the ratio of fuel and air or the mixing of the reactants, and, in case of FGM or FPI, a reaction progress variable, describing the development of the chemical reaction from the unburnt mixture to burnt products. This progress variable is typically a product mass fraction, a necessary constraint for its choice is the strict monotonicity throughout the flame. For hydrocarbons, such a typical product is CO_2, but also linear combinations with CO or H2O are conceivable.

The filtered transport equations that need to be solved additionally to the Navier-Stokes equations in the LES context for mixture fraction ξ and progress variable Y read as follows:

$$\frac{\partial}{\partial t}\left(\bar{\rho}\tilde{\xi}\right) + \frac{\partial}{\partial x_j}\left(\bar{\rho}\tilde{u}_j\tilde{\xi}\right) = \frac{\partial}{\partial x_j}\left(\bar{\rho}\left(\tilde{D}_\xi \frac{\partial \tilde{\xi}}{\partial x_j} + J^{sgs}_{\xi,j}\right)\right)$$

$$\frac{\partial}{\partial t}\left(\bar{\rho}\tilde{Y}\right) + \frac{\partial}{\partial x_j}\left(\bar{\rho}\tilde{u}_j\tilde{Y}\right) = \frac{\partial}{\partial x_j}\left(\bar{\rho}\left(\tilde{D}_Y \frac{\partial \tilde{Y}}{\partial x_j} + J^{sgs}_{Y,j}\right)\right) + \bar{\dot{\omega}}_Y$$

To include interaction of turbulence and chemistry on the non-resolved subgrid scales different approaches are possible. At first, the most common approach is the use of so-called Presumed Probability Density Function models. In that context, a subgrid distribution for mixture fraction and sometimes also the progress variable is assumed. The data in the chemistry table is then pre-integrated before the computation to account for the subgrid variations of the mixture fraction, for example. A typical example for the mixture fraction is the use of a β-PDF [10] depending on mean and variance of the mixture fraction, which will be explained in more detail later on. In the context of this work, the reaction progress variable is modeled using a δ-PDF, only depending on the mean value within a computational cell. For reacting scalars like the progress variable not many presumed shapes for the PDF are known. Furthermore, statistical independence of the normalized progress variable and the mixture fraction are assumed, following Landenfeld et al. [11]. The second possibility to account for turbulence chemistry interaction is the transport of the PDF, where no shape needs to be assumed beforehand.

However, additional modeling is required to use FGM tabulated chemistry in the context of LES as even the main species act on subgrid dimensions, as the flames are thin compared to typical grid sizes. Within this project, this modeling includes (1) PDF modeling, (2) Artificially Thickened Flames (ATF) modeling for premixed flames which are even thinner than non-premixed flames, and (3) additional models for pollutant prediction, in this case NO_x and CO, as they cannot be sufficiently described by one single progress variable.

6.3 Transported Euler Monte Carlo PDF

Looking at one computational cell, one observes a certain value of the mixture fraction. However, this information is not complete, as the distribution of the mixture within this cell is not resolved and therefore unknown. As an example, a cell with a mean mixture fraction of 0.5 could either have a constant distribution with zero variance or it could be filled half and half with mixture fractions 1 and 0. This is illustrated in Fig. 6.1.

It shows the possible mixtures on the left, and the respective subgrid distributions in the middle. On the right hand side of Fig. 6.1, a smoother and hence more realistic representation of the subgrid distribution is shown.

As chemical reactions tend to be non-linear, these different subgrid distributions have a strong influence on whether reaction occurs or not. The mean value of the mixture fraction is not sufficient to predict the chemical state within a cell, as the following inequality applies:

$$\phi\left(\tilde{\xi}\right) \neq \tilde{\phi}\left(\xi\right)$$

Therefore, the subgrid distribution of the mixture fraction is needed and is included via an integration of the PDF, following

$$\tilde{\phi}\left(\xi\right) = \int_{0}^{1} \phi\left(\xi\right) P\left(\xi\right) d\xi$$

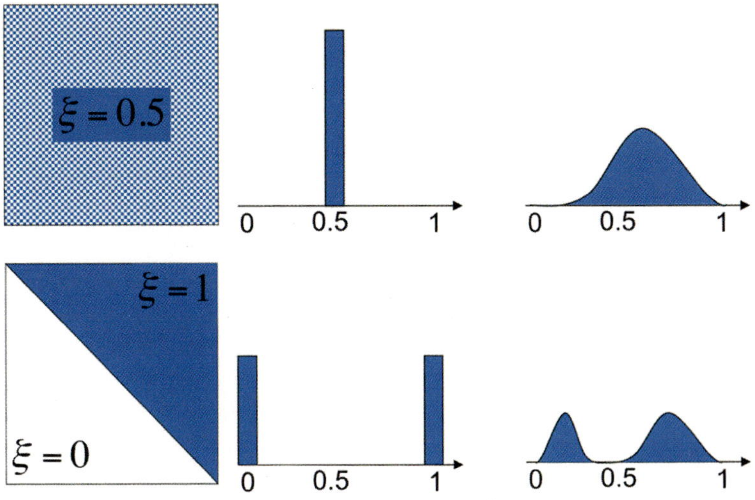

Fig. 6.1 Subgrid distribution of the mixture fraction

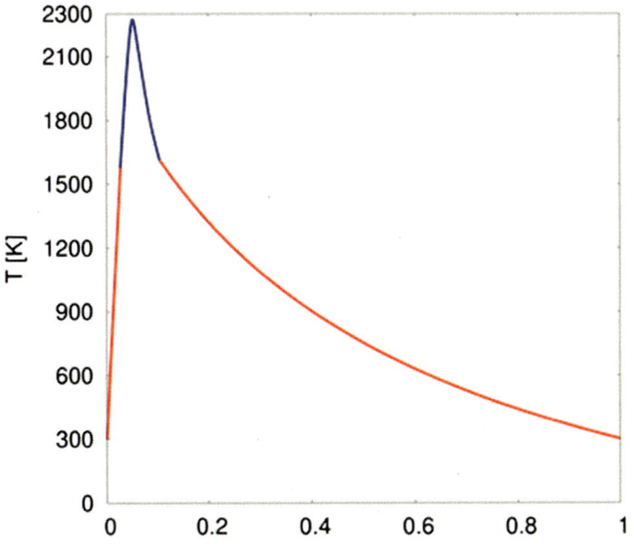

Fig. 6.2 Equilibrium temperature depending on mixture fraction

The first term in the integral, $\phi(\xi)$, covers the reaction kinetics, whereas the second term, $P(\xi)$, describes the PDF distribution within the computational cell. Figure 6.2 further illustrates the connection between subgrid distribution and the occurrence of chemical reactions. The figure shows the adiabatic flame temperature of methane/hydrogen combustion with air depending on the mixture fraction. The blue region denotes the flammable region, the red lines show the extrapolation towards pure fuel and oxidizer, respectively. If the distribution shown in the top right of Fig. 6.1 is present, one ends up in the non-burning region, with the distribution shown in the bottom right, one has part of the cell consisting of a flammable mixture, allowing for chemical reactions. Which shape of the subgrid PDF occurs depends on the variance of the mixture fraction. If it is small, one obtains a rather homogeneous distribution; if it becomes larger, one tends towards unmixed states.

As mentioned above, the standard approach now computes a presumed shape of the PDF, depending on mixture fraction and variance, e.g. assumes a β-shape, as given in the equation below

$$P\left(\xi;\tilde{\xi},\tilde{\xi}''^{2}\right)=\frac{\xi^{\alpha-1}(1-\xi)^{1-\beta}}{\int_0^1\xi^{\alpha-1}(1-\xi)^{1-\beta d\xi}}$$

with the coefficients α and β themselves depending on mean and variance. The disadvantage of this model is that the real shape of the PDF can only be approximated and not all possible subgrid distributions can be represented by this equation. For example, no separate double or even triple peaks are possible.

Therefore, a transport equation for the distribution function $P(\xi)$ can be derived.

$$\frac{\partial}{\partial t}\left(\bar{\rho}\tilde{P}\right)+\frac{\partial}{\partial x_i}\left(\bar{\rho}\tilde{P}u_i\right)+\frac{\partial}{\partial\psi_\alpha}\left(\bar{\rho}\tilde{P}S_\alpha\right)=\frac{\partial}{\partial x_i}\left[\bar{\rho}\left(\tilde{D}+D_t\right)\frac{\partial\tilde{P}}{\partial x_i}\right]+E\left(\psi,x,t\right)$$

This allows now for any possible PDF shape, but a significantly higher computational cost. It is very useful for either improving simulation results where the β-PDF does not suffice or for the assessment of existing models, to evaluate the realistic PDF shape locally. A solution of the above equation using the Finite Volume approach as for all other equations is too expensive. Therefore, it was discretized using stochastic algorithms. The distribution was represented by a discrete number of particles per computational cell. It was found for this work that 50 particles per cell were adequate to represent the distribution correctly and to keep the computational cost reasonable. By the transport equation given above, the particles exchange their properties, as symbolically shown in Fig. 6.3.

From the thus obtained information on the distribution, an ensemble average per cell can be used to compute mean and variance of the mixture fraction in each computational cell. These can then be used to either access the chemistry table or to compare with the shape of a β-PDF having the same mean and variance, as shown in Fig. 6.4. The thin lines below the distribution denote the mixture fraction of the individual particles, the solid line the PDF reconstructed from them, and the dashed line represents the β-PDF having same mean and variance.

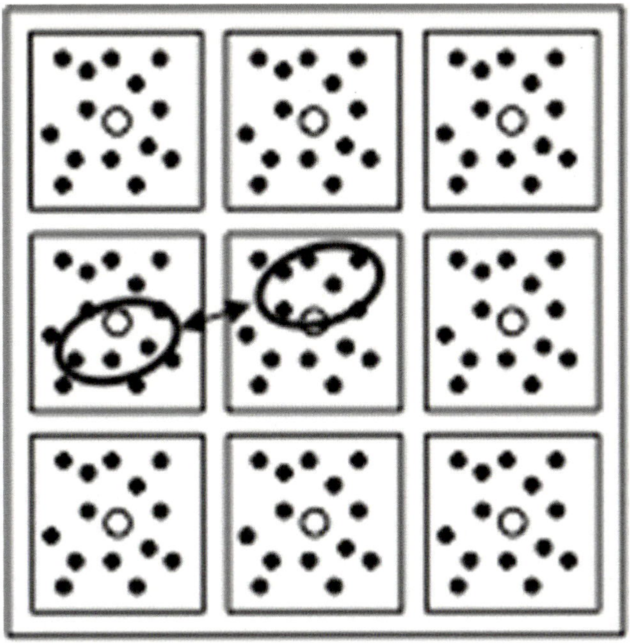

Fig. 6.3 Exchange of PDF information between cells

Fig. 6.4 Mixture fraction PDF for one computational cell

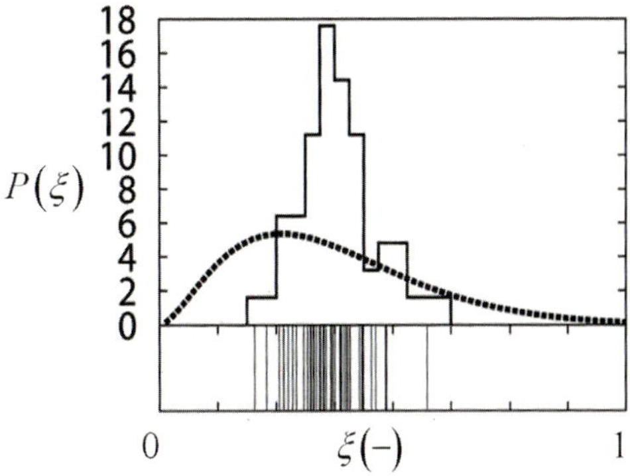

6.3.1 *Application*

The modeling described above was applied to a turbulent non-premixed bluff body stabilized flame of the Sydney flame series, developed by the group of Masri at the University of Sydney. These flames are target flames of TNF workshop, they are well investigated experimentally and provide a large amount of validation data for numerical simulations. The flame considered here consists of a rotationally symmetric bluff body with a diameter of $D_{bb} = 50$ mm. The fuel is a 1:1 by volume mixture of natural gas (CNG) and hydrogen (H2) at ambient conditions ($p = 1$ bar, $T = 300$ K). It enters the combustion chamber through a central pipe with $D_j = 3.6$ mm with a velocity of $u_j = 108$ m/s and a Reynolds number of $Re_j = 14{,}600$. The stoichiometric mixture fraction of the fuel is $\xi_{st} = 0.05$. The bluff body is located in a square duct of width 150 mm with co-flowing air at $u_{air} = 35$ m/s and ambient conditions as well. Measurements of velocities and scalars were taken by Dally and coworkers [12, 13]. The flame is stabilized behind the bluff body by a recirculation zone consisting of two counter rotating vortices. The inner vortex transports fuel to the outside, while the outer vortex entrains air and hot gases into the recirculation zone. The fuel is preheated in the recirculation zone which enhances mixing and leads to a nearly uniform temperature distribution close to the bluff body.

The configuration is shown in (Fig. 6.5), together with an instantaneous representation of the hydroxyl (OH) mass fraction to illustrate the position and shape of the flame. The simulations of the HM1e configuration were performed on a numerical domain with the dimensions 0.25 m in axial, 0.22 m in radial and 2π in azimuthal direction. The computational grid consists of approximately 1.2 million control volumes, which are given by 400 nodes for the axial resolution, 94 for radial and 32 for the tangential direction. The radial distribution of the nodes is equidistant for the region from the centerline to the outer edge of the bluff body and non-equidistant beyond the bluff body. In order not to start the computational domain directly at the upper edge of the bluff body, which can cause instabilities of the simulations, parts of the bluff body are modeled using an immersed boundary technique. The modeled bluff body penetrates the domain by 8.75 mm, which is sufficient to observe the characteristic flow patterns around the bluff body and additionally increases the stability of the simulation. The chemistry is represented employing a steady flamelet approach with a constant strain rate of $a = 10\text{s}^{-1}$.

The chemistry table is pre-calculated using the one-dimensional flamelet code CHEM1D [14] and additionally pre-integrated applying a presumed β-shaped PDF approach. The filtered mixture fraction space is resolved with 200 and the normalized sub-grid variance with 100 equally distributed sampling points, respectively. A detailed analysis of this flame regarding the Monte Carlo PDF method can be found in the work of Kuehne [1, 2]. A short excerpt from that work is shown here, showing the evaluation of the β-PDF for the mixture fraction.

Three different positions above the bluff body at 45, 65 and 90 mm radially located in the flame front are shown. Figure 6.6 shows the comparison between the individual particles' mixture fraction, the β-PDF and the actual PDF calculated from the particles. As can be seen in the two lowest plots in Fig. 6.6, at this position a very good agreement between the PDF shapes can be observed. The shortcomings of the β-PDF can be found in the two higher positions. At 65 mm, the two central plots in Fig. 6.6, the β-PDF overestimates the amount of air. Furthermore, at 90 mm above the bluff body, the double peaks of the PDF cannot be represented by the β-PDF, conceptually.

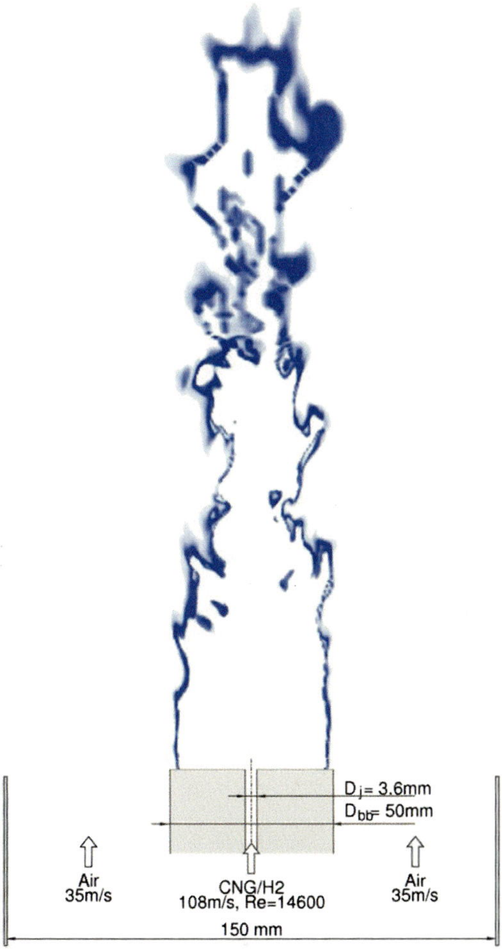

Fig. 6.5 Sketch of the bluff body configuration

After having investigated the PDF of the mixture fraction, dependent variables which are important to value the combustion process were studied. Two of those being of special interest are the temperature and the mass fraction of the hydroxyl (OH) radical. The results of their radial distributions of the time averaged mean values are shown in Fig. 6.7. Three different heights above the bluff body are shown, 65, 90 and 120 mm. The solid line denotes the standard finite volume approach, the dashed line the Monte Carlo PDF method, the symbols show the experimental data. One can observe that the temperature only slightly improves by the more complicated Monte Carlo PDF approach, wheras more significant changes in the OH mass fraction can be seen. The amount of OH produced in the flame and the shape and position of the profiles match the experimental data better. This can be explained by the lack of the β-PDF to correctly predict the subgrid distribution of the mixture fraction at all positions.

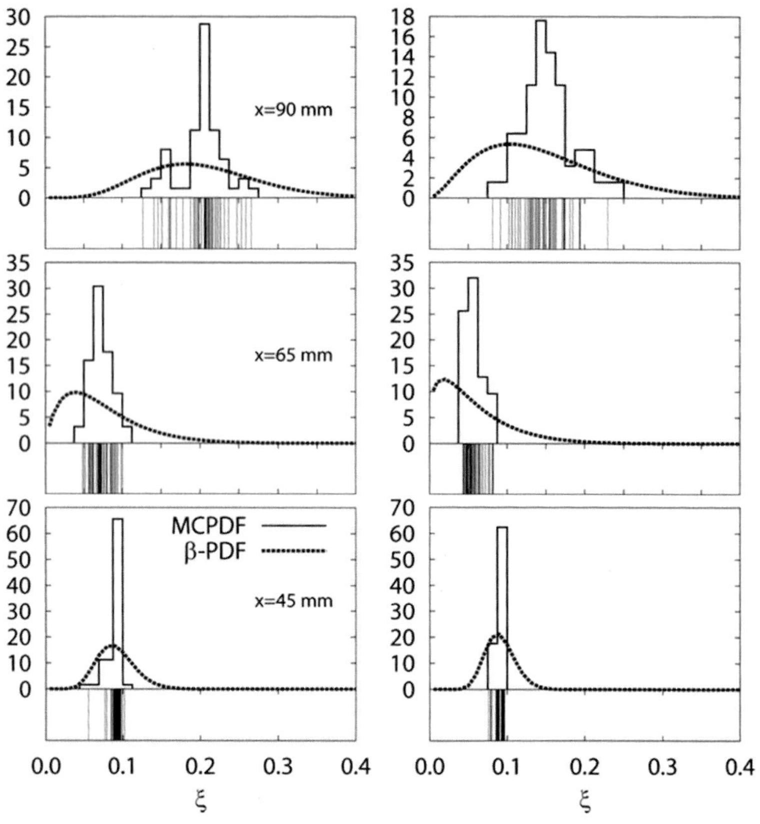

Fig. 6.6 PDF shapes at different heights above the bluff body

6.4 Premixed Flame Modeling

In the modeling of premixed flames the correct prediction of the flame propagation is a key issue. As a flammable mixture exists at many positions in the computational domain, not only where fuel and oxidizer meet with the correct amount, the correct flame speed is essential to predict the flame position and shape.

The flame velocity s_l can be deduced directly from the species transport equation, for example based on the source term and the mass fraction of CO2 in the burnt gases and the density in the fresh mixture, as shown in the equation below.

$$s_l = \frac{1}{\rho_u y_{CO_2,b}} \int_{-\infty}^{\infty} \dot{\omega}_{CO_2} dx \approx s_{l\Delta} = \frac{1}{\rho_u y_{CO_2,b}} \sum_i \dot{\omega}_{CO_2,i} \Delta_{x,i}$$

The differential value $s_{l\Delta}$ obtained on a numerical grid can be approximated accordingly. Figure 6.8 shows the source term and mass fraction of CO_2 from detailed simulations in CHEM1D [14] and from the LES code FASTEST on

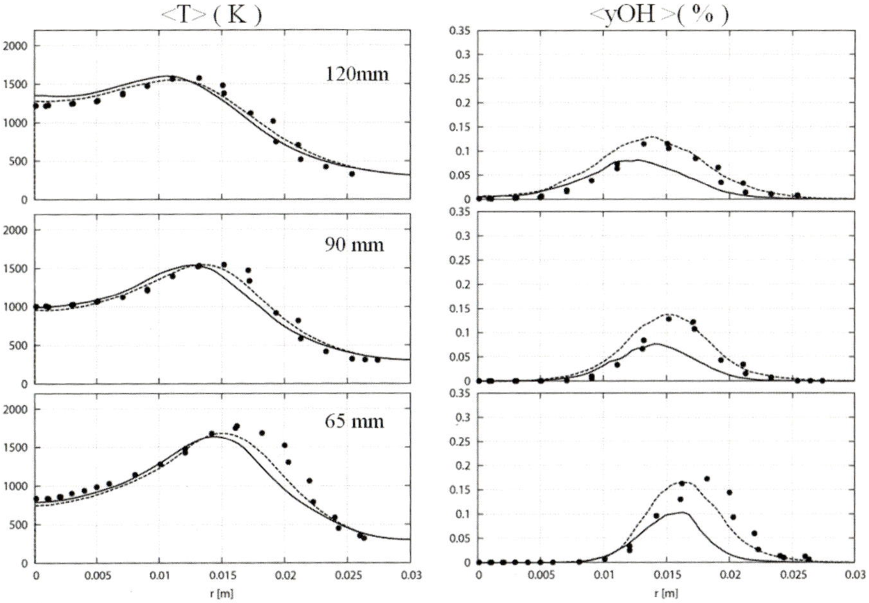

Fig. 6.7 Radial profiles of time averaged temperature and hydroxyl mass fraction

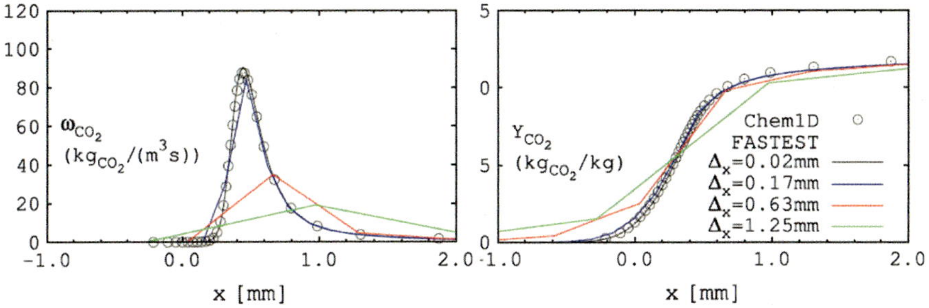

Fig. 6.8 CO_2 source term and mass fraction depending on grid size

different numerical grid sizes. It can be observed that at some point, the grid gets too coarse and the correct shape cannot be reproduced anymore.

If one computes the flame speed according to the equation given above for the different grids, one obtains Fig. 6.9. The simulations were carried out using two different inflow velocities 0.2 and 0.5 m/s, one below, the other one well above the correct flame velocity for the mixture.

For small cell sizes, the correct value of the flame speed is obtained, whereas for typical LES grid sizes, marked with the blue box in Fig. 6.9, the flame speed cannot be predicted correctly and tends towards the inflow velocity. Obviously, additional modeling is required for premixed flames in the LES context. The approach chosen in this work is the combination of Flamelet Generated Manifolds as chemistry model combined with the Artificially Thickened Flames (ATF) model [15, 16] to correctly capture the thin flame front and hence the flame velocity.

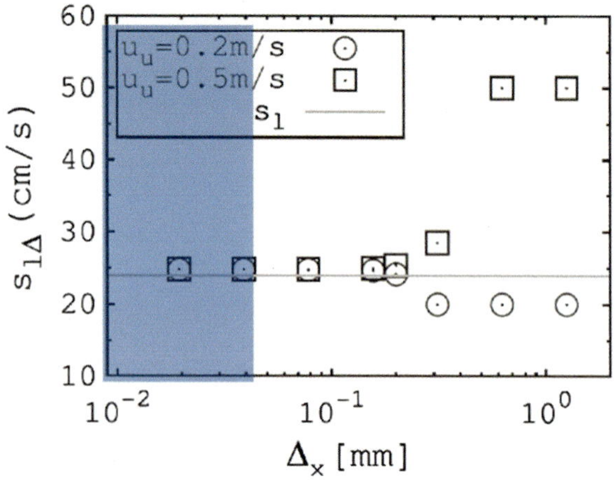

Fig. 6.9 Flame speed depending on grid size

The basic idea behind the ATF model is to thicken the flame front such that it can be resolved on the numerical grid. For that purpose, the diffusion term and the source term in the species transport equation are multiplied respectively divided by a thickening factor F. By the thickening procedure some of the flame wrinkling gets lost, therefore, an additional efficiency function E needs to be introduced to correct the mass burning rate within a flame segment. The modified transport equation then looks as follows:

$$\frac{\partial \rho Y_k}{\partial t} + \frac{\partial}{\partial x_i}(\rho u_i Y_k) = \frac{\partial}{\partial x_i}\left(\rho F \varepsilon D_k \frac{\partial Y_k}{\partial x_i}\right) + \frac{\varepsilon}{F}\dot{\omega}_k$$

As the source term is reduced and the flame is thickened with this procedure, the one-dimensional flame structure finally looks as seen in Fig. 6.10. All further details on the modeling approach can be found in Kuenne et al. [3].

6.4.1 *Application*

The combined FGM/ATF modeling was applied to a swirl stabilized premixed flame, the TECFLAM swirl burner, shown in Fig. 6.11. The methane-air mixture ($T = 300$ K, equivalence ratio $= 0.83$) enters the radial and tangential channels where the swirl is generated. Hereafter, the fuel moves upward again, passes the annulus around the bluff body and enters the unconfined section. Here, the Reynolds number based on the bulk velocity (5 m/s) and the hydraulic diameter ($D_h = 30$ mm) is Re = 10,000. The swirl number, representing the ratio of azimuthal and axial momentum at the nozzle exit, is 0.7. The nozzle is placed concentrically inside a coflow of pure air issuing with 0.5 m/s. The flame, which stabilizes by the recirculation of hot gases above the bluff body, has a thermal power of 30 kW. A block structured grid containing approx. 3.2 million control volumes was used. The

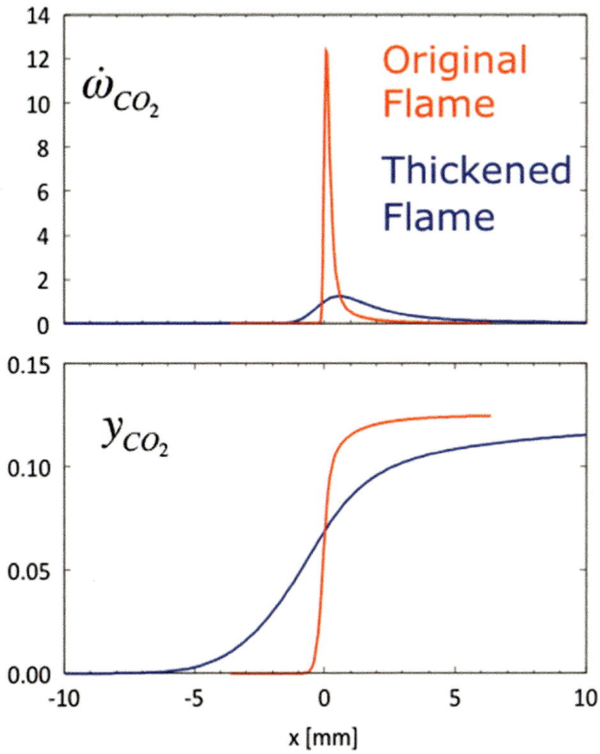

Fig. 6.10 Source term and mass fraction of CO2 with (*blue*) and without (*red*) artificial thickening

Fig. 6.11 Schematic of TECFLAM swirl burner

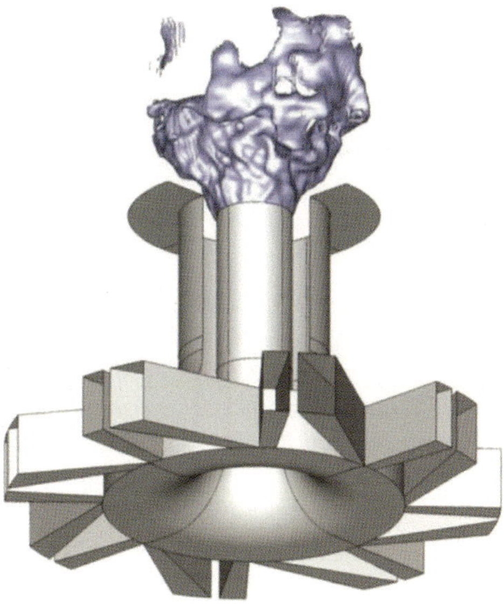

grid was elliptically smoothed to obtain a better orthogonality and is refined towards the near nozzle region, whereas it gets coarser with increasing distance. A spatially constant thickening factor of $F_{max} = 5$ was used.

Figure 6.12 shows the CO_2 mass fraction and temperature temporal mean and fluctuations at different heights above the bluff body. The symbols denote the experimental data by [17, 18], the lines show two different efficiency function models. As can be seen, the simulation performs very well for both CO_2 and temperature. The reduction of the temperature above the bluff body in the lower measuring planes cannot be reproduced by this model, as this is due to a water cooled bluff body and so far, no heat transfer was included in the FGM model.

6.5 Pollutant Prediction

As has been shown before, the main characteristics of a flame, whether premixed or non-premixed can be well described by a single progress variable in the context of FGM modeling. However, the prediction of pollutant formation is still complicated. An example is the slow formation of nitric oxide (NO) whose time scales are much longer than those of the main species. Therefore, its formation does not have a direct relation to the progress variable, which is generally based on main species. With the use of the standard FGM approach, the NO mass fraction is heavily underestimated, the profiles where it is built in a flame cannot be predicted either. Another example of a pollutant important in the development of gas turbines that is difficult to predict is carbon monoxide (CO). If premixed flamelets are used for the generation of the FGM table being then applied to a non-premixed flame in the context of regime-independent modeling, the CO mass fraction can be largely overestimated, due to the different chemistry in rich premixed flames compared to that in rich parts of non-premixed flames. In order to improve the prediction of these two gas turbine-related pollutants, additional modeling in the FGM context was introduced, based on additional transport equations for the respective species.

6.5.1 NO Prediction

As mentioned above, due to the different chemical time scales, NO cannot be well described by the progress variable. Figure 6.13 shows the development of NO and CO_2 along the flamelet. One can well observe how long the NO needs to develop when CO_2 has long reached its equilibrium value. This indicates that an additional transport equation for the NO mass fraction with the source term being tabulated depending on mixture fraction and progress variable and the subgrid variance of the mixture fraction. This equation is given below.

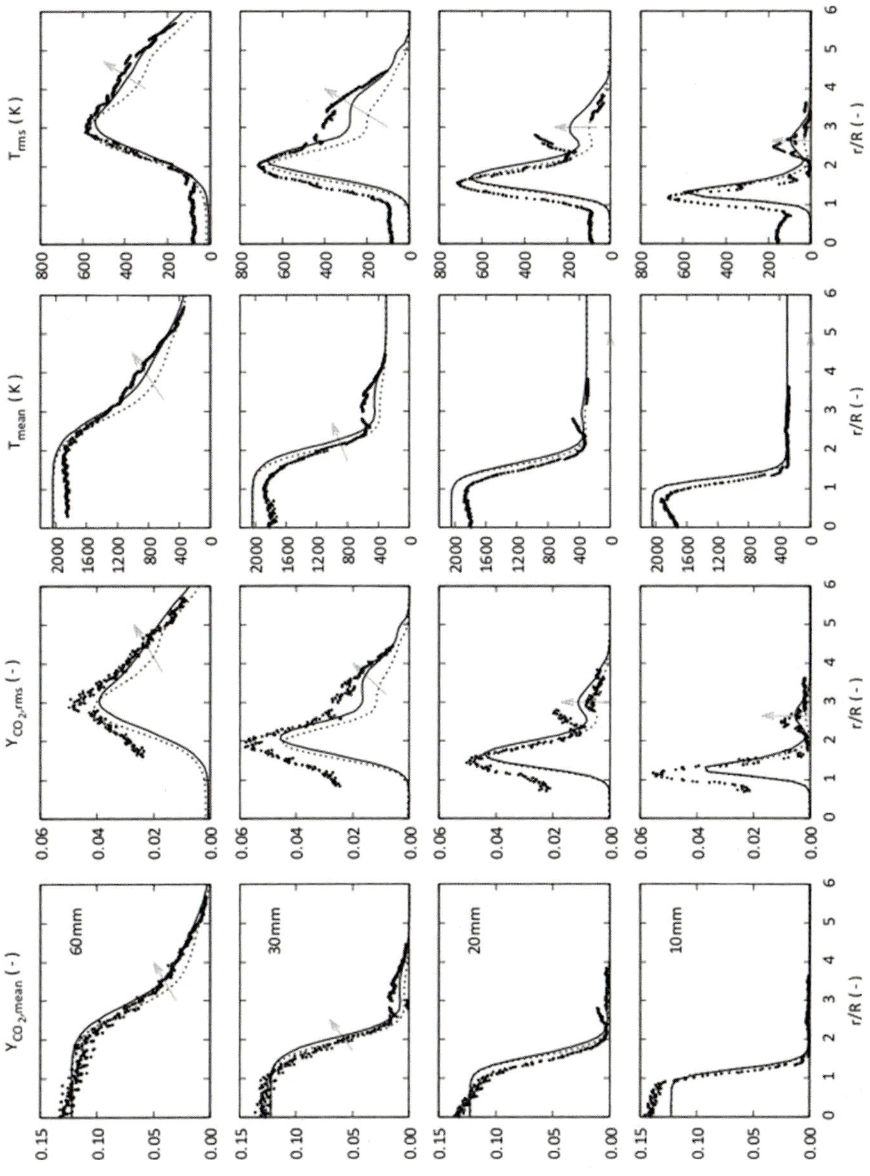

Fig. 6.12 Mean and Fluctuation of CO_2 mass fraction and temperature the TECFLAM swirl burner

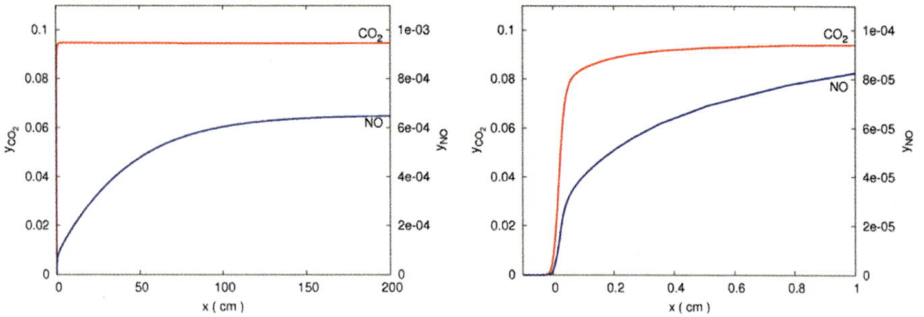

Fig. 6.13 CO_2 (*red*) and NO (*blue*) mass fraction along the flamelet for different ranges in physical space

$$\frac{\partial}{\partial t}\left(\bar{\rho}\tilde{y}_{NO}\right) + \frac{\partial}{\partial x_j}\left(\bar{\rho}\tilde{u}_j\tilde{y}_{NO}\right) = \frac{\partial}{\partial x_j}\left(\bar{\rho}\left(\tilde{D}_{NO}\frac{\partial \tilde{y}_{NO}}{\partial x_j} + J_{NO,j}{}^{sgs}\right)\right) + \bar{\omega}_{NO}\left(\tilde{\xi}, \tilde{\xi}''^2, \tilde{Y}\right)$$

A possible problem that might occur during the LES due to strong convective processes is that the NO concentration in the flamelet used to compute the source term strongly differs from the one transported in the actual flame. Therefore, an additional distinction between NO production and consumption was included in the model. The source term was split accordingly and the NO mass fraction needed for the consumption term was taken from the transport equation. The final transport equation thus reads as:

$$\frac{\partial}{\partial t}\left(\bar{\rho}\tilde{y}_{NO}\right) + \frac{\partial}{\partial x_j}\left(\bar{\rho}\tilde{u}_j\tilde{y}_{NO}\right) = \frac{\partial}{\partial x_j}\left(\bar{\rho}\left(\tilde{D}_{NO}\frac{\partial \tilde{y}_{NO}}{\partial x_j} + J^{sgs}_{NO,j}\right)\right) + \bar{\omega}^P_{NO} + \bar{\omega}^C_{NO}\cdot\tilde{y}_{NO}$$

6.5.1.1 Application

The modeling was then applied to the non-premixed bluff body stabilized flame already used in Sect. 6.3 of this chapter. The LES code used was the same as for the premixed flame in Sect. 6.4. The grid used to compute the flame consisted of approximately 1.7 m cells. Again, an O-grid structure was used to refine the cells in the fuel jet and around the bluff body. The computational domain covered one bluff body diameter upstream. As this was not sufficient to develop turbulent structures in the coflow, artificial turbulent inflow using the method of Klein et al. [19] was used.

Although the NO mass fraction itself is not covered by CO_2, its source term is. More than 90% of its total amount can be represented by the progress variable, as can be seen in Fig. 6.14.

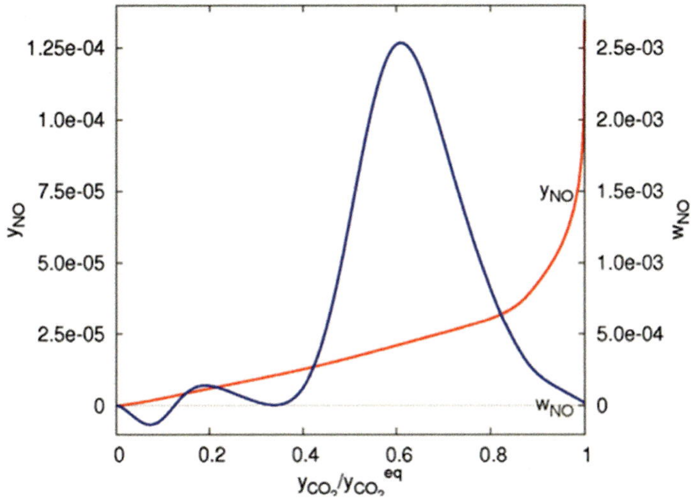

Fig. 6.14 NO mass fraction (*red*) and source term (*blue*) along the flamelet in progress variable space

The results are discussed in detail in the work of Ketelheun et al. [4], therefore, only the radial profiles of the NO mass fraction and its fluctuations are shown here in (Fig. 6.15). A strong improvement can be observed for the models using the additional transport equation compared to the standard FGM model. Although the amount of NO is still not perfectly predicted, the shape of the profiles matches. The discrepancies at higher positions above the bluff body can be attributed to incorrect predicting the jet breakup behavior.

6.5.2 CO Modeling

The additional modeling used for the prediction of CO was originally developed by Ansys in the context of Bray-Moss-Libby models in RANS simulations, see the work by Wegner et al. [20]. It is based on the separation of time scales of fast and slow chemical reactions.

The basic assumption of the model is that the source term of CO is not fully represented within the burnout zone of the flamelets where the progress variable reaches its maximum. Therefore, an additional progress variable for the post flame zone is defined.

The source terms of CO are derived differently before and in the flame front compared to behind the flame. If the progress variable has not yet reached its maximum, the source terms are taken from the standard FGM table based on mixture

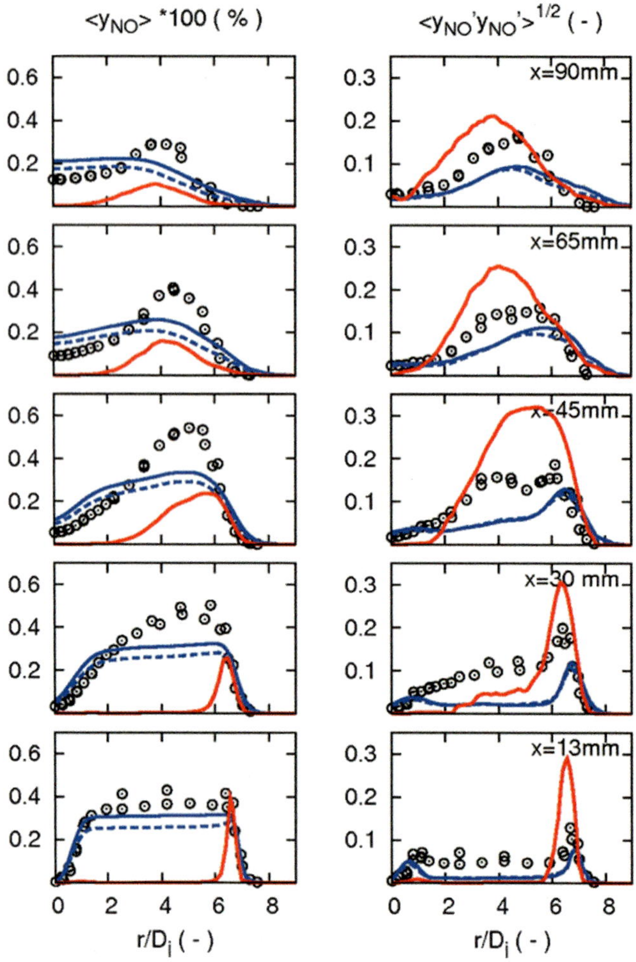

Fig. 6.15 Mean (*left*) and fluctuations (*right*) of the NO mass fraction. Standard FGM model (*red*), full source term (*solid blue*), splitted source term (*dashed blue*) models, compared to experiments (*symbols*)

fraction and progress variable. Starting at the point where the main parts of the flame finished reaction (i.e. $Y = Y_{max}$), an additional progress variable is defined, following

$$\zeta = \frac{\tilde{y}_{CO} - y_{CO}^{ff}}{y_{CO}^{eq} - y_{CO}^{ff}}$$

In this equation, y_{CO}^{ff} denotes the mass fraction of CO in the flame front, y_{CO}^{eq} the mass fraction at chemical equilibrium behind the flame. The source terms are then tabulated according to mixture fraction and this new progress variable ζ.

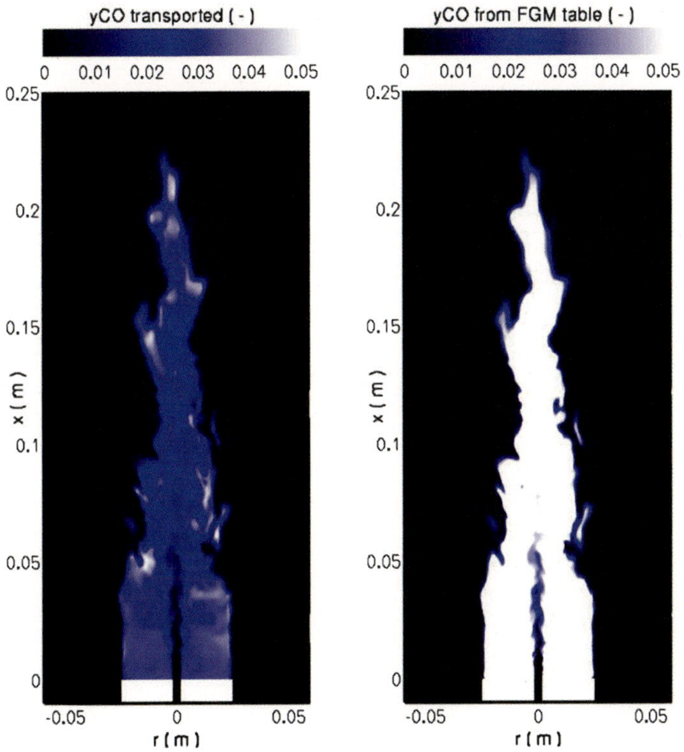

Fig. 6.16 CO mass fraction transported (*left*) and from the FGM table (*right*)

During the LES, the progress variable is constantly evaluated in order to decide which CO source terms to choose. All other variables are stored in the standard table, they are assumed to remain constant behind the flame. Details on the modeling can be found in Ketelheun et al. [5].

6.5.2.1 Application

This modeling was again applied to the bluff body flame described above, using the same code and grid. Figure 6.16 shows instantaneous snapshots of the CO mass fraction in the flame. The right plot shows the CO mass fraction as directly taken from the chemistry database, as in the standard FGM model. The left plot shows the transported mass fraction of CO, using the additional modeling. It can be clearly seen that the amount of CO produced is strongly reduced by the new modeling approach. To get a better quantitative comparison, Fig. 6.17 shows radial profiles of the CO mass fraction and its fluctuation.

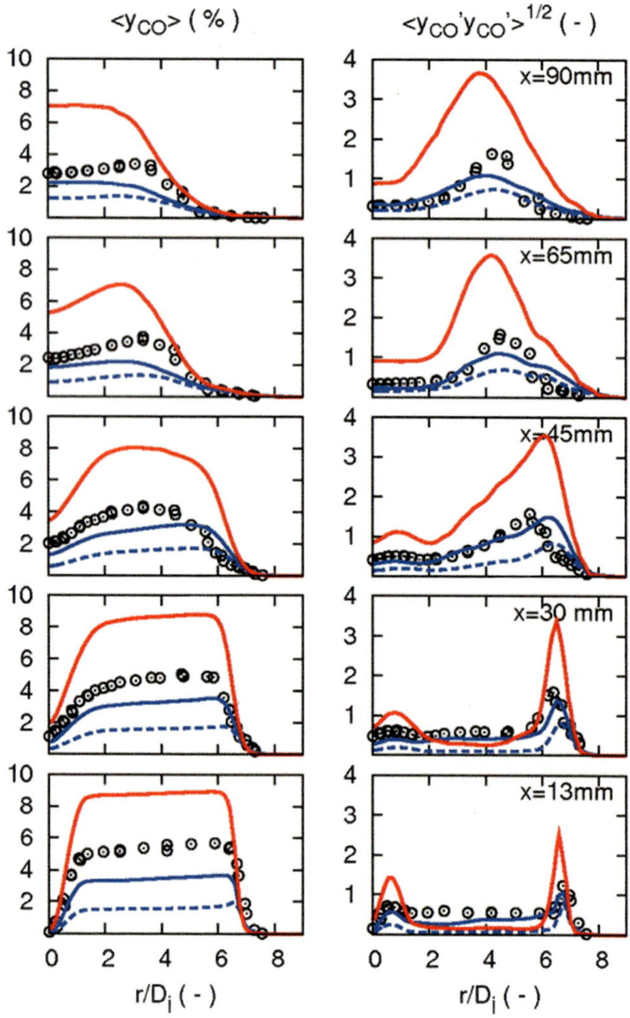

Fig. 6.17 Mean (*left*) and fluctuations (*right*) of the CO mass fraction. Standard FGM model (*red*), transport equation (*solid blue*) and time scale separation (*dashed blue*) models, compared to experiments (*symbols*)

Again, a strong improvement by the additional transport equation can be observed. After the mass fraction was too high with the standard FGM model, it now drops too low. This indicates that additional modeling will be required, considering bimolecular reactions further reducing the amount of consumption present in the flame. The additional separation of time scales gives no advantages; this indicates that the (rather thick) non-premixed flame is sufficiently resolved and that the flamelets sufficiently capture the CO distribution. This indicates that the model is not needed in this context; however, it has its justification for infinitely thin flame front models like BML- or G-equation models.

6.6 Summary and Conclusions

Three aspects of combustion modeling using tabulated chemistry in LES were presented in this chapter. First, an investigation of the subgrid structure using transported Euler Monte Carlo PDF methods was shown. The computation of the subgrid PDF instead of presuming it improves the simulation results and allows for the analyses of the shortcomings of other methods. The second aspect covered is the simulation of premixed flames. As LES grids are typically too coarse, an artificial thickening of the flame was introduced. This enables the use of FGM tabulated chemistry in the context of premixed combustion. Very good agreement with the experimental data could be obtained. Third, additional transport equations to better model pollutant formation in non-premixed flames were introduced. Additionally, the source term of the respective species was treated separately. This improved the prediction especially for NO, but also for CO improvement was observed. All modeling approaches were applied to configurations with flow features that can also be found in gas turbines. In future work, the modeling approaches will be combined, e.g. via the prediction of NO in premixed flames. Furthermore, the inclusion of heat transfer processes into the FGM context will be focused.

Acknowledgements The authors acknowledge the financial support from the German Research Council (DFG) through the SFB568.

References

Project-Related Publications

1. Kuehne, J.: Analysis of combustion LES using an Eulerian Monte Carlo PDF method. PhD thesis, TU Darmstadt (2011)
2. Kuehne, J., Ketelheun, A., Janicka, J.: Analysis of sub-grid PDF of a progress variable approach using a hybrid LES/TPDF method. Proc. Combust. Inst. **33**, 1411–1418 (2011)
3. Kuenne, G., Ketelheun, A., Janicka, J.: LES modeling of premixed combustion using a thickened flame approach coupled with FGM tabulated chemistry. Combust. Flame **158**, 1750–1767 (2011)
4. Ketelheun, A., Olbricht, C., Hahn, F., Janicka, J.: NO prediction in turbulent flames using LES/FGM with additional transport equations. Proc. Combust. Inst. **33**(2), 2975–2982 (2011)
5. Ketelheun, A., Aschmoneit, K., Janicka, J.: CO prediction in LES of turbulent flames with additional modeling of the chemical source term. ASME Turbo Expo, Copenhagen, Denmark, accepted for publication, 11–15 June 2012 (2012)

Other Publications

6. Smith, G.P., Golden, D.M., Frenklach, M., Moriarty, N.W., Eiteneer, B., Goldenberg, M., Bowman, C.T., Hanson, R.K., Song, S., Gardiner Jr., W.C., Lissianski, V.V., Qin, Z.: GRI3.0: http://www.me.berkeley.edu/grimech/ (2008). Accessed Apr 2008
7. Peters, N.: Laminar diffusion flamelet models in non-premixed turbulent combustion. Prog. Energy Combust. Sci. **10**, 319–339 (1984)
8. van Oijen, J.A., de Goey, L.P.H.: Modelling of premixed laminar flames using Flamelet-Generated Manifolds. Combust. Sci. Technol. **161**, 113–137 (2000)
9. Gicquel, O., Darabiha, N., Thévenin, D.: Laminar premixed hydrogen/air counterflow flame simulations using flame prolongation of ILDM with differential diffusion. Proc. Combust. Inst. **28**, 1901–1908 (2000)
10. Rhodes, R.P.: A probability distribution function for turbulent flows. In: Murthy, S.N.B. (ed.) Turbulent Mixing in Nonreactive and Reactive Flows, pp. 235–241. Plenum Press, New York (1975)
11. Landenfeld, T., Sadiki, A.: A turbulence-chemistry interaction model based on a multivariate presumed beta-pdf method for turbulent flames. Flow Turbul. Combust. **68**(2), 111–135 (2002)
12. Dally, B.B., Fletcher, D.F., Masri, A.R.: Flow and mixing fields of turbulent bluff-body jets and flames. Combust. Theory Model. **2**, 193–219 (1998)
13. Dally, B.B., Masri, A.R., Barlow, R.S., Fiechtner, G.J.: Instantaneous and mean compositional structure of bluff-body stabilized nonpremixed flames. Combust. Flame **114**, 119–148 (1998)
14. Chem1D. A one-dimensional laminar flame code, developed at Eindhoven University of Technology, http://www.combustion.tue.nl/chem1d (2008). Accessed Apr 2008
15. Butler, T., O'Rourke, P.: A numerical method for two dimensional unsteady reacting flows. In: Proceedings of the 16th Symposium (International) on Combustion, pp. 1503–1515 (1977)
16. Colin, O., Ducros, F., Veynante, D., Poinsot, T.: A thickened flame model for large eddy simulations of turbulent premixed combustion. Phys. Fluids **12**, 1843–1863 (2000)
17. Schneider, C., Dreizler, A., Janicka, J.: Fluid dynamical analysis of atmospheric reacting and isothermal swirling flows. Flow Turbul. Combust. **74**, 103–127 (2005)
18. Gregor, M., Seffrin, F., Fuest, F., Geyer, D., Dreizler, A.: Multi-scalar measurements in a premixed swirl burner using 1D Raman/Rayleigh scattering. Proc. Combust. Inst. **32**, 1739–1746 (2009)
19. Klein, M., Sadiki, A., Janicka, J.: A digital filter based generation of inflow data for spatially developing direct numerical or large eddy simulations. J. Comput. Phys. **186**, 652–665 (2003)
20. Wegner, B., Gruschka, U., Krebs, W., Egorov, Y., Forkel, H., Ferreira, J., Aschmoneit, K.: CFD prediction of partload CO emissions using a two-timescale combustion model. J. Eng. Gas Turbines Power **133**(7), 071502 (2011)

Chapter 7
A Simplified Model for Soot Formation in Gas Turbine Combustion Chambers

J. Marquetand, M. Fischer, I. Naydenova, and U. Riedel

Abstract Soot resulting from combustion processes is known to have a negative impact on health and environment. Soot also may lead to material damages especially in gas turbines. Therefore, a deeper insight in the processes leading to the formation and consumption of the soot precursors and soot particles is needed. For describing these processes a reaction kinetical model is needed representing the different steps. The processes happening in the gas phase are described by elementary chemical reactions, whereas for the particle phase a more complex formalism is needed due to the enormous number of different particles possible. In this work the processes in the particle phase are represented by the detailed soot model. A kinetical description of the processes is developed along with a mathematical representation of the model. This is done for shock tube conditions that can be assumed to be spatially homogeneous, allowing to focus on models of the chemical processes. In a next step this model is implemented for laminar flame conditions. Here additional transport phenomena are considered which helps improving gas phase and particle reaction models for representing soot formation in practical combustion devices. Besides giving a deeper insight into soot formation processes, the detailed soot model is needed for calibrating and adjusting a simplified soot model. This is used for simulating soot formation in complex technical systems like gas turbines.

Keywords Polyaromatic hydrocarbons • Soot • Modelling

J. Marquetand • M. Fischer • I. Naydenova • U. Riedel (✉)
Reactive Flows Research Group, Interdisciplinary Center for Scientific Computing, University of Heidelberg, Im Neuenheimer Feld 368, 69120 Heidelberg, Germany
e-mail: uwe.riedel@dlr.de

J. Janicka et al., *Flow and Combustion in Advanced Gas Turbine Combustors*, Fluid Mechanics and Its Applications 102, DOI 10.1007/978-94-007-5320-4_7,

7.1 Introduction

Soot resulting from combustion processes is a well-known phenomenon. On one hand this soot is used as a product for example in tires; on the other hand it is known to have a negative impact on health and environment. Additionally soot also may lead to material damages especially in gas turbines. Therefore, a deeper insight in the processes leading to the formation and consumption of the soot precursors and soot particles is needed.

The main processes happening during soot formation are understood and may be considered as different steps which, according to recent work [1–4] can be described by special sub-models: At the onset, there is the build-up of the first aromatic ring from species in the gas phase, such as acetylene and propargyl [5], which result from decomposing of the fuel molecules. This ring formation is regarded as a rate-limiting step in the evolution of soot particles [6].

The first aromatic ring gives rise to the build-up of polyaromatic hydrocarbons (PAHs): The number of rings increases by reactions with gas-phase species. Further growth of the PAHs via joining and ongoing formation of aromatic rings finally leads to the build-up of soot particles [5].

The size of these primary particles increases when mainly acetylene from the gas phase reacts with their respecitive surfaces [5]. Using the assumption of chemical similarity, this surface growth can be described by the kinetic model of the hydrogen abstraction and acetylene-addition pathway (HACA) [5]. Another effect leading to larger soot particles is the coagulation of two particles [5], theoretically described by the Smoluchowsky equation [7].

The process reducing particle sizes or even leading to complete decomposition of the particles is oxidation. Here the surface of the particles is reduced by reactions with gas-phase species like oxygen or OH [8–10].

For describing these processes a reaction kinetical model is needed representing the different steps. The processes happening in the gas phase are described by elementary chemical reactions, whereas for the particle phase a more complex formalism is needed due to the enormous number of different particles possible. In this work the processes in the particle phase are represented by the detailed soot model. A kinetical description of the processes is developed along with a mathematical representation of the model. This is done for shock tube conditions that can be assumed to be spatially homogeneous, allowing to focus on models of the chemical processes. In a next step this model is implemented for laminar flame conditions. Here additional transport phenomena are considered which helps improving gas phase and particle reaction models for representing soot formation in practical combustion devices.

Besides giving a deeper insight into soot formation processes the detailed soot model is needed for calibrating and adjusting the simplified soot model. This is used for simulating soot formation in complex technical systems like gas turbines. Under these conditions computational effort for using the detailed soot model is prohibitive. The most complex conditions this detailed model can be used for are laminar premixed flames mentioned above. Therefore the level of detail in describing the single processes needs to be reduced for CFD simulations. This is accomplished in our simplified soot model by having just two additional rate equations describing soot concentration and soot volume fraction.

So the steps towards the simplified model are: (1) developing a detailed kinetical model for the processes in the gas phase leading to soot precursors and particles, (2) developing a detailed model representing the formation and consumption of soot particles, (3) developing and implementing a method for simulating soot formation with the detailed model, and (4) improving and implementing a simplified model allowing simulations of soot formation in complex technical combustion devices including prior model validation.

The following chapters explain the methods used for accomplishing these tasks and they show the results obtained for the different soot models. The last chapter gives a conclusion and an outlook of future tasks resulting from the work presented here.

7.2 Methods

7.2.1 *Detailed Reaction Kinetics Model*

Numerous models exist for soot formation simulation in flames, whereas such models are rarely available for shock tube experiments, related to specific problems. The goal of the proposed work is the development of a detailed kinetic model of soot formation based on the comprehensive models of PAH formation and growth [2, 4, 11–14]. The concept of soot particle nucleation is described through a combination of the different pathways of PAH formation and growth, and soot particle inception published in [12, 14–16], and the traditional H-abstraction-C_2H_2-addition (HACA) route of PAH and soot particle growth [2, 17, 18]. The model is initially developed and applied for soot formation simulation in shock tube pyrolysis and oxidation of toluene, n-heptane, and toluene/alcohol mixtures [1]. It is further extended and applied for soot formation simulation of methane, ethylene, benzene and n-heptane pyrolysis and oxidation of methane and propane in conditions typical of shock tube experiments. The main idea taken from the models listed above is that the particle mass increases by their reactions with the gaseous species simultaneously with particle size growth by collision reactions among the PAH species.

Two different detailed kinetic models (*Model-1* and *Model-2*) are developed. These models describe gas-phase chemistry of soot precursor and particle formation in terms of different pathways. The formation and evolution of soot particles in both models differ with respect to the type of the species leading to soot particle inception and growth. In the following sections the gas-phase kinetics and the formation of soot precursers and particles for both models are described in detail.

7.2.2 Gas-Phase Reaction Mechanism

7.2.2.1 Description of the Detailed Kinetic Model 1 (Model-1)

A detailed kinetic model of soot formation during pyrolysis of various hydrocarbons in shock tube experiments is developed [19]. The model is further extended with a part of the mechanism of n-heptane oxidation [20] and applied for soot formation simulation in CH_4, C_3H_8, and n-C_7H_{16} rich-oxidation behind a reflected shock wave [21]. The gas-phase reaction mechanism includes a complete set of reactions of polyaromatic hydrocarbon (PAH) formation, described in [6], for laminar premixed acetylene and ethylene flames with all modifications presented in [2]. The PAH nomenclature from [6] is adopted in the mechanism. The notation A_i describes a PAH molecule, where the index i denotes the number of benzene rings included in the PAH molecule.

The gas-phase kinetic mechanism of polyaromatic hydrocarbon formation describes the pyrolysis and oxidation of C_2 species, the formation of higher linear hydrocarbons up to C_6 species, the formation of benzene and further reactions leading to pyrene, and the oxidation pathways of the aromatic species. Benzene and phenyl molecules are formed by reactions of C_4H_X species with acetylene, by cyclization reactions of C_6H_X species, and the combination of C_3H_3 propargyl. The formation pathway of pyrene, starting with benzene, follows the HACA reaction sequence, along with the ring-ring condensation, via the formation of biphenyl. The further growth of higher aromatic compounds proceeds via acetylene addition reactions and the ring-forming addition of vinylacetylene C_4H_4 to aromatic radicals. The vinylacetylene addition channel contributes mostly to the production of naphthalene, whereas phenanthrene is primarily formed by ring-ring condensation reactions.

Model-1 also contains a reaction mechanism of acetylene pyrolysis [22, 23] and a set of the gas-phase reactions of the mechanism of small pure carbon clusters formation up to C_{30} [24, 25] are included.

In addition the gas-phase reaction mechanism described as the polyyne pathway by Krestinin [26, 27] is considered. The most important pathway of polyyne growth is the subsequent increase of the carbon chain by a reaction of substitution $C_{2n}H + C_{2m}H_2 = C_{2n+2m}H_2 + H$, $n,m = 1,2,\ldots n + m < 6$, whose rate coefficients are close to the gas-kinetic collision limit. Acetylene and its radical C_2H play the key role in the polyyne growth process. The appropriate chemical composition that consists of H, C_2H, C_2H_3 and C_4H_3 radicals, responsible for the polyyne growth depends on the type and the structure of the fuel molecules. The polyyne molecules grow up to $C_{12}H_2$ and the average size of the polyyne radicals increases. Large polyyne radicals react with each other or with higher homologues, forming a variety of aggregates, which may contain ring closures.

As a result, the detailed kinetic scheme of the gas phase in *Model-1* consists of 1,850 unidirectional elementary reactions between 172 species including 691 reverse reactions counted separately.

7.2.2.2 Description of the Detailed Kinetic Model 2 (Model-2)

Recent progress in the development and understanding of the chemistry of polycyclic aromatic hydrocarbons provided foundation for further improvement of *Model-1*.

The gas-phase reaction mechanism in *Model-2* is a combination of *Model-1* and the recently evaluated mechanism of C_1-C_4 hydrocarbon oxidation [28] and a set of reactions of C_3-, C_5-, C_6- and C_7- hydrocarbons presented in [12, 13, 20]. Some of those reactions lead to formation of benzene, phenyl and higher PAH.

Also additional reaction paths of PAH formation and growth for PAH between benzene and pyrene (A_1–A_4) published in [4], are included in the reaction scheme. Large polyaromatics up to coronene A_4–A_7 are described by a few reactions paths adopted also from [4]. As a result, several pathways of PAH formation and growth are incorporated in the reaction mechanism: the alternating H-abstraction/C_2H_2-addition (HACA) route, resulting in a successive growth of PAH, the combination reactions of phenyl with benzene, the cyclopentadienyl recombination, the ring closure reactions of aliphatic hydrocarbons, and the recombination reactions between A_iR, an aromatic structure A_i with an aliphatic chain R and aliphatic radicals.

The developed kinetic model consists of a gas phase reaction mechanism, which describes the pyrolysis and oxidation of the parent hydrocarbons, and the formation and growth of PAH through different reaction pathways up to coronene.

The variety of reaction paths of PAH formation and growth included in *Model-2*, made it possible to describe the process of soot formation in case of pyrolysis and oxidation of hydrocarbons with both aliphatic and aromatic structures, without the necessity to include the polyynes in the particle inception process.

Compared to *Model-1* a set of reactions of polyyne hydrogenation and decomposition is included [11] into *Model-2*. This strategy provided different behavior of the polyynes whose concentration decreases approximately by one order of magnitude with increasing the number of carbon atoms in the polyyne molecule. The higher polyynes ($C_{10}H_2$ and $C_{12}H_2$) are excluded from the scheme because of lack of thermodynamic data.

The elementary reactions for the processes described in *Model-2* sum up to 2,228 between 204 gas-phase species. For simulations with the simplified soot model the reactions for the C1-C4-chemistry are updated with latest results for the reaction kinetics [28]. With these modifications the reaction mechanism contains 2,379 direct and reverse elementary reactions between 213 different species. The corresponding thermodynamic data for the gas-phase species is taken from available literature sources [2, 4, 29].

7.2.3 Soot Precursors and Particle Inception, Surface Growth, Coagulation and Oxidation

The processes described above give rise to soot particles by way of chemical species from the gas phase transforming into particulate matter. A key aspect of this soot formation process is the deposition of soot mass through reactions of gaseous species with the soot particle surface. Frenklach and Wang [17], suggested a mechanism for soot particles surface growth which is based on the postulate of the chemical similarity between analogous surface and gas-phase reactions of carbonaceous species. For soot particles this means that chemical reactions at the soot particle surface are similar to those of large polycyclic aromatic hydrocarbons (PAH). Following this idea all soot precursors and particles are considered effectively as large molecules.

The different processes contributing to the formation of soot may be represented with one reaction for every intermediate step in the evolution of a soot particle. This description leads to a huge number of chemical reactions and of chemical species, with an according number of differential equations to solve, resulting in an enormous computational effort.

Within this work the particles are treated as polymer species and take part in elementary reactions for polymers similar to those of regular gas-phase species. Each polymer species $P[n]$ represents a large number of molecules with common characteristics but a differing size, where n denotes the size of the polymers. The size of the polymer may be represented by characteristics like mass, volume, or the number of monomers. Using this formalism, a infinite number of molecules or particles $P[1]$, $P[2]$, ... ,$P[n_{\max}]$ is reduced to just one species $P[n]$.

Representing the precursers and particles by polymers the huge number of similar reactions repeating with molecules of different size can be combined to polymer reactions. The many different rate coefficients associated with the explicit reactions need to be replaced by just one coefficient for the polymer reaction, which leads to a size-independent formulation of the reaction rate. As an example for a polymer reaction

$$P[n] + X \xrightarrow{k} P[n+1] \tag{7.1}$$

shows a gas-phase species X reacting with a polymer $P[n]$ and thereby increasing the size of the polymer to $n+1$.

As each polymer is not just one species but represents different sizes for each polymer there is a size distribution giving the number of molecules with a corresponding size. As an example a log-normal size distribution of a polymer is given in Fig. 7.1.

Although formulated like simple chemical reactions the polymers and their reactions cannot be solved as regular elementary reactions. They need a special mathematical formalism as not just one concentration needs to be computed but the whole size distributions. These methods increase computational effort compared to

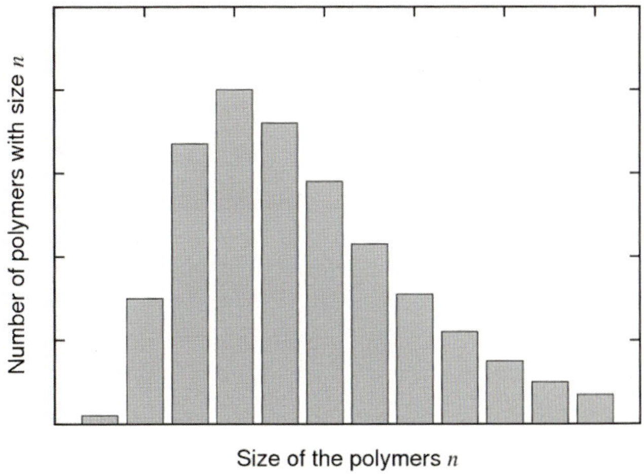

Fig. 7.1 Log normal size distribution of polymers over a general size axis. This may represent characteristics like number of monomers, mass, or volume

elementary reactions. But overall considering just one polymer reaction decreases computational cost significantly compared to computations needed for the huge number species if not grouped into polymers.

The next two sections below focus on the reaction kinetical description of the soot precursor and particle formation processes and the two models used.

7.2.3.1 Description of the Detailed Kinetic Model 1 (Model-1)

In *Model-1*, only the basic mechanism of surface growth presented in [2] is considered. Following [2], the first precursors are incepted in condensation reactions of pyrene molecules. As an extension to this process, the reactions of pyrene, phenanthrene and biphenyl molecules and their radicals [19] are included. Two different types of particles are considered for the precursors formed from the PAH-particles with dehydrogenated C* sites (active) and saturated particles with C-H sites (inactive). The formation and consumption of active sites on the soot particle surface occur in the reactions with H_2, H, and $H_2O/OH/O_2$ species. Surface growth is provided by the reactions of soot particles with active sites with C_2H_2 [2], accompanied by H abstraction, and by the reactions of pyrene, phenanthrene, and naphthalene condensation on the active soot particles [19]. The soot particles with active sites participate in coagulation reactions, where only a free-molecular regime of coagulation is considered.

In the polyyne pathway, soot precursors are formed in polymerization reactions of higher polyynes $C_{12}H_2$, $C_{10}H_2$. In *Model-1*, only the high temperature path of initiation reactions between neutral polyyne molecules, and reactions of particle growth with participation of polyyne molecules and their radicals, are adopted from the original model proposed by Krestinin et al. [26, 27]. Revised thermodynamic data for several polyyne molecules included in the gas-phase part of the mechanism, are taken from the literature [30] or calculated using group additivity techniques. The surface growth in the polyyne pathway occurs via reactions of soot precursors and particles with the most reactive gas-phase species C_2H_2, C_2H, C_2, C_4H_4, C_4H_2, C_4H, C_4, C_6H_2, C_6H, C_6, C_8H_2, C_8H, C_8, $C_{10}H_2$, $C_{10}H$, C_{10}, $C_{12}H_2$, $C_{12}H$, and C_{12}. According to the main assumption of the polyyne model, a growing particle generates active sites in each act of interaction with the gas-phase species. Therefore only one type of active soot particles is considered in the model. Soot particles formed through the polyyne pathway react also with PAH and coagulate with each other.

In *Model-1* there are 94 polymer reactions between four polymer species, describing the processes in the particle phase. These four species represent the precursors and the soot particles in an active and inactive form, each.

7.2.3.2 Description of the Detailed Kinetic Model (Model-2)

The large PAH are considered as soot precursors. Their mean diameter is between 1 and 2 nm. These data are in agreement with the experiential measurements of young soot particle size reported in [4, 29]. The C/H ratios of the soot precursors formed during these reactions are close to the upper boundary of the C-H-diagram reported in [31]. In *Model-2* the precursors are formed mainly by radical-molecule reactions of different PAH including species with five- and six-membered rings [4, 12, 16]. The radical-radical reactions played a minor role. Therefore they are reduced to minimum. These reactions result in the formation of polyaromatic molecules (soot precursors) containing from 25 to 46 carbon atoms, which are stabilized by the formation of new chemical bonds.

The precursors grow by the HACA mechanism, where the initiation of PAH radicals occurs in H-abstraction reactions with H and OH radicals described in details in [4]. Termination reactions with H radicals and H_2, and H_2O molecules are also included in the model. The rate coefficients of these reactions are adopted from the literature [4]. The reaction flow analysis showed that at about 70% of the surface growth is due to C_2H_2 addition by the reaction described in the model of Appel et al. [2], with the same rate coefficient. Species like C_4H_2, C_6H_2, and different PAH are measured during pyrolysis of aliphatic and aromatic hydrocarbons behind shock wave [32, 33] and are found in rather high concentrations. Therefore, growth reactions with the participation of those aliphatic species, and various PAH molecules and their radicals, are introduced in the model. The particle mass increases by reactions with gaseous species. Simultaneously particle size growth by collision reactions among the PAH species.

Oxidation takes place in reactions with O and OH radicals. All types of particles participate in coagulation reactions, for the case of free-molecular regime. The processes in particle phase for *Model-2* are represented by 105 polymer reactions between 4 polymer species. As in *Model-1* these species represent the precursors and the soot particles in an active and inactive form.

7.3 Detailed Soot Model

The reaction kinetical mechanism of soot precursor and particle formation and degradation described above is developed using a revised and substantially modified version of the software package MACRON [34]. This version was used before for modeling the formation of solid carbon particles during C_3O_2 thermal decomposition [24, 25] and soot formation simulation during pyrolysis and oxidation of different hydrocarbons behind shock wave [1, 19, 21, 35–39]. The program MACRON is the first implementation combining the solution of the ordinary differential equations generated by the chemical kinetics in the gas phase with the framework of a discrete Galerkin Method suggested by Deuflhard and Wulkow [34, 40] for solving polymer reaction problems.

This approach gives the advantages to preserve the discrete character of any elementary transformations of heterogeneous particles and to describe them as elementary chemical reactions for the particles of all sizes. The heterogeneous particles can also react with the gas-phase species, and thus, a connection with the gas-phase chemistry of soot particles is provided during the whole calculation.

As a drawback the method uses predefined size distributions and solving those is relatively time consuming and may result in unrealistic distribution functions. Therefore the *Compartment Method* is newly developed. This method is a combination of the commonly used Method of Moments [41, 42] and Sectional Methods [43–45]. It uses two moments $\mu_0(t)$ and $\mu_1(t)$ for every polymer species to account for the size distribution function representing the distribution of the according polymer's size and the size axis of the distribution function is separated into a selectable number of intervals k with a discrete size.

The statistical moments μ_m describing the size distributions are generally defined as

$$\mu_m = \sum\nolimits_{n=0}^{\infty} n^m c_n, \tag{7.2}$$

with m defining the actual moment, n for the particle's size and c_n for the concentration of particles with that size [42]. Some of these moments correspond to physical characteristics of the distribution that can easily be compared to data measured in experiments. The zeroth moment

$$\mu_0 = \sum\nolimits_{n=0}^{\infty} c_n \tag{7.3}$$

gives the polymer concentration, here the concentration of the precursors or particles and the first moment μ_1 denotes the summed concentration of the embedded monomers, this means the total number of atoms contained in the related polymer species.:

$$\mu_1 = \sum\nolimits_{n=0}^{\infty} n c_n \tag{7.4}$$

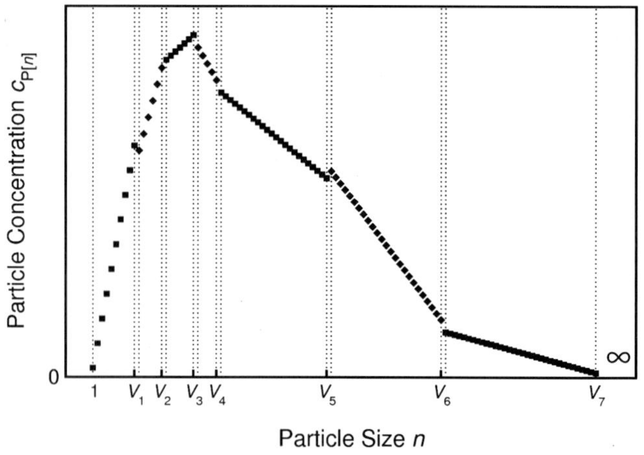

Fig. 7.2 General scheme for discrete compartments: every interval $V_{k-1} + 1, \dots V_k$ on the particle size axis features its own independent linear approximation of the discrete size distribution, calculated from the conserved quantities μ_0^k, $\mu_{1,i}^k$

The precursors and the particles are assumed to be build up by C-, H- and O-atoms. These represent the monomer species building a polymer P. So for each species there is a first moment $\mu_{1,C}^P$, $\mu_{1,H}^P$, $\mu_{1,O}^P$ representing the number of corresponding atoms within the polymer. Together with the zeroth moment this results in four equations to solve for each polymer species and each compartment in the model.

These moments are considered for calculating the concentration and mass of the particles or the polymers, respectively. They are used to algebraically reconstruct locally linear approximations of the size distribution such that the latter reproduce the moment's values (see Fig. 7.2). The first interval starts from one and reaches to the given size N_1, the next interval ranges from $N_1 + 1$ to N_2. This repeats for the given number of intervals and ends with size N_n. Within every interval the values of μ_0^k and μ_1^k are calculated which corresponds to a local linear approximation within the interval. A size distribution resulting from this is shown in Fig. 7.2.

Within the *Compartment Method* the particles of all sizes may be grouped together leading to just one compartment on the size axis of the distribution. Thereby computational effort is reduced as just four equations per polymer species need to be considered. Additionally inter-compartment reactions can be neglected, further reducing the simulation time.

In many applications the mean values of the moments, resulting from a single compartment, are adequate. As this work only considers soot characteristics that are calculated from mean values, like soot volume fraction, all simulations with the detailed soot model are carried out with a single compartment.

Using just one compartment leads to size-independent reaction kinetics simplifying the formulation of the kinetical mechanism. In that case rate equations for the moments $\mu_0, \mu_{1,i}$ are essentially as simple as the rate equations for the gas-phase species. As an example the reaction rate for the surface growth of particle P[n] by acetylene (C_2H_2) from the gas phase is given in detail:

$$P\left[n_C\right] + C_2H_2 \rightarrow P\left[n_C + 2\right], \tag{7.5}$$

where n_C represents the number of C-atoms with in the polymer. For H- and O-atoms the formulation is about the same.

From this reaction the rate equation of acetylene can be derived following the formalism of elementary reactions. This leads to

$$\frac{d c_{C_2H_2}}{dt} = -k \mu_0^P c_{C_2H_2}, \tag{7.6}$$

with the concentration of acetylene denoted by $c_{C_2H_2}$and that of the polymer P[n] by the zeroth moment μ_0^P. As for all the polymer reactions, the rate constant k is calculated from Arrhenius parameters resulting from the reaction mechanism described above.

For the zeroth and first moments the rate equations result in

$$\frac{d\mu_0^P}{dt} = 0 \tag{7.7}$$

$$\frac{d\mu_{1,C}^P}{dt} = 2k\mu_0^P c_{C_2H_2} \tag{7.8}$$

$$\frac{d\mu_{1,H}^P}{dt} = 2k\mu_0^P c_{C_2H_2} \tag{7.9}$$

$$\frac{d\mu_{1,O}^P}{dt} = 0. \tag{7.10}$$

As for the surface growth process the total number of particles stays constant, there is no change for the zeroth moment (see Eq. 7.7). While reacting with acetylene the particle grows by two C- and H-atoms (see Eqs. 7.8 and 7.9). Therefore, the reaction rate is twice as fast as for acetylene. Oxygen does not contribute in the reaction, so there is no change at all.

For every polymer species the rate equations of $\mu_0(t)$ and $\mu_{1,i}(t)$ are summed up over all reactions that contain the considered species. These differential equations do not need a special numerical solution. They are, together with the rate equations for the normal gas phase species, part of the overall equation system and can be computed with a solver for stiff differential equations. With four moments per polymer and 4 different polymer species in the reaction kinetical model described above overall this leads to 16 additional equations to solve for the polymers in the detailed soot model.

The values of μ_0 and $\mu_{1,i}$ resulting from the integration of the equation system allow the calculation of the mean number of C-, H- and O-atoms per particle n_i with

$$n_i = \frac{\mu_{1,i}^P}{\mu_0^P}. \tag{7.11}$$

As the molar mass of the monomers $M_{\mu_{1,i}}^P$ is known, the mean mass of the particles $M_{\mu_0}^P$ can be derived from

$$M_{\mu_0}^P = \sum n_i M_{\mu_{1,i}}^P \tag{7.12}$$

using n_i. The molar mass of all the polymer species P and their concentration μ_0^P can be used for the calculation of the soot volume fraction

$$f_V = \frac{\mu_0^P M_{\mu_0}^P}{\rho_{soot}}. \tag{7.13}$$

In this work the density of soot ρ_{soot} is assumed to be $1.8g\ /cm^3$, a value commonly used in literature [4, 18].

Treating the particles as spheres, another commonly used assumption, it is possible to calculate the mean particle from molar mass $M_{\mu_0}^P$ and soot density ρ_{soot}

$$D_{soot} = \sqrt[3]{\frac{6M_{\mu_0}^P}{\pi \rho_{soot} N_A}}. \tag{7.14}$$

The soot yield Y_{soot} is calculated from the concentration of C atoms in the particle represented by the first moment $\mu_{1,C}^P$ and the initial concentration of C atoms in the fuel $c_{C,fuel}^0$

$$Y_{soot} = \frac{\mu_{1,C}^P}{c_{C,fuel}^0}. \tag{7.15}$$

Another value often measured in experiments and giving information about the inner structure of the particles is the C/H-ratio. This can easily be derived from the first moments of C-atoms and H-atoms by

$$\frac{C}{H} = \frac{\mu_{1,C}^P}{\mu_{1,H}^P}. \tag{7.16}$$

The characteristics described above are used to compare the results of the detailed soot model with the corresponding results from experiments.

The compartment method for one interval is implemented in a C++-program named POLYKARP. It interacts with the HOMREACT-suite, a simulation program for homogeneous conditions [46]. This program is written in FORTRAN code and uses available stiff integrators DASSL or LIMEX [47]. The two programs communicate by interchanging different files like the reaction scheme, the thermodynamic data and a control file.

Furthermore, another program is implemented for the simulations of soot formation under laminar premixed flame conditions [48]. It is based on the program MIXFLAME [49–51], but the code is updated to FORTRAN 90 and extended to use another DAE solver [52]. The new program also contains routines to simulate soot formation applying the Compartment Method described above. The code is build up by two parts: SIIN (Simulation Interface) written in C++ and doing the preprocessing, like reading the simulation conditions, the reaction mechanism and the thermodynamic data. Additionally some preliminary computations concerning the transport parameters of the species are done in that part of the program.

The simulation of the flame itself is implemented in the second part of the program FLASI (Flame Simulation). It is written in FORTRAN 90 and uses the preprocessed data from SIIN, that is transferred to FLASI via FORTRAN 90 modules. Within FLASI the conservation equations for mass and energy are solved at every timestep. The flame is spatially resolved by a predefined number of gridpoints that may be rearranged within the simulation.

Compared to regular gas-phase species not the reaction rates are more complicated for the polymer species. When regarding transport phenomena as necessary in flame simulations an additional effect of the polymer species needs to be considered: They do not have a constant mass as gas-phase species but it changes during the simulation due to processes like surface growth and coagulation. This variation needs to be taken into account for the calculation of the polymer's transport parameters, as these often depend on the molecule's mass.

For initial calculation of the transport parameters in SIIN a predefined molar mass $\hat{M}_{\mu_0}$ of the polymers is used as the real value for the mass is not known at that point of the simulation. These pre-calculated transport parameters are corrected within the flame simulation FLASI, where all the size changing processes from the polymer reactions are taken into account for the molar mass of the polymers.

The first step in correcting the transport parameters is calculating the actual molar mass of the polymers at the current time and position in the flame. As given in Eq. (7.12) the molar mass of a polymer can be derived from the moments and the molar masses of the monomers.

The varying mass of the polymers results in a special treatment for their conservation equations. For regular gas-phase species, because of their constant molar mass, the mass conservation equation also implies the preservation of molecule's number. In contrast for polymers separate equations need to be solved for conserving their mass and number. The equation of the zeroth moment is solved for representing the particles concentration whereas the first moments assure the conservation of the mass and the number of C-, H-, and O-atoms in the system.

7.4 Simplified Soot Model

As described above there are numerous interacting reactions and processes happening while a particle is formed and grows in reactions with gas phase species. Within the simplified soot model all these reactions are grouped into four main processes: nucleation, coagulation, surface growth, and oxidation.

These soot formation processes are represented by two rate equations: One considering the soot concentration which gives an information of the number of particles (Eq. 7.17). The second one, the rate equation of the soot volume fraction, corresponds to the volume or the mass, respectively (Eq. 7.18).

$$\frac{dC_{soot}}{dt} = \underbrace{\alpha_1 c^2_{C_2H_2} + \alpha_2 c^2_{C_6H_5}}_{nucelation\ \tilde{\alpha}_1, \tilde{\alpha}_2} - \underbrace{\beta c^2_{soot}}_{coagulation\ \tilde{\beta}} \quad . \tag{7.17}$$

The nucleation and coagulation both influence the number of particles and therefore are part of the soot concentration rate equation. Within the simplified model it is assumed that nucleation is depending on the species concentrations of propargyl and phenyl. These are considered as most important species in forming the first aromatic ring [5] leading to further growth of PAHs and soot particles, finally. The parameters α_1 and α_2 in the nucleation part of Eq. (7.17) represent the different contributions of the respective species to the nucleation of the soot particles. The representation of the complicated particle nucleation from the gas phase depending on just two species concentrations is a strong simplification of the real processes. Therefore it is not possible to determine the values of the parameters α_1 and α_2 directly from theory. In this work these values are adjusted to be consistent with the experimental results.

Coagulation, the process of two particles colliding and sticking together, depends on the soot concentration itself and the collision number β corresponding to the reaction of two soot particles with each other [53]. This decreases the number of particles in the system, thus reduces the soot concentration, but does not affect the soot volume fraction at all.

The surface growth process is assumed to depend on the concentration of acetylene as the main growth species [18]. The volume of the particles is decreased by oxidation. Here OH is assumed to be the dominant species reacting with the particle surface [9].

Following the common assumption of spherical particles and applying the kinetic theory of gases assuming a first order growth law [8, 54], the rate equation of the soot volume fraction results in

$$\frac{df_V}{dt} = \underbrace{\delta_1 c^2_{C_2H_2} + \delta_2 c^2_{C_6H_5}}_{nucelation} + \underbrace{\gamma c_{C_2H_2} c^{1/3}_{soot} f^{2/3}_V \left(1 - \frac{f_V}{f_{V,\infty}}\right)}_{surfacegrowth\ \tilde{\gamma}} - \underbrace{\gamma c_{OH} c^{1/3}_{soot} f^{2/3}_V}_{oxidation\ \tilde{\varepsilon}} . \tag{7.18}$$

This equation depends on nucleation, surface growth and oxidation, as these influence the volume of the particles.

Within Eq. (7.18) δ_1 and δ_2 represent the contribution of propargyl and phenyl to the nucleation, as described for the soot concentration rate.

But here the unit transformation

$$\delta_i = \frac{\alpha_i \cdot M_{soot}}{\rho_{soot}} \tag{7.19}$$

of the parameters is needed to represent the nucleation process in the volume fraction rate equation. The density of soot is chosen to be $\rho_{soot} = 1.8\text{g}/\text{cm}^3$ [18] and the mass of a nucleating particle is assumed to be ten-times the mass of a benzene molecule $M_{soot} = 781,2\text{g}/\text{mol}$. The model parameters γ and ε corresponding to surface growth and oxidation represent characteristic properties like the mass of the gas-phase species C_2H_2 and OH.

The factor $1 - \frac{f_V}{f_{V,\infty}}$ in Eq. (7.18) limits the contribution of the surface growth to an upper empirical value, the maximum volume fraction $f_{V,\infty}$, which can be measured in experiments. The simplified model is very sensitive to that value but is often not given for a specific experiment. Therefore values resulting from simulations with the detailed soot model are used, when there are no experimental values available.

To represent the temperature dependence of the values from the detailed model within the simplified model Eq. (7.20) is used which by adjusting the parameters a_i is fit to the bell-shaped results of the detailed model. In some cases a single bell-shaped curve is not sufficient to represent the temperature dependence of $f_{V,\infty}$ from the detailed model. Then a second bell-shaped curve is considered in addition, having parameters $b_i \neq 0$

$$f_{V,\infty}^{fit}(T) = \frac{a_1}{a_2} \cdot \exp\left(\frac{-(T - a_3)^2}{a_2}\right) + \frac{b_1}{b_2} \cdot \exp\left(\frac{-(T - b_3)^2}{b_2}\right) \tag{7.20}$$

As mentioned before, spherical particles are assumed. So the particle diameter D_{soot} can be calculated from the soot concentration C_{soot} and the volume fraction f_V by

$$D_{soot} = \left(\frac{6 f_V}{\pi C_{soot} N_A}\right)^{\frac{1}{3}} \tag{7.21}$$

with N_A representing Avogadro's number.

The soot yield Y_{soot} is calculated from $C_{C,total}$, the total carbon atom concentration in the system. That is determined by the initial concentration of the fuel and the number of carbon atoms within and $C_{C,soot}$ representing the concentration of C-atoms appearing in the soot particles

$$Y_{soot} = \frac{C_{C,soot}}{C_{C,total}}. \tag{7.22}$$

Here $C_{C,soot}$ is calculated from

$$C_{C,soot} = \frac{N_{\mathrm{A}} f_V \rho_{\mathrm{soot}}}{M_{\mathrm{C}}}, \tag{7.23}$$

with M_C beeing the molar mass of a C-atom and the soot density ρ_{soot} mentioned above.

The concentrations of the gas-phase species (C_2H_2, C_3H_3, C_6H_5, OH) used in the two-equation soot model are calculated from the reaction model of the gas-phase. The feedback from the simplified soot model back to the detailed gas phase chemistry is given by

$$\begin{aligned}
\frac{dC_{C_3H_3}}{dt} &= \left[\frac{dC_{C_3H_3}}{dt}\right]_{gas} - 2 \cdot \tilde{\alpha}_1 \\
\frac{dC_{C_6H_5}}{dt} &= \left[\frac{dC_{C_6H_5}}{dt}\right]_{gas} - 2 \cdot \tilde{\alpha}_2 \\
\frac{dC_{C_2H_2}}{dt} &= \left[\frac{dC_{C_2H_2}}{dt}\right]_{gas} - \tilde{\gamma}\frac{\rho_{soot}}{m_{C_2H_2} N_A} \\
\frac{dC_{OH}}{dt} &= \left[\frac{dC_{OH}}{dt}\right]_{gas} - \tilde{\varepsilon}\frac{\rho_{soot}}{m_{OH} N_A}.
\end{aligned} \tag{7.24}$$

Here the equations in square brackets represent the pure gas-phase processes. The processes of nucleation, surface growth, and oxidation are represented by the variables $\tilde{\alpha}_1, \tilde{\alpha}_2, \tilde{\gamma}$, and $\tilde{\varepsilon}$ as formulated in Eqs. (7.17) and (7.18).

7.5 Results and Discussion

The detailed and the simplified soot model are used for simulating soot formation at different conditions. For both models simulations at shock-tube conditions are performed. Additionally the detailed model is used for simulations of laminar premixed flames. The simplified model is not applied for flame conditions, as it is not yet implemented in a flame simulation program.

7.5.1 Detailed Soot Model

The detailed model is developed and used to obtain a deeper insight in soot formation and consumption. For the simulations in this chapter the gas-phase mechanism described above is used together with the polymer reactions. The

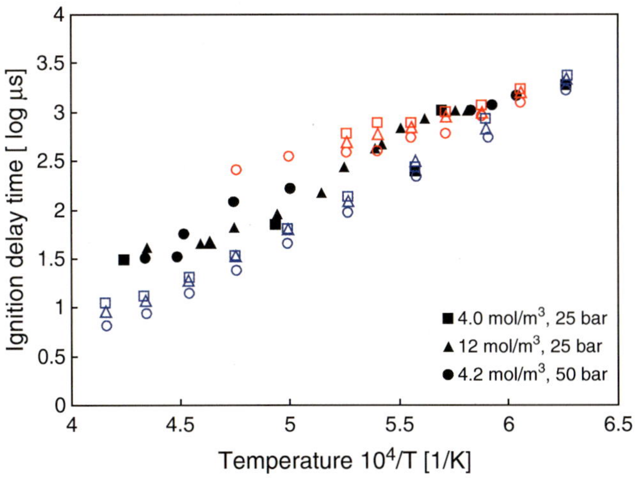

Fig. 7.3 Temperature dependence of the ignition delay time for the pyrolysis of ethylene with different C-atom concentrations in Ar at 50 bar (*filled*: experiment [55], *open*: simulation; *blue*: *Model-1*, *red*: *Model-2*)

simulations under spatially homogenous conditions are carried out for developing the reaction kinetical scheme. Under these conditions transport can be neglected and the investigations can focus on the chemical processes. For further development of the detailed model towards CFD-simulations it is implemented for flame conditions, additionally regarding transport processes. The results for both systems are shown here.

7.5.1.1 Shock Tube Simulations

The newly developed gas-phase mechanism is validated for different conditions. This is necessary as without proper description of the processes in the gas phase a good representation of the soot formation and consumption processes is impossible. Within this project two different models are developed containing different chemical reaction pathways. Both models are compared to experiments to show the progress achieved in mechanism development. The simulations for shock tube conditions are carried out with the simulation program Macron, described above.

As a first test the results of the two models are compared to experimental ignition delay times at different conditions. In Fig. 7.3 the results for the pyrolysis of ethylene with different C-atom concentrations in argon at 50 bar are plotted.

For all temperatures and concentrations *Model-1* underestimates the ignition delay time. Better results can be achieved with *Model-2* that reproduces the ignition delay times at lower temperatures. But at higher temperatures the delay times are overpredicted, in contrast to *Model-1*.

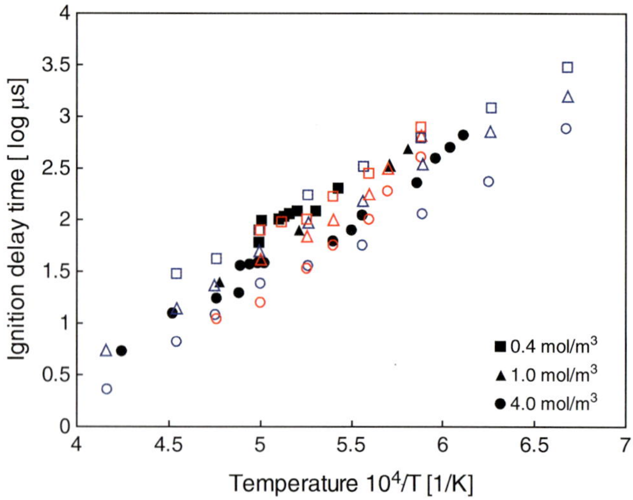

Fig. 7.4 Temperature dependence of the ignition delay time for the pyrolysis of benzene with different C-atom concentrations in Ar at 50 bar (*filled*: experiment [55], *open*: simulation; *blue*: *Model-1*, *red*: *Model-2*)

For the pyrolysis of benzene with different C-atom concentrations at 50 bar (see Fig. 7.4) there is no such a trend for the models. Under these conditions they both represent the measured values within an acceptable range.

Compared to the integral ignition delay time regarding the species profiles over time gives much more information on the evolution of the chemical system. These additional simulations are carried out for *Model-2* to get deeper insights on its predictive capabilities. In Fig. 7.5 the time evolution of the OH concentration profiles during oxidation of toluene are given at different temperatures. In both cases there is a good agreement with experiment [56] concerning the level of OH. For the lower temperature and the higher pressure case the increase is predicted a little earlier then measured.

In addition to the gas-phase species the two models are tested for their capability in predicting the soot formation processes. Therefore simulations of the temperature dependence of the soot yield are carried out.

These results show the progress achieved in developing the reaction kinetical model for gas-phase chemistry and soot formation. Under all conditions the results of *Model-2* better represent the experiments then *Model-1*. Therefore, in the following simulations just *Model-2* is used.

7.5.1.2 Laminar Premixed Flame Simulations

Simulations with laminar premixed flames give a deeper insight and represent a harder test of the model as under these conditions a wide range of temperatures is

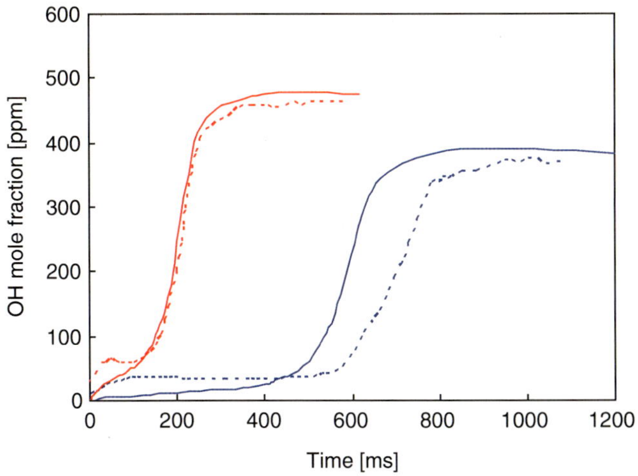

Fig. 7.5 Measured (*solid*, [56]) and simulated (*dashed*) OH-concentration profiles during toluene oxidation ($\Phi = 1$, 0.1% $C_6H_5CH_3$, 0.9% O_2; *red*: 1,689 K, 1.79 atm; *blue*: 1,586 K, 1.9 atm)

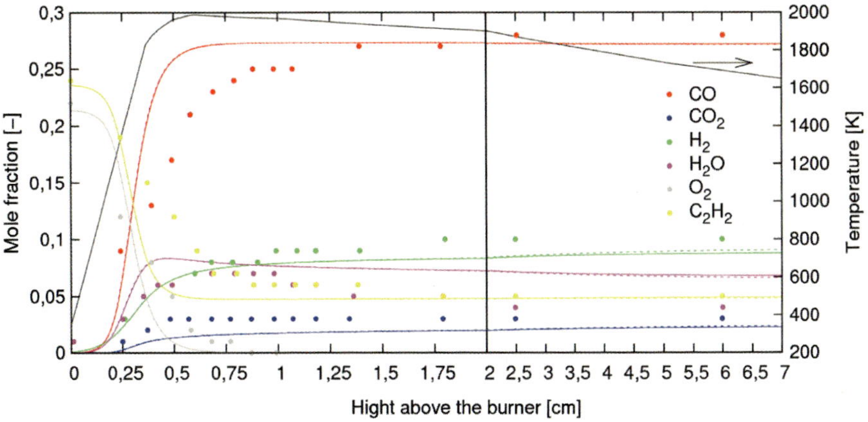

Fig. 7.6 Mole fractions of main combustion species and temperature in a burner stabilized acetylene, measured (*symbols*, [57]) and simulated (*lines*) just for gas phase (*dashed*) and additionally considering particle phase (*solid*) (C/O = 1.1, $x_{Ar} = 0.55$, p = 120 mbar, $v_{Gas} = 20.4$ cm/s)

considered. Generally, these temperatures reach from about room temperature up to the adiabatic temperature at 1,500–2,000 K. In addition, under these conditions transport processes need to be considered. This is the next step towards simulating technical devices.

In Fig. 7.6 the simulation results for the main combustion species in a burner stabilized laminar acetylene flame are shown together with measurements [58]. The simulation temperature profile is fixed to the experimental values for neglecting radiation losses.

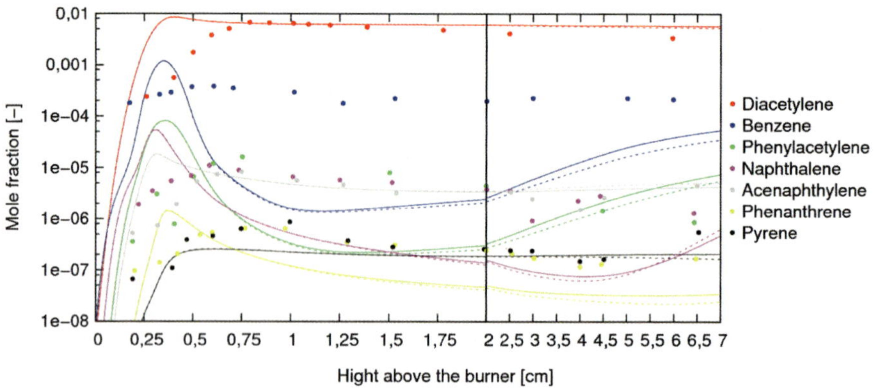

Fig. 7.7 Mole fractions of soot precursor species in a burner stabilized acetylene flame, measured (*symbols*, [57]) and simulated (*lines*) for gas phase only (*dashed*) and additionally considering the particle phase (*solid*) (C/O = 1.1, $x_{Ar} = 0.55$, p = 120 mbar, $v_{Gas} = 20.4$ cm/s)

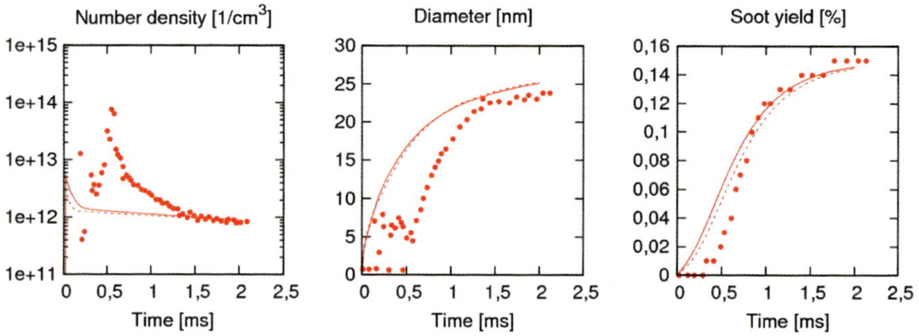

Fig. 7.8 Time evolution of number density, diameter and soot yield for the oxidation of n-heptane in argon, measured (*symbols*, [58]) and simulated (*lines*) with (*solid*) and without (*dashed*) nucleation from phenyl ($c_C = 7.89$ mol/m^3, $\Phi = 5$, T = 1,750 K, p = 25 bar, $t_r = 2.0$ ms)

Simulation predicts faster combustion and a thinner flame front. H_2 is predicted very good for lower flame regions and shows about the same deviation from the experiments for higher flame regions as CO and CO_2. Both show bigger differences within the flame front. But taking into account that up to now development of the reaction mechanism in *Model-2* is just done for shock tube conditions the overall agreement between experiment and simulation is very good.

The experimental [57] and simulated mole fractions of the soot precursors in the gas phase are shown in Fig. 7.7 for the same flame conditions. Here the general agreement between simulation and experiment is good, too. But for the aromatic species benzene and naphthalene, regarded as main nucleation species, there is a big difference compared to measurements. For both species the simulated values are lower by at least one magnitude compared to experiment. This effect is not related to the soot model as it can be seen with and without considering soot formation in the simulations (Fig. 7.8). It is an effect resulting from the gas phase reaction scheme.

7.5.2 Simplified Soot Model

Simulations with the simplified soot model are carried out for shock-tube conditions. The simulations were done using the gas-phase mechanism of *Model-2* together with the two additional differential equations for soot concentration and volume fraction.

The simplified model is tested for aliphatic and aromatic fuels. As a first test the simulation results for the time evolutions of some typical soot characteristics during oxidation of n-heptane are compared to values measured by Kellerer et al. [58]. In this case the value of the maximum volume fraction $f_{V,\infty}$ can be calculated from given experimental values with the help of Eq. (7.21).

There is some difference for the number density and the particle diameter within the first millisecond. This is caused by the strong simplification of the nucleation process within the simplified model. The soot yield is not depending on the number of particles that much and therefore is in better agreement with the measurements. After about 1.5 ms the simulation reaches the level of the experiments and at the end of the simulation time there is good agreement. This is important, as for the following simulations only the values at the end of the simulation time are considered.

For a deeper insight in the simulation the influence of the soot model on the gas phase is plotted in Fig. 7.9. The gas-phase species included in the soot model are shown together with some precursor species for the simulation with and without activating the simplified model.

There is nearly no influence on the concentrations of the species responsible for nucleation (C_3H_3, C_6H_5) and oxidation (OH) when using the simplified soot model. In contrast the model has a strong influence on the concentration of acetylene and the precursor species pyrene and coronene.

Acetylene is reduced as it is considered as growth species and therefore a large fraction of it reacts to soot. The precursors are reduced as the acetylene reacting to soot is no longer available for forming these species and so less precursors are formed.

The model parameters used for all the simulations are given in Table 7.1. They are a compromise for the different conditions considered here. A better agreement between the model and experiments could be achieved by readjusting the parameters for specific conditions.

As described above for the previous simulation the maximum volume fraction $f_{V,\infty}$ can be calculated from experimental results. For the following simulations this is not possible as only the soot yield is given and no values for deriving the volume fraction. Therefore $f_{V,\infty}$is found by doing simulations with the detailed soot model, as described above. The resulting volume fractions from the detailed model are shown in Fig. 7.10 for the oxidation of n-heptane in argon together with the fit curve resulting from Eq. (7.20).

There is a very good agreement between the simulation results and the fit for considering just one curve under these conditions. Also for other conditions of n-heptane oxidation considering just one curve is sufficient.

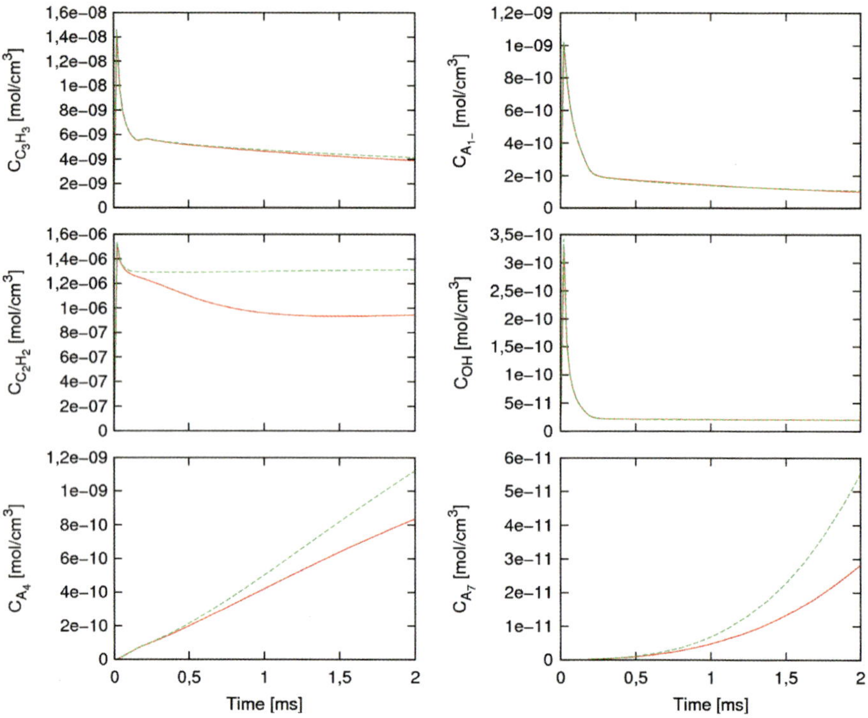

Fig. 7.9 Time evolution of the number of gas-phase species included in the simplified soot model and soot precursors during the oxidation of n-heptane in argon, with (*red*) and without (*green*) considering the simplified soot model ($c_C = 7.89$ mol/m^3, $\Phi = 5$, T = 1,750 K, p = 25 bar, $t_r = 2.0$ ms)

Table 7.1 Parameters for the simplified soot model with and without considering nucleation from phenyl

Parameter	Without phenyl	With phenyl
α_1	$6.0 \cdot 10^3$	$5.0 \cdot 10^3$
α_2	0.0	$2.0 \cdot 10^{13} T^{-2.05} \exp\left(\frac{12{,}133}{RT}\right)$
β	$1.0 \cdot 10^9$	$1.0 \cdot 10^9$
γ	99.82	99.82
ε	$3.34 \cdot 10^3$	$3.34 \cdot 10^3$

In contrast for the simulations carried out for oxidation of toluene two curves are needed for a good agreement to the simulation results of the detailed model. In Fig. 7.11 these curves are plotted together with the resulting fit and the simulation results of the detailed soot model for the oxidation of toluene.

These fit curves are generated for all the conditions simulated below. The resulting fit parameters are given in Table 7.2. The values are used within the simulations to calculate the maximum volume fraction for the corresponding temperature.

In Fig. 7.12 the temperature dependence of the soot yield for the oxidation of aliphatic fuel n-heptane in argon at different pressures is given. The results are

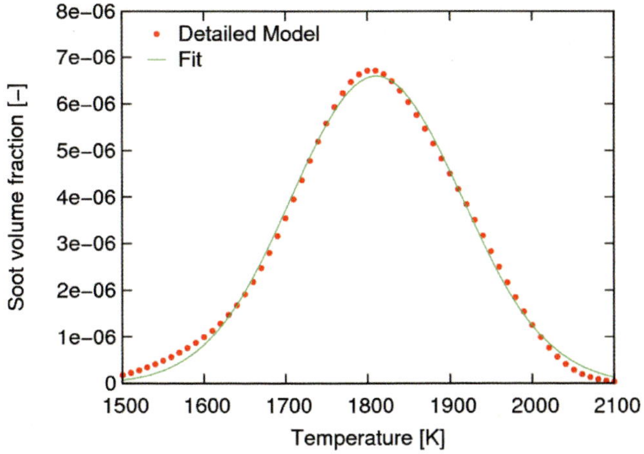

Fig. 7.10 Maximum soot volume fraction for the oxidation of n-heptane in argon resulting from the detailed soot model (*symbols*) and the corresponding fit curve for these conditions ($x_{Ar} = 0.55$, $\Phi = 5$, $p = 40$ bar, $t_r = 2.0$ ms)

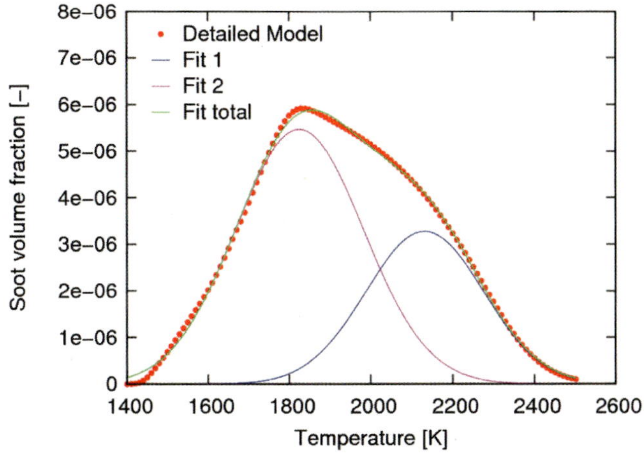

Fig. 7.11 Maximum soot volume fraction for the oxidation of toluene in argon resulting from the detailed soot model (*symbols*) and the corresponding fit curve for this conditions ($x_{toluene} = 1.5\%$, $x_{O2} = 1.5\%$, $t_r = 2.0$ ms)

plotted for experiments [58] and simulations with the detailed and the simplified soot model. The latter one is used with and without considering nucleation from phenyl.

Between the two models there is just a slight difference at the peak values of the soot yield for the different pressures and they both agree well to the experimental results. No difference can be seen for the simplified model with and without considering nucleation from phenyl under these conditions.

Table 7.2 Parameters for calculating the temperature dependence of the maximum volume fractions (Eq. 7.20) within the simplified soot model at different conditions

Conditions	a_1	a_2	a_3	b_1	b_2	b_3
n-Heptane						
30 bar	0.0718658	20,765.2	1,796.70			
40 bar	0.141515	21,442.0	1,811.08			
50 bar	0.238233	23,243.6	1,822.82			
Toluene						
0% O_2	0.290462	53,215.2	2,160.22	0.405129	44,739.6	1,831.01
1.5% O_2	0.135201	41,232.7	2,132.48	0.270689	49,448.2	1,825.43
2.5% O_2	0.058832	34,696.2	2,138.64	0.223786	57,054.1	1,835.88

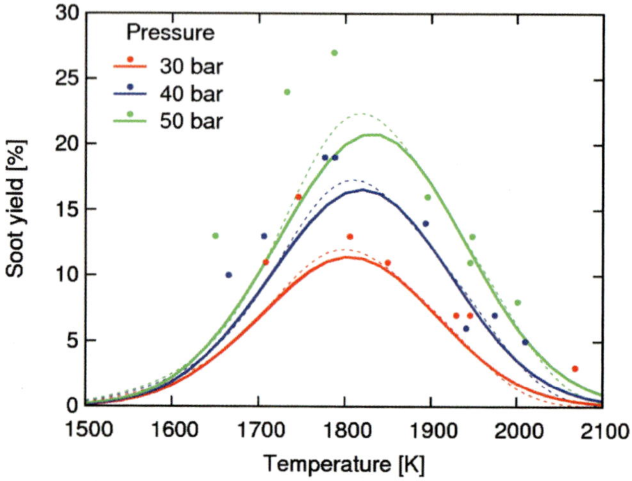

Fig. 7.12 Temperature dependence of the soot yield during oxidation of n-heptane in argon at different pressures, measured (*symbols*, [58]) and simulated (*lines*) with the detailed model (*dashed*) and the simplified model (*solid*) with (*thick*) and without (*thin*) nucleation from phenyl ($x_{Ar} = 0.99$, $\Phi = 5$, $t_r = 2.0$ ms)

7.6 Conclusion

A detailed reaction kinetical description for the formation and consumption of soot is developed within this work. Two models are implemented containing different reaction paths. The processes in gas phase are described by elementary reactions leading to the formation of polyaromatic soot precursors up to coronene. The processes in the particle phase are modeled with the help of polymere reactions. Here about 100 polymer reactions of 4 species representing soot precursors and particles are formulated to cover the main processes nucleation, coagulation, surface growth, and oxidation. The two models differ in considering polyynes forming soot particles. In *Model-1* they can form particles directly whereas in *Model-2* they only form precursor species. This modification is done as for the thermo kinetical- and transport-data of the bigger polyyne some more research is needed.

To simulate the processes in the particle phase, implemented as polymer reactions, the Compartment Method is developed. It combines aspects of the Methods of Moments and the Sectional Methods. The new method is capable of simulating particle size distributions. It is implemented in simulation programs for spatially homogenous and 1D-conditions, as assumed for shock tubes and premixed laminar flames. Up to now a version is used considering just one size compartment and therefore calculationg mean values for the particle species. This is sufficient for comparisons with experiments here.

The two reaction kinetical models are validated against experiments at different conditions. They show good agreement with ignition delay times and species profiles measured in shock tubes. Besides this gas-phase characteristics the model is also validated for soot characteristics like soot yield. There is a bigger difference between the two models as for the gas phase. The results of *Model-2* better agree with experiments.

The detailed soot model is also tested for premixed laminar flames and generally showed a good agreement with experiments although up to now all the mechanism development is carried out for shock tube conditions. The flame simulation show some new aspects for the future development of gas phase kinetics. This can also be seen for the results of soot formation in particle phase. They generally agree well with measurements, but also show some aspects that need to be improved in future versions of the mechanism.

Besides the detailed kinetical models a simplified soot model is further developed and enhanced. It describes soot formation by just two additional equations: One for the soot concentaration and one for soot volume fraction. The small additional computational effort therefore allows the use of the model in simulations of complex three dimensional systems. As a first test of the simplified model it is implemented in a simulation program for spatially homogeneous conditions, representing shock tubes.

The original version of the simplified model considered nucleation of the soot particles from propargyl only. This results in problems for aromatic fuels as there the pathways forming particles differ from aliphatic fuels. To overcome this problem in the current version nucleation from phenyl is implemented additionally. This version of the model is also capable of using a temperature dependent value for the maximum soot volume fraction, a parameter very important in the model. It limits contribution of surface growth to soot formation to an upper limit. The values for the volume fraction used in the model may be derived from experiments, or from simulations with the detailed model, if no measured values are available for the corresponding conditions. For the simplified model two set of parameters are determined, one for the model with nucleation from phenyl and one for model not regarding phenyl as a nucleation species.

The two versions of the model are compared to experiments with aliphatic and aromatic fuels. For the aliphatic fuel n-heptane there is nearly no difference between the results of both versions. But for the aromatic toluene the nucleation from phenyl has a big influence on the soot yield and gives much better agreement with experiments than without including it. The variable maximum volume fraction also improved the agreement with experiments. So overall there is a good agreement of the extended simplified soot model to the experiments despite the strong simplifications implemented.

Although the simulations with the detailed soot model are in good agreement with experiments for shock tubes, in further development of the reaction kinetical model the results from flame simulations need to be considered additionally. They showed bigger differences to experiments especially for species like benzene that are regarded to be important in soot formation processes. With the simulations of flames there is an additional focus on the thermo data and transport properties of the species.

Up to now all the soot characteristics are regarded as mean values. But from experiments it is known that characteristics of soot particles and their impact on health strongly depend on the size. Therefore the Compartment Method needs to be implemented in its full form. This allows simulations giving the size distribution of the soot particles for different conditions.

With the updated reaction model in future work the simplified soot model needs to be recalibrated for the latest mechanism and results from the detailed model. For future applications with respect to technical systems an automated finding of the model parameters is needed.

References

1. Agafonov, G.L., Naydenova, I., Vlasov, P.A., Warnatz, J.: Detailed kinetic modeling of soot formation in shock tube pyrolysis and oxidation of toluene and n-heptane. Proc. Combust. Inst. **31**, 575–583 (2007)
2. Appel, J., Bockhorn, H., Frenklach, M.: Kinetic modeling of soot formation with detailed chemistry and physics: laminar premixed flames of C_2 hydrocarbons. Combust. Flame **121**, 122–136 (2000)
3. D'Anna, A., Violi, A.: Detailed modeling of the molecular growth process in aromatic and aliphatic premixed flames. Energy Fuel **19**, 79–86 (2005)
4. Richter, H., Granata, S., Green, W.H., Howard, J.B.: Detailed modeling of PAH and soot formation in a laminar premixed benzene/oxygen/argon low-pressure flame. Proc. Combust. Inst. **30**, 1397–1405 (2005)
5. Frenklach, M.: Reaction mechanism of soot formation in flames. Phys. Chem. Chem. Phys. **4**, 2028–2037 (2002)
6. Wang, H., Frenklach, M.: A detailed kinetic modeling study of aromatics formation in laminar premixed acetylene and ethylene flames. Combust. Flame **110**, 173–221 (1997)
7. Smoluchowski, M.V.: Versuch einer mathematischen Theorie der Koagulationskinetik kolloider Loesungen. Z. Phys. Chem. **92**, 129–168 (1917)
8. Haynes, B.S., Wagner, H.G.: Soot formation. Progr. Energ Combust. Sci. **7**, 229–273 (1981)
9. Neoh, K.G., Howard, J.B., Sarofim, A.F.: Effect of oxidation on the physical structure of soot. Proc. Combust. Inst. **20**, 951–957 (1985)

10. Lucht, R.P., Sweeney, D.W., Laurendeau, N.M.: Laser-saturated fluorescence measurements of hydroxyl radical in atmospheric pressure methane/oxygen/nitrogen flames under sooting and non-sooting conditions. Combust. Sci. Tech. **42**, 259–281 (1985)
11. Frenklach, M., Clary, D.W., Yuan, T., Gardiner, W.C., Stein, S.E.: Mechanism of soot formation in acetylene-oxygen mixtures. Combust. Sci. Technol. **50**, 79–115 (1986)
12. Richter, H., Howard, J.B.: Formation and consumption of single-ring aromatic hydrocarbons and their precursors in premixed acetylene, ethylene and benzene flames. Phys. Chem. Chem. Phys. **4**, 2038–2055 (2002)
13. Skjøth-Rasmussen, M.S., Glarborg, P., Østberg, M., Johannessen, J.T., Livbjerg, H., Jensen, A.D., Christensen, T.S.: Formation of polycyclic aromatic hydrocarbons and soot in fuel-rich oxidation of methane in a laminar flow reactor. Combust. Flame **136**, 91–128 (2004)
14. D'Alessio, A., D'Anna, A., Minutolo, P., Sgro, L.A., Violi, A.: On the relevance if surface growth in soot formation in premixed flames. Proc. Combust. Inst. **28**, 2547–2554 (2000)
15. Violi, A., Sarofim, A.F., Voth, A.F., Voth, G.A.: Kinetic Monte Carlo-molecular dynamics approach to model soot inception. Combust. Sci. Technol. **176**, 991–1005 (2004)
16. Violi, A.: Modeling of soot particle inception in aromatic and aliphatic premixed flames. Combust. Flame **139**, 279–287 (2004)
17. Frenklach, M., Wang, H.: Detailed modeling of soot particle nucleation and growth. Proc. Comb. Inst. **23**, 1559–1566 (1991)
18. Frenklach, M., Wang, H.: In: Bockhorn, H. (ed.) Soot formation in combustion, pp. 165–192. Springer, Berlin/London (1994)
19. Vlasov, P.A., Warnatz, J.: Detailed kinetic modeling of soot formation in hydrocarbon pyrolysis behind shock waves. Proc. Combust. Inst **29**, 2335–2341 (2002)
20. Correa, C., Niemann, H., Schramm, B., Warnatz, J.: Reaction mechanism reduction for higher hydrocarbons by the ILDM mehod. Proc. Combust. Inst. **28**, 1607–1614 (2000)
21. Vlasov, P.A., Warnatz, J., Naydenova, I.: Kinetic Modeling of the soot formation mechanism during rich oxidation of methane, propane, and n-heptane behind shock waves. Chem. Phys. Rep. **23**, 36–43 (2004)
22. Kiefer, J.H., Sidhu, S.S., Kern, R.D., Xie, K., Chen, H., Harding, L.B.: The homogeneous pyrolysis of acetylene II: the high temperature radical chain mechansim. Combust. Sci. Technol. **82**, 101–130 (1992)
23. Kruse, T., Roth, P.: Kinetics of C_2 reactions during high-temperature pyrolysis of acetylene. J. Phys. Chem. A **101**, 2138–2146 (1997)
24. Agafonov, G.L., Nullmeier, M., Vlasov, P.A., Warnatz, J., Zaslonko, I.S.: Kinetic modeling of solid carbon particle formation and thermal decomposition during carbon suboxide pyrolysis behind shock waves. Combust. Sci. Technol. **174**, 185–213 (2002)
25. Sojka, J., Warnatz, J., Vlasov, P.A., Zaslonko, I.S.: Kinetic modeling of carbon suboxide thermal decomposition and formation of soot-like particles behind shock waves. Combust. Sci. Technol. **158**, 439–460 (2000)
26. Krestinin, A.V.: Polyyne model of soot formation process. Proc. Combust. Inst. **27**, 1557–1563 (1998)
27. Krestinin, A.V.: Detailed modeling of soot foormation in hydrocarbon pyrolysis. Combust. Flame **121**, 513–524 (2000)
28. Heghes, C.: C_1-C_4 hydrocarbon oxidation mechanism. Dissertation, University of Heidelberg, Germany (2006)
29. Burcat, A., Ruscic, B.: Third Millennium Ideal Gas and Condensed Phase Thermochemical Database for Combustion with updates from Active Thermochemical Tables. ANL-05/20 and TAE 960. Technion-IIT, Aerospace Engineering, and Argonne National Laboratory Chemistry Division (2005)
30. Burcat, A.: Third Millennium ideal gas and condensed phase thermochemical database for combustion. Techion Aerospace Engineering (TAE) report (2001)
31. Keller, A., Kovacs, R., Homann, K.: Large molecules, radicals, ions, and small soot particles in fuel-rich hydrocarbon flames. Part IV: large polycyclic aromatic hydrocarbons and their radicals in a fuel-rich benzene-oxygen flame. Phys. Chem. Chem. Phys. **2**, 1667–1675 (2000)

32. Kern, R.D., Singh, H.J., Esslinger, M.A., Winkeler, P.W.: Product profiles observed during the pyrolysis of toluene, benzene, butadiene and acetylene. Proc. Comb. Inst. **19**, 1351–1358 (1982)
33. Mathieu, O., Frache, G., DjebaÃ-li-Chaumeix, N., Paillard, C., Krier, G., Muller, J., Douce, F., Manuelli, P.: Characterization of adsorbed species on soot formed behind reflected shock waves. Proc. Combust. Inst. **31**, 511–519 (2007)
34. Ackermann, J., Wulkow, M.: MACRON – A program package for macromolecular kinetics. Preprint SC 90–14. Konrad-Zuse-Zentrum fuer Informationstechnik, Berlin (1990)
35. Agafonov, G.L., Naydenova, I., Nullmeier, M., Vlasov, P.A., Warnatz, J.: Detailed kinetic modeling of PAH growth and soot formation in shock tube pyrolysis of benzene and ethylene. In: Proceedings of the 20th ICDERS, Montreal, Canada (2005)
36. Naydenova, I., Nullmeier, M., Vlasov, P.A., Warnatz, J.: Detailed kinetic modeling of soot formation in pyrolysis of benzene, acetylene and benzene/acetylene mixtures behind shock waves. In: Proceedings of the 2nd International Workshop on Trends in Numerical and Physical Modeling for Turbulent Processes in Gas Turbine Combustors (and Automotive Engines), Heidelberg (2004)
37. Naydenova, I., Nullmeier, M., Warnatz, J., Vlasov, P.A.: Detailed kinetic modeling of soot formation during shock-tube pyrolysis of C_6H_6: direct comparison with the results of time-resolved laser-induced incandescence (LII) and CW-laser extinction measurements. Combust. Sci. Technol. **176**, 1667–1703 (2004)
38. Naydenova, I., Vlasov, P.A., Warnatz, J.: Detailed kinetic modeling of soot formation in pyrolysis of benzene/acetylene/argon mixtures. ECM, Louvain-la-Neuve (2005)
39. Sojka, J.: Simuation der Rußbildung unter homogenen Verbrennungsbedingungen. Dissertation, Ruprecht-Karls-Universitaet Heidelberg, Interdisziplinaeres Zentrum fuer Wissenschaftliches Rechnen (2001)
40. Deuflhard, P., Wulkow, M.: Computational treatment of polyreaction kinetics by orthogonal polynomials of a discrete variable. Impact Comput. Sci. Eng. **1**, 269–301 (1989)
41. Seinfeld, J.H., Pandis, S.N.: From air pollution to climate change. In: Atmospheric chemistry and physics. Wiley, New York (1998)
42. Frenklach, M., Harris, S.J.: Aerosol dynamics modeling using the method of moments. J. Colloid Interface Sci. **118**, 252–261 (1987)
43. Colket, M.B., Hall, R.J.: Successes and uncertainties in modeling soot formation in Laminar, premixed flames. In: Bockhorn, H. (ed.) Soot formation in combustion, pp. 442–470. Springer, Berlin/London (1994)
44. Gelbard, F., Seinfeld, J.H.: Simulation of multicomponent aerosol dynamics. J. Colloid Interface Sci. **78**, 485–501 (1980)
45. Gelbard, F., Tambour, Y., Seinfeld, J.H.: Sectional representation for simulating aerosol dynamics. J. Colloid Interface Sci. **76**, 541–556 (1980)
46. Maas, U., Warnatz, J.: Ignition processes in carbon-monoxide-hydrogen-oxygen mixtures. Proc. Combust. Inst. **22**, 1695–1704 (1988)
47. Deuflhard, P., Nowak, U., Wulkow, M.: Recent developments in chemical computing. Comput. Chem. Eng. **14**, 1249–1258 (1990)
48. Jens Marquetand.: Simulation der Rußbildung in Flammen und Stoßrohren mit einem detaillierten und einem semi-empirischen Modell, Dissertation, Ruprecht-Karls-Universitaet Heidelberg, Interdisziplinaeres Zentrum fuer Wissenschaftliches Rechnen Heidelberg (2010)
49. Warnatz, J.: Calculation of the structure of laminar flat flames I: flame velocity of freely propagating ozone decomposition flames. Ber. Bunsenges. Phys. Chem. **82**, 193–200 (1978)
50. Warnatz, J.: Calculation of the structure of laminar flat flames II: flame velocity of freely propagating hydrogen-air and hydrogen-oxygen flames. Ber. Bunsenges. Phys. Chem. **82**, 643–649 (1978)
51. Warnatz, J.: Calculation of the structure of laminar flat flames III: structure of burner-stabilized hydrogen-oxygen and hydrogen-fluorine flames. Ber. Bunsenges. Phys. Chem. **82**, 834–841 (1978)

52. Deuflhard, P., Nowak, U.: Efficient numerical simulation and identification of large chemical reaction systems. Ber. Bunsenges. Phys. Chem. **90**, 940–946 (1986)
53. Haynes, B.S., Wagner, H.G.: Z. Phys. Chem. **133**, 201–213 (1982)
54. Atkins, P.W.: Physical chemistry. Oxford University Press, New York (2002)
55. Tanke, D.: Rußbildung in der Kohlenwasserstoffpyrolyse hinter Stoßwellen. Dissertation, Universitaet Goetingen (1994)
56. Vasudevan, V., Davidson, D.F., Hanson, R.K.: Shock tube measurments of toluene ignition tumes. Proc. Combust. Inst. **30**, 1155–1163 (2005)
57. Bockhorn, H., Fetting, F., Wenz, H.W.: Investigation of the formation of high molecular hydrocarbons and soot in premixed hydrocarbon-oxygen flames. Ber. Bunsenges. Phys. Chem. **87**, 1067–1073 (1983)
58. Kellerer, H., Mueller, A., Bauer, H., Wittig, S.: Soot formation in a shock tube under elevated pressure conditions. Combust. Sci. Technol. **113–114**, 67–80 (1996)

Chapter 8
Discontinuous Galerkin Methods for Premixed Combustion Multiphase Problems

M. Oberlack and F. Kummer

Abstract Employing the low Mach number approximation, premixed combustion can be modeled as a two-phase flow with the flame modeled as a non-material interface. This will induce jumps in velocity and pressure at the interface, which depend on density ratio. In classical numerical methods for single-phase flows, such as finite volume, it is difficult to represent jumps accurately and stable, i.e. without oscillations, at the same time. Within this work we propose an extension to the Discontinuous Galerkin method which is able to exactly represent jumps, with sub-cell accuracy.

Keywords Discontinuous Galerkin • G-Equation • Level set • Premixed combustion

8.1 Introduction

8.1.1 Premixed Combustion and Multiphase Flows

Since in premixed combustion the flame thickness is very thin, it can be modeled as a discontinuity such as a two-phase flow with non-material interface. If physical

M. Oberlack (✉) • F. Kummer
Department of Mechanical and Processing Engineering, Technische Universität Darmstadt, Petersenstr. 30, 64287 Darmstadt, Germany

Graduate School Computational Engineering, Technische Universitüt Darmstadt, Dolivostr. 15, 64287 Darmstadt, Germany

M. Oberlack
Center of Smart Interface, Technische Universität Darmstadt, Petersenstr. 32, 64287 Darmstadt, Germany
e-mail: oberlack@fdy.tu-darmstadt.de

J. Janicka et al., *Flow and Combustion in Advanced Gas Turbine Combustors*, Fluid Mechanics and Its Applications 102, DOI 10.1007/978-94-007-5320-4_8,

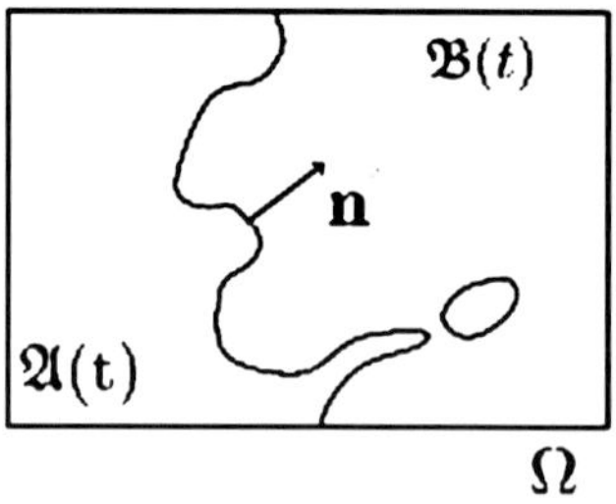

Fig. 8.1 Illustration of multiphase problem setting; phase $A(t)$, phase $B(t)$ and the interface $I(t)$

properties, namely density and viscosity, jump at the interface between the two fluids and if there is mass flux across the interface, the pressure and velocity field will contain jumps. (Rem.: For material interfaces, the velocity is always continuous, but it will contain a kink; the pressure field contains a kink at the interface, which could be derived from the Navier-Stokes equation itself, and a jump due to surface tension models.) In [27], it was proposed to use a Level Set approach to model this situation. For premixed combustion, it should be mentioned that the Level Set field and the associated evolution equation are called G-field and G-equation, respectively. However, in this work, adopting the established name in numerics, we use φ to notate the Level Set field.

(1) Notation (Phases)**:** For the usual D – dimensional computational domain $\Omega \in \mathrm{IR}^D$ the time-dependent disjoint decomposition

$$\mathbb{R} \geq 0 \ni t \mapsto (A(t),\ I(t),\ B(t)) \in \text{Pot } (\Omega)^3 \quad \text{with} \quad A(t) \cup I(t) \cup B(t) \forall t$$

is given. The sets A and B which correspond to unburned and burned Gas, respectively, are open, not necessarily connected, but each decomposes into a finite disjoint decomposition of simply connected parts, with D – dimensional volume of each part greater than 0. The set $I(t) = \bar{A}(t) \cap \bar{B}(t)$ represents the premixed flame, or more general the phase interface, between A and B and is a $(D-1)$ – dimensional manifold. This setup is illustrated in Fig. 8.1. The normal field on I, is oriented such that "it points from A to B", i.e. $\mathbf{n}_I = \mathbf{n}_A$ for the outer normal field $\mathbf{n}_A$ of A.

(2) Definition (Jump-Operator)**:** For a function that is continuous everywhere but on I we define

$$f_A(\mathbf{x}) := \lim_{\substack{\xi \to X \\ \xi \in A}} f(\xi) \text{ and } f_B(\mathbf{x}) : \lim_{\substack{\xi \to X \\ \xi \in B}} f(\xi)$$

and the usual jump notation is given as

$$[\![f]\!] (\mathbf{x}) := f_B(\mathbf{x}) - f_A(\mathbf{x}).$$

The Navier-Stokes equations for immiscible, incompressible, two-phase flow are given as

$$\frac{\partial}{\partial t}\mathbf{u} + \mathrm{div}\,(\mathbf{u} \otimes \mathbf{u}) + \nabla p = \frac{1}{\mathrm{Re}}\Delta \mathbf{u} \tag{8.1}$$

$$\mathrm{div}\,(\mathbf{u}) = 0 \tag{8.2}$$

within the bulk-phase $\Omega \backslash I$ together with the Jump conditions

$$[\![-s \cdot \rho + \rho \cdot \mathbf{u} \cdot \mathbf{n}]\!] = 0 \tag{8.3}$$

$$[\![\mathbf{n} \cdot \left(-s \cdot \rho \cdot \mathbf{u} + \rho \cdot (\mathbf{u} \otimes \mathbf{u}) + p \cdot \mathbf{n} - \mu \cdot \left(\nabla \mathbf{u} + \nabla \mathbf{u}^T\right)\right)]\!] = 0 \tag{8.4}$$

$$[\![\mathbf{t} \cdot \mathbf{u}]\!] = 0 \tag{8.5}$$

on the interface I. The field s denotes the interface speed, which is defined in §4. For premixed combustion, it consists of an advection part and a model for the self-propagation of the flame:

$$s = \mathbf{u}_A \cdot \mathbf{n}_I + s_l. \tag{8.6}$$

The flame speed s_l usually is given by a model which depends on local curvature and thermodynamic parameters. The jump condition (Eq. 8.3) could be derived from the conservation of mass, while (Eq. 8.4) could be derived from momentum conservation, see e.g. [26]; Eq. 8.5 enforces a no-slip condition in the interface tangent plane, with **t** denoting the tangent vector. It is necessary to close the system.

The Eqs. 8.1 and 8.2 – together with the jump conditions (Eqs. 8.3, 8.4 and 8.5) – are called singular, because the jump conditions can be translated to singular distributions and vice-versa. The relation between jumps and Delta – distributions is given in §25.

The numerical treatment of singular equations requires two ingredients. At first, the position of singularity – i.e. the position of the "front", or the "interface" – must be described, which is presently done by the well-known Level Set method. Second, the singularities – or discontinuities – of the field quantities must be represented in the numerics. This is achieved by the modulation of the DG basis by the characteristic function of the subdomains, or phases to be shown in §12.

8.1.2 Discontinuous Galerkin Methods

According to Cockburn [4], the Discontinuous Galerkin method was invented by Reed and Hill in 1973, [17], aiming on neutron transport problems.

Cockburn and Shu developed Runge-Kutta DG schemes (RKDG) in 1989 [5] for scalar hyperbolic problems. In order to work, these methods usually require total-variation-diminishing (TVD) Runge-Kutta schemes and generalized slope limiters. Later, these results have been extended to hyperbolic systems in multiple dimensions [6, 7].

For an incompressible Navier-Stokes solver, usually a Poisson equation has to be solved to compute the pressure, resp. to enforce conservation of mass. Furthermore, also the treatment of the viscous terms requires some discretization of the Laplace operator. In 1982, Douglas Arnold developed the so-called Interior Penalty discretization [1], see §9, which is essential for the solvers that are presented in Sects. 8.5 and 8.6.

8.1.3 Extended Finite Element

The extended finite element (XFEM) method was initially proposed by Moës, Dolbow, and Belytschko [15], in order to model solid mechanic problems which contain cracks and other discontinuities induced from material defects. Several authors have extended XFEM to two-phase flows, see e.g. [12, 13] and [8, 11].

In single-phase settings it is commonly accepted that fractional-step – approaches like the Projection method, but also approaches like Pressure correction, SIMPLE or PISO offer better performance than 'overall'-schemes, which assemble a large nonlinear, differential-algebraic system from the Navier-Stokes equations. Form this point of view, it seems promising to extend the ideas of such schemes to multiphase solvers. For XFEM, this has successfully been done in [3].

Because of the similarities between XFEM and XDG, it should be possible to extend these ideas to XDG. In this work, the building blocks of a fractional-step multiphase XDG solver will be constructed.

This article is organized as follows: within the Sects. 8.2 and 8.4 the Extended DG (XDG) method is introduced and formally constructed. After defining the XDG method, we address the question of how to determine the additional degrees-of-freedom of the method. This is demonstrated for linear, scalar steady-state and transient examples. For such problems, to fix the additional degrees-of-freedom, we propose what we call the "patched-decomposition" – approach. This approach is introduced for the Poisson equation with jump, which stands as a prototype for a steady-state problem (Sect. 8.5). Subsequently, these ideas are extended to transient equations. In order to explain the effect of the interface movement onto the $\partial/\partial t$ – operator, some results from Distribution theory are summed up (Sect. 8.6.1). Given that, the patched-decomposition approach is extended to a singular Heat equation (Sect. 8.6.2).

8.2 The Level-Set – Framework

(3) Definition (Level-Set – function)**:** A continuous function

$$\mathbb{R}_{\geq 0} \times \Omega \ni (t, \mathbf{x}) \mapsto \varphi(t, \mathbf{x}) \in \mathbb{R}$$

is called *Level Set function* if

$$\begin{cases} A(t) = \{\mathbf{x};\ \varphi(t,\mathbf{x}) < 0\} \\ I(t) = \{\mathbf{x};\ \varphi(t,\mathbf{x}) = 0\} \\ B(t) = \{\mathbf{x};\ \varphi(t,\mathbf{x}) > 0\} \end{cases}$$

and φ fulfils the signed-distance property, i.e.

$$|\varphi(t,\mathbf{x})| = \min\{|\mathbf{x}-\mathbf{y}|;\ \mathbf{y} \in I(t)\}.$$

Consequently,

$$\operatorname{sign}(\varphi(t,\mathbf{x})) = \begin{cases} -1 & \mathbf{x} \in A(t) \\ 1 & \mathbf{x} \in B(t) \end{cases}.$$

(4) Remark (Level Set speed)**:** The connection between the Level-Set function and the surface speed in normal direction is given by

$$s = \frac{\partial \varphi}{\partial t} \cdot \frac{1}{|\nabla \varphi|},$$

but only on the interface itself. This equation even holds for Level Set functions which do not fulfill the signed-distance property, i.e. where $|\nabla\varphi| \neq 0$.

In premixed combustion, the surface speed model (Eq. 8.6) is obviously only physically motivated on the interface, i.e. on the flame itself. Since the level set function has to be evolved in the whole domain Ω, s must be somehow extended. This is indeed a nontrivial problem, but will not be addressed here; hence we refer to the textbooks [16, 21, 22] and the works [19, 20] for further reading. For a DG-specific Level-Set evolution algorithm we refer to [10].

8.3 Formal Definition of Classical Discontinuous Galerkin Method

At first, the numerical grid needs to be defined. The usual inner product of two functions f and g is notated as: $<f,g> := \int_\Omega f \cdot g \, d\mathbf{x}$.

(5) Definition (Numerical Grid)**:** Given is a set $G = \{K_1, \ldots, K_J\}$ of cells K_j. By notating the cells with an index j, an index-mapping, i.e. a bijective mapping

$$\{1, \ldots, J\} \ni j \leftrightarrow K_j \in G$$

is induced. These cells fulfill the usual requirements for a numerical grid, i.e. they do not overlap and their union covers the whole domain Ω. For each cell K_j, a diffeomorphism T_j is given so that K_j can be given as the image of a reference element K_{ref}, which is usually a triangle or a quadrilateral (for $D = 2$) or a tetrahedron or a hexahedron (for $D = 3$), i.e. $K_j = T_j\ (K_{\text{ref}})$.

(6) Definition (Discontinuous Galerkin Approximation)**:** The *discontinuous Galerkin space*, or *DG-space* of degree p is defined as

$$DG_p(G) := \left\{ f \in L^2(\Omega); \forall K \in G : f|_K \text{ is polynomial and deg } (f|_K) \leqslant p \right\}.$$

(7) Definition (DG – basis)**:** In subsequence let N_p be the number of polynomials up to degree p in D variables. For $DG_p(G)$, an orthonormal basis

$$\Phi = \left(\Phi_{1,\,1},\ \ldots,\ \Phi_{1,N_p}, \cdots\cdots, \Phi_{J,\,1},\ \ldots,\ \Phi_{J,\,N_p}\right)$$

is given. For the basis it is further assumed that the support of each basis element is limited to exactly one cell, i.e.

$$\text{supp}\left(\Phi_{j,n}\right) = \overline{K}_j.$$

An element u in $DG_p(G)$ can be represented as $u = \underline{\Phi} u^*$, with the DG-coordinate vector

$$u^* = \left(u^*1, 1, \ldots, u^*{}_{1,N_p}, \cdots\cdots, u^*{}_{J,1}, \ldots, u^*{}_{J\,N_p}\right)^T \in \mathbb{R}^{J \cdot Np}.$$

(8) Definition and Remark (DG – projection)**:** The projector

$$L^2(\Omega) \ni f \mapsto Proj_p(f) = \Phi \cdot f* \to DG_p(G),$$

is given by the property $< f - Proj_p(f), g > = 0$ for all g in $DG_p(G)$. For the *orthonormal* basis functions $\Phi_{j,n}$, the DG-coordinates of $Proj_p(f)$ are given by

$$f_{j,n}{}^* = < \Phi_{j,n}, f >$$

Since this work only focuses on the Poisson- and Heat-equation, we will only give the DG discretization of the Laplace operator by the so-called Interior Penalty method [1, 2].

(9) Corollary and Definition (Interior Penalty DG discretization of the Laplace operator)**:** The DG-discretization of the Laplace operator

$$H^2(\Omega) \ni u \mapsto \Delta u \in L^2(\Omega),$$

together with boundary conditions

$$\begin{aligned} u|\Gamma_{\text{Diri}} &= g_{\text{Diri}} \quad \text{and} \\ (\mathbf{n} \cdot \nabla u)\,|\Gamma_{\text{Neu}} &= g_{\text{Neu}} \end{aligned}$$

for the partition $\partial\Omega = \Gamma_{\text{Neu}} \cup \Gamma_{\text{Diri}}$ of the domain boundary by the Interior Penalty method is denoted by

$$DG_p(G) \ni u \mapsto \mathbf{L}\Delta(u) \in DG_p(\Omega).$$

For the boundary of some cell K in G, and some u in $L^2(\Omega)$, we define $u^{\text{in}}(\mathrm{x}) := \lim_{\alpha\to 0}(u(\mathbf{x} - \mathbf{n}\,\alpha))$ and $u^{\text{out}}(\mathrm{x}) := \lim_{\alpha\to 0}(u(\mathbf{x} + \mathbf{n}\,\alpha))$, for an outer normal $\mathbf{n}$ on ∂K, if these limits exist.

Given a linear Riemannian for the *primal variable u*, namely $G^R(u^{\text{in}}, u^{\text{out}})$ and a linear Riemannian for the *derivative variable* ∇u, namely $\sigma^R(u^{\text{in}}, u^{\text{out}}, \nabla u^{\text{in}}\mathbf{n}, \nabla u^{\text{out}}\cdot\mathbf{n}, \mathbf{n})$, $\mathbf{L}_\Delta(u)$ can be specified by the equation

$$\langle \mathbf{L}_\Delta(u), \Phi_{j,n} \rangle = \int_{\partial K_j} \left(\sigma^R \cdot \Phi_{j,n} - \left(G^R - u\right) \cdot \nabla\Phi_{j,n}\right) \cdot \mathbf{n}\,\mathrm{dS} - \int_{K_j} \left(\nabla\Phi_{j,n} \cdot \nabla u\right) d\mathbf{x}$$

for all j,n.

The Riemannians are given as

$$\begin{aligned} G^R\left(u^{\text{in}}, u^{\text{out}}\right) &= \frac{1}{2}\left(u^{\text{in}} + u^{\text{out}}\right) \quad \text{and} \\ \sigma^R\left(u^{\text{in}}, u^{\text{out}}, \nabla u^{\text{in}} \cdot \mathbf{n}, \nabla u^{\text{out}} \cdot \mathbf{n}, \mathbf{n}\right) &= \frac{\nu}{2}\left(\nabla u^{\text{in}} + \nabla u^{\text{out}}\right) - \mu \cdot \mathbf{n} \cdot \left(u^{\text{in}} - u^{\text{out}}\right) \end{aligned}$$

on the interior edges of the grid. On Γ_{Diri} the Rimannians are

$$\begin{aligned} G^R\left(u^{\text{in}}\right) &= g_{\text{Diri}} \quad \text{and} \\ \sigma^R\left(u^{\text{in}}, \nabla u^{\text{in}} \cdot \mathbf{n}, \mathbf{n}\right) &= \nabla u^{\text{in}} - \mu \cdot \mathbf{n} \cdot \left(u^{\text{in}} - g_{\text{Diri}}\right) \end{aligned}$$

and on Γ_{Neu} they are defined as

$$\begin{aligned} G^R\left(u^{\text{in}}\right) &= u^{\text{in}} \quad \text{and} \\ \sigma^R\left(u^{\text{in}}, \nabla u^{\text{in}} \cdot \mathbf{n}, \mathbf{n}\right) &= \mathbf{n} \cdot g_{\text{Neu}}. \end{aligned}$$

The operator $\mathbf{L}_\Delta$ is affine-linear, i.e. for u, v in $DG_p(G)$ one gets the equality

$$\mathbf{L}\Delta(u + v) = \mathbf{L}\Delta(u) + \mathbf{L}\Delta(v) - \mathbf{L}\Delta(0).$$

For $= \underline{\Phi} \cdot u^* \in DG_p(G)$, let be $\mathbf{L}_\Delta(u) =: w = \underline{\Phi} w^*$. The matrix M_Δ and the vector $c' \in \mathrm{IR}^{J \cdot Np}$ denote the linear and the affine coordinate representation of $\mathbf{L}_\Delta$, i.e.

$$w* = M\Delta \cdot u * + c'.$$

It can be shown that the Interior Penalty operator is symmetric, i.e. $M_\Delta = {M_\Delta}^T$.

(10) Remark (about the penalty parameter)**:** The so-called penalty parameter μ in §9 must be chosen large enough to ensure coercivity of the operator matrix; However, a larger penalty parameter increases the condition number of the matrix and therefore increases the cost of running an iterative sparse solver. For computing μ, a lower-bound estimation from [23] is used.

8.4 The XDG Framework

Ideas similar to the ones presented below have been proposed by many authors in the context of Finite Element methods, summarized under the term "extended Finite Element", or XFEM. Consequently, the method presented here is called "extended DG" or XDG.

(11) Definition (cut cells)**:** Given the definitions from §1, for a Level Set function $\varphi(\mathbf{x})$, the set of *cut cells* is defined as:

$$Cut(\varphi) := Near(\varphi, 0) := \{K \in G;\ K \cap I \neq \{\}\}.$$

(12) Definition (XDG space): The extended DG space is defined as

$$XDG_p(\varphi, G) := DG_p(G) \oplus \bigoplus_{\substack{j \in J_c \\ n \in N_s}} 1_{A(t)} \cdot \phi_{j,n} \cdot \mathbb{R}$$

where the set $N_S \subseteq \{0, \ldots, N_p - 1\}$ is the set of separate degrees-of-freedom (DOF) for the cut-cells and the indices of all cut cells $J_C = \{j;\ K_j \text{ in } Cut(\varphi)\}$. $\mathbf{1}_X$ denotes the characteristic function of the set X, i.e. on X, $\mathbf{1}_X = 1$ and everywhere else $\mathbf{1}_X = 0$. The integer N_p is the number of DOF in an uncut cell, i.e. $N_p = \dim(DG_p(\{K_{\mathrm{ref}}\}))$.
(13) Definition (Species Subgrids)**:** We introduce two subgrids $G(A)$ and $G(B)$, which contain all cells that are partly or fully occupied by species A resp. B:

$$G(A) = \{K \in G;\ \mathrm{Vol}_D(A \cap K) > 0\}.$$

(14) Definition (Species Continuation)**:** An isomorphism

$$\begin{aligned} DG_p\,(G(A)) \times DG_p\,(G(B)) &\cong XDG_p(\varphi, G) \\ (u_A, u_B) &\mapsto 1_A \cdot u_A + 1_B \cdot u_B \\ (Cont_A(u), Cont_B(u)) &\leftarrow u \end{aligned} \tag{8.7}$$

is given. By an trivial injection $DG_p(G(A)) \to DG_p(G)$ the mappings $Cont_A(u)$ and $Cont_B(u)$ are continued onto $DG_p(G)$.

8.5 The Singular Poisson Equation

(15) Notation (restricted differential operator)**:** For an open set $X \subset \Omega$, a function $f \in H^n(\Omega)$ we define, for a differential operator ∂^α,

$$\partial^\alpha|_X f := \begin{cases} \partial^\alpha\,(f|_X) & \text{in } X \\ 0 & \text{in } \Omega \backslash X \end{cases}$$

the restriction of ∂^α onto X.

(16) Example: Let H be the Heaviside function. Then

$$\frac{\partial}{\partial x}\,|_{\mathbb{R}\backslash\{0\}}\, H(x) = 0,$$

while $\partial_x H(x)$ is not defined at $x = 0$, at least not in the classical sense.

The Poisson problem with jump is given as

$$\begin{cases} \Delta\,|_{\Omega\backslash I}\,\Psi & = f \quad \text{in } \Omega \\ [\![\Psi]\!] & = g_1 \quad \text{on } I \\ [\![\nabla\Psi \cdot \mathbf{n}_I]\!] & = g_2 \quad \text{on } I \\ \Psi & = g_{\text{Diri}} \text{ on } \Gamma_{\text{Diri}} \\ \nabla\Psi \cdot \mathbf{n}_{\partial\,\Omega} & = g_{\text{Neu}} \text{ on } \Gamma_{\text{Neu}} \end{cases}, \tag{8.8}$$

with the disjoint decomposition $\partial\Omega = \Gamma_{\text{Diri}} \cup \Gamma_{\text{Neu}}$, We assume the problem to be well-posed. It may stand as a prototype problem for a linear PDE with jumps. For solving Eq. 8.8, we propose a procedure which contains two ingredients that we refer to as "jump subtraction" and "patching". At first, these procedures should be motivated and explained on the continuous level. Afterwards, the XDG discretization will be given.

Jump subtraction: We assume to know an Ansatz function $\Psi_A \in H^2\,(\Omega \backslash I)$ that fulfills the jump condition, i.e.

$$\begin{cases} [\![\Psi_A]\!] = g_1 & \text{on } I \\ [\![\nabla\Psi_A \cdot \mathbf{n}_I]\!] = g_2 & \text{on } I \end{cases}$$

and is sufficiently smooth everywhere else. Then, instead of solving Eq. 8.8, one could define the decomposition

$$\Psi = \Psi_S + \Psi_A,$$

which immediately yields $\Psi_S \in H^2(\Omega\backslash I)$, and solve the equation

$$\begin{cases} \Delta\Psi_S = f - \Delta\big|_{\Omega\backslash I}\Psi_A & \text{in } \Omega \\ \Psi_S = g_{\text{Diri}} - \Psi_A & \text{on } \Gamma_{\text{Diri}} \\ \nabla\Psi_S \cdot \mathbf{n}_{\partial\Omega} = g_{\text{Neu}} - \nabla\Psi_A \cdot \mathbf{n}_{\partial\Omega} & \text{on } \Gamma_{\text{Neu}} \end{cases}. \tag{8.9}$$

Patching: The remaining problem is the construction of the Ansatz function Ψ_A; while this is a challenging problem in the whole domain Ω it seems, assuming the presence of a signed-distance Level Set function, quite easy in a close neighborhood of I. Since for the Poisson equation the interface is stationary, it is sufficient to know the Ansatz function within the cut cells only.

Their union will be referred to as *patching domain* in subsequence:

$$K_C := \bigcup_{K \in Cut(\varphi)} K.$$

Hence, within K_C the decomposed equation, (Eq. 8.9), is discretized, while outside, i.e. on $\Omega\backslash K_C$, where Ψ_A is unknown, Eq. 8.8 is discretized. An example for the construction of the Ansatz function within K_C is given in §20.

We define the patched decomposition

$$\Psi_+ := \Psi - \mathbf{1}_{K_C} \cdot \Psi_A, \tag{8.10}$$

where the jump subtraction is only done within the patching domain K_C, for a given Ansatz function $\Psi_A \in XDG_p(\varphi,G)$. Since Ψ_+ contains an artificial jump on the boundary of the patching domain, see Fig. 8.2, these jumps will "activate" the penalty terms in the Laplace discretization $\mathbf{L}_\Delta$. These artificial penalties are compensated by a so-called patching-correction field c_+, which is specified in subsequence. An explicit form of c_+ is difficult to give, therefore we define it by its characteristic property; an explicit formula can be gained from the definition of $\mathbf{L}_\Delta$, see §9, by "manipulating" the fluxes like illustrated in Fig. 8.2.

(17) Corollary and Definition (characteristic property of patching decomposition)**:** For an Ansatz function $\Psi_A \in XDG_p(\varphi,G)$ with no jump and no kink on the boundary of the patching domain, there exists a $c_+ (\Psi_A \cdot \mathbf{1}_{KC} \in DG_p(G))$ so, that for all $\Psi_S \in DG_p(G)$ the equation

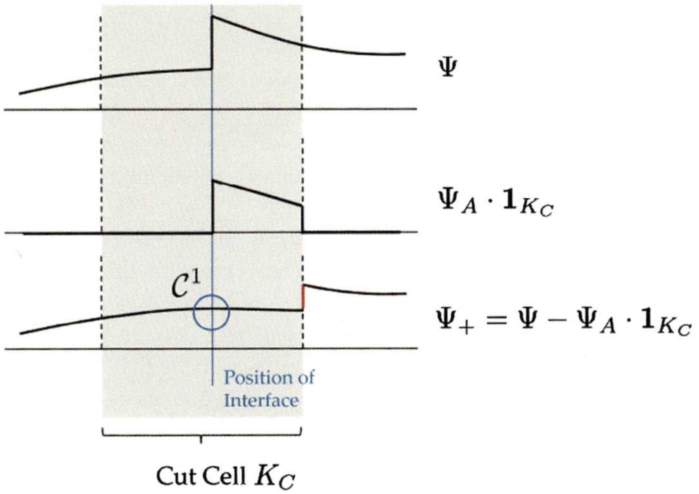

Fig. 8.2 Illustration of patching, 1D example. One may assume that the solution of some singular problem looks like Ψ. When the patched Ansatz $\Psi_A \cdot \mathbf{1}_{KC}$ – which is 0 outside of the patching domain K_C – is subtracted from Ψ, we observe two things: at first, jump and kink at I is removed, i.e. Ψ_+ and its first derivative are continuous. Second, another artificial jump in Ψ_+ arises, which is located exactly at a cell boundary. The idea of patching is to correct this jump in the flux functions, by the patching-correction vector c_+

$$\mathbf{L}_\Delta(\Psi_+) + c_+(\Psi_A \cdot \mathbf{1}_{K_C}) = \mathbf{L}_\Delta(\Psi_S) + \sum_{p \in \{A,B\}} \mathbf{1}_{p \setminus K_C} \cdot \left(\mathbf{L}_\Delta\left(Cont_{p\setminus K_C}(\Psi_A)\right) - \mathbf{L}_\Delta(0)\right). \tag{8.11}$$

holds and that

$$\operatorname{supp}(c_+(\Psi_A \cdot \mathbf{1}_{K_C})) = \bigcup_{\substack{K \in G \\ \mathrm{Vol}_D - 1(\partial K_C \cap K) \neq 0}} K. \tag{8.12}$$

Equation 8.11 could be interpreted as

$$\begin{aligned} \mathbf{L}\Delta(\Psi_+) + c_+(\Psi_A \cdot 1_{K_C}) &\approx \Delta\Psi_S + \sum p \in \{A,B\}\, 1_{p\setminus_{K_C}} \cdot \Delta\Psi_A \\ &= \Delta\Psi_S + 1_{\Omega\setminus K_C} \cdot \Delta\Psi_A \\ &= \begin{cases} \Delta\Psi_+ & \text{in } K_C \\ \Delta\Psi & \text{elsewhere, i.e. in } \Omega\setminus K_C \end{cases} . \end{aligned}$$

Here, it is assumed that

$$\mathbf{1}_{p \setminus K_C} \cdot \left(\mathbf{L}\Delta \left(Cont_p \left(\Psi_A \right) \right) - \mathbf{L}\Delta(0) \right) \approx \mathbf{1}_{p \setminus K_C} \cdot \Delta \Psi_A,$$

i.e. that the left-hand-side of the "equation" above is, for species S, an approximation to the Laplacian of the Ansatz function outside of the patching domain. Therefore it is required that the stencil of the operator $\mathbf{L}_\Delta$ is only one cell. If the stencil of the Laplacian would be larger, Eq. 8.11 would not hold, at least for narrow patching domains that just consist of the cut cells. The reason for this is the following: obviously, the continuation $Cont_A(\Psi_A)$ of the Ansatz function will feature a jump at $\partial K_C \cap B$ (the part of the boundary of the patching domain which is located in B). Because of the penalization, the Laplacian in all cells that are adjacent to the jump will show high oscillations. But since no cell in domain of interest (regarding Eq. 8.11), namely $A \setminus K_C$, bounds to the jump in $Cont_A(\Psi_A)$, these oscillations wont reach the domain of interest. Equation 8.12 describes that the support of c_+ is limited to cells that share an edge with the boundary of the patching domain K_C.

It is still an issue to approximate $\Delta|_{\Omega \setminus I} \Psi_A$ on the patching domain K_C, since Ψ_A must be assumed to be unknown outside of K_C.

(18) Corollary and Definition (restricted evaluation of $\mathbf{L}_\Delta$)**:** (i) For $\Psi_A \in DG_p(G)$, we define

$$\mathbf{L}_{rst} \left(\Psi_A \cdot \mathbf{1}_{K_C} \right) := \mathbf{L}\Delta \left(\Psi_A \cdot \mathbf{1}_{K_C} \right) + c_+ \left(\Psi_A \cdot \mathbf{1}_{K_C} \right).$$

(ii) For $\Psi_A \in XDG_p(\varphi, G)$, we define

$$\mathbf{L}_{rst} \left(\Psi_A \cdot \mathbf{1}_{K_C} \right) := \sum_{p \in \{A,B\}} \mathbf{1}_p \cdot \mathbf{L}_{rst} \left(Cont_{\,p} \left(\Psi_A \right) \cdot \mathbf{1}_{K_C} \right)$$

Interpretation: The purpose of $\mathbf{L}_{rst}$ $(\Psi_A \cdot \mathbf{1}_{KC})$ is to be an approximation to $\Delta|_{\Omega \setminus I} \Psi_A$ on the domain K_C, although Ψ_A is practically not known outside of K_C, i.e. only $\Psi_A \cdot \mathbf{1}_{KC}$ is known. We illustrate that by the diagram:

$$\begin{array}{ccc} \Psi_A & \xrightarrow{\text{patch.}} & \Psi_A \cdot \mathbf{1}_{K_C} \\ \mathbf{L}\Delta \downarrow & & \downarrow \mathbf{L}_{rst} \\ \mathbf{L}\Delta(\Psi_A) - \mathbf{L}\Delta(0) \mapsto & (\mathbf{L}\Delta(\Psi_A) - \mathbf{L}\Delta(0)) \cdot \mathbf{1}_{K_C} & \approx \mathbf{L}_{rst}\left(\Psi_A \cdot \mathbf{1}_{K_C}\right) \end{array}$$

This diagram is commutative, i.e. the $\approx$ in the lower right corner turns into an equality, if the jump of Ψ_A and $\nabla \Psi_A$ on ∂K_C are zero.

At this point we are in the position to formulate the central result of this section, the patched-decomposition scheme for the Poisson equation; it is illustrated in Fig. 8.4.

(19) Corollary (XDG scheme for a Poisson equation with an additive jump)**:** Given is a discretization $\Psi_A \cdot \mathbf{1}_{KC} \in XDG_p\,(\varphi,G)$ of the Ansatz function Ψ_A within the patching domain. A consistent discretization of the discontinuous Poisson problem defined in Eq. 8.8 is given by

$$\mathbf{L}_\Delta\left(\Psi_+\right) + c_+\left(\Psi_A \cdot \mathbf{1}_{K_C}\right) = Proj_p(f) - Proj_p\left(\mathbf{L}_{rst}\left(\Psi_A \cdot \mathbf{1}_{K_C}\right)\right). \quad (8.13)$$

The numerical solution $\Psi \in XDG_p\,(\varphi,G)$ of the discontinuous Poisson problem is given as

$$\Psi = \Psi_+ + \Psi_A \cdot \mathbf{1}_{K_C}.$$

Rationale: Given is the Poisson equation

$$\Delta|_{\Omega\backslash I}\ \Psi = f.$$

Patching and Jump-subtraction yields:

$$\Delta|_{\Omega\backslash\partial K_C} \Psi_S = f - \mathbf{1}_{K_C} \cdot \Delta|_{\Omega\backslash I} \Psi_A.$$

Equation 8.13 is just the DG- approximation of this.

(20) Remark (Construction of Ansatz function)**:** Given are g_1, g_2 in $L^2(I)$; It is further assumed that the curvature of I is bounded, which is the case if $\Delta\varphi$ is bounded. Then an Ansatz Ψ_A function with

$$\begin{cases} [\![\Psi_A]\!] = g_1 & \text{on } I \\ [\![\nabla\Psi_A \cdot \mathbf{n}_I]\!] = g_2 & \text{on } I \end{cases}$$

is given in a neighbourhood of I by

$$\Psi_A := \begin{cases} 0 & \text{for } \mathbf{x} \in A \\ \varphi(\mathbf{x}) \cdot g_2(\mathbf{x}_0) + g_1(\mathbf{x}_0) & \text{for } \mathbf{x} \in B \end{cases}$$

with $\mathbf{x}_0 = \mathbf{x} - \varphi(\mathbf{x})\,(\nabla\varphi)(\mathbf{x})$ for a signed-distance Level-Set – function φ.

(21) Example (Poisson equation with additive jump condition)**:** Given is a domain $\Omega = (0,10)^2$ discretized by a Cartesian equidistant grid with 64×64 equidistant cells. The polynomial degree of the DG interpolation is 2 and there are no common modes for both phases; the phases are explicitly given as

$$A = \{x \in \Omega;\ |x - (5,5)| > 3\},\ \ I = \partial A,\ \ \text{and } B = \Omega\backslash(A \cup I).$$

The Ansatz function is explicitly given as

$$\Psi_A(x) = \begin{cases} 0 \text{ if } \mathbf{x} \in A \\ \frac{x}{5} \text{ if } \mathbf{x} \in B \end{cases},$$

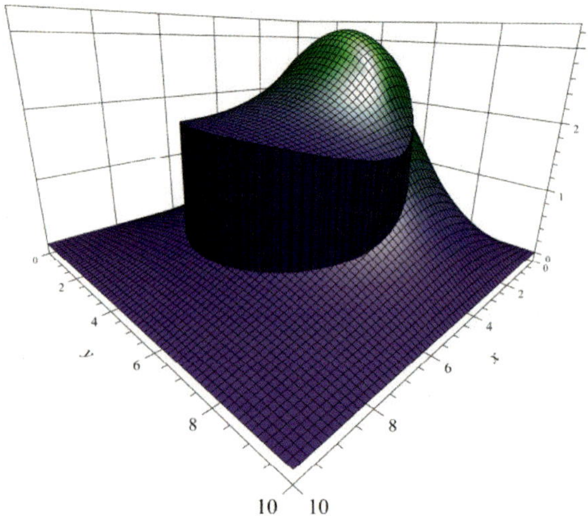

Fig. 8.3 Poisson problem $\Delta|_{\Omega \setminus I}\Psi = f$ with jump $[[\Psi]] = x/5$ and kink $[[\mathbf{n}_I \nabla\Psi]] = (x-5)/17.5$ on the interface I, which is a circle around point (5,5). The right-hand-side f is equal to -1 for $0 < x < 4$ and3 $< y < 7$. On the boundary of the domain $(0,10)^2$, discretized by 64×64 equidistant cells, homogeneous Dirichlet boundary conditions are used

which coincides with the jump conditions

$$[\![\Psi]\!] = \frac{x}{5}$$

$$[\![\mathbf{n}_I \cdot \nabla\Psi]\!] = \frac{x-5}{17.5}.$$

On the boundary, homogeneous Dirichlet boundary conditions are assumed; the right-hand-side of the Poisson problem is given as

$$f = \begin{cases} -1 \text{ if } |x-2| < 2 \text{ and } |y-5| < 2 \\ 0 \quad\quad\quad\quad\quad \text{otherwise} \end{cases}.$$

Numerical results are shown in Fig. 8.3, the Ansatz function and the Patching are illustrated in Fig. 8.4.

Notes on Performance: The presented XDG method may be most advantageous in the case of moving interfaces, i.e. if the Poisson problem is solved multiple times with a different interface position at each time. In this case the matrix of the system remains unchanged and no re-meshing is necessary when the interface position is changed. Exactly the opposite would be the case, if the grid would be re-meshed and adapted to the interface position every time it changes.

The XDG performance is compared to a Poisson problem without any jump condition, i.e. $\Delta\Psi = \mathrm{f}$ with homogeneous Dirichlet boundary conditions, using the same grid as for the cut-cell problem.

A conjugate gradient (CG) algorithm (see [14], page 120), without any preconditioning was used for solving the linear system. The termination criterion is set to an

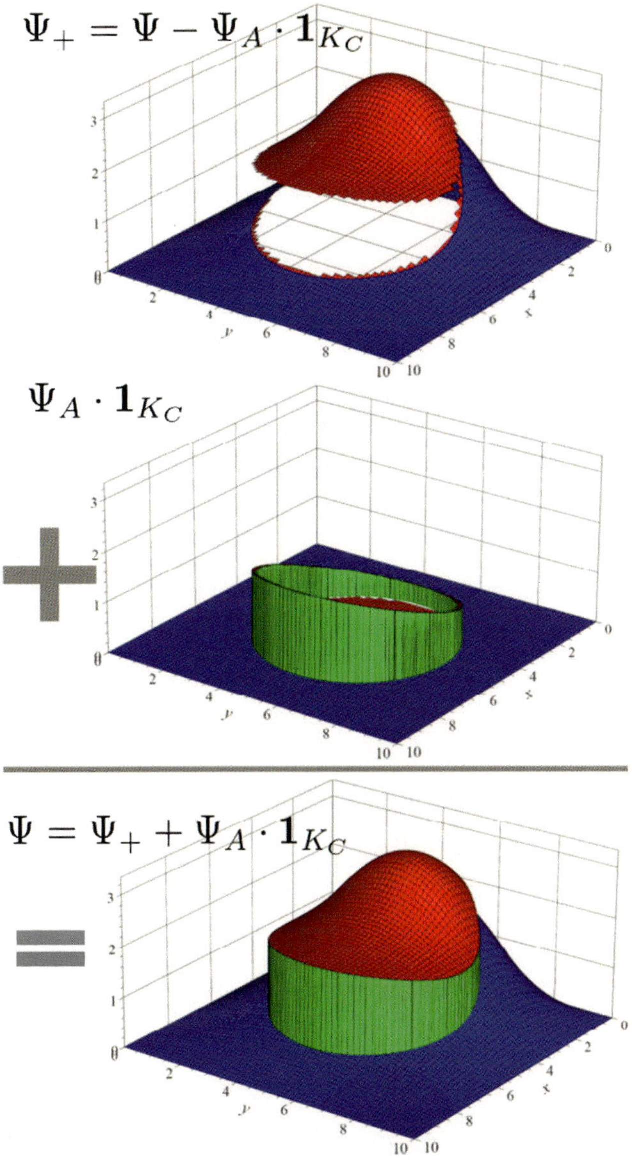

Fig. 8.4 Illustration of patching, 2D example: The solution Ψ_+ of the patched-decomposed equation (Eq. 8.13) shows an "artificial" jump on the boundary of the patching domain, which compensates with the jump in the patched Ansatz $\Psi_A \cdot \mathbf{1}_{KC}$. In sum, this gives the numerical solution of the singular problem specified in Fig. 8.3 and §21

absolute residual 2-norm of 10^{-7}, i.e. the algorithm terminates if $|| M\,x - b\,|| < 10^{-7}$ for the linear system $M\,x = b$. One CPU core (AMD Phenom II X4 940 3 GHz) was used to run the solver, using the Hypre-implementation [9].

The overhead, in terms of CPU time, can be broken down into three individual parts:

- the computation of the Ansatz function Ψ_A. In this example, it is explicitly known, but this may vary from case to case; therefore, the effect on runtime is not further investigated for this example.
- the computation of the patching-correction vector c_+: runtime is 0.015 (below 0.5% of overall solver runtime) seconds. It should be noted that at the present point our implementation contains extensive software optimization potential in comparison to the rest of the code.
- the solution of the linear system: runtime is 3.6 seconds (approximately 80% of total runtime) for 1,978 iterations.

The runtime of the matrix assembly – which is exactly the same for both, the discontinuous and the smooth problem – is approximately 0.8 s (18% of total runtime). However, this may heavily depend on problem dimension (2D, 3D), DG polynomial degree and many other factors. The "cost" of computing c_+ seems to be quite small in comparison to matrix assembly and negligible in comparison to the solver runtime.

The solver runtime heavily depends on preconditioning, but due to a lack of effective preconditioning techniques this was not further investigated. For the smooth problem the solver runtime, for the same residual threshold is 2.9 s for 1,604 CG iterations, which is approximately 20% faster than for the discontinuous problem.

8.6 The Singular Heat Equation

For transient problems, it is more convenient to notate the time-dependent phases in a time-space fashion. Therefore we introduce the time-space – versions of the phases and the interface:

(22) Definition (time-space formulation)**:** We define

$$\begin{aligned} \Omega^\times &:= \mathbb{R}_{>0} \times \Omega\,, \\ A^\times &:= \{(t,\mathbf{x}) \in \Omega^\times; \mathbf{x} \in A(t)\}\,, \\ I^\times &:= \{(t,\mathbf{x}) \in \Omega^\times; \mathbf{x} \in I(t)\}\,, \\ B^\times &:= \{(t,\mathbf{x}) \in \Omega^\times; \mathbf{x} \in B(t)\}\,. \end{aligned}$$

The time-space normal, which is only reasonable in an dimensionless setting, could be found from the surface speed s and the space-normal $\mathbf{n}_I$ by basic geometric considerations:

$$\mathbf{n}^\times := (-s, \mathbf{n}_I) \cdot \frac{1}{\sqrt{1+s^2}}$$

The full initial-boundary value problem with is given as

$$\begin{cases} \frac{\partial}{\partial t} u\big|_{\Omega^\times \setminus I^\times} - \nu \cdot \Delta|_{\Omega^\times \setminus I^\times} u & = 0 \qquad in\, \Omega^\times \\ [\![u]\!] & = g_1 \qquad on\, I^\times \\ [\![\nabla u \cdot \mathbf{n}_I]\!] & = g_2 \qquad on\, I^\times \\ u & = g_{\text{Diri}}\ on\, \mathbb{R}_{>0} \times \Gamma_{\text{Diri}} \\ \nabla u \cdot \mathbf{n}_{\partial\Omega} & = g_{\text{Neu}}\ on\, \mathbb{R}_{>0} \times \Gamma_{\text{Neu}} \\ u & = u^0 \qquad on\ \{0\} \times \Omega \end{cases}, \tag{8.14}$$

with the disjoint decomposition of the boundary into Dirichlet- and Neumann-region and a function $\nu(\mathbf{x})$.

8.6.1 Brief Overview About Theory of Distributions

In order to construct an implicit Euler scheme for the Heat equation with jump, Eq. 8.14 needs to be integrated over time. Therefore, one has to clarify the right-hand-side of

$$\int_{t^0}^{t^1} \frac{\partial}{\partial t} u\bigg|_{\Omega^\times \setminus I^\times} dt = ?.$$

Subsequently we show that it is advantageous to evaluate this integral by means of distribution theory. It was developed by Laurent Schwartz in 1944 [18] who himself was heavily influenced by the works of Jacques Hadamard, Paul Dirac and Sergei Lwowitsch Sobolew. In the year 1951, Schwartz received the Fields medal for his work.

We premise knowledge about the definition of the test-function space $D(\Omega)$ and the special definition of convergence within this space, within the domain of distribution theory. Details on the definitions may be taken from [24, 25]. One key goal of distribution theory is to explain derivatives which are not defined in the classical way by the infinitesimal difference quotients, e.g. the derivative of the Heaviside function. This is achieved by identifying functions with linear functionals on $D(\Omega)$ and shifting the burden of the differentiation to the C^∞- test functions. It can be shown that, for regular distributions that are differentiable in the classical sense, the distributional derivative coincides with the classical one.

(23) Definition/Notation (Distributions and Derivatives)**:** (i) *Distributions*: A linear functional T, notated as

$$D(\Omega) \ni \chi \mapsto (T, \chi) \in \mathbb{R}$$

is called to be a distribution if for all series χ_i with $\chi_i \to 0$ (in $D(\Omega)$) also $(T,\chi_i) \to 0$ (in IR). The set of all distributions is denoted as $D'(\Omega)$.

(ii) The *derivative of a Distribution*, for a multi-index α, of a distribution T is given by

$$(\partial^\alpha T,\ \chi) := (-1)^{|\alpha|} \cdot\ (T,\ \partial^\alpha\ \chi).$$

(24) Definition (special distributions)**:** (i) *regular distributions*: locally integrable functions f are called *regular distributions* and embedded into D'(Ω) by

$$(f, \chi) := \int_\Omega f \cdot \chi\ d\mathbf{x}.$$

(ii) *the (Dirac-) delta – distribution:* For the $(D-1)$ – dimensional manifold I and a locally integrable function f, one defines

$$(\delta_I \cdot f, \chi) := \int_I f \cdot \chi\ dS.$$

For the problems investigated in this work, i.e. certain PDE's with jumps at the interface, we have to calculate distributional derivative of functions with jumps at the interface I. The following two paragraphs create a direct link between the jump operator and the δ – distribution, resp. its derivative.

(25) Corollary (derivative of functions in $C^1(\Omega\backslash I)$)**:** For u in $C^1(\Omega\backslash I)$

$$\frac{\partial}{\partial x_i} u = \left.\frac{\partial}{\partial x_i}\right|_{\Omega\backslash I} u + \delta_I \cdot \mathbf{n} \cdot \mathbf{e}_i \cdot [\![u]\!]$$

Proof:

$$\begin{aligned}
\left(\frac{\partial}{\partial x_i} u, \chi\right) &=^{\text{Def.}} -\left(u, \frac{\partial}{\partial x_i}\chi\right) \\
&= -\int_{\text{supp}(\chi)\cap A} u.\text{div}(e_i\ \chi)\ d\mathbf{x} - \int_{\text{supp}(\chi)\cap B} \%\ d\mathbf{x} \\
&=^{\text{Gauss}-\int-\text{thm.}} \int_{\text{supp}(\chi)\cap A} \nabla u \cdot \mathbf{e}_i \cdot \chi\ d\mathbf{x} + \int_{\text{supp}(\chi)\cap B} \%\ d\mathbf{x} \\
&\quad - \underbrace{\int_{\partial\ \text{supp}(\chi)\cap A} \mathbf{e}_i \cdot \mathbf{n} u \cdot \chi\ dS}_{=0} - \underbrace{\int_{\partial\text{supp}(\chi)\cap B} \% dS}_{=0} \\
&\quad + \int_I \mathbf{e}_i \cdot \mathbf{n}_I (u_B - u_A) \cdot \chi\ dS \\
&= \left(\left.\frac{\partial}{\partial x_i}\right|_{\Omega\backslash I} u, \chi\right) + (\delta_I \cdot \mathbf{n} \cdot \mathbf{e}_i \cdot [\![u]\!]\,, \chi)
\end{aligned}$$

End of Proof.

(26) Definition (definite integrals of distributions)**:** (i) Given is a distribution T in D'(Ω); for a Lebesgue – measurable, bounded subset K of Ω let be

$$\int_K T\,d\mathbf{x} := \lim_{i\to\infty} (T, \chi_i)$$

for a sequence $(\chi_1,\chi_2,\ldots)$ in D(Ω) with $\chi_i \to \mathbf{1}_{KC}$. (in $L^2(\Omega)$).(ii) Given is a distribution T in $D(\Omega \times \Omega_2)$; for a Lebesgue – measurable, bounded subset K of Ω_2 the distribution $\int_K T\,d\mathbf{y}$ in D'(Ω) is defined

$$\left(\int_K T\,d\mathbf{y},\ \chi\right) := \lim_{i\to\infty} (T, (\mathbf{x},\mathbf{y}) \mapsto \chi(\mathbf{x}) \cdot \rho_i(\mathbf{y}))$$

for a test function χ in D(Ω) and a sequence $(\rho_1, \rho_2, \ldots)$ in $D(\Omega_2)$ with $\rho_i \to \mathbf{1}_K$ (in $L^2(\Omega)$).

By time integration of the right-hand-side of

$$\left.\frac{\partial}{\partial t}\right|_{\Omega^\times \setminus I^\times} u = \frac{\partial}{\partial t} u - \delta_I\ [\![u]\!] \cdot \frac{-s}{\sqrt{1+s^2}}$$

we are now able to answer the question from the beginning of this section:

$$\int_{t=t^0}^{t^1} \left.\frac{\partial}{\partial t}\right|_{\Omega^\times \setminus I^\times} u\,dt = u(t^1,-) - u(t^0,-) - \int_{t=t^0}^{t^1} \delta_I\ [\![u]\!] \cdot \frac{-s}{\sqrt{1+s^2}}\,dt. \quad (8.15)$$

This relation is used in the rationale of §28, in order to construct an implicit Euler scheme for the Heat equation with a jump. It should be mentioned that it is important for the construction of the numerical schemes to understand the difference between Eq. 8.15 and

$$\int_{t=t^0}^{t^1} \frac{\partial}{\partial t} u\,dt = u(t^1,-) - u(t^0,-).$$

Whichever of them applies depends on the problem that should be solved.

8.6.2 Implicit Euler XDG Schemes for the Heat Equation

(27) Definition and Remark: For some ν in $C^\infty(\Omega)$ we notate the matrix of the linear mapping

$$DG_P(G) \ni f \mapsto Proj_p(\nu \cdot f) \in DG_p(G)$$

as M_ν. For $0 \le j < J$ and $0 \le n,m < N_p$

$$[M_\nu]_{j \cdot N_p+m,\, j \cdot N_p+n} = \langle \nu \cdot \phi_{j,n},\ \phi_{j,m} \rangle .$$

All other entries of M_ν are 0, therefore M_ν is of block-diagonal shape with $N_p \times N_p$ – blocks and therefore, *numerically easy to invert*, since N_p is usually small (below 100).

Rationale:

$$\langle Proj_p(\nu \cdot f),\ \phi_{j,m} \rangle = \left\langle \nu \cdot \left(\sum\nolimits_{j,n} f^*_{j,n} \cdot \phi_{j,n} \right),\ \phi_{j,m} \right\rangle = \sum\nolimits_{j,n} f^*_{j,n} \cdot \langle \nu \cdot \phi_{j,n}, \phi_{j,m} \rangle .$$

At this point, the foundation for the formulation of an implicit Euler scheme for a scalar Heat equation with jump has been laid out; It is important to notice that the initial value u^0 at time t^0 and the numerical solution u^1 at time t^1 are in different XDG spaces, since the jump has moved, i.e. u^1 is in $XDG_p(\varphi^1,G)$ but u^0 is in $XDG_p(\varphi^0,G)$. In consequence, any nontrivial linear combination of u^1 and u^0 has *two* jumps, at I^1 and at I^0 and is therefore neither a member of $XDG_p(\varphi^1,G)$ nor $XDG_p(\varphi^0,G)$. Since an implicit Euler scheme requires linear combinations between initial and new solution, this is "lifted" by the projection onto the standard DG space, i.e. by taking linear combinations of the form $\alpha\, u^1 + \beta\, Proj_p(u^0)$ in $XDG_p(\varphi^1,G)$.

(28) Corollary (Implicit Euler scheme for a scalar Heat equation with additive jump condition)**:** Given is the initial-boundary value problem defined in Eq. 8.14; w.l.o.g. let be $t^0 = 0$ and $t^1 > 0$. Then we define/recall, for i in $\{0,1\}$:

- the initial value at time t^0 and the solution at time t^1: $u^i = u(t^i, -)$.
- an Ansatz function $u_A \in H^2(\Omega^\times \backslash I^\times)$ that fulfills the jump condition, especially $u^i{}_A \in H^2(\Omega \backslash I(t^i))$.
- the decomposition into smooth part and Ansatz, i.e. $u^i{}_S = u^i - u^i{}_A$.
- the Patching domain

$$K_C := \left(\bigcup_{K \in Cut(\varphi^0)} K \right) \cup \left(\bigcup_{L \in Cut(\varphi^1)} L \right),$$

  consisting of the cut cells at times t^0 and t^1.
- the patched decomposition: $u^i{}_+ = u^i - \mathbf{1}_{KC} \cdot u^i{}_A$
- the patching correction vector $c_+ = c_+\ (u^1{}_A \cdot \mathbf{1}_{KC}),\ \Phi^T$ in the sense of §17, for the patching domain K_C and Ansatz function $u^1{}_A$ at time t^1.

An implicit Euler discretization is given by

$$\left(\frac{1}{\Delta t} M_\nu^{-1} \cdot M_\Delta \right) \cdot u_+^{*1} = c' + c_+ + \langle Proj_p \left(\mathbf{L}_{rst} \left(u_A^1 \cdot \mathbf{1}_{K_C} \right) \right),\ \Phi^T \rangle$$
$$+ M_\nu^{-1} \cdot \left(\frac{1}{\Delta t} u_+^{*0} - \left\langle Proj_p \left(\frac{I_{K_C}}{\Delta t} \cdot (u_A^1 - u_A^0) - \int_{t^0}^{t^1} \delta_{I^\times} \cdot \frac{-s}{\sqrt{1+s^2}} [\![u_A]\!]\ dt \right),\ \Phi^T \right\rangle \right) \tag{8.16}$$

(Rem.: by $< Proj_p(f), \underline{\Phi}^T >$, we denote the DG coordinates of $Proj_p(f)$) If M_ν is symmetric, also the matrix of the linear system, i.e. $(M_\nu{}^{-1}/\Delta t - M_\Delta)$ is symmetric.

Rationale: Given is the Heat equation, i.e.

$$\left.\frac{\partial}{\partial t}\right|_{\Omega^\times \setminus I^\times} u = \nu \cdot \Delta|_{\Omega^\times \setminus I^\times} u.$$

Integration of the left-hand-side over the interval (t^0, t^1) and an implicit Euler Ansatz for the right-hand-side yields:

$$u^1 - u^0 - \int_{t^0}^{t^1} \delta_{I^\times} \cdot \frac{-s}{\sqrt{1+s}^2} [\![u]\!]\, dt = \Delta t \cdot \nu \cdot \Delta|_{\Omega^\times \setminus I^\times} u^1. \tag{8.17}$$

Inserting the decomposition $u^i = u^i{}_S + u^i{}_A$ into Eq. 8.17 yields

$$u_S^1 - u_S^0 + u_A^1 - u_A^0 - \int_{t^0}^{t^1} \delta_{I^\times} \cdot \frac{-s}{\sqrt{1+s}^2} [\![u]\!]\, dt = \Delta t \cdot \nu \cdot \left(\Delta u_S^1 + \Delta|_{\Omega^\times \setminus I^\times} u_A^1\right). \tag{8.18}$$

A DG discretization of Eq. 8.18 yields

$$\left(\frac{1}{\Delta t} I - M_\nu \cdot M_\Delta\right) \cdot u_S^{*1} = M_\nu \cdot \left(c' + \left\langle Proj_p \left(\Delta|_{\Omega^\times \setminus I^\times} u_A^l\right), \Phi^T\right\rangle\right)$$
$$+ \frac{1}{\Delta t} u_S^{*0} - \left\langle Proj_p \left(\frac{1}{\Delta t} \cdot \left(u_A^l - u_A^0\right) - \int_{t^0}^{t^l} \delta_{I^\times} \cdot \frac{-s}{\sqrt{1+s^2}} [\![u_A]\!]\, dt\right), \Phi^T \right\rangle \tag{8.19}$$

Note that, the matrix of the linear system to solve, in Eq. 8.19 is not necessarily symmetric; to get a symmetric system, Eq. 8.19 is multiplied by $M_\nu{}^{-1}$. Afterwards, patching may be applied and the term $Proj_p(\Delta|_{\Omega \setminus I} u_A{}^1)$ may be approximated by the same method that was used for the Poisson equation, see §20. This finally gives Eq. 8.16. End of rationale.

(29) Remark (Simplified implicit Euler scheme for a scalar Heat equation with additive jump condition)**:** If the time-dependent Ansatz function of §28 can be written as

$$u_A(t, \mathbf{x}) = \begin{cases} u_1(\mathbf{x}) \text{ if } \mathbf{x} \in A(t) \\ u_2(\mathbf{x}) \text{ if } \mathbf{x} \in B(t) \end{cases}$$

with *time-independent* functions u_1 and u_2. Then Eq. 8.16 simplifies to

$$
\begin{aligned}
&\left(\frac{1}{\Delta t} M_\nu^{-1} - M_\Delta\right) \cdot u_+^{*1} = c' + c_+ \\
&\quad + Proj_p \left(\mathbf{1}_A \cdot \mathbf{L}_{rst}\left(Cont_A\left(u_A^l\right) \cdot \mathbf{1}_{K_C}\right) + \mathbf{1}_B \cdot \mathbf{L}_{rst}\left(Cont_B\left(u_A^l\right) \cdot \mathbf{1}_{K_C}\right)\right) \\
&\quad + M_\nu^{-1} \cdot \left(\frac{1}{\Delta t} u_+^{*0}\right).
\end{aligned} \tag{8.20}
$$

Rationale: Because

$$
\left.\frac{\partial}{\partial t}\right|_{\Omega^\times \setminus I^\times} u_A = 0,
$$

it follows that

$$
\frac{1_{K_C}}{\Delta t} \cdot \left(u_A^1 - u_A^0\right) - \int_{t^0}^{t^1} \delta_{I^\times} \cdot \frac{-s}{\sqrt{1+x^2}} [\![u_A]\!]\, dt = \int_{t^0}^{t^1} \left.\frac{\partial}{\partial t}\right|_{\Omega^\times \setminus I^\times} u_A\, dt = 0.
$$

(30) Example (Scalar Heat equation with additive jump condition)**:** Consider the notation of §28 and §29. Given is a domain $\Omega = (-1,1)^2$ discretized by a Cartesian equidistant grid with 64×64 cells. The polynomial degree of the DG interpolation is 2 and there are no common modes for both phases, which are explicitly given as

$$
A(t) = \{\mathbf{x} \in \Omega;\ |\mathbf{x}| < 0.25 + t\},\ I(t) = \partial A(t), \text{ and } B(t) = \Omega \setminus (A(t) \cup I(t)).
$$

The Ansatz function is explicitly given as

$$
u_A(t, \mathbf{x}) = \begin{cases} 1 + 0.3 \cdot x & \text{if } \mathbf{x} \in A(t) \\ 0 & \text{if } \mathbf{x} \in B(t) \end{cases}.
$$

On $\partial\Omega$, homogeneous Dirichlet boundary conditions are assumed; the initial value is given as

$$
u^0(\mathbf{x}) = (1 - x^2) \cdot (1 - y^2) + u_A(0, \mathbf{x}).
$$

Because of the structure of the Ansatz function, the discretization presented in §29 can be used. Numerical results are shown in Fig. 8.5.

8.7 Summary and Outlook

Presently we developed schemes for the solution of singular linear, scalar problems such as the Poisson (§19) and the Heat equation (§28). It should be mentioned that for those problems only affine jump conditions in the from $[[u]] = g$ have been

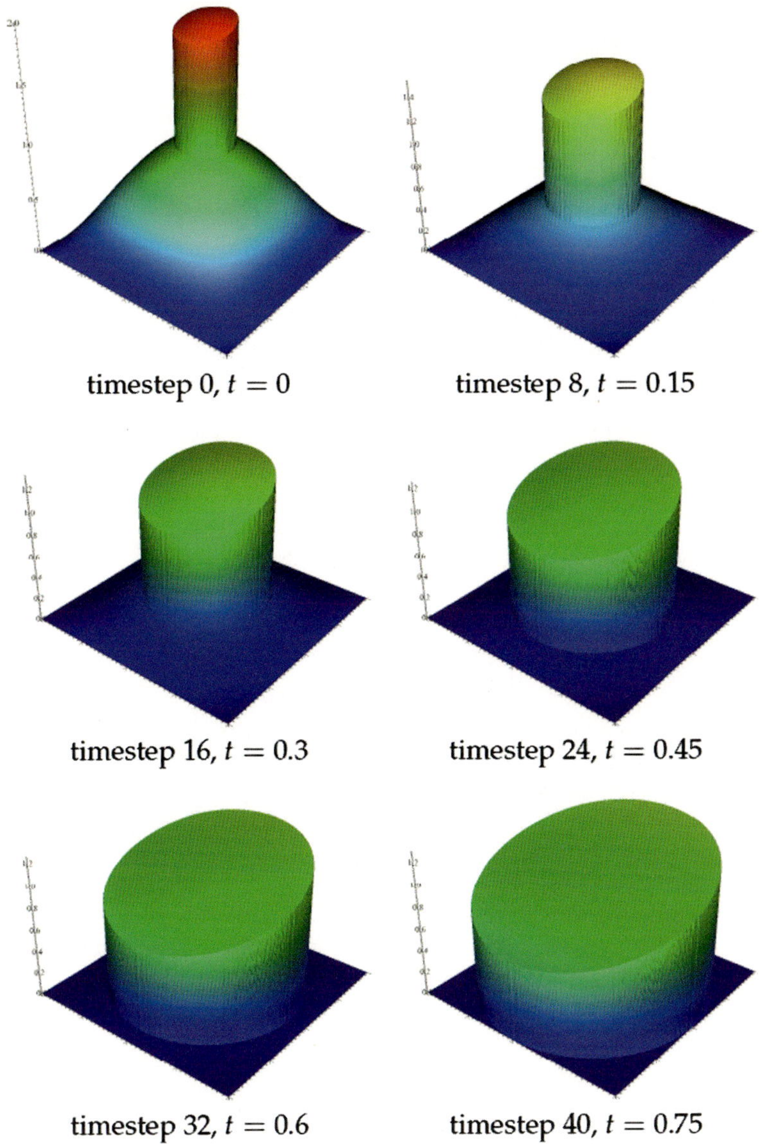

Fig. 8.5 Scalar heat equation with time-dependent jump

investigated. For jump conditions of this type it is clear that the solution is in an affine linear manifold within the XDG space, i.e. u is in $u_A + DG_p(G)$, for an Ansatz function u_A with $[[u_A]] = g$.

More complicated jump conditions would be affine-linear ones like e.g.

$$[\![\varpi \cdot u]\!] = g \text{ with } [\![\varpi]\!] \neq 0 \text{ on } I.$$

First experiments show that such Jump conditions could be approximated in an iterative procedure as follows. Given is a problem which formally reads as

$$\begin{cases} Op\ u = f & \text{in } \Omega \setminus I \\ \beta \cdot u_B + \alpha \cdot u_A = \gamma & \text{on } I \end{cases} \tag{8.21}$$

with a linear operator Op. This problem may be substituted by *a sequence* of problems with affine jump conditions

$$\begin{cases} Op\ \nu^n = f & \text{in } \Omega \setminus I \\ [\![\nu^n]\!] = c^n & \text{on } I \end{cases} \qquad n = 1, 2, 3, \ \ldots$$

which can be solved in the way that we propose for the Poisson equation (see Sect. 8.5). It converges to the solution of the original problem (Eq. 8.21), i.e.

$$\begin{cases} \nu^n \ \rightarrow \ u \\ \beta \cdot \nu_B^n + \alpha \cdot \nu_A^n \ \rightarrow \gamma \end{cases} \quad \text{for } n \ \rightarrow \ \infty.$$

Here, the relation between the real jump condition and the c^n is

$$c_n = \frac{-\gamma + \nu_A^{n-1} \cdot (\beta + \alpha) \cdot (1 - \vartheta) + \nu_B^{n-1} \cdot (\beta + \alpha) \cdot \vartheta}{-\beta \cdot (1 - \vartheta) - \alpha\vartheta}$$

with a scalar ϑ in (0,1). The basic idea behind that formula is the assumption that

$$\begin{aligned} \nu_A^n &\approx \vartheta \cdot \nu_A^{n-1} + (1 - \vartheta) \cdot \nu_B^{n-1} - \theta \cdot c^n \\ \nu_B^n &\approx \vartheta \cdot \nu_A^{n-1} + (1 - \vartheta) \cdot \nu_B^{n-1} + (1 - \theta) \cdot c^n \end{aligned} \quad \text{on } I.$$

As this is still work-in-progress, it will not be discussed further.

A second critical issue seems to be how the jump and kink conditions, which have been derived from conservation laws and additional constitutive laws (Eqs. 8.3, 8.4 and 8.5), can be integrated into the fractional-step schemes, and the numerical stability of these algorithms. This will be the issue of future works.

Acknowledgements The authors acknowledge the financial support from the German Research Council (DFG) through the SFB568.

References

1. Arnold, D.N.: An interior penalty finite element method with discontinuous elements. SIAM J. Numer. Anal. **19**(4), 742 (1982)
2. Arnold, D.N., Brezzi, F., Cockburn, B., Marini, L.D.: Unified analysis of discontinuous galerkin methods for elliptic problems. SIAM J. Numer. Anal. **39**(5), 1749–1779 (2002)

3. Chen, T., Minev, P.D., Nandakumar, K.: A projection scheme for incompressible multiphase flow using adaptive eulerian grid. Int. J. Numer. Methods Fluids **45**, 1–19 (2004)
4. Cockburn, B., Karniadakis, G.E., Shu, Chi-Wang.: Discontinuous Galerkinmethods: Theory, Computation and Applications. Number 11 in Lecture Notes in Computational Science and Engineering. Springer, Berlin (2000)
5. Cockburn, B., Shu, Chi-Wang.: TVB Runge-Kutta local projection discontinuous galerkin finite element method for conservation laws II: general framework. Math. Comput. **52**(186), 411 (1989)
6. Cockburn, B., Shu, Chi-Wang.: The p1-rkdg method for two-dimensional euler equations of gas dynamics. In: NASA Contractor Report187542, ICASE Report 91–32, p. 12 (1991)
7. Cockburn, B. Shu, Chi-Wang.: The Runge-Kutta discontinuous galerkin method for conservation laws v. J. Comput. Phys., **141**(2), 199–224 (1998)
8. Esser, P., Grande, J., Reusken, A.: An extended finite element method applied to levitated droplet problems. Int. J. Numer. Methods Eng. **84**(7), 757–773 (2010)
9. Falgout, R. Jones, J., Yang, Ulrike.: The design and implementation of hypre, a library of parallel high performance preconditioners. In: Barth, T.J., Griebel, M., Keyes, D.E., Nieminen, R.M., Roose, D., Schlick, T., Bruaset, A. M., Tveito, A. (eds.) Numerical solution of partial differential equations on parallel computers, volume 51 of Lecture notes in computational science and engineering, pp. 267–294. Springer, Berlin/Heidelberg (2006). 10.1007/3-540-31619-1_8
10. Grooss, J., Hesthaven, J.: A level set discontinuous galerkin method for free surface flows. Comput. Methods Appl. Mech. Eng. **195**(25–28), 3406–3429 (2006)
11. Groß, S., Reusken, A.: An extended pressure finite element space for two-phase incompressible flows with surface tension. J. Comput. Phys. **224**(1), 40–58 (2007)
12. Marchandise, E., Geuzaine, P., Chevaugeon, N., Remacle, J.-F.: A stabilized finite element method using a discontinuous level set approach for the computation of bubble dynamics. J. Comput. Phys. **225**, 949–974 (2007)
13. Marchandise, E., Remacle, J.-F.: A stabilized finite element method using a discontinuous level set approach for solving two phase incompressible flows. J. Comput. Phys. **219**(2), 780–800 (2006)
14. Meister, A.: Numerik linearer Gleichungssysteme: eine Einführung in moderne Verfahren. Vieweg, Wiesbaden (2005)
15. Moës, N., Dolbow, J., Belytschko, T.: A finite element method for crack growth without remeshing. Int. J. Numer. Methods Eng. **46**, 131–150 (1999)
16. Osher, S.J., Fedkiw, R.P.: Level Set Methods and Dynamic Implicit Surfaces. Springer, New York (2002)
17. Reed, W.H., Hill, T.R.: Triangular mesh methods for the neutron transport equation. In: National Topical Meeting on Mathematical Models and Computational Techniques for Analysis of Nuclear Systems, CONF-730414–2; LA-UR–73-479, p. 23. Los Alamos Scientific Laboratory, New Mexico (USA) (1973)
18. Schwartz, L.: Théorie des distributions. Hermann, Paris (1966)
19. Sethian, J.A.: Evolution, implementation, and application of level set and fast marching methods for advancing fronts. J. Comput. Phys. **169**(2), 503–555 (2001)
20. Sethian, J.A., Smereka, P.: Level set methods for fluid interfaces. Annu. Rev. Fluid Mech. **35**, 341–372 (2003)
21. Sethian, J.A.: Level Set Methods: Evolving Interfaces in Geometry, Fluid Mechanics, Computer Vision, and Materials Science. Cambridge University Press, Cambridge (1996)
22. Sethian, J.A.: Level Set Methods and Fast Marching Methods. Cambridge University Press, Cambridge (1999)
23. Shahbazi, K.: An explicit expression for the penalty parameter of the interior penalty method. J. Comput. Phys. **205**, 401–407 (2005)
24. Triebel, H.: Höhere Analysis. Verlag Harri Deutsch, Thun und Frankfurt am Main (1980)
25. Walter, W.: Einführung in die Theorie der Distributionen. BI Wissenschaftsverlag, Mannheim (1973)

26. Wang, Y., Oberlack, M.: A thermodynamic model of multiphase flows with moving interfaces and contact line. Contin. Mech. Thermodyn. **23**, 409–433 (2011)
27. Williams, F.A.: Turbulent combustion. In: Buckmaster, J.D. (ed.) The Mathematics of Combustion, volume 2 of Frontiers in Applied Mathematics, pp. 267–294. Society for Industrial Mathematics (SIAM), Philadelphia (1985)

Part III
Interaction and Fluid-Mechanical Processes

In gas turbine combustors, the thermal load of combustor walls in terms of wall temperature profiles represents an important aspect of the total combustor concept. Its correct capturing is essential, bearing in mind that the efficiency and longevity of the gas turbine are strongly dependent on the maximum temperature, which a combustor wall can withstand. The characteristics of turbulent flow adjacent to the solid boundaries differ substantially from those accounted in the core region as investigated in the other parts of this book. This relates primarily to the enhanced viscosity influence but also to the non-viscous wall-blockage effects expressed in terms of both Reynolds stress and stress dissipation anisotropies. In the high-temperature reactor cases the steep temperature gradients due to wall heating have also to be accounted for. The most important variations of the fluid properties, i.e. viscosity and density, are concentrated in the immediate wall vicinity. The strongest modification of the flow structure occurs in the inner part of the temperature layer. All these features, especially in conjunction with phenomena, such as swirl and separation, invalidate the use of the wall models based on the assumptions of a local equilibrium. Correct capture of these phenomena can be achieved only by integration of the governing equations up to the wall using the exact boundary conditions.

Chapter 9 proposes a solution for both the physical modeling and numerical implementation of the wall-boundary conditions. This should permit the advantageous use of the second-moment and related turbulence closures within RANS and Hybrid LES/RANS frameworks for prediction of flows near solid boundaries for complex and technically relevant applications. The feasibility of the methods is illustrated in various wall-bounded turbulent flows also associated with scalar transport under constant and variable property conditions.

Chapter 9
Computational Modelling of Flow and Scalar Transport Accounting for Near-Wall Turbulence with Relevance to Gas Turbine Combustors

S. Jakirlić, R. Jester-Zürker, G. John-Puthenveettil, B. Kniesner, and C. Tropea

Abstract An overview is given of the activities in the framework of the German Collaborative Research Center "Flow and Combustion in Future Gas Turbine Combustion Chambers" (Sonderforschungsbereich SFB 568) concerning the development of computational models in the framework of the conventional RANS method with special focus on the near-wall turbulence and a method combining a near-wall RANS models in the wall vicinity with the conventional LES in the core flow and their applications to the flow separating from sharp-edged and continuous surfaces and different swirl combustor configurations under the conditions of constant and variable fluid properties.

Keywords RANS • Hybrid LES/RANS • Reynolds stress and scalar-flux transport models • Near-wall turbulence • Flow separation • Swirling flows • Variable fluid properties

---

S. Jakirlić (✉) • R. Jester-Zürker • G. John-Puthenveettil • B. Kniesner • C. Tropea
Department of Mechanical Engineering, Institute of Fluid Mechanics and Aerodynamics, Technische Universität Darmstadt, Petersenstr. 17, 64287 Darmstadt, Germany

S. Jakirlić • C. Tropea
Center of Smart Interfaces, Technische Universität Darmstadt, Petersenstr. 17, 64287 Darmstadt, Germany
e-mail: s.jakirlic@sla.tu-darmstadt.de

R. Jester-Zürker
Voith Hydro Holding GmbH & Co, KG-tts, Alexanderstrasse 11, 89522 Heidenheim, Germany

B. Kniesner
Astrium Space Transportation, TP24 System Analysis, 81663 Munich, Germany

J. Janicka et al., *Flow and Combustion in Advanced Gas Turbine Combustors*, Fluid Mechanics and Its Applications 102, DOI 10.1007/978-94-007-5320-4_9,

9.1 Introduction

The largest majority of the flow configurations encountered in nature and technical practice is bounded by the solid walls: earth boundary layer, flows around cars and aircrafts, turbomachinery flow, flow in combustors, etc. Up to 60% of the total velocity change in a wall boundary layer is situated in the viscous sub-layer and transition zone making only 0.6% of the boundary layer thickness. About 50% of the total aerodynamic drag on a modern passenger aircraft is allotted to its boundary layer. The most intensive turbulence production due to strong shear originating from the kinematic blocking is pertinent to the immediate wall vicinity. A prominent representative of the flow configurations influenced by the high-temperature gradients is a combustion chamber. The thermal load of combustor walls in terms of the wall temperature development represents an important aspect of the total combustor concept. Its correct capturing is of the great importance, bearing in mind that the efficiency and longevity of the gas turbine are strongly dependent on the maximal temperature, which a combustor wall can withstand. These are just a few examples illustrating the necessity of studying flow behaviour in the near-wall region.

The characteristics of turbulent flow adjacent to the solid boundaries differ substantially from those accounted in the core region. This relates primarily to the enhanced viscosity influence but also to the non-viscous wall-blockage effects expressed in terms of both Reynolds stress and stress dissipation anisotropies. In the high-temperature reactor cases the steep temperature gradients due to wall heating have also to be accounted for. The most important variations of the fluid properties (viscosity and density) are concentrated in the immediate wall vicinity. The strongest modification of the flow structure occurs in the inner part of the temperature layer. All these features (especially in conjunction with some other phenomena, such as swirl and separation) invalidate the use of the wall models based on the assumptions of local equilibrium. Correct capture of these phenomena can be achieved only by integration of the governing equations up to the wall using the exact boundary conditions. However, the latter approach requires very fine near-wall resolution of the order of $\Delta x^+ \approx 50$, $y^+ \leq 0.5-1.0$ and $\Delta z^+ \approx 15-20$. It results in the fact (keeping in mind that the near-wall resolution becomes progressively important with the Reynolds number - $N \propto \mathrm{Re}_\tau^{1.76}$) that almost 50% of the total number of the numerical nodes should be situated in the viscous sublayer and the buffer layer [39]. It is very demanding to fulfil in the entire flow domain of complex three-dimensional flows, as e.g. in car or aircraft configurations. All that makes a wall-resolved LES (Large-Eddy Simulation) too costly. Furthermore, the grid refinement in the normal-to-wall direction causes extremely elongated near-wall grid cells with higher aspect ratio leading to slower convergence of numerical procedures. Such non-uniform grids with higher cell aspect ratio, especially in the wall vicinity, are usually encountered in everyday Computational Fluid Dynamics. It should be noted that the RANS models are less sensitive against such an irregular grid topology. Accordingly, a solution is in coupling a low-Reynolds-number RANS (Reynolds-Averaged Navier–Stokes) model with the conventional LES in the framework of a two-layer scheme.

In the present work, we propose a solution for both the physical modelling and numerical implementation of the wall-boundary conditions, which should permit the advantageous use of the second-moment and related turbulence closures within RANS and Hybrid LES/RANS frameworks for prediction of flows near solid boundaries for complex and technically relevant applications. The feasibility of the method proposed in the frameworks of both RANS and hybrid LES/RANS simulations will be illustrated against the available reference databases in different wall-bounded turbulent flows separating from sharp-edged and continuous surfaces and in swirl combustor configurations associated also with scalar transport under constant and variable property conditions.

9.2 Computational Model

The continuity $\partial \overline{\rho}/\partial t + \partial\left(\overline{\rho}\widetilde{U}_j\right)/\partial x_j = 0$, momentum and energy equations governing the flow and heat transfer under the variable property conditions read:

$$\frac{\partial(\overline{\rho}\widetilde{U}_i)}{\partial t} + \frac{\partial\left(\overline{\rho}\widetilde{U}_j\widetilde{U}_i\right)}{\partial x_j} = -\frac{\partial \overline{P}}{\partial x_i} + \frac{\partial}{\partial x_j}\left(\overline{\tau}_{ij}^{\mu} + \overline{\tau}_{ij}^{t}\right) \tag{9.1}$$

$$\frac{\partial(\overline{\rho}C_p\widetilde{\Theta})}{\partial t} + \frac{\partial\left(\overline{\rho}C_p\widetilde{U}_j\widetilde{\Theta}\right)}{\partial x_j} = \frac{\partial}{\partial x_j}\left(\overline{q}_j^{\mu} + \overline{q}_j^{t}\right) \tag{9.2}$$

Here, $\overline{\tau}_{ij}^{\mu}$ $(= 2\overline{\mu}\widetilde{S}_{ij} - 2\overline{\mu}\widetilde{S}_{kk}\delta_{ij}/3;\quad \widetilde{S}_{ij} = 0.5\left(\partial\widetilde{U}_i/\partial x_j + \partial\widetilde{U}_j/\partial x_i\right))$ and $\overline{q}_j^{\mu}$ $(= \overline{\lambda}\partial\widetilde{\Theta}/\partial x_j$; with $\overline{\lambda} = C_p\mu/\mathrm{Pr})$ represent viscous stress tensor and viscous heat flux, whereas turbulent stress tensor $\overline{\tau}_{ij}^{t}$ and turbulent heat $\overline{q}_j^{t}$ are to be modeled (see the following subsections). It is noted, that the term $\overline{\tau}_{ij}^{\mu}\widetilde{S}_{ij}$ denoting the (viscous) dissipation function is omitted in the energy equation. Its contribution is negligible at low Mach numbers applied in the present work. In these equations the overbar $(\overline{\Phi})$ and the tilde $(\widetilde{\Phi})$ denote the standard (Reynolds) and the mass weighted (Favre) averages $(\widetilde{\Phi} \equiv \overline{\rho\Phi}/\overline{\rho})$, respectively. The temperature dependence on viscosity μ and heat conductivity λ is defined via a power-law formulation, while Prandtl number Pr and specific heat at constant pressure Cp were kept constant.

$$\overline{\mu} = \mu_{ref}\left(\widetilde{\Theta}/\Theta_{ref}\right)^{0.71} \quad \overline{\lambda} = \lambda_{ref}\left(\widetilde{\Theta}/\Theta_{ref}\right)^{0.71} \tag{9.3}$$

Density is evaluated from the equation for ideal gas $\overline{\rho} = \overline{P}/\left(R\widetilde{\Theta}\right)$, with R denoting the universal gas constant.

The most important features of the turbulence models used presently in the frameworks of both RANS and hybrid LES/RANS are outlined in the next subsections. For detailed specification of both models readers are referred to Jakirlić and Hanjalić [1], Jakirlić and Jester-Zürker [8], Jakirlić et al. [4, 5, 9, 10] and Jakirlić and Jovanović [6].

9.2.1 Near-Wall Second-Moment Closure (SMC) Model

The present model implies the solution of modeled transport equation for the Reynolds stress tensor $\widetilde{u''_i u''_j}$ ($\overline{\tau}^t_{ij} = -\overline{\rho}\widetilde{u''_i u''_j}$ in Eq. 9.1) and the equation governing a new scale-providing variable, referred to as the 'homogeneous dissipation rate' $\widetilde{\varepsilon^h}$ (Fig. 9.1), Jakirlić and Hanjalić [1]:

$$\frac{\partial\left(\bar{\rho}\widetilde{U}_k\widetilde{u''_i u''_j}\right)}{\partial x_k} = \frac{\partial}{\partial x_k}\left[\left(\frac{1}{2}\overline{\mu}\delta_{kl} + C_s\bar{\rho}\frac{\widetilde{k}}{\widetilde{\varepsilon^h}}\widetilde{u''_k u''_l}\right)\frac{\partial\widetilde{u''_i u''_j}}{\partial x_l}\right] + \bar{\rho}\left(P_{ij} - \widetilde{\varepsilon^h_{ij}} + \Phi_{ij} + \Phi^w_{ij}\right) \quad (9.4)$$

$$\frac{\partial\left(\bar{\rho}\widetilde{U}_k\widetilde{\varepsilon^h}\right)}{\partial x_k} = \frac{\partial}{\partial x_k}\left[\left(\frac{1}{2}\overline{\mu}\delta_{kl} + C_\varepsilon\bar{\rho}\frac{\widetilde{k}}{\widetilde{\widetilde{\varepsilon^h}}}\widetilde{u''_k u''_l}\right)\frac{\partial\widetilde{\varepsilon^h}}{\partial x_l}\right] + \bar{\rho}\left(C_{\varepsilon,1}P_k - C_{\varepsilon,2}\widetilde{\widetilde{\varepsilon^h}}\right)\frac{\widetilde{\varepsilon^h}}{\widetilde{k}} + C_{\varepsilon,3}\overline{\mu}\frac{\widetilde{k}}{\widetilde{\widetilde{\varepsilon^h}}}\widetilde{u''_j u''_k}\frac{\partial^2\tilde{U}_i}{\partial x_j\partial x_l}\frac{\partial^2\tilde{U}_i}{\partial x_k\partial x_l} \quad (9.5)$$

with $P_{ij} = -\widetilde{u''_i u''_k}\partial\widetilde{U}_j/\partial x_k - \widetilde{u''_j u''_k}\partial\widetilde{U}_i/\partial x_k$ and $P_k = 0.5P_{ii}$ representing the stress production term and the production rate of the kinetic energy of turbulence respectively and $\widetilde{\widetilde{\varepsilon^h}} = \widetilde{\varepsilon^h} - \left(\partial\widetilde{k}^{1/2}/\partial x_l\right)^2$. The main features of the model are an anisotropic formulation of the dissipation correlation $\widetilde{\varepsilon^h_{ij}}$, a quadratic formulation of the pressure strain model term Φ_{ij} and a wall-normal-free formulation of the Gibson and Launder (1978) wall reflection term Φ^w_{ij}. The quantity $\widetilde{\varepsilon^h}$ differs from the conventional dissipation rate $\widetilde{\varepsilon}\,(= \widetilde{\varepsilon^h} + 0.5D^\nu_k)$ by a non-homogeneous part, which is active only in the immediate wall vicinity up to $y^+ \approx 20$. This 'inhomogeneous' part corresponds exactly to one half of the molecular diffusion of the kinetic energy of turbulence ($0.5D^\nu_k$, see Fig. 9.1) and, thus, it needs no modeling. Such an approach offers a number of convenient advantages: the dissipation equation (Eq. 9.5) retains the same basic form, the proper near-wall behaviour of $\widetilde{\varepsilon}$ is recovered without any additional terms, and the correct asymptotic behaviour of the stress dissipation components $\widetilde{\varepsilon}_{ij} = \widetilde{\varepsilon^h} + 0.5D^\nu_k$ when a solid wall is approached is fulfilled automatically without necessity for any wall geometry-related parameter.

The applied turbulence model for the variable fluid property (i.e. compressible) cases is a straight-forward adaptation of the incompressible version of the turbulence model.

A model describing the unknown turbulent heat fluxes ($\overline{q}^t_i = -\bar{\rho}C_p\widetilde{u''_i\theta}$), appearing in Eq. 9.2, is required when solving the turbulent energy equation. A common approach is the prescription of a constant Prandtl number in the framework

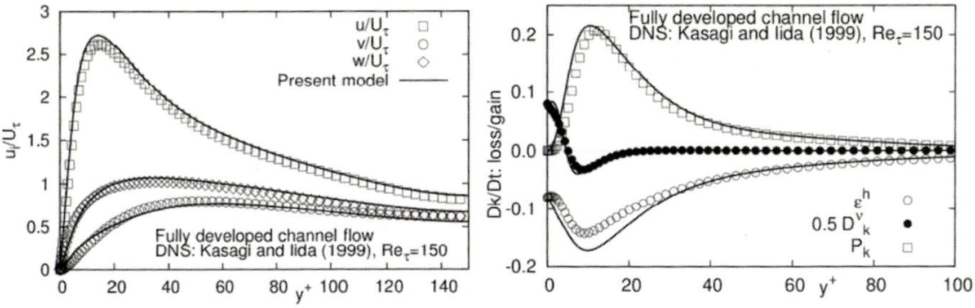

Fig. 9.1 Turbulence intensity components (*left*), terms in the budget of the equation for the kinetic energy of turbulence (*right*) in a channel flow with constant fluid properties (DNS: [33]) obtained with the present turbulence model

of the simple gradient diffusion hypothesis (SGDH) or the general gradient diffusion hypothesis (GGDH), the latter employing the Reynolds stress tensor and turbulent time scale in the diffusion coefficient. More sophisticated models solve additional differential equations either for temperature variance $\widetilde{\theta^2}$ and scalar dissipation $\widetilde{\varepsilon}_\theta$ or directly for the turbulent heat fluxes $\widetilde{u''_i\theta}$:

$$\frac{\mathbf{D}\left(\overline{\rho}\widetilde{u''_i\theta}\right)}{\mathbf{D}t} = D^v_{i\theta} + D^t_{i\theta} + \overline{\rho}P_{i\theta} + \overline{\rho}\Phi_{i\theta} - \overline{\rho}\varepsilon_{i\theta} \tag{9.6}$$

with $P_{i\theta} = -\widetilde{u''_i u''_j}\partial\widetilde{\Theta}/\partial x_j - \widetilde{u''_j\theta}\partial\widetilde{U}_i/\partial x_j$. The latter model group is in the focus of the present work. Three turbulent heat flux models were applied presently, differing mainly in the formulation of the pressure-temperature gradient term $\Phi_{i\theta}$. In conjunction with the differential Reynolds-stress turbulence model described above, only the near-wall heat transfer models were taken into account.

The basic model, denoted as model **I** throughout the work, is the proposal of Lai and So (1990).

Model **II** is an extension of the basic model to account for the mean scalar gradient, a proposal introduced by Jones (1992) in the context of a high Reynolds number scalar transport model.

Model **III** is a recent development of Seki et al. (2003) being characterized by a very complex definition of the pressure-temperature-gradient correlation. This model formulation accounts for contributions from both the mean velocity and the mean scalar gradients. The model for the turbulent transport $D^t_{i\theta}$ is the gradient diffusion model due to Daly and Harlow (1970). The viscous dissipation model adopted in the present work is the proposal of Lai and So (1990). Interested readers are referred to the work of Jakirlić and Jester-Zürker [8] for the model specifications.

The predictive capabilities of the three heat flux models described were tested by computing a channel flow with constant-fluid properties (passive scalar transport) at bulk Reynolds number $\mathrm{Re}_m = 4560$, for which the reference DNS has been performed by Kasagi and Iida [33]. The corresponding Reynolds stress

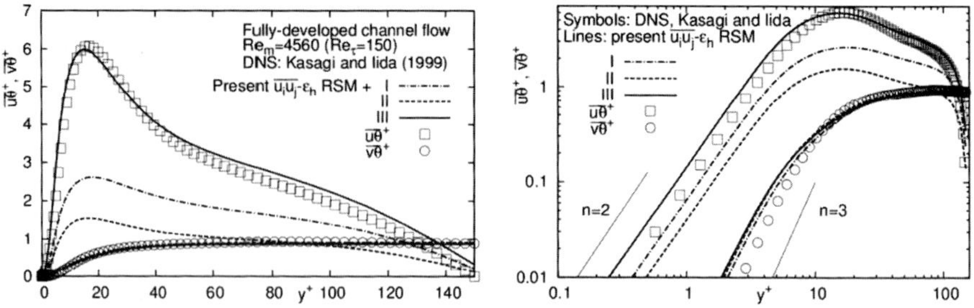

Fig. 9.2 Heat flux components $u''\theta^+$ and $v''\theta^+$ in a channel flow with constant fluid properties

components, production and dissipation rates of the turbulent kinetic energy are displayed in Fig. 9.1. The thermal boundary conditions correspond to a constant wall temperature. The ratio of the bottom wall temperature to the temperature of the top wall is chosen small enough providing the velocity field being not affected. The heat flux components obtained are displayed in Fig. 9.2 (note the correct near-wall asymptotic behaviour in accordance to: $\theta = b_\theta y^2 + c_\theta y^3 + \ldots$, $\overline{u\theta} = \overline{b_1 b_\theta} y^2 + \ldots$ and $\overline{v\theta} = \overline{c_2 b_\theta} y^3 + \ldots$). All models predicted well the $\widetilde{v''\theta}$ component (being the only reminding heat flux component in the energy Eq. 9.2, simplified under conditions of fully-developed flow), unlike the streamwise heat flux $\widetilde{u''\theta}$, which could be correctly reproduced only by the model *III*. Obviously that the non-linear description of the interaction between the heat flux and the turbulent Reynolds stresses in the expression for $\Phi_{i\theta}$, including both the mean velocity and mean temperature gradients, contributed strongly to the correct capturing of the component $\widetilde{u''\theta}$. The $\widetilde{u''\theta}$ profiles obtained with the models *I* and *II* follow the same tendency as the DNS result, but their intensities exhibit a severe underprediction.

9.2.2 Hybrid LES/RANS (HLR) Model

In the present zonal hybrid LES/RANS formulation the RANS model covers the near-wall region and the LES model the remainder of the flow domain. Both methods share the same temporal resolution. The mass-weighted equations governing the velocity (Eq. 9.1) and temperature (Eq. 9.2) fields operate as the Reynolds-averaged Navier–Stokes equations in the near-wall layer ($\widetilde{U}_i$ und $\widetilde{\Theta}$ represent the mass-weighted, ensemble-averaged velocity and temperature fields U_i and Θ) or as the filtered Navier–Stokes equations in the outer layer (with $\widetilde{U}_i$ und $\widetilde{\Theta}$ representing the mass-weighted, spatially filtered velocity and temperature fields). The turbulent stress tensor $\overline{\tau}^t_{ij}$ in Eq. 9.1 representing either the subgrid-stress tensor ($-\overline{\rho}\overline{\tau}_{ij}$) or the Reynolds-stress tensor ($-\overline{\rho}\widetilde{u''_i u''_j}$) is expressed in terms of the mean strain tensor via Boussinesq's relationship:

$$\overline{\tau}_{ij}^t \equiv -\overline{\rho\tau}_{ij} = \overline{\mu}_t \left(\frac{\partial \widetilde{U}_i}{\partial x_j} + \frac{\partial \widetilde{U}_j}{\partial x_i} - \frac{2}{3} \frac{\partial \widetilde{U}_k}{\partial x_k} \delta_{ij} \right) - \frac{1}{3} \overline{\rho\tau}_{kk} \delta_{ij} \tag{9.7}$$

$$\overline{\tau}_{ij}^t \equiv -\overline{\rho}\widetilde{u''_i u''_j} = \overline{\mu}_t \left(\frac{\partial \widetilde{U}_i}{\partial x_j} + \frac{\partial \widetilde{U}_j}{\partial x_i} - \frac{2}{3} \frac{\partial \widetilde{U}_k}{\partial x_k} \delta_{ij} \right) - \frac{2}{3} \overline{\rho} k \delta_{ij} \tag{9.8}$$

with $\widetilde{k} = 0.5\widetilde{u''_i u''_i}$. The turbulent heat flux $\overline{q}_i^t$ in the equation governing the temperature field (Eq. 9.2) is modeled by using the simple gradient diffusion hypothesis

$$\overline{q}_i^t = -\overline{\rho} C_p \widetilde{u''_i \theta} = \overline{\lambda}_t \frac{\partial \widetilde{\Theta}}{\partial x_i} \quad \text{with } \overline{\lambda}_t = \frac{\overline{\mu}_t C_p}{\mathrm{Pr}_t} \tag{9.9}$$

The equations governing the velocity and temperature field in the hybrid LES/RANS framework are:

$$\frac{\mathbf{D}\left(\overline{\rho}\widetilde{U}_i\right)}{\mathbf{D}t} = -\frac{\partial \overline{p}^*}{\partial x_i} + \frac{\partial}{\partial x_j} \left[(\overline{\mu} + \overline{\mu}_t) \left(\frac{\partial \widetilde{U}_i}{\partial x_j} + \frac{\partial \widetilde{U}_j}{\partial x_i} - \frac{2}{3} \frac{\partial \widetilde{U}_k}{\partial x_k} \delta_{ij} \right) \right] \tag{9.10}$$

$$\frac{\mathbf{D}\left(\overline{\rho}\widetilde{\Theta}\right)}{\mathbf{D}t} = \frac{\partial}{\partial x_j} \left[\left(\frac{\overline{\mu}}{\mathrm{Pr}} + \frac{\overline{\mu}_t}{\mathrm{Pr}} \right) \frac{\partial \widetilde{\Theta}}{\partial x_j} \right] \tag{9.11}$$

Conventionally the isotropic parts of the stress tensors ($\overline{\rho\tau}_{kk}\delta_{ij}/3$ and $2\overline{\rho}k\delta_{ij}/3$) are grouped together with the pressure in the equations of motion: $\overline{p}^* = \overline{p} + \overline{\rho\tau}_{kk}/3$ (if Eq. 9.1 operate as the filtered Navier–Stokes equations in the outer layer) and $\overline{p}^* = \overline{p} + 2\overline{\rho}k/3$ (if Eq. 9.1 operate as the Reynolds-averaged Navier–Stokes equations in the near-wall layer).

In order to illustrate important issues related to the merging of a near-wall RANS model with conventional LES at a discrete interface the results of the simulations of the flow in a plane channel subjected to wall heating ($q_w = 100\ W/m^2$) at a moderate Reynolds number $\mathrm{Re}_m \approx 22000$ ($\mathrm{Re}_\tau = 640$; DNS: [17]) are interactively discussed in the following section. The dimensions of the computational domain adopted were $(L_x, L_y, L_z) = (2\pi h, 2h, \pi h)$ with $2h$ being the channel height. Periodic boundary conditions were used in both homogeneous directions x and z. The computations with the reference LES (using dynamic Smagorinsky model) have been performed using the grid $(N_x, N_y, N_z) = (96, 128, 144)$. The computations with the hybrid model and complementary LES (using the standard Smagorinsky model) have been performed on the eight times coarser grid (48, 64, 72) and the 32 times coarse grid (24, 64, 36).

In the present two layer hybrid LES/RANS scheme the coupling of the instantaneous LES field and the ensemble-averaged RANS field at the interface is realized via the turbulent viscosity, which makes it possible to obtain solutions using one system of equations. This means practically that the governing equations (Eqs. 9.10

and 9.11) are solved in the entire solution domain irrespective of the flow sub-region (LES or RANS). Depending on the flow zone, the hybrid model implies the determination of the turbulent viscosity $\overline{\mu}_t$ either from a $k-\varepsilon$ RANS model: $\overline{\mu}_t = \overline{\rho} C_\mu f_\mu k^2/\varepsilon$ or from the LES formulation: $\overline{\mu}_t = \overline{\mu}_{SGS} = \overline{\rho}(C_S\Delta)^2 |\widetilde{S}|$. The Smagorinsky constant C_S takes the value of 0.1. $\Delta = (\Delta_x \times \Delta_y \times \Delta_z)^{1/3}$ represents the filter width and $|\widetilde{S}| = (\widetilde{S}_{ij}\widetilde{S}_{ij})^{1/2}$ the strain rate modulus.

The near-wall variation of the turbulent viscosity $\overline{\mu}_t$ is obtained from a $k-\varepsilon$ RANS model implying solving of the following two transport equations:

$$\frac{\mathbf{D}\left(\overline{\rho}\widetilde{k}\right)}{\mathbf{D}t} = \frac{\partial}{\partial x_j}\left[\left(\overline{\mu} + \frac{\overline{\mu}_t}{\sigma_k}\right)\frac{\partial \widetilde{k}}{\partial x_j}\right] + \overline{\rho}P_k - \overline{\rho}\widetilde{\varepsilon} \tag{9.12}$$

$$\frac{\mathbf{D}\left(\overline{\rho}\widetilde{\varepsilon}\right)}{\mathbf{D}t} = \frac{\partial}{\partial x_j}\left[\left(\overline{\mu} + \frac{\overline{\mu}_t}{\sigma_\varepsilon}\right)\frac{\partial \widetilde{\varepsilon}}{\partial x_j}\right] + \overline{\rho}\frac{C_{\varepsilon,1}P_k - f_\varepsilon C_{\varepsilon,2}\widetilde{\varepsilon}}{\tau} + P_{\varepsilon,3} \tag{9.13}$$

with $\tau = \widetilde{k}/\widetilde{\varepsilon}$. The near-wall and viscous damping functions (f_μ and f_ε) and the production term $P_{\varepsilon,3}$ (Eqs. 9.12 and 9.13), are presently modeled (other combinations including the anisotropy-reflecting $\zeta - f$ RANS model of [28], as well as the one-equation SGS model of [48], have also been tested, see [15] for more details) in line with the proposal of Chien [22] and Launder and Sharma [35].

The computational time increase compared to an LES (using the standard Smagorinsky model) performed on the same grid amounts about 30 and 50% respectively. In both turbulence models the scale supplying variable represents the so-called "quasi-homogeneous" part of the total viscous dissipation rate $\widetilde{\widetilde{\varepsilon}} = \widetilde{\varepsilon} - 2\nu\left(\partial k^{1/2}/\partial x_n\right)^2$, taking zero value at the wall.

Because $\widetilde{k}$ and $\widetilde{\varepsilon}$ are not provided (in the case of the subgrid-scale (SGS) model of Smagorinsky) within the LES sub-domain, their SGS values are estimated using the proposal of Mason and Callen [36]:

$$k_{SGS} = \frac{(C_S\Delta)^2|\widetilde{S}|^2}{0.3} \quad \text{with} \quad \varepsilon_{SGS} = (C_S\Delta)^2|\widetilde{S}|^3 \tag{9.14}$$

The RANS equations for $\widetilde{k}$ and $\widetilde{\varepsilon}$ are solved in the entire flow field, but with the discretization coefficients taking zero values in the LES sub-region. By manipulating appropriately the source terms, the numerical solution of these equations in the framework of the finite volume method provides the interface values of the $\widetilde{k}_{RANS}$ and $\widetilde{\varepsilon}_{RANS}$ being equal to the corresponding SGS values. By doing so, the boundary condition at the LES/RANS interface (*ifce*) implying the equality of the modeled turbulent viscosities (by assuming the continuity of their resolved contributions across the interface, [45]) at both sides of the interface:

$$\overline{\mu}_{t,ifce}\big|_{RANS-side} = \overline{\mu}_{t,ifce}\big|_{LES-side} \tag{9.15}$$

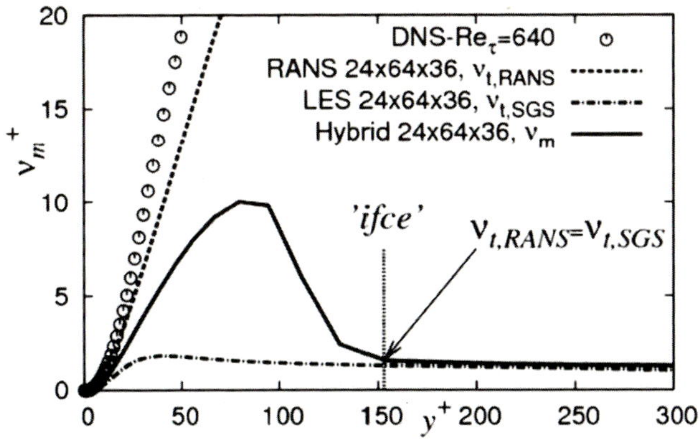

Fig. 9.3 Variation of modeled turbulent viscosity across the interface in a fully-developed channel flow

is implicitly imposed without any further adjustment, see Fig. 9.3 for illustration. In such a way a smooth transition of the turbulent viscosity is ensured.

One of the advantages of a zonal approach is the possibility to predefine the LES-RANS interface. However, in unknown flow configurations, this could be a difficult issue. Therefore, a certain criteria expressed in terms of a control parameter should be introduced. Presently, the following control parameter

$$k^* = \left\langle \frac{k_{\text{mod}}}{k_{\text{mod}} + k_{res}} \right\rangle \tag{9.16}$$

is adopted, representing the ratio of the modeled (SGS) to the total turbulent kinetic energy in the LES region, averaged over all grid cells in homogeneous direction at the interface on the LES side. As soon as this value exceeds 20%, the interface is moved farther from the wall and in opposite direction when the value goes below 20%. This additionally ensures that in the limit of a very fine grid (very low level of the residual turbulence) LES is performed in the most of the solution domain. In contrast, in the case of a coarse grid, RANS prevails. As the interface separates the near wall region from the reminder of the flow, it would be suitable to choose a wall-defined parameter for denoting the interface location. In the present study, the dimensionless wall distance y^+ was adopted. Despite possible difficulties in respect to the definition of y^+ in flow domains where the wall shear stress approaches zero, as e.g. in separation and reattachment regions, no problems in the course of the computations have arisen (one may recall that the same non-dimensional wall distance y^+ is regularly used in the Van Driest's wall-damping of $\overline{\mu}_t$ also in LES of separating and reattaching flows). It is noted, that the interface y^+ doesn't represent a model parameter in the HLR method. It only denotes the computational nodes at which the prescribed value of k^* is obtained. Figures 9.4 and 9.5 illustrate exemplarily the behaviour of the interface position in the flow over a 2D-hump (baseline and flow-control-by-steady-suction configurations).

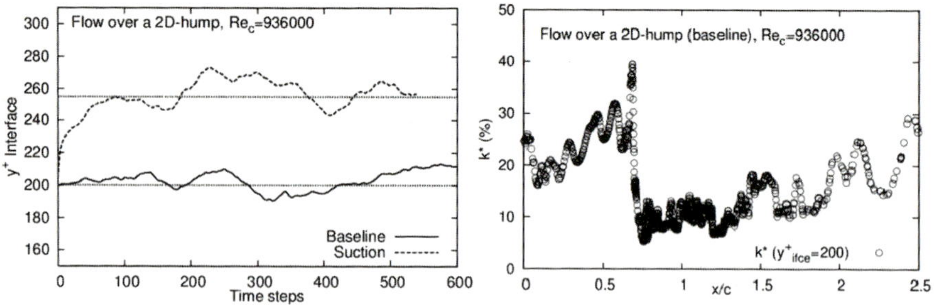

Fig. 9.4 Temporal variation of the interface position in terms of dimensional wall distance y^+ (*left*) and local variation of the control parameter $k*$ along the interface at $y^+ = 200$ (*right*) corresponding to the lower deflected wall (in the central vertical plane of the baseline configuration) in the flow over a 2-D hump

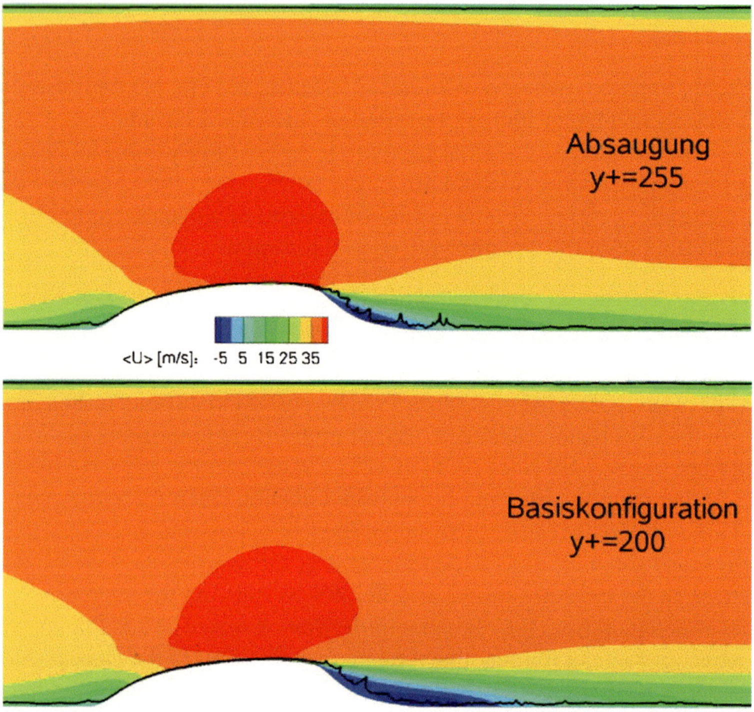

Fig. 9.5 Time-averaged mean velocity field and corresponding evolution of the interface at $y^+ = 200$ (baseline configuration) and $y^+ = 255$ (flow-control-by-steady-suction configuration) along lower and upper walls in the periodic flow over a 2D hump at $Re_C = 936{,}000$

A snapshot of the instantaneous velocity field with the corresponding spatial development of the averaged interface value along the upper wall and the lower wall, its temporal evolution in terms of y^+ as well as the local variation of the control parameter k* along the interface is displayed.

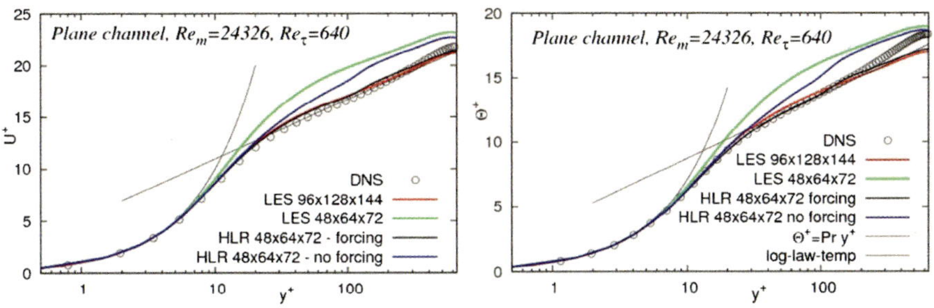

Fig. 9.6 Semi-log plot of mean velocity and mean temperature profile across a fully-developed channel flow: effect of turbulence forcing in the interface region

A typical outcome of the application of a zonal (it is also pertinent to some seamless methods, as e.g. DES method – Detached Eddy Simulation, e.g. [37]) hybrid LES/RANS method is the formation of a kind of buffer layer aligned with the region around interface separating the RANS sub-region from the LES one. Closely connected to this phenomenon is an unphysical bump at the mean velocity profile, as well as at the mean temperature profile, at the distance corresponding to the interface position. This is especially visible in computations of wall-attached flows where there is no dominating forcing (e.g. channel flows), Fig. 9.6 (blue lines). From the physical point of view the velocity mismatch is caused by the fact that due to the relatively high turbulent viscosity in the RANS region, the fluctuations are strongly damped and do not recover until some distance behind the interface. Within this so-called buffer region, the flow is neither RANS nor LES, and the corresponding turbulent viscosity is not exactly adjusted to the flow. The idea of the forcing method, which was introduced by Piomelli et al. [38] and followed by Davidson and Dahlström [23], is to generate fluctuations at or around the interface to accelerate their recovery.

Hereby, the important issues are the type and intensity of these fluctuations. Preliminary computations showed that only stochastic fluctuations are not sufficient. A method originating from a digital-filter-based generation of inflow data for spatially developing DNS and LES due to Klein et al. [34] was adopted in the present work. Its use increases the computational costs to some extent. However, it should be emphasized that the fluctuations are computed only at the interface. The steps to be performed are summarized as follows: (a) creation of random fluctuations, (b) filtering of fluctuations in space and time, (c) adjustment to local Reynolds stresses and (d) introduction into momentum equations through a source term. The strength of the forcing depends on the interface position. The closer the interface is to the wall, the less forcing is needed. Two important observations emerged. First, it seems sufficient to introduce forcing only into the equations governing the velocity field, which for instance also reduces the bump in the mean temperature profile in the case of the conjugate heat transfer computations (not considered here). Second, the forcing is needed only in the direction normal to the interface. Herewith, the

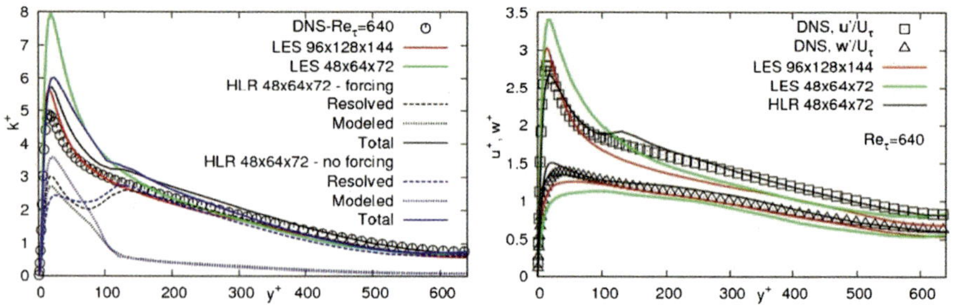

Fig. 9.7 Kinetic energy of turbulence and normal Reynolds stress components profiles across a fully-developed channel flow: effect of turbulence forcing in the interface region

recovery of the fluctuations on the LES side of the interface is accelerated and the afore-mentioned velocity and temperature bumps are eliminated to a largest extent, Fig. 9.6 (red lines).

Figure 9.7 shows the profiles of the total k, which were obtained by summing up the resolved and modeled parts. The modeled part is determined by the model used and the resolved part resulted from the averaging of the instantaneous velocity fields over a certain time period. Due to unsteady treatment of the RANS layer and the RANS field excitation by the adjacent instantaneous LES field through the interface, one can observe a resolved part also in this region. The modeled part clearly diminishes when crossing the interface at about $y^+ \approx 120-150$. Figure 9.7 illustrates also the effects of the interface turbulence forcing on the profile of the kinetic energy of turbulence. The reduction of the modeled fraction and increase of the resolved turbulent kinetic energy within the RANS layer are obvious. Simply said, the damping effect of the RANS layer on the LES sub-region is appropriately weakened.

One can still note a certain kink at the both profiles (noticeable also at Reynolds stress component profiles, Fig. 9.7), which originates by largest extent from the resolved part. However, this fact does not influence the computational procedure.

All three normal Reynolds stress components, depicted in Fig. 9.8, are correctly captured (apart of the incorrect asymptotic behaviour of the normal-to-wall component, not shown here), despite the use of the linear $k-\varepsilon$ model due to Chien [22] in the near-wall RANS layer, known to result in a completely isotropic situation in the fully-developed channel flow: $\overline{u^2} = \overline{v^2} = \overline{w^2} = 2k/3$.

9.2.3 Numerical Method

All HLR and LESc computations were performed by using the in-house code FASTEST (Flow Analysis by Solving Transport Equations Simulating Turbulence) relying on a finite volume method for block-structured, body-fitted, non-orthogonal, hexahedral meshes [24]. Block interfaces are treated in a conservative manner, consistent with the treatment of inner cell-faces. Cell centred (collocated) variable

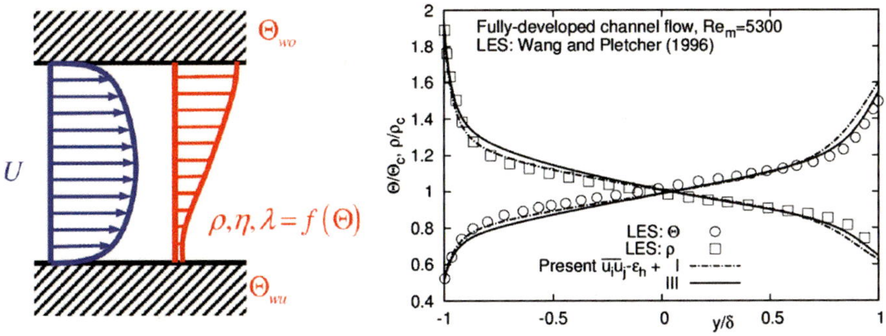

Fig. 9.8 Schematic of the channel flow with variable flow properties (*left*) and corresponding temperature and density profiles (*right*)

arrangement and Cartesian vector and tensor components are used. The equations are linearised and solved sequentially using an iterative method. The velocity–pressure coupling is ensured by the pressure-correction method based on the SIMPLE algorithm (appropriately modified to account for the compressibility effects) which is embedded in a geometric multi-grid scheme with standard restriction and prolongation and an ILU for smoothing, Durst et al. (1996). To avoid decoupling of pressure and velocity on the collocated grid the selective interpolation method proposed by Rhie and Chow (1983) is applied. To enable simulations with a large number of control volumes the code is strictly parallelised. The parallelisation is based on a domain decomposition, which is directly related to the block structure of the spatial discretisation. By using an automatic partitioning tool, an optimal load balancing of the processors can be achieved, even for complex geometries with hundreds of blocks, Sternel et al. (2004). Communication between the processors is organised via the Message Passing Interface (MPI) technique. The SMC computations were performed by an in-house computer code FAN2D (Flow Analysis Numerically) based on the Finite Volume numerical method for solving RANS-equations on the orthogonal computational grids. The closest-to-the-wall grid point was located at $y^+ \leqslant 0.5$. The convective transport of all variables was discretised by a second-order, central differencing scheme. In the case of the equations for turbulent quantities some upwinding is used by applying the so called "flux blending" technique. Time discretisation in HLR and LESc simulations was accomplished applying the Crank-Nicolson scheme.

9.3 Validation and Application of the Present SMC-RANS and Hybrid LES/RANS Methods: Results and Discussion

The developments presented in both RANS and Hybrid LES/RANS frameworks have been thoroughly validated in a number of turbulent flow configurations dealing with flow, heat and mass transfer under the conditions of constant and variable

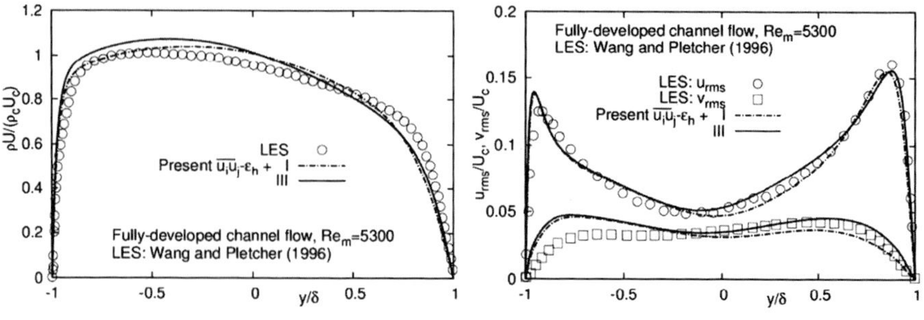

Fig. 9.9 Channel flow with variable fluid properties – mass flow rate (*left*) and turbulent intensities (*right*)

fluid properties: in channel and pipe (e.g. laminarizing flow due to strong heating) geometries, separating flows from sharp-edged and curved continuous surfaces at low and high Reynolds numbers, flow in real three-dimensional configurations, as e.g. flow over a 3D hill, flow in a 3D diffuser (characterized by secondary currents and corner separation), thermal mixing in a 3D T-shape junction geometry as well as flow and mixing in swirl combustor configurations. Here only a selection of the results obtained by both modeling strategies will be shown in order to illustrate their predictive performances.

In addition, the rationale and performances of the newly proposed wall boundary conditions, the so-called "hybrid wall-functions", to be used in the LES framework are outlined.

9.3.1 *Second-Moment Closure*

A selection of results highlighting the model features and its predictive capabilities is presented in different non-equilibrium flows with intensive wall heating causing significant modification of the mean flow and turbulence structure. For more detailed insight into the modeling and computational issues and more extensive result presentation interested readers are referred to Jakirlić and Hanjalić [1], Jakirlić et al. [2, 3], Jester-Zürker et al. [11], Jester-Zürker [12], Jakirlić and Jovanović [6] and Jakirlić and Jester-Zürker [8].

9.3.1.1 Channel Flow with Variable Fluid Properties

A channel flow at $\mathrm{Re_m} = 5{,}400$ in conjunction with the scalar transport under conditions of variable fluid properties (LES: [47]) was considered, focusing on the configuration with the ratio of the upper wall temperature to the lower wall temperature being varied up to 3, Fig. 9.8.

Figure 9.9-left displays the mass flow rate $\overline{\rho}\widetilde{U}$ across the channel. The temperature gradient and consequently the density variation in the channel cross-section

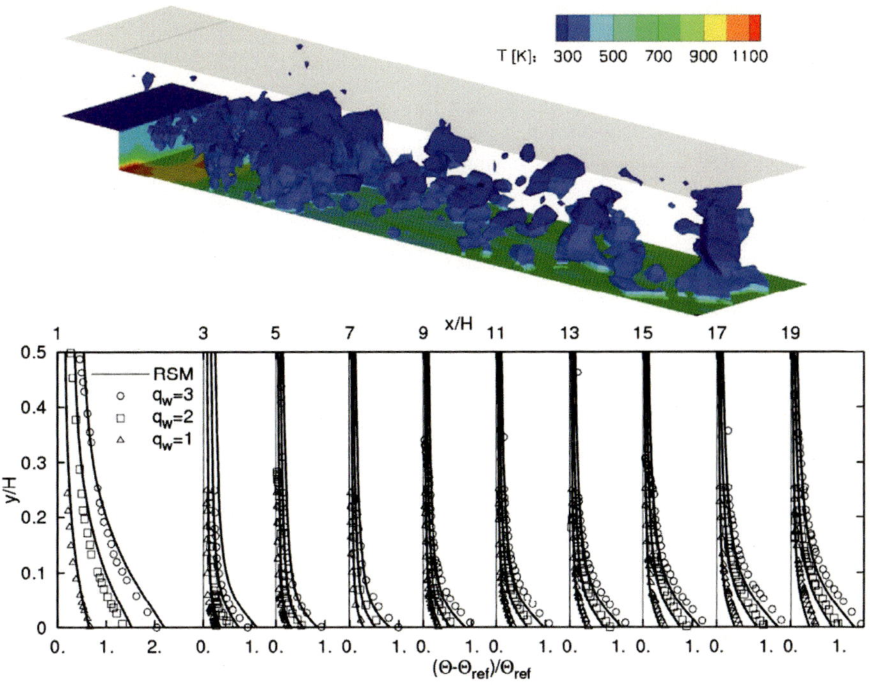

Fig. 9.10 Flow over a backward-facing step with variable flow properties – HLR-results of the pressure fluctuation contours for the $q_w = 3$ kW/m^2 case coloured by temperature (*upper*) and development of the mean temperature profile (*lower*)

cause the asymmetric velocity profile, unlike in the constant density flows. The basic profile is obtained in reasonable agreement with the DNS data. The slopes of the mass flow rate profile at both the bottom and top walls are well captured. Figure 9.9-right displays the streamwise and normal-to-the-wall Reynolds stress components u_{rms} and v_{rms}. The main characteristic of both profiles is their asymmetric shape. Apart of the certain deviation of the normal component in the lower channel half both Reynolds stress intensity components agree reasonably well with the reference LES data.

9.3.1.2 Backward-Facing Step Flow with Variable Fluid Properties

The case with the flow Reynolds number based on the step height and the upstream centerline velocity is $Re_H = 5{,}540$ (H = 0.041 m) was computed next. The conditions upstream of the step correspond to fully-developed flow in a channel of height $2h$ providing the expansion ratio of $ER = 1.5$. The bottom wall downstream of the step was heated by a uniformly supplied heat flux. Three cases with increasing heat flux ($q_w = 1$, 2 and 3 kW/m^2; reference LES by [18]) were computed in addition to the isothermal flow (Exp.: [32]). Figure. 9.10 depict the

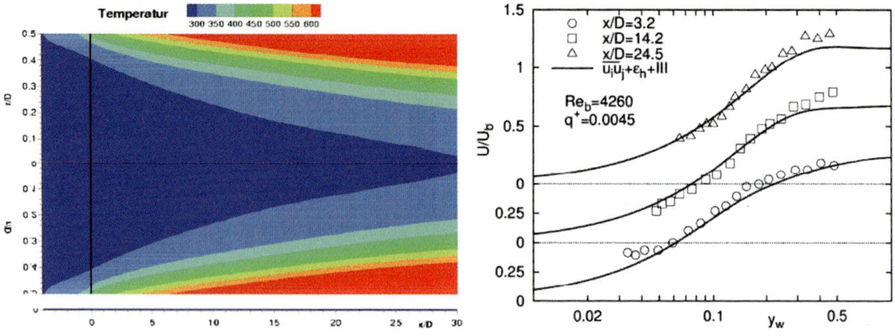

Fig. 9.11 Iso-contours of the mean temperature field (*left*) and velocity profile development in the "laminarizing" pipe configuration (*right*)

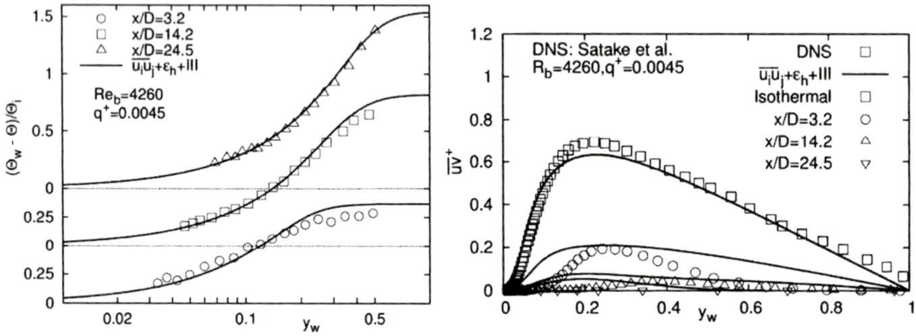

Fig. 9.12 Mean temperature (*left*) and shear stress component (*right*) profile development in the "laminarizing" pipe configuration

pressure fluctuation contours for the $q_w = 3$ kW/m^2 case coloured by temperature (obtained by HLR-LS method) for the case with the highest wall heat flux level $q_w = 3$ kW/m^2 and the mean temperature evolution. The wall temperature takes the values slightly below 1,000 K in the region of the secondary recirculation and associated reattachment region illustrating strong temperature variation. The influence of the strong temperature gradient is visible also in the region far downstream of the step; the values up to 700 K are measured.

9.3.1.3 Laminarizing Flow in a Pipe

Following flow configuration represents a vertical circular tube with air flowing in upward direction investigated experimentally by Shehata and McEligot [44], Figs. 9.11 and 9.12. After a portion of a fully-developed pipe flow (length = 4D up to the solid line in Fig. 9.11-left; with D being the pipe diameter) with constant wall temperature Θ_w the air enters a 30D-long pipe subjected to intensive heating (with negligible buoyancy effects). The thermal boundary conditions correspond to a constant heat flux. Three different heating rates were considered $q_i^+ = 0.0018$,

0.0035 and 0.0045. In the latter case for which the DNS database obtained by Satake et al. [43] was also available, some laminarization phenomena have been observed.

The evolution of the mean velocity profile (Fig. 9.11-right, the profiles at three selected locations were shown: x/D = 3.2, x/D = 14.2 and x/D = 24.5) indicates clearly the laminarizing features of the upward flow in the vertically positioned pipe. Whereas at the first position investigated a short logarithmic region still exists the profiles at the remaining locations exhibit laminar-like shapes. The same is valid for the mean temperature profiles displayed in Fig. 9.12-left. The substantial growth of the thermal boundary layer thickness obtained with the present model follows closely the experimental results. In Fig. 9.12-right the profiles of the shear stress components including also those corresponding to the isothermal flow situation with respect to the inlet section are presented. Here, the results are compared with the DNS data. The behaviour of shear stress component is in accordance with a severe suppression of the turbulence intensity due to local acceleration caused by strong viscosity increase.

9.3.2 *Hybrid LES/RANS*

The feasibility of the method was checked against the available experiments, DNS (Direct Numerical Simulation), fine-grid and coarse-grid LES results. For more details with respect to the case descriptions and associated physics, further modeling and computational issues and more detailed result presentation interested readers are referred to Kniesner [15], Jakirlić et al. [4, 5, 9, 10], Jakirlić and Kniesner [7], Šarić et al. [16] and John-Puthenveettil et al. [13].

9.3.2.1 Periodic Flow Over an Axisymmetric 2D Hill

The flow in a plane channel with a periodic arrangement of smoothly contoured hills mounted on the bottom wall at two hill-height-based Reynolds numbers $Re_H = 10{,}600$ and $Re_H = 37{,}000$ has been chosen as the next test case, Fig. 9.13.

The highly-resolved large eddy simulations performed at grids comprising about 4.6 Mio. cells [25] and 13 Mio. cells [19] as well as the recent experimental work [41, 42] were taken as reference.

The computational grid used by the present HLR model comprises only 240,000 ($80 \times 100 \times 30$) computational cells. The HLR results are assessed comparatively with the results of the LES simulations obtained by applying the same grid resolution (denoted by LESc).

Selected results obtained by the present zonal Hybrid LES/RANS method are shown in Figs. 9.14, 9.15, and 9.16 for both Reynolds numbers. The results presentation includes the friction factor development at the lower wall, streamline patterns (Fig. 9.14) and the profiles of mean axial velocity and shear stress components at selected streamwise locations. The profiles of all quantities exhibit qualitatively very similar shape in the Reynolds number range considered. The model

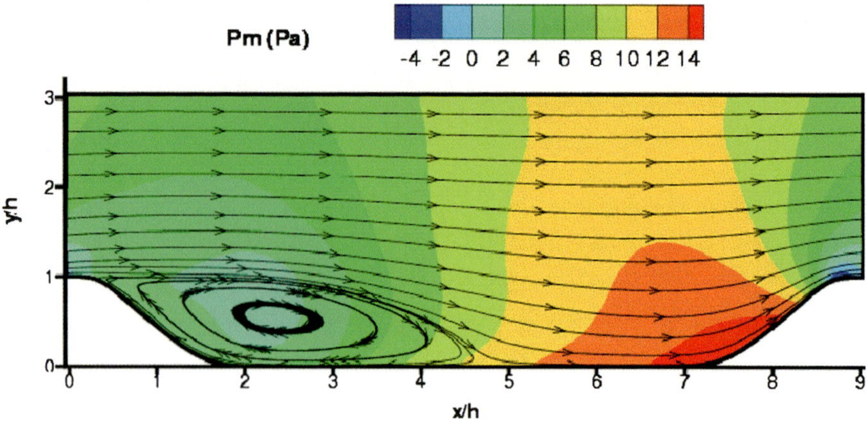

Fig. 9.13 Periodic flow over a 2-D hill at $Re_H = 10{,}600$ – the time-averaged pressure field obtained by HLR-LS

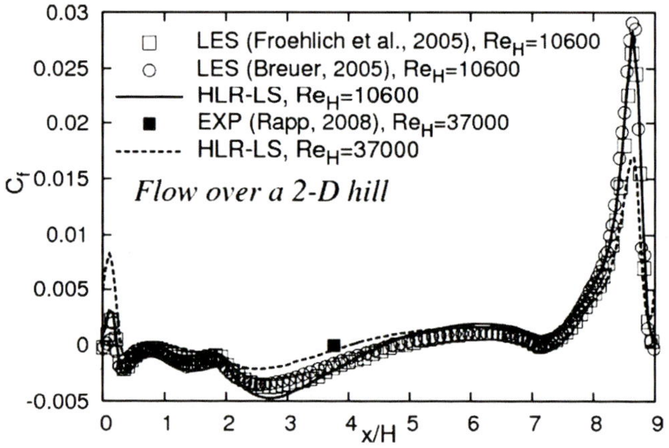

Fig. 9.14 Friction coefficient evolution along the lower deflected wall

performances concerning the Reynolds number dependence on the flow development differ primarily with respect to the turbulence enhancement in the separated shear layer (Fig. 9.16) and the consequent separation bubble shortening (Figs. 9.14 and 9.15).

9.3.2.2 Flow Over a 3D Wall-Mounted Hump

Turbulent flow over a smoothly-contoured, wall-mounted hump (Figs. 9.5 and 9.17) situated in a plane channel (height 0.909c; c = 0.42 m) represents next flow configuration of interest, this especially due to high chord-based Reynolds number $Re_c = 936{,}000$. This flow was experimentally examined at the NASA Langley Research Center [27].

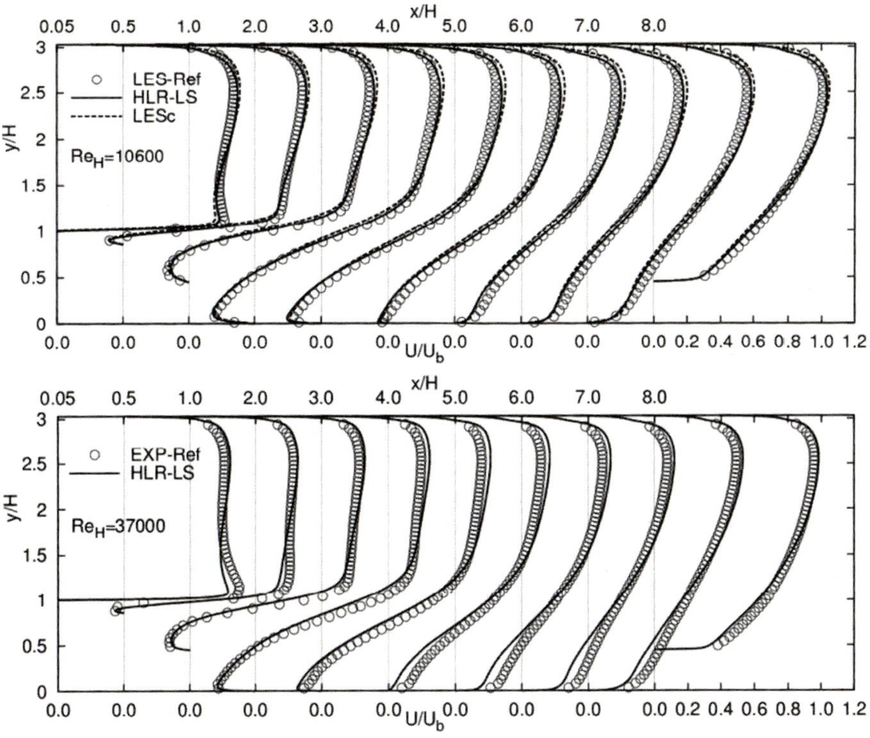

Fig. 9.15 Evolution of the axial mean velocity profiles in the periodic flow over a 2D hill at $Re_H = 10{,}600$ and $Re_H = 37{,}000$

The effect of the suction on the flow pattern over the hump geometry is clearly visible in Fig. 9.17. The position of both the separation and reattachment points for the flow configurations without (no flow control) and with suction (flow control) are in good agreement with experimental results.

Figure 9.18 shows exemplarily the axial velocity and shear stress profile development in all characteristic flows regions in the steady-suction configuration. The HLR-LS method, unlike the LESc simulation, despite a relatively coarse mesh for such a high Reynolds number, wall-bounded flow, was capable of capturing important effects of the flow control by steady suction qualitatively and quantitatively.

9.3.2.3 Separating Flow in 3D-Diffuser Configurations

Flow in a three-dimensional diffuser with the upper and one side walls being appropriately deflected, for which the experimentally obtained reference database was provided by Cherry et al. [20, 21], represents the next test case, see Fig. 9.19. Such diffuser configuration has also a high practical relevance. It mimics a diffuser

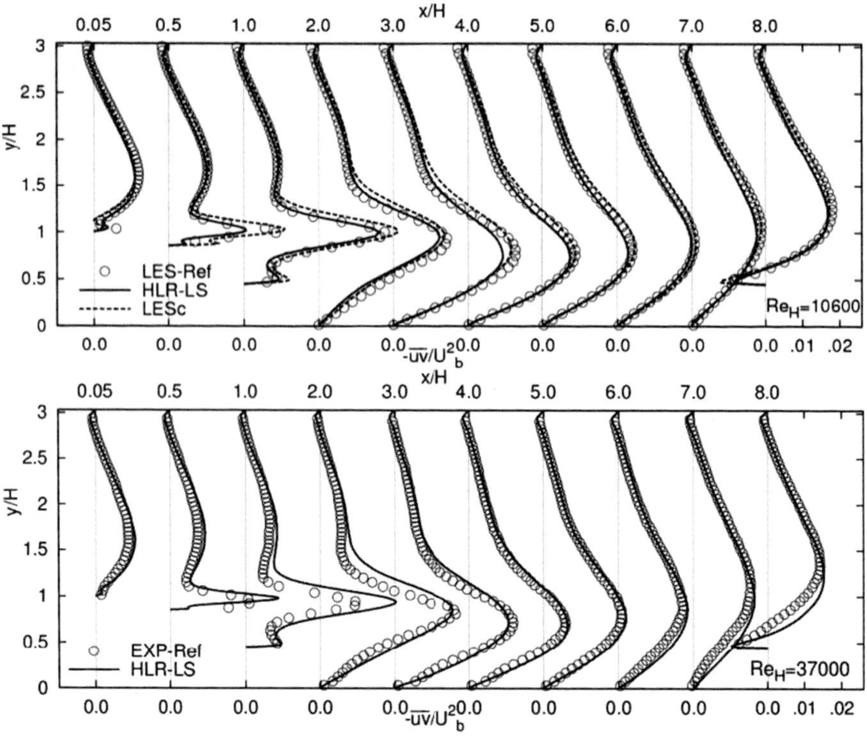

Fig. 9.16 Evolution of the shear stress profiles in the periodic flow over a 2D hill at $Re_H = 10{,}600$ and $Re_H = 37{,}000$

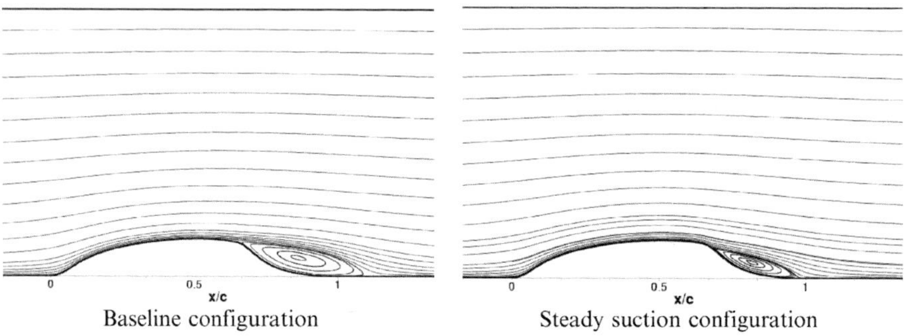

Fig. 9.17 Separation pattern in flow over a 2D hump

situated between a compressor and the combustor chamber in a jet engine. Its task is to decelerate the flow discharging from the compressor over a very short distance to the velocity field of the combustor section. Typically, a uniform inlet profile over the diffuser outlet is desirable. Such a flow situation is associated by a strong pressure increase, which may result in flow separation. In case of separation, the flow in the diffuser is characterized by a complex three-dimensional unsteady separation

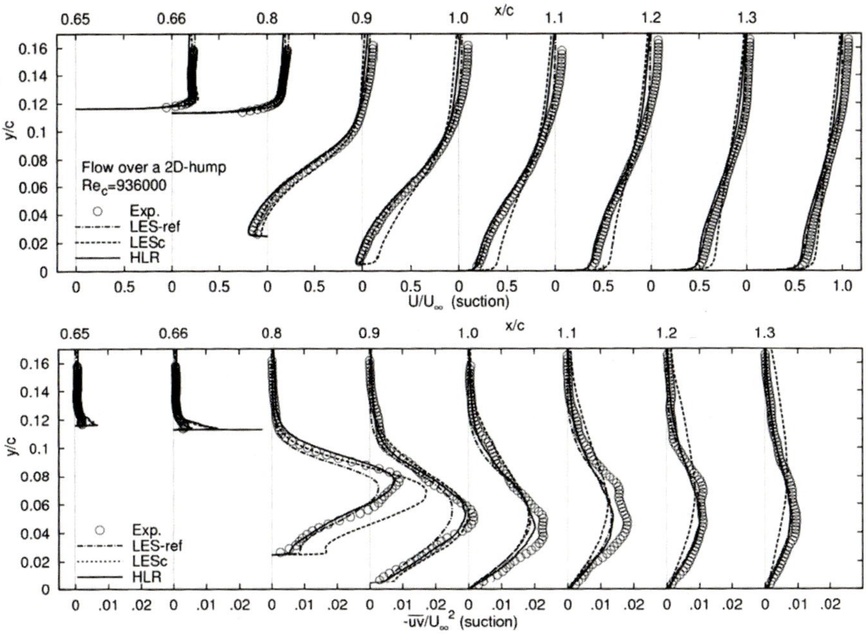

Fig. 9.18 Mean velocity and shear stress profile development in the flow over a 2D hump with steady suction flow control

pattern (associated with corner separation and corner reattachment). Furthermore, the flow in a 3D diffuser is featured by a strong secondary motion induced by the Reynolds stress anisotropy. Two diffuser configurations were considered characterized by the same fully-developed flow in the inlet duct (height $h = 1$ cm, width $B = 3.33$ cm) but slightly different expansion geometries:

The upper-wall expansion angle is reduced from 11.3^o (Diffuser 1) to 9° (Diffuser 2) and the side-wall expansion angle is increased from 2.56^o (Diffuser 1) to 4° (Diffuser 2). These slight modifications in the diffuser geometry led to substantial changes in the flow structure with respect to the onset, location, shape and size of the three-dimensional separation pattern associated with the corner separation and corner reattachment, Figs. 9.19 and 9.20. Whereas the separation zone spreads over the upper wall in the diffuser 1, it occupies the deflected side wall in the diffuser 2. The results obtained by the present HLR method exhibit qualitatively and quantitatively a fairly close agreement with the spatial development of the three-dimensional back-flow region, both in size and shape.

9.3.2.4 Thermal Mixing in a T-Shaped Junction

The flow and heat transfer in a T-junction configuration investigated experimentally by Hirota et al. [29] was served as the next test case, Fig. 9.21. Two flow streams characterized by different velocities and different temperatures originating from the

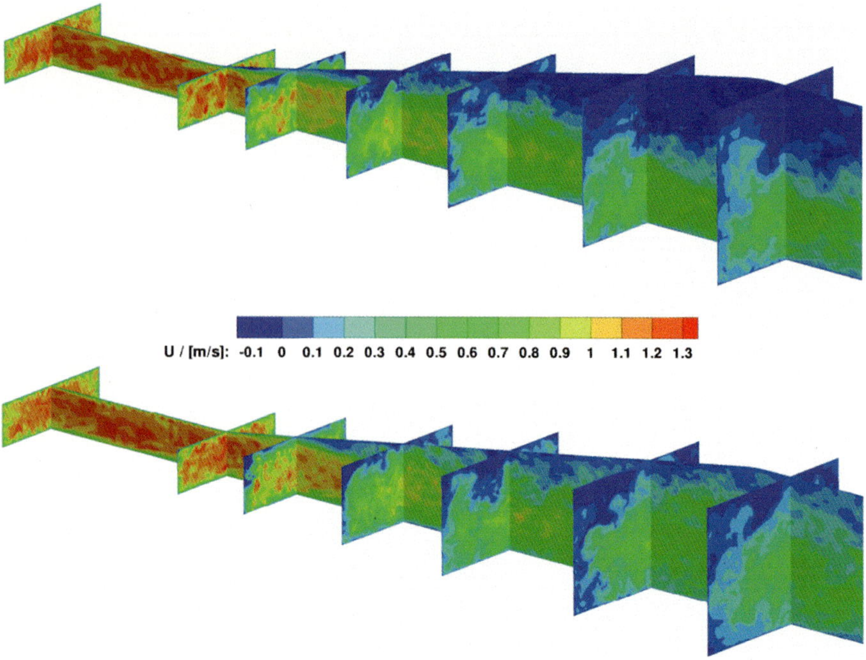

Fig. 9.19 Instantaneous velocity field in both diffuser configurations, diffuser 1 (*upper*) and diffuser 2 (*lower*), obtained by a zonal hybrid LES/RANS model illustrating different flow separation patterns

main horizontal channel ("cold" stream at $T_c = 12°C$) and the vertically positioned branch channel ("hot stream" at $T_h = 60°C$) cross, i.e. impinge onto each other in the T-junction region creating a recirculation zone which extends up to $x/H = 4$ in the right branch of the main channel. The flow Reynolds number based on the characteristics of the main channel is $Re_H = 15{,}000$. First impression of the flow structure could be gained from the following figure showing the instantaneous velocity field obtained by the present HLR.

Figure 9.22 shows the velocity vectors indicating the shape and size of the flow separation pattern and the evolution of the velocity and temperature profiles along with the experimental data.

9.3.2.5 Flow and Mixing in a Swirl Combustor

The present HLR method was furthermore applied in a generic mixing chamber model, representing an unfolded segment of a simplified Rich-Quick-Lean (RQL) combustion chamber operating under isothermal, non-reacting conditions at ambient pressure. The in-house PIV measurements have been performed providing

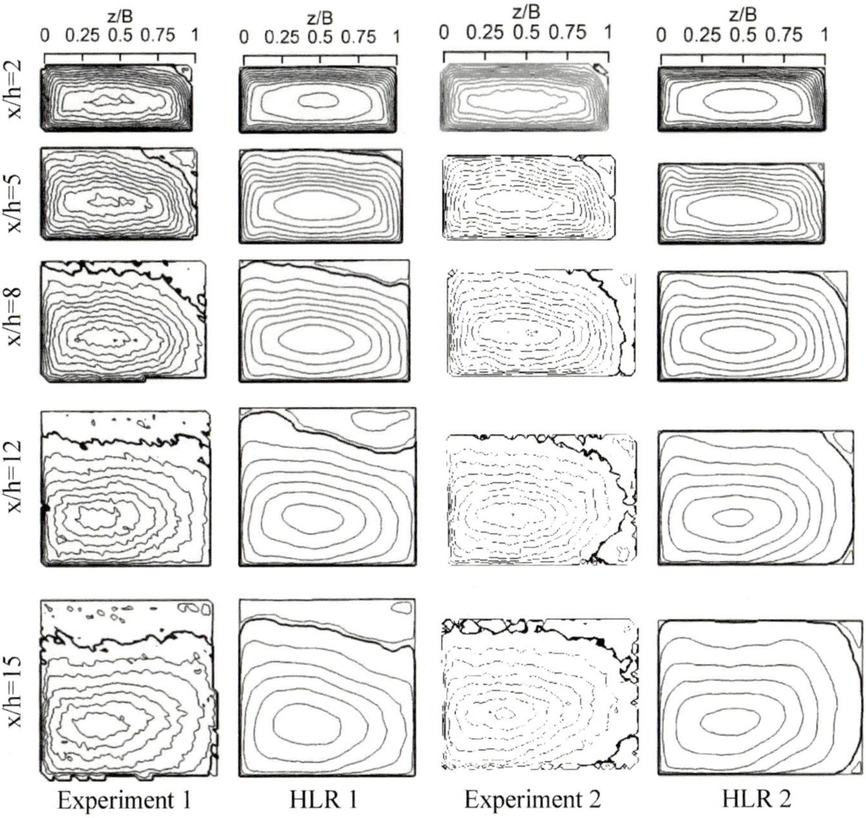

Fig. 9.20 Experimental vs. HLR-results, contours of streamwise velocity at cross-sections x/h = 2, 5, 8, 12 and 15 in both diffuser configurations

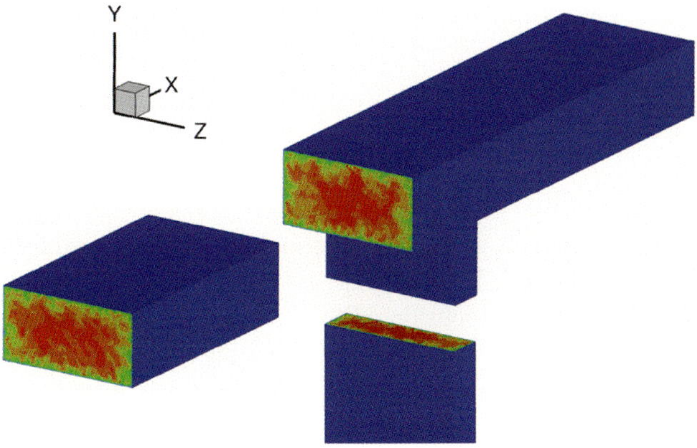

Fig. 9.21 Part of the flow domain considered indicating the instantaneous velocity field

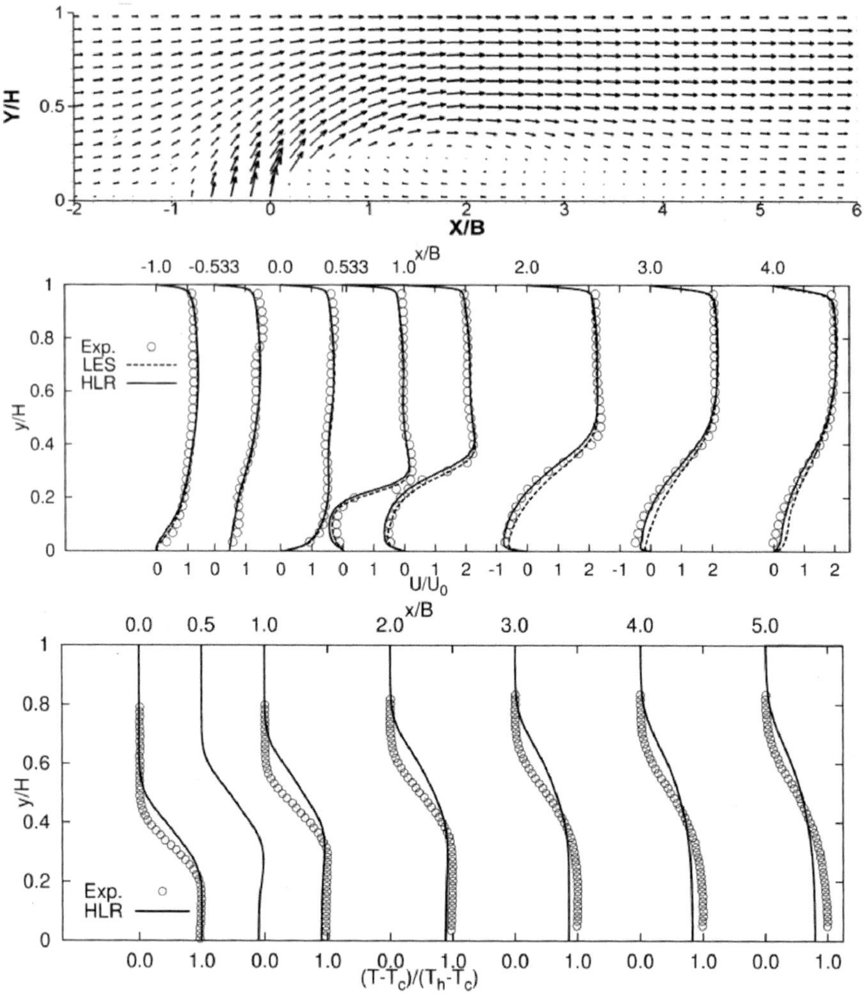

Fig. 9.22 Comparison between experimentally and computationally obtained mean velocity and temperature fields in the *x-y* ($z = 0$) plane

profiles of all velocity and Reynolds-stress components at selected locations within the combustor, Gnirß and Tropea [26]. Two configurations without and with secondary air injection were considered. Figures 9.23 and 9.24 illustrate the flow topology in the present combustor configurations by displaying the time-averaged streamlines and the iso-contours of the axial velocity component.

A strong influence of the swirling jets resulting in a closed recirculation zone in the core flow can be observed. The velocity profile evolution exhibiting regions with negative velocity values in the mixing chamber core and positive peak values coinciding with the shear layer regions is a typical representative of the flow in

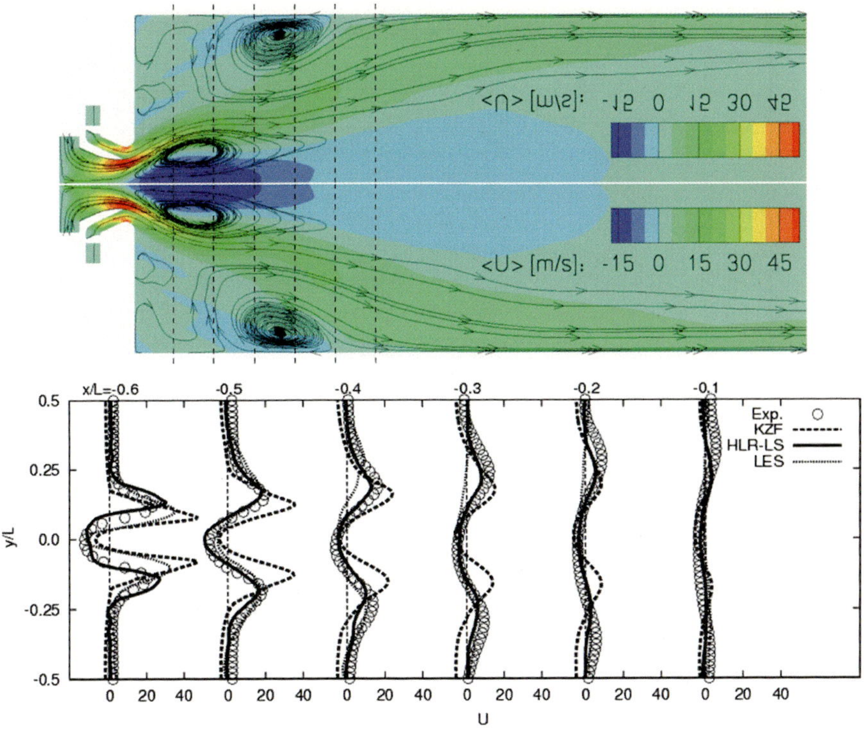

Fig. 9.23 Computationally (HLR-LS) obtained time-averaged streamlines and associated axial velocity contours and mean velocity profile evolution in the horizontal plane (z = 0) of the mixing chamber for the case without secondary air injection

a swirl combustor. As it can be seen from the previous figures the HLR model reproduced all important flow features, in particular with regard to the interaction of the secondary air jets with the swirled main flow and swirl-induced free recirculation zone evolution in good agreement with experimental results.

9.3.3 Wall Modelling in LES by Hybrid Wall-Functions (HWF)

A Large Eddy Simulation of high Reynolds number flows request a very fine mesh to be used. The higher the Reynolds number is the thinner the boundary layer and the lower the size of the smallest scales near the wall. Accordingly, the numerical grid resolving all these scale up to a relevant extent has to be refined not only in the wall-normal direction but also in the streamwise and spanwise directions aiming at capturing appropriately the structural characteristics of the flow. Therefore, an LES is not reasonable for many technical configurations at high Reynolds numbers. Instead of carrying out a wall-resolved LES one can use a hybrid LES/RANS

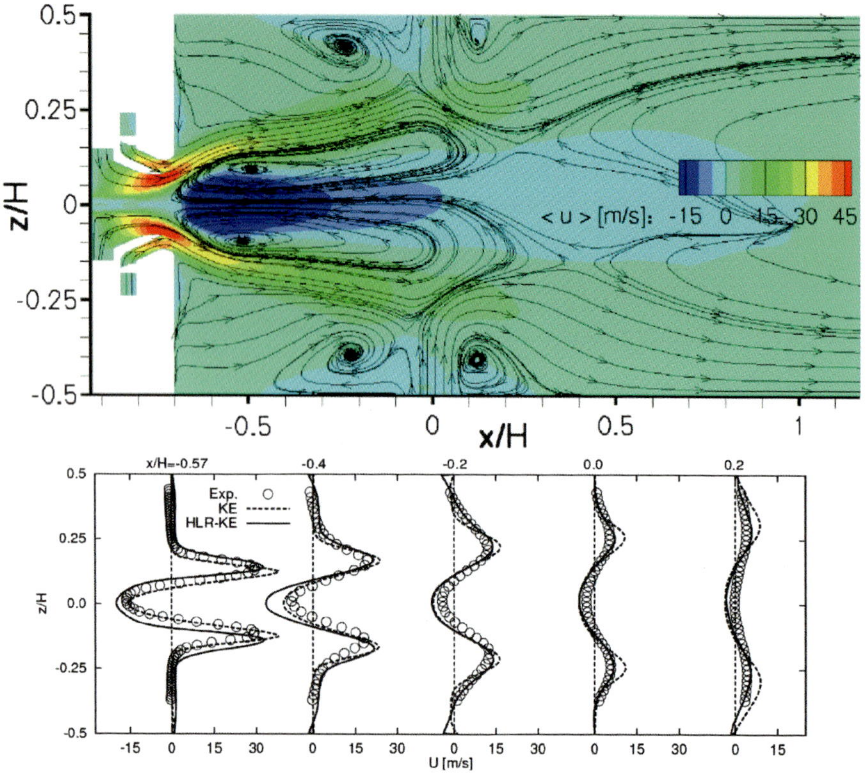

Fig. 9.24 Computationally (HLR) obtained time-averaged streamlines and axial velocity contours and mean velocity profile evolution in the vertical plane (y = 0) of the mixing chamber plane for the case with secondary air injection

method (see previous section) or the "wall-functions" for the near-wall treatment. In both cases the grid resolution can be reduced to a great extent.

Wall functions are to return the correct instantaneous wall-shear stress matching the instantaneous tangential velocity at the wall-nearest grid point. The calculated wall-shear stress is then applied as boundary condition at the wall. When using the standard wall function relying on the law-of-the-wall for velocity field

$$U^+ = \frac{1}{\kappa} \ln \left(E y^+ \right) \tag{9.17}$$

it has to be ensured that the first grid point off the wall is placed in the logarithmic layer. In cases that involve separation and reattachment y^+ of the wall-nearest grid point varies throughout the flow and as a consequence it is hard to assure that the first grid point is positioned in the log-law layer. By taking a wall function expression blending between linear law (exact velocity profile pertinent to the viscous sublayer) and "equilibrium" wall function that is valid in all three zones of the boundary layer, i.e. viscous sublayer, buffer layer and logarithmic layer independent of the position of the wall-adjacent computational node, this problem can be redressed. The so-called "compound wall-function" (CWF) formulation proposed by Popovac and Hanjalic [40] in the RANS framework was presently adopted for LES. By changing the exponents in the CWF-formulation a better agreement in the buffer layer in comparison to DNS could be realized. Another feature of the CWF method is that it includes a factor ψ that accounts for non-equilibrium effects in the flow. Presently, $\psi = 1$ is adopted implying the use of equilibrium wall functions (Eq. 9.17) as the hybrid wall-function's (HWF) upper bound (the term hybrid refers to the fact that the three regions in the boundary layer are captured by one expression).

The modified HWF equation reads, [14]:

$$U^+ = y^+ e^{-\Gamma_1} + \left[\frac{1}{\kappa\psi}\ln\left(E y^+\right) U^+\right] e^{-1/\Gamma_2} \tag{9.18}$$

with $\Gamma_1 = 0.0045\left(y^+\right)^{4.5} / \left(15 + y^+\right)$ and $\Gamma_2 = 0.018\left(y^+\right)^4 / \left(15 + 5y^+\right)$. The Kármán constant being $\kappa = 0.41$ and $E = 8.3$. The non-equilibrium factor was set to $\psi = 1$. For the temperature the wall function expression by Kader [31] without any modification was applied

$$\Theta^+ = \Pr y^+ e^{-\Gamma_\Theta} + \left[\alpha \ln\left(y^+\right) + \beta\left(\Pr\right)\right] \Theta^+ e^{-1/\Gamma_\Theta} \tag{9.19}$$

where $\Gamma_\Theta = 0.01\left(\Pr y^+\right)^4 / \left(1 + 5\Pr^3 y^+\right)$, $\beta = \left(3.85\Pr^{1/3} - 1.3\right)^2 + 2.12\ln\left(\Pr\right)$ and $\alpha = 2.12$.

By iteratively solving Eq. 9.18 for the friction velocity U_τ the wall-shear stress can be deduced. After inserting U_τ in Eq. 9.19 the dimensionless temperature Θ^+ can be finally calculated.

The wall functions for velocity and temperature have been tested in different plane channel configurations and in the flow over a backward-facing step. Figure 9.25-left illustrates the functional principle of the proposed wall boundary treatment.

Independent of the position of the wall-closest numerical node, be it in the buffer layer or in the logarithmic region, the LES simulation using the present hybrid wall function returned correctly the velocity profile in the logarithmic region agreeing reasonably with the available DNS data [30] for the channel flow at $Re_\tau = 2003$. In Fig. 9.25-right the dimensionless temperature profile close to the wall is displayed for channel flow $Re_\tau = 640$ (DNS: [17]) with a constantly introduced heat flux. The first grid point near the wall is positioned in the buffer layer.

Figure 9.26 shows the velocity and temperature distributions for the backward-facing step at $Re_H = 28{,}000$ with external constant wall heating [46]. y^+ of the wall-nearest grid point in the recirculation zone after the backstep was situated in the buffer layer and in the rear part in the logarithmic layer.

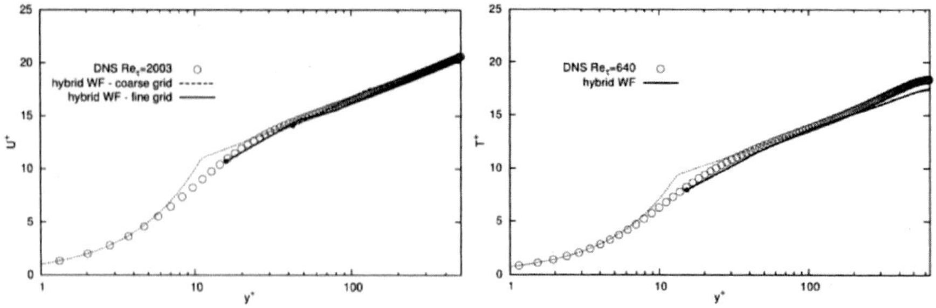

Fig. 9.25 Illustration of the hybrid wall-function performance in computing a plane channel flow with heat transfer

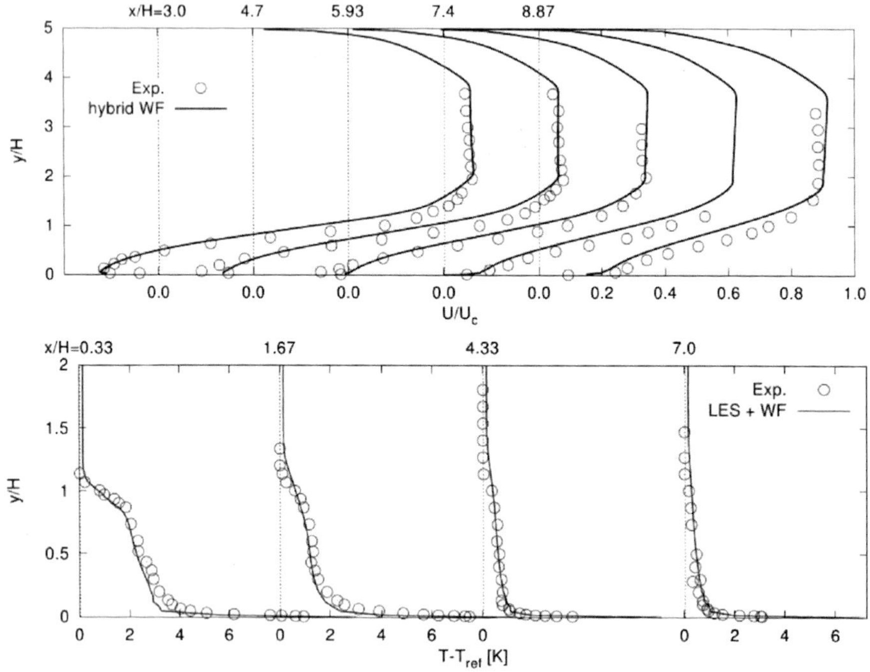

Fig. 9.26 Development of the mean velocity and mean temperature profiles in the flow over a backward-facing step computed by using the present hybrid-wall function

9.4 Conclusions

The potential of the near-wall Second-Moment Closure model and a hybrid LES/RANS model scheme was illustrated by computing a series of wall-bounded flow configurations featured by separation, reattachment and swirl effects in a broad range of Reynolds numbers. Very promising results with respect to the structural characteristics of the instantaneous flow field, the mean velocity field and associated integral parameters (friction coefficient, pressure coefficient, flow spreading rate) as well as the turbulence quantities demonstrate the model feasibility and applicability in a broad range of complex, wall-bounded turbulent flows.

Acknowledgements The authors acknowledge the financial support from the German Research Council (DFG) through the SFB568.

References

Project-Related Publications

1. Jakirlić, S., Hanjalić, K.: A new approach to modelling near-wall turbulence energy and stress dissipation. J. Fluid Mech. **539**, 139–166 (2002)
2. Jakirlić, S., Jester-Zürker, R., Tropea, C.: Joint effects of geometry confinement and swirling inflow on turbulent mixing in model combustors: a second-moment closure study. J. Prog. CFD **4**(3–5), 198–207 (2004)
3. Jakirlić, S., Eisfeld, B., Jester-Zürker, R., Kroll, N.: Near-wall, Reynolds-stress model calculations of transonic flow configurations relevant to aircraft aerodynamics. Int. J. Heat Fluid Flow **28**(4), 602–615 (2007)
4. Jakirlić, S., Kniesner, B., Kadavelil, G., Gnirß, M., Tropea, C.: Experimental and computational investigations of flow and mixing in a single-annular combustor configuration. Flow Turbul. Combust. **83**(3), 425–448 (2009)
5. Jakirlić, S., Kadavelil, G., Kornhaas, M., Schäfer, M., Sternel, D.C., Tropea, C.: Numerical and physical aspects in LES and hybrid LES/RANS of turbulent flow separation in a 3-D diffuser. Int. J. Heat Fluid Flow **31**(5), 820–832 (2010)
6. Jakirlić, S., Jovanović, J.: On unified boundary conditions for improved prediction of near-wall turbulence. J. Fluid Mech. **656**, 530–539 (2010)
7. Jakirlić, S., Kniesner, B.: Near-wall RANS modelling in LES of heat transfer in backward-facing step flows under conditions of constant and variable fluid properties. In: ASME 3rd Joint U.S.-European Fluids Engineering Summer Meeting: Symposium on "DNS, LES and Hybrid RANS/LES Methods", Montreal, Quebec, Canada, Paper No. FEDSM-ICNMM2010-30354, 1–5 August 2010
8. Jakirlić, S., Jester-Zürker, R.: Convective heat transfer in wall-bounded flows affected by severe fluid properties variation: a second-moment closure study. In: ASME 3rd Joint U.S.-European Fluids Engineering Summer Meeting: "7th Symposium on Fundamental Issues and Perspectives in Fluid Mechanics", Montreal, Quebec, Canada, Paper No. FEDSM-ICNMM2010-30729, 1–5 August 2010
9. Jakirlić, S., Chang, C.-Y., Kadavelil, G., Kniesner, B., Maduta, R., Sarić, S., Basara, B.: Critical evaluation of some popular hybrid LES/RANS methods by reference to flow separation at a curved wall (invited lecture). In: 6th AIAA Theoretical Fluid Mechanics Conference, Honolulu, HI, Paper No. AIAA-2011-3473, 27–30 June 2011
10. Jakirlić, S., Kniesner, B., Kadavelil, G.: On interface issues in LES/RANS coupling strategies: a method for turbulence forcing. JSME J. Fluid Sci. Technol. **6**(1), 56–72 (2011)
11. Jester-Zürker, R., Jakirlić, S., Tropea, C.: Computational modelling of turbulent mixing in confined swirling environment under constant and variable density conditions. Flow Turbul. Combust. **75**(1–4), 217–244 (2005)
12. Jester-Zürker, R.: Reynolds-Spannungsmodellierung des Skalartransports unter Bedingungen variabler Stoffeigenschaften in Drallbrennerkonfigurationen (Second-Moment Closure

modelling of scalar transport in swirl combustors under variable flow property conditions). PhD thesis, Technische Universität Darmstadt. Shaker Verlag, Aachen. ISBN 978-3-8322-6742-1 (2006)
13. John-Puthenveettil, G., Jia, L., Reimann, T., Jakirlić, S., Sternel, D.C.: Thermal mixing of flow-crossing streams in a T-shaped junction: a comparative LES, RANS and Hybrid LES/RANS study. In: 7th International Symposium on Turbulence, Heat and Mass Transfer, Palermo, Italy, 24–27 September 2012
14. John-Puthenveettil, G.: Computational modelling of complex flows using eddy-resolving models accounting for near-wall turbulence. PhD thesis, Technische Universität Darmstadt (2012)
15. Kniesner, B.: Ein hybrides LES/RANS Verfahren für konjugierte Strömung, Wärme- und Stoffübertragung mit Relevanz zu Drallbrennerkonfigurationen (A hybrid LES/RANS method for conjugated flow, heat and mass transfer with relevance to swirl combustor configurations). PhD thesis, Technische Universität Darmstadt, Darmstadt. http://tuprints.ulb.tu-darmstadt.de/950/ (2008)
16. Šarić, S., Jakirlić, S., Čavar, D., Kniesner, B., Altenhöfer, P., Tropea, C.: Computational study of mean flow and turbulence structure in inflow system of a swirl combustor. In: Tropea, C., et al. (eds.) Notes on numerical fluid mechanics and multidisciplinary design, vol. 96, pp. 462–470. Springer, Berlin/New York (2007). ISBN 978-3-540-74458-0

Other Publications

17. Abe, H., Kawamura, H., Matsuo, Y.: Surface heat-flux fluctuations in a turbulent channel flow up to $Re_\tau = 1020$ with $Pr = 0.025$ and 0.71. Int. J. Heat Fluid Flow **25**, 404–419 (2004)
18. Avancha, R.V.R., Pletcher, R.H.: Large eddy simulation of the turbulent flow past a backward-facing step with heat transfer and property variations. Int. J. Heat Fluid Flow **23**, 601–614 (2002)
19. Breuer, M., Peller, N., Rapp, C., Manhart, M.: Flow over periodic hills – numerical and experimental study in a wide range of Reynolds numbers. Comput. Fluids **38**, 433–457 (2009)
20. Cherry, E.M., Elkins, C.J., Eaton, J.K.: Geometric sensitivity of three-dimensional separated flows. Int. J. Heat Fluid Flow **29**, 803–811 (2008)
21. Cherry, E.M., Elkins, C.J., Eaton, J.K.: Pressure measurements in a three-dimensional separated diffuser. Int. J. Heat Fluid Flow **30**, 1–2 (2009)
22. Chien, K.-Y.: Predictions of channel and boundary-layer flows with a low-Reynolds-number turbulence model. AIAA J. **20**(1), 33–38 (1982)
23. Davidson, L., Dahlström, S.: Hybrid LES/RANS: computation of the flow around a three-dimensional hill. In: 6th International Symposium on Engineering Turbulence Modelling and Measurements, Sardinia, Italy, 23–25 May 2005
24. FASTEST-Manual.: Institute of Numerical Methods in Mechanical Engineering, Department of Mechanical Engineering, Technische Universität Darmstadt, Germany (2005)
25. Fröhlich, J., Mellen, C.P., Rodi, W., Temmerman, L., Leschziner, M.A.: Highly resolved large-eddy simulation of separated flow in a channel with streamwise periodic constrictions. J. Fluid Mech. **526**, 19–66 (2005)
26. Gnirß, M., Tropea, C.: Simultaneous PIV and concentration measurements in a gas-turbine combustor model. Exp. Fluids **45**(4), 643–656 (2008)
27. Greenblatt, D., Paschal, K.B., Yao, C.S., Harris, J., Schaeffler, N.W., Washburn, A.E.: A separation control CFD validation test case, Part1: baseline and steady suction, In: AIAA Paper No. 2004–2220 (2004)
28. Hanjalić, K., Popovac, M., Hadziabdić, M.: A robust near-wall elliptic-relaxation eddy-viscosity turbulence model for CFD. Int. J. Heat Fluid Flow **25**, 1047–1051 (2004)

29. Hirota, M., Mohri, E., Asano, H., Goto, H.: Experimental study on turbulent mixing process in cross-flow type T-junction. Int. J. Heat Fluid Flow **31**(5), 776–784 (2010)
30. Hoyas, S., Jimenez, J.: Scaling of the velocity fluctuations in turbulent channels up to $Re_\tau = 2003$. Phys. Fluids **18**, 011702 (2006)
31. Kader, B.A.: Temperature and concentration profiles in fully turbulent boundary layers. Int. J. Heat Mass Transf. **24**, 1541–1544 (1981)
32. Kasagi, N., Matsunaga, A.: Three-dimensional particle-tracking velocimetry measurements of turbulence statistics and energy budget in a backward facing step flow. Int. J. Heat Fluid Flow **16**, 477–485 (1995)
33. Kasagi, N., Iida, O.: Progress in direct numerical simulation of turbulent heat transfer. In: Proceedings of the 5th ASME/JSME Joint Thermal Engineering Conference, San Diego, CA, 15–19 March 1999
34. Klein, M., Janicka, J., Sadiki, A.: A digital filter based generation of inflow data for spatially developing direct numerical or large-eddy simulations. J. Comput. Phys. **186**, 652–665 (2003)
35. Launder, B.E., Sharma, B.I.: Application of the energy dissipation model of turbulence to the calculation of flow near a spinning disc. Lett. Heat Mass Transf. **1**, 131–138 (1974)
36. Mason, P.J., Callen, N.S.: On the magnitude of the subgrid-scale eddy coefficient in large-eddy simulation of turbulent channel flow. J. Fluid Mech. **162**, 439–462 (1986)
37. Nikitin, N.V., Nicoud, F., Wasistho, B., Squires, K.D., Spalart, P.R.: An approach to wall modelling in large-eddy simulations. Phys. Fluids **12**(7), 1629–1632 (2000)
38. Piomelli, U., Balaras, E., Pasinato, H., Squires, K.D., Spalart, P.R.: The inner-outer layer interface in large-eddy simulations with wall-layer models. Int. J. Heat Fluid Flow **24**(4), 538–550 (2003)
39. Pope, S.: Turbulent flows. Cambridge University Press, Cambridge (2000). ISBN 0-521-59886-9
40. Popovac, M., Hanjalic, K.: Compound wall treatment for RANS computation of complex turbulent flows and heat transfer. Flow Turbul. Combust. **78**, 177–202 (2007)
41. Rapp, Ch.: Experimentelle Untersuchung der turbulenten Strömung über periodische Hügel. PhD thesis, Technical University Munich, Germany (2008)
42. Rapp, C., Manhart, M.: Flow over periodic hills: an experimental study. Exp. Fluids **51**, 247–269 (2011)
43. Satake, S., Kunugi, T., Shehata, A.M., McEligot, D.M.: Direct numerical simulation for laminarization of turbulent forced gas flows in circular tubes with strong heating. Int. J. Heat Fluid Flow **21**, 526–534 (2000)
44. Shehata, A.M., McEligot, D.M.: Mean structure in the viscous layer of strongly-heated internal gas flows. Measurements. Int. J. Heat Mass Transf. **41**, 4297–4313 (1998)
45. Temmerman, L., Hadžiabdić, M., Leschziner, M.A., Hanjalić, K.: A hybrid two-layer URANS-LES approach for large eddy simulation at high Reynolds numbers. Int. J. Heat Fluid Flow **26**, 173–190 (2005)
46. Vogel, J.C., Eaton, J.K.: Combined heat transfer and fluid dynamic measurements downstream of a backwards-facing step. ASME J. Heat Transf. **107**, 922–929 (1985)
47. Wang, W.-P., Pletcher, R.: On the large eddy simulation of a turbulent channel flow with significant heat transfer. Phys. Fluids **8**(12), 3354–3366 (1996)
48. Yoshizawa, A., Horiuti, K.: A statistically-derived subgridscale kinetic energy model for the large-eddy simulation of turbulent flows. J. Phys. Soc. Jpn. **54**, 2834–2839 (1985)
49. Daly, B.J., Harlow, F.H.: Transport equations in turbulence. Phys. Fluids **13**, 2634–2649 (1970)
50. Durst, F., Schäfer, M.: A parallel block-structured multigrid method for the prediction of incompressible flows. Int. J. Numer. Methods Fluids **22**, 549–565 (1996)
51. Gibson, M.M., Launder, B.E.: Ground effects on pressure fluctuations in the atmospheric boundary layer. J. Fluid Mech. **86**, 491–511 (1978)
52. Lai, Y.G., So, R.M.C.: Near-wall modelling of turbulent heat fluxes. Int. J. Heat Mass Trans. **33**(7), 1429–1440 (1990)
53. Jones, W.P.: Turbulence modelling for combustion flows. In: Modelling for Combustion and Turbulence. Lecture series 1992–03. von Karman Institute for Fluid Dynamics, Rhode Saint Geneèse (1992)

54. Rhie, C.M., Chow, W.L.: A numerical study of the turbulent flow past an isolated airfoil with trailing edge separation. AIAA J. **21**, 1525–1532 (1983)
55. Seki, Y., Kawamoto, N., Kawamura, H.: Proposal of turbulent heat flux model and its application to turbulent channel flow with various thermal boundary conditions. In: Hanjalic et al. (eds.) Turbulence, Heat and Mass Transfer 4, pp. 569-576. Begell House Inc., New York (2003)
56. Sternel, D.C., Junglas, D., Martin, A., Schäfer, M.: Optimisation of partitioning for parallel flow simulation on block structural grids. In: Topping, B.H.V., Mota Soares, C.A. (eds.) Proceedings of the 4th International Conference on Engineering Computational Technology. Stirling, paper 93. Civil-Comp Press, Stirling (2004)

Part IV
Cross-Sectional Projects

An essential prerequisite for an efficient overall integral model for the simulation of combustion chambers is, on the one hand, the efficiency of sub-models, and on the other hand, the development and efficient integration of numerical methods. The contributions dealing with this topic constitute the content of this part. Adequate optimization algorithms are developed that allow for reaching the proximity of the technical optimum with a relatively small number of complex simulations.

Chapter 10 presents a gradient-based optimization strategy which is employed involving a parallel multigrid solver for the flow and sensitivity equations.

In Chap. 11, another approach for the optimization of turbulent flows is followed by incorporating multilevel optimization algorithms. With this kind of algorithms, different levels describing a problem can be efficiently used for the optimization. Typical examples are discretization levels or models of different physical fidelity. Especially a discrete adjoint approach is applied here. The numerical results that show the efficiency of the adjoint mode and the optimization algorithms include shape optimization and boundary control examples for the Navier–Stokes Equations (NSE), Large Eddy Simulation (LES) and Reynolds Averaged Navier–Stokes (RANS) Equations.

Since LES was recognized as an attractive approach for combustor simulation due to its demonstrated superiority over classical RANS, quality of LES results has to be addressed. In Chap. 12, quality assessment studies for LES that have been carried out in simple configurations are extended to complex geometries following a dynamic quality control technique. Thereby an efficient computation of the adjunct flow equations is necessary along with efficient methods of model reduction such as Proper Orthogonal Decomposition (POD) and Centroidal Voronoi Tessellations (CVT).

The last chapter in this part demonstrates the feasibility of the overall integral model for the simulation of complex combustion chambers.

Chapter 10
Efficient Numerical Schemes for Simulation and Optimization of Turbulent Reactive Flows

J. Siegmann, G. Becker, J. Michaelis, and M. Schäfer

Abstract An approach for the efficient simulation and optimization of turbulent reactive flow problems is presented. A gradient-based optimization strategy is employed involving a parallel multigrid solver for the flow and sensitivity equations. The geometry variation is realized using NURBS surfaces providing a large scale of possible deformations with a small number of design variables. The sensitivity-based computation of the gradient of the objective function is systematically verified by comparisons with finite-difference approximations. The efficiency of the multigrid method and the parallelization is investigated. The functionality of the optimization approach is illustrated by results for representative test cases.

Keywords Optimization • Continuous sensitivity equation • NURBS • Multigrid • Parallelization

10.1 Introduction

Computational fluid dynamics (CFD) is a major subject in various engineering applications, e.g., aerodynamics in aeronautical and automotive engineering, which has led to the development of sophisticated CFD codes. With those and increasing computational power numerical flow optimization becomes increasingly interesting besides the pure flow simulation. The objectives for the optimization can be miscellaneous, one can think of aerodynamic improvements via the cooling of an engine block to, as of interest here, the design improvement of a combustion chamber.

J. Siegmann (✉) • G. Becker • J. Michaelis • M. Schäfer
Institute for Numerical Methods in Mechanical Engineering, Mechanical Engineering, Technische Universität Darmstadt, Dolivostr. 15, 64293 Darmstadt, Germany
e-mail: siegmann@fnb.tu-darmstadt.de; becker@fnb.tu-darmstadt.de; michaelis@fnb.tu-darmstadt.de; schaefer@fnb.tu-darmstadt.de

J. Janicka et al., *Flow and Combustion in Advanced Gas Turbine Combustors*, Fluid Mechanics and Its Applications 102, DOI 10.1007/978-94-007-5320-4_10,

This work considers the implementation and verification of a sensitivity-based flow and shape optimization method for arbitrary surfaces. The governing state equations as well as the resulting sensitivity equations are solved with our in-house code FASTEST. This solver applies a fully conservative finite-volume discretization to solve the incompressible Navier–Stokes equations on a non-staggered, block-structured and cell-centered grid [1]. A highly efficient computation can be obtained due to the close relationship of solving sensitivities and flow governing equations.

Common methods for the resolution of minimization problems can be divided into strategies with and without derivative determination of the cost functional. In the first case the cost functional is approximated via a polynomial ansatz that can then be differentiated at low cost. Although this approach requires no additional gradient information a large number of expensive flow calculations is needed to reach the aspired minimum of the cost functional. Due to this disadvantage gradient-free optimization techniques like, for instance, evolutionary algorithms are not efficient for the considered type of problems. This can already be shown with according test calculations for rather simple configurations [2, 3].

Other approaches are based on the gradient evaluation of the cost functional. Hence, there are several ways to calculate the gradient. One option is the finite-difference method, which can become very expensive due to the fact that one has to calculate two complete flow states for each design parameter. Furthermore the gradient can be obtained via the solution of the adjoint equation or – as in this work – by the evaluation of sensitivities [4]. While the adjoint problem has to be solved backward in time [5], the sensitivities can be calculated forward in time. For each design variable an equation system similar to the Navier–Stokes problem, but linear in its unknowns, has to be solved. The numerical solution of this problem can be obtained with already well-established procedures such as a block-structured finite-volume discretization, a pressure-correction scheme, multigrid methods and domain decomposition. The sensitivities for the different design variables can be calculated independently, which gives an additional approach for an efficient parallelization strategy. The resulting gradient is then transferred to the optimization algorithm, which allows the determination of the cost functionals minimum.

The sensitivities are calculated with a differentiate-then-discretize approach presented by Borggaard and Burns [6], i.e., the sensitivity equations are obtained via differentiation of the non-discretized partial differential equations followed by discretization. The resulting PDE-system is then solved with a finite-volume approach. Particular attention to the boundary conditions is required, which have to be differentiated as well. Thus, for the optimization with shape parameters the derivative of the flow at the boundary as well as the derivative of the boundary describing functional has to be calculated. In our case the surfaces are described with NURBS surfaces, i.e., we need their derivative with respect to the shape parameters, which can be obtained easily. Furthermore the NURBS surfaces provide a large scale of possible deformations witch a rather low amount of design variables.

The basic functionality of the approach considered is verified via the comparison with the finite-difference method. Furthermore we show how the approach performs within a shape optimization environment. As a test case the flow over a hill [7] is considered in order to verify the implementation of the algorithm. Furthermore, the functionality of the approach is illustrated for a non-isothermal scenario relevant for gas turbine combustion chambers. Additionally we investigate the efficiency of the implemented acceleration strategies.

10.2 Theoretical Background

In this section we present the basic equations in the context of flow sensitivity analysis. The continuous sensitivity equations are derived from the Navier–Stokes equations system. We discuss the properties of different optimization methods and the involvement of the sensitivity equation system. Furthermore, we present a surface approximation with NURBS obtaining parameter dependent geometry parts. In order to achieve a high numerical efficiency a geometric multigrid method and parallelization techniques are introduced.

10.2.1 Governing Equations

10.2.1.1 Navier–Stokes Equations

The steady incompressible flows of interest are modeled by the conservation equations of mass, momentum and energy. The energy conservation equation can be simplified to a scalar transport equation for the temperature, while neglecting the work performed by pressure and friction forces and assuming further that the specific heat is constant. Considering Newtonian fluids the Navier–Stokes equations system and the equation for temperature can be written as

$$\frac{\partial u_j}{\partial x_j} = 0, \tag{10.1}$$

$$\frac{\partial \left(\rho u_i u_j\right)}{\partial x_j} - \frac{\partial \left(\mu \tau_{ij}\right)}{\partial x_j} = -\frac{\partial p}{\partial x_i} + \rho f_i, \tag{10.2}$$

$$\frac{\partial \left(\rho c_p u_j T\right)}{\partial x_j} - \frac{\partial}{\partial x_j}\left(\kappa \frac{\partial T}{\partial x_j}\right) = \rho q, \tag{10.3}$$

with

$$\tau_{ij} = \frac{\partial u_i}{\partial x_j} + \frac{\partial u_j}{\partial x_i}. \tag{10.4}$$

Therein ρ and μ represent the fluid density and dynamic viscosity, respectively. The flow velocity is denoted by u_i for the i-th component, p is the pressure, and f_i is the i-th component of a body force. In the temperature equation κ denotes the heat conductivity, T the temperature and q possibly present heat sources. This system of partial differential equations is solved inside the flow domain Ω surrounded by the boundary Γ. The equation system is closed with proper boundary conditions. At an outlet we assume a zero gradient Neumann boundary condition, which is sufficient when the outflow region is far enough from the main region of interest.

In case of parameter dependent flows the solution of the partial differential equations system can be influenced by external design variables. These variables can control body forces or heat sources inside Ω, e.g. the gravitational field in a centrifuge, depending on the parameter controlled rotational speed. Values at the boundary Γ might depend on design variables as well, either as a Dirichlet or a Neumann boundary condition. However, even the entire boundary Γ can be modified by changing the control parameters.

10.2.1.2 The k–ε-Model of Turbulence

In the case of a turbulent flow it is no longer feasible to calculate the flow states directly from the Navier–Stokes equations since this would take by far too much computational effort. In order to achieve acceptable results within moderate computing times statistical models can be used, where the original set of equations is averaged in time. This transforms Eqs. (10.1) and (10.2) into:

$$\frac{\partial \bar{u}_j}{\partial x_j} = 0, \tag{10.5}$$

$$\frac{\partial \left(\rho \bar{u}_i \bar{u}_j\right)}{\partial x_j} - \frac{\partial \left(\left(\mu + \mu_t\right) \tau_{ij}\right)}{\partial x_j} = -\frac{\partial p}{\partial x_i} + \rho f_i, \tag{10.6}$$

with

$$\tau_{ij} = \frac{\partial \bar{u}_i}{\partial x_j} + \frac{\partial \bar{u}_j}{\partial x_i}. \tag{10.7}$$

As an example, the $k-\varepsilon$-model of turbulence, being part of the class of eddy viscosity models, is used to calculate the eddy viscosity μ_t. It can be expressed via the turbulence kinetic energy k and its rate of dissipation ε. The closure of the mathematical system (10.1 and 10.2) is realized using the standard $k-\varepsilon$-model of Launder and Spalding [9]. Furthermore, in turbulent flows particular attention is needed regarding the area near the wall as in this region the assumptions for the turbulence models are no longer correct. In this so-called viscous sublayer we come across a viscosity dominated flow. As an example, we use here the low-Re approach of Chien [10]. The introduced wall dissipation makes a correction of the equations for k and ε necessary.

The k-equation results into

$$\frac{\partial}{\partial x_j}\left[\rho\bar{u}_j k - \left(\mu + \frac{\mu_t}{\partial_k}\right)\frac{\partial k}{\partial x_j}\right] = G - \rho\varepsilon - 2\mu\frac{k}{y^2}, \tag{10.8}$$

using a modified eddy viscosity

$$\mu_t = C_\mu f_\mu \rho\frac{k^2}{\varepsilon} \tag{10.9}$$

with

$$f_\mu = 1 - e^{-0.0115y^+} \tag{10.10}$$

in which y^+ denotes the normal wall distance. The equation for ε results into

$$\frac{\partial}{\partial x_j}\left[\rho\bar{u}_j\varepsilon - \left(\mu + \frac{\mu_t}{\partial_\varepsilon}\right)\frac{\partial\varepsilon}{\partial x_j}\right] = C_{\varepsilon 1}G\frac{\varepsilon}{k} - C_{\varepsilon 2}\rho\frac{\varepsilon^2}{k}f_2 - 2\mu\frac{\varepsilon}{y^2}e - 0.5^{y+} \tag{10.11}$$

with

$$f_2 = 1 - 0.22e\frac{(R_T)^2}{6}, \quad \text{and} \quad R_T = \frac{\rho k^2}{\mu\varepsilon}. \tag{10.12}$$

Using this formulation we can ensure a correct gradient of the velocities needed for the sensitivity equations.

10.2.2 Continuous Sensitivity Equations

For computing the sensitivity equations two general approaches can be distinguished [4]. On the one hand the discretize-then-differentiate method where the Navier–Stokes equations are discretized in advance and later differentiated with respect to the design parameters, resulting in the discrete sensitivity equations (DSE). This can be realized by automatic differentiation software [11], working on the source code of the flow solver. However, these tools must be controlled manually, since they need to differentiate the complete code, including limiters, blending functions, or other non-differentiable terms. Furthermore, it is necessary for parameter dependent surfaces to compute the mesh sensitivities, i.e., differentiating the sophisticated mesh deformation routines.

On the other hand the continuous approach, that we employ here, starts from the averaged analytical flow Eqs. (10.5) and (10.6) [12]. Note that the procedure is identical for the non-averaged equations. Implicit differentiation of the system results in the continuous sensitivity equations (CSE). The numerical discretization in this framework is more flexible and can differ from that of the state equations, for instance, another meshing could be used. Although it is convenient to use the same discretization scheme and to reuse major parts of the state solver. Shape parameters appear by implicit differentiation as additional terms in the boundary conditions of the CSE and their treatment needs special attention, however, the explicit computation of mesh sensitivities becomes unnecessary [13]. The same approach is used for the additional state equations for energy, turbulent kinetic energy and dissipation.

For sake of simplicity we introduce the following notation:

$$\frac{\partial \phi}{\partial a} - s_{\phi}^{a} \tag{10.13}$$

Having (10.13) the CSE can be written as

$$\frac{\partial s_{\bar{u}_j}^{a}}{\partial x_j} - 0, \tag{10.14}$$

$$\frac{\partial}{\partial x_j}\left(\rho s_{\bar{u}_i}^{a} \bar{u}_j\right) - \frac{\partial}{\partial x_i}\left(\mu s_{\tau_{ij}}^{a}\right) = -\frac{\partial s_p^{a}}{\partial x_i} - \frac{\partial}{\partial x_j}\left(\rho \bar{u}_i s_{u_j}^{a} + s_{\mu_t}^{a} \tau_{ij}\right). \tag{10.15}$$

with

$$s_{\tau_{ij}}^{a} = \frac{\partial s_{u_i}^{a}}{\partial x_j} + \frac{\partial s_{u_j}^{a}}{\partial x_i} \quad \text{and} \quad \tau_{ij} = \frac{\partial \bar{u}_i}{\partial x_j} + \frac{\partial \bar{u}_j}{\partial x_i}. \tag{10.16}$$

assuming that the parameter a has no influence on the fluid properties ρ and μ, as well as on the body force f_i. For the sensitivity of the turbulent kinetic energy we get:

$$\frac{\partial}{\partial x_j}\left(\rho \bar{u}_j s_k^{a} - \frac{\partial}{\partial x_j}\left(s_{\mu_t}^{a} k - (\mu + \mu_t)\, s_k^{a}\right)\right) = G_k - \rho s_{\varepsilon}^{a} - 2\mu \frac{s_k^{a}}{y^2} - \frac{\partial}{\partial x_j} s_{\bar{u}_j}^{a} \tag{10.17}$$

with

$$G_k = s_{\mu_t}^{a} \tau_{ij} \frac{\partial \bar{u}_j}{\partial x_j} + \mu_t \tau_{ij} \frac{\partial s_{\bar{u}_j}^{a}}{\partial x_j} + \mu_t s_{\tau_{ij}}^{a} \frac{\partial \bar{u}_j}{\partial x_j} \tag{10.18}$$

and

$$s_{\mu_t}^{a} = C_{\mu} f_{\mu} \rho \left(\frac{2 s_k^{a} k \varepsilon - k^2 s_{\varepsilon}^{a}}{\varepsilon^2}\right) + C_{\mu} s_{f_{\mu}}^{a} \rho \frac{k^2}{\varepsilon} \tag{10.19}$$

as well as

$$s^a_{f_\mu} = 0.0115 e^{-0.0115 y^+} s^a_{y+} \quad \text{with} \quad s^a_{y+} \frac{\rho s^a_{u_\tau} y}{\mu}. \tag{10.20}$$

For the sensitivity of the dissipation rate we obtain:

$$\begin{aligned}
&\frac{\partial}{\partial x_j}\left(\rho \bar{u}_j s^a_\varepsilon - \frac{\partial}{\partial x_j}\left(s^a_{\mu_t}\varepsilon - (\mu+\mu_t) s^a_\varepsilon\right)\right) = -\frac{\partial}{\partial x_j}\left(\rho s^a_{\bar{u}_j} k\right) + C_{\varepsilon 1} G_k \frac{\varepsilon}{k} \\
&+ C_{\varepsilon 1} G\left(\frac{s^a_\varepsilon k - s^a_k \varepsilon}{k^2}\right) - C_{\varepsilon 2}\rho\left(\frac{2 s^a_\varepsilon \varepsilon k - s^a_k \varepsilon^2}{k^2}\right) + C_{\varepsilon 2}\rho \frac{\varepsilon^2}{k} s^a_{f_2} - 2\mu \frac{s^a_\varepsilon}{y^2} e^{-0.5 y^+} \\
&+ \mu \frac{\varepsilon}{y^2} e^{-0.5 y^+} s^a_{y+}
\end{aligned} \tag{10.21}$$

with

$$s^a_{f_2} = \frac{1}{3} 0.22 e^{\frac{-R_T^{-2}}{6}} R_T s^a_{R_T} \quad \text{and} \quad s^a_{R_T} = \frac{\rho}{\mu}\left(\frac{2 s^a_k k \varepsilon - s^a_\varepsilon k^2}{e^2}\right) \tag{10.22}$$

Furthermore, it is necessary to close the CSE system. Thus the boundary conditions need to be differentiated with respect to the control parameter a as well. The zero gradient condition is retained at the outlet. Dirichlet boundary conditions are implicitly differentiated like the state equations. Having a shape parameter a, the boundary geometry itself is affected by changes of the control variable. For this the Dirichlet boundary condition must be stated more precisely as $u_i\,(x(a), y(a), z(a), a) = \bar{u}_i\,(x(a), y(a), z(a), a)$ on $\Gamma_D(a)$. Implicit differentiation of this condition yields

$$\frac{\mathrm{d}u_i}{\mathrm{d}a} = \frac{\mathrm{d}\bar{u}_i}{\mathrm{d}a} \quad \text{on } \Gamma_D(a), \tag{10.23}$$

$$\Leftrightarrow \quad \frac{\partial u_i}{\partial a} + \frac{\partial u_i}{\partial x_j}\frac{\mathrm{d}x_j}{\mathrm{d}a} = \frac{\mathrm{d}\bar{u}_i}{\mathrm{d}a} \quad \text{on } \Gamma_D(a), \tag{10.24}$$

where d/da denotes the total derivative with respect to the shape parameter a.

Equation (10.24) leads to the boundary condition for the sensitivity $s^a_{u_j}$ at the manipulable boundary

$$s^a_{u_i} = \frac{\mathrm{d}\bar{u}_i}{\mathrm{d}a} - \frac{\partial u_i}{\partial x_j}\frac{\mathrm{d}x_j}{\mathrm{d}a} \quad \text{on } \Gamma_D(a). \tag{10.25}$$

The sensitivity of the shape position $\partial x_j/\partial a$ appears, as well as the gradient of the flow velocities at the boundary $\partial u_i/\partial x_j$. For a closer insight into the topic we refer to [14].

10.2.3 Optimization Methods

10.2.3.1 General Optimization Problem

According to Jacobi et al. [15] a basic optimization problem consists of a given *cost* or *objective functional* $\mathcal{J}$ depending on a set of variables with the possibility to add some additional *constraints* F of any kind. The variables are separated in *state variables* $\phi \in \mathcal{R}^n$ and *control variables* $\alpha \in \mathcal{R}^m$. State variables are internal values and cannot be manipulated directly. They depend on the control variables $\phi = \phi\ (a)$. These so called design parameters can be controlled from the outside. The constraints must be satisfied by all candidate sets of state and control variables. The challenge is to find the internal variables and the corresponding set of design parameters which minimize the given cost functional and additionally satisfy the constraints. We examine a minimization problem by default. In mathematical notation this can be written as

$$\min \quad \mathcal{J}\ (\phi(a).\,a) \quad \text{subject to} \quad F\ (\phi(a).\,a) = 0. \tag{10.26}$$

In our context the state variables are flow properties like velocity, pressure, temperature, turbulent kinetic energy, etc. The control variables, for instance, are the inflow velocity, external heat insulation or geometry variations. The last case is also known as shape optimization [16]. In a flow optimization problem the main constraints are always the partial differential equations that describe the flow field presented in Sect. 10.2.1. Most objective functionals used in practice do not depend explicitly on the design parameters, thus unbounded optimal controls are possible. This means an optimal solution of the objective functional depending only on the state variables is reached for an infinite great value of any design parameter. To avoid these unrealistic solutions the control size should be limited somehow in such a case. That changes the basic problem Eq. (10.26) into

$$\begin{aligned} \min \ \mathcal{J}\ (\phi(a)), \quad \text{subject to} \quad & F\ (\phi(a).a) = 0. \\ & a_i \le \kappa, \qquad \text{for some constant } \kappa. \end{aligned} \tag{10.27}$$

10.2.3.2 Gradient-Based Optimization

A gradient-based optimization strategy is chosen to find optimal design parameters. The algorithms in this category use the functional evaluations as well as information about the gradient of the functional. Of course, it requires additional computational effort to calculate the gradient of the functional. A detailed comparison between a gradient-based and a derivative-free optimization technique is given in [17]. A simple and straightforward approach for computing the gradient is the difference quotient. However this requires n additional functional evaluations to get

$\nabla \mathscr{J} \in \mathcal{R}^m$. A more sophisticated approach is the usage of the sensitivity equations to calculate the gradient. No matter how the gradient is finally calculated, this additional input of information about the slope usually results in a much smaller number of iterations of the optimization algorithm.

The trust-region method is a representative example of an optimization algorithm. Using the gradient allows to build a quadratic model of the functional. The trust-region method defines an adaptive region in which the correlation between the model and the original functional is trustful. The crucial idea is to optimize a simpler model of the objective functional instead of to struggle with the minimization of the original functional. The optimization of a quadratic sub-problem is quite well explored and we can choose between several established optimization strategies. An approved approach is the usage of quasi-Newton methods presented in [18].

10.2.4 NURBS Surface Approximation

For the optimization of arbitrary shape optimization problems a flexible surface representation is needed. Hence, the method of choice is the usage of non-uniform rational B-spline surfaces (NURBS surfaces) as they provide the required features with a manageable number of design variables. Regarding our problem the first step is the approximation of a given surface grid with a NURBS surface according to the algorithm presented in Becker et al. [19]. The resulting surface can be described as

$$\mathscr{S}(s, t) = \sum_{i=1}^{\hat{n}} \sum_{j=1}^{\hat{m}} \cdot \mathscr{N}_i^p(s) \mathscr{N}_j^q(t) \mathbf{P}_{i,j}, \qquad s, t \in [0, 1], \tag{10.28}$$

where $P_{i,j}$ stands for the control points of the NURBS surfaces which are driven by the design parameters in our problem. The resulting movement of the volume grid is realized with the grid generation routines described in Schäfer et al. [20].

10.2.5 Numerical Techniques

10.2.5.1 Discretization and Solution Methods

In numerical calculations accuracy, efficiency and robustness are key requirements. To this respect numerous strategies have been investigated and implemented. An improved multidimensional interpolation provided a distinct gain in accuracy during calculations on irregular grids [21–23]. The modification of the pressure correction equation enlarged the convergence area of the relaxation parameters leading to a significant improvement within the robustness [22]. Furthermore, in order to reduce the need of computational costs as well as needed memory, strategies for local grid refinement have been investigated [24–26]. In the case of turbulent flows a step-length control for the multigrid method (see net section) led to significant improvements [21, 22].

10.2.5.2 Multigrid Methods

Conventional iterative solvers usually show a drastic increase of the computational effort with decreasing grid spacing. In particular, for 3-dimensional problems the cost swells rapidly. Using a Fourier series expansion of the error it is possible to show that basic iterative solvers remove efficiently errors with a wavelength of about the characteristic control volume size h (see, e.g., [27]). Error components with longer wavelengths are only slowly eliminated.

Multigrid algorithms are based on the idea to solve an equation system involving a hierarchy of meshes (see, e.g., [28, 29]). We employ a geometric multigrid algorithm where the different grid levels are generated by coarsening steps from the finest grid.

The linear equation system (10.14) and (10.15) can be written shortly as

$$\mathbf{L}^h \mathbf{s}^h = \mathbf{b}^h, \tag{10.29}$$

where $\mathbf{s}$ summarizes the unknown sensitivities and the superscript h emphasizes the spatial discretization. Although this is a linear system we employ the full approximation scheme (FAS), a non-linear multigrid algorithm. For linear problems the linear multigrid algorithm and the FAS are analytically equivalent. However, in contrast to linear multigrid algorithms, FAS does not approximate the error on different grid levels, but improves directly the solution of the sensitivities on the different grids. In future studies, this can be used for adaptive optimization methods, wherein the gradient is computed simultaneously with low computational effort on the coarse grid levels. Utilizing error estimation it is possible to extend the multigrid cycle to finer grids during the optimization process. Thereby, more accurate gradient approximations can be obtained near the optimum. Furthermore some advanced techniques like τ-extrapolation based on the FAS approach can be used, see [30]. Hence, we write Eq. (10.29) in non-linear notation

$$\mathbf{L}^h \left(\mathbf{s}^h\right) = \mathbf{b}^h. \tag{10.30}$$

The SIMPLE algorithm yields after m iterations towards an approximation of the solution, which satisfies Eq. (10.30) up to a residual vector $\mathbf{r}^h$. The non-linear defect equation reads:

$$\begin{aligned} & \mathbf{L}^h \left(\mathbf{s}^h\right) = \mathbf{b}^h \\ \Leftrightarrow \quad & \mathbf{L}^h \left(\mathbf{s}^{m,h} + \mathbf{s}^h - \mathbf{s}^{m,h}\right) - \mathbf{L}^h \left(\mathbf{s}^{m,h}\right) = \mathbf{b}^h - \mathbf{L}^h \left(\mathbf{s}^{m,h}\right) \\ \Leftrightarrow \quad & \mathbf{L}^h \left(\mathbf{s}^{m,h} + \mathbf{e}^h\right) - \mathbf{L}^h \left(\mathbf{s}^{m,h}\right) = \mathbf{r}^h, \end{aligned} \tag{10.31}$$

where $e^h = s^h - s^{m,h}$ is the current error on grid level h. The idea is to solve the defect Eq. (10.31) on the coarser grid levels. The computed fine grid values are restricted by a restriction operator $\mathcal{R}_h^{2h}$ to the next coarser grid level $(2\,h)$. For building up the new discretization $\mathbf{L}^{2h}$ on the coarser grid we can make use of the conservativity of the finite volume method, i.e., the flux over the faces of the assembled coarse grid CV is equal to the fluxes over the faces of the eight underlying fine grid CVs. After the restriction we transform Eq. (10.31) into

$$\mathbf{L}^{2h}\left(\mathscr{R}_h^{2h}\left(\mathbf{s}^{m,h}\right) + \mathbf{e}^{2h}\right) = \mathscr{R}_h^{2h}\left(\mathbf{r}^h\right) + \mathbf{L}^{2h}\left(\mathscr{R}_h^{2h}\left(\mathbf{s}^{m,h}\right)\right), \tag{10.32}$$

wherein the coarse grid variable $s^{2h} = \mathcal{R}_h^{2h}\left(s^{m,h}\right) + e^{2h}$ is defined and the known terms on the right hand side are merged. We obtain the new coarse grid defect equation

$$\mathbf{L}^{2h}\left(\mathbf{s}^{2h}\right) = \mathbf{b}^{2h}, \tag{10.33}$$

which is similar to Eq. (10.30) except for the grid spacing. Thus, it is possible to use in turn coarser meshes to solve this defect equation. After finding an approximate solution s^{2h} for Eq. (10.33), by solving directly or employing further recursions using more grid levels, it is necessary to adapt the error correction back to the fine grid.

An interpolation operator I_{2h}^h prolongates the found coarse grid defect $\mathbf{e}^{2h}$ to the fine grid

$$\mathbf{e}^h = \mathscr{I}_{2h}^h\left(\mathbf{e}^{2h}\right) \approx \mathscr{I}_{2h}^h\left(\tilde{\mathbf{s}}^{2h} - \mathscr{R}_h^{2h}\left(\mathbf{s}^{m,h}\right)\right). \tag{10.34}$$

which can be used iteratively to update the previously computed fine grid solution

$$\mathbf{s}^h - \mathbf{s}^{m,h} + \mathbf{e}^h. \tag{10.35}$$

This updated solution is iterated again by the SIMPLE algorithm to smooth high-frequency errors appearing from the interpolations between the different grid levels. We use V-cycles to move through the grid hierarchy. For further details on the topic see [8, 31].

10.2.5.3 Parallelization Strategy

Despite the improvements in computer technique and the usage of efficient algorithms size, number of parameters or the complexity of many problems make it impractical or impossible to solve them on a single processor. Parallel computing offers a solution for this challenge. Two generic message-passing strategies for parallelization of the sensitivity equations are reasonable.

The first possibility is the separate calculation of each parameter sensitivity on a different processor in parallel after previously determining the state variables once.

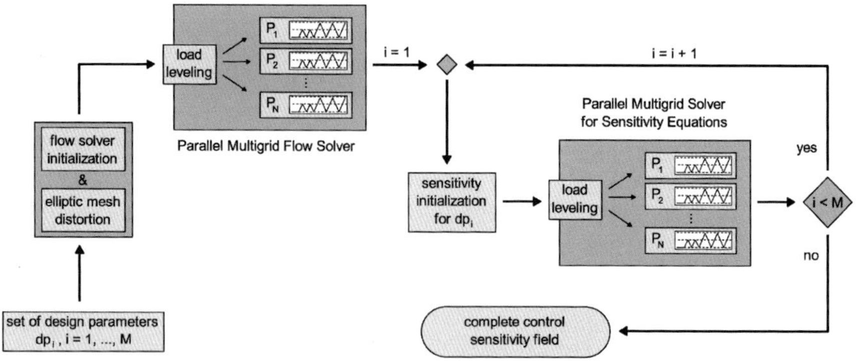

Fig. 10.1 Schematic work-flow for parallel computation of the control sensitivity information within the entire flow field for M design parameters (*dpi*) on N processors (P_j)

This strategy benefits from the fact that each system has almost the same structure and differs only in the boundary constraints, but is absolutely independent from the other parameter sensitivity systems. This fits perfectly the requirements of an efficient parallelization. Drawbacks are the strong coupling between the number of parameters and the required number of processors and the risk that the sensitivity computations of different design parameters have variable duration leading to idling processors.

The other strategy is based on the idea of solving each sensitivity system on several processors in parallel, before computing successively the next sensitivity. In this setting the NURBS surface fitting and mesh distortion is computed before the state solution. Thereafter the multiple solver runs for the sensitivity equations due to several design parameters are done in a serial manner, but each intrinsic computation is done in parallel, splitted onto several processors solving the actual sensitivity system simultaneously. Figure 10.1 illustrates this procedure. The advantages of this methodology are the rather loose coupling between the number of parameters and the number of processors as well as the immunity to varying computational cost for the different sensitivities.

We focus on the second approach described, e.g., in [32]. The computational domain is subdivided into non-overlapping subdomains which are each delegated to a single processor. In principal this decomposition is independent from the splitting used for the parallel flow solver.

Due to the local properties of the discretization schemes, using only the adjacent neighbors, the processors can solve their problem domain mostly independent from the other processors. Only at the interface boundaries an information exchange is needed, which is realized by an additional layer of ghost cells, containing the values from the adjacent control volume in the next subdomain. The content of these ghost cells is interchanged at fixed synchronization points, after the initialization, following each SIMPLE iteration and finally at the end of the program execution.

Besides these local exchanges some global information exchanges are needed as well, e.g., for convergence checking. For the processor communication we utilize the Message Passing Interface (MPI) [33].

The NURBS surface initialization is done on a predefined processor as well as the computation of the related sensitivity boundary conditions. These calculations have relatively small computational cost and the results are broadcasted hereafter to the relevant processors.

10.3 Results

In this section, numerical results are presented in three steps. First we verify our implementation of the continuous sensitivity equation method and compare the obtained results with those of the (expensive) finite difference method. In a second step we present the usage of the sensitivities in a shape optimization environment for the straightforward example flow over a 2-dimensional bump. At last we present the results obtained for a generic heat exchanger where we combine flow and temperature sensitivities with shape optimization.

10.3.1 Verification Sensitivities

10.3.1.1 Verification of Flow Sensitivities

As a straightforward test case for verification we use a two-dimensional Poiseuille flow between two infinite flat plates. For this case an analytical solution exist which enables as to verify our flow solution and by that the results of our finite difference method. This solution then provides the base for the comparison with the results from the sensitivity calculation. The Poiseuille flow is set as inlet condition for the channel (see Fig. 10.2). For the given setup it can be described analytically bywith

$$\hat{u}(y) = \bar{u}a \frac{y\,(3h - y)}{(1.5h)^2}, \tag{10.36}$$

where ū is the maximum velocity in x-direction and $3h$ denotes the height of the channel. The Reynolds number results to $\mathrm{Re} = 1$ for $\bar{u} = 1$ and $a = 1$. For the finite-difference method two calculations are performed with a difference between the two inlet velocities of $\bar{u}_{\text{mod}} - \bar{u} = 0.0001$ m/s. The sensitivity for the velocity in x-direction each field point hence results to:

$$dq_t = \frac{\bar{u}_{\text{mod}} - \bar{u}}{0.0001}. \tag{10.37}$$

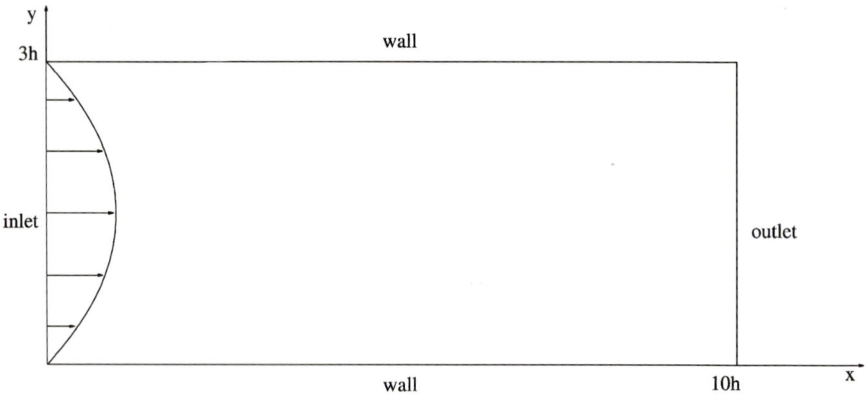

Fig. 10.2 Setup for the calculation of flow between two infinite parallel plates

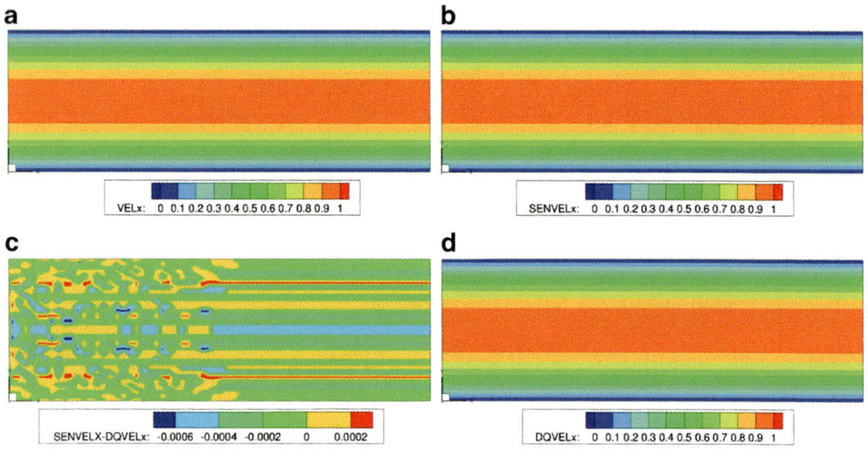

Fig. 10.3 Results for the flow between two infinite parallel plates. (**a**) Velocity in x-direction, (**b**) Sensitivity for the velocity in x-direction, (**c**) Difference between sensitivity and finite-difference, (**d**) Finite-difference for the velocity in x-direction

The boundary condition at the inlet for the sensitivity calculation can be obtained easily via straightforward differentiation:

$$s_u^a(y) = \frac{y\,(3h - y)}{(1.5h)^2}. \tag{10.38}$$

The obtained results are illustrated in Fig. 10.3. As expected, the solver provides the analytical solution for the given problem. The resulting sensitivity field looks exactly like the velocity field as we use the same boundary conditions for the two problems.

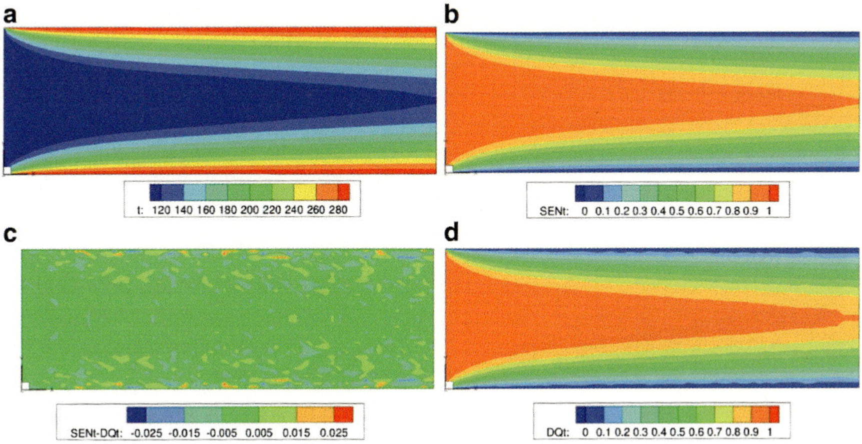

Fig. 10.4 Results for the temperature between two infinite parallel plates. (**a**) Temperature, (**b**) Sensitivity for the temperature, (**c**) Difference between sensitivity and finite-difference, (**d**) Finite-difference for the temperature

Comparing the solution of the finite-difference method with the sensitivity computation (see Fig. 10.3c) shows that there are only minor differences between the two results which are randomly distributed over the field. In summary the obtained results lead us to the conclusion, that the sensitivity equations for the Navier–Stokes equations have been derived and implemented correctly. Furthermore, in comparison to the solution with the finite-difference method, a computation of the sensitivities only takes about 10–15% of the time needed for the second flow calculation.

10.3.1.2 Verification of Temperature Sensitivities

To verify the derivation and the implementation of the heat transfer we again refer to the test case introduced in Fig. 10.2. As boundary condition for the flow we use a very low inlet velocity of $\bar{u} = 1 \cdot 10^{-6}\text{m/s}$ in order to add a convective component to the energy transport. For the energy equation we set the inlet temperature to $T_{inlet} = a$ with $a = 100°\text{C}$ and the temperature of the walls to $T_{wall} = 300°\text{C}$. Note, that only the inlet temperature is dependent on the design parameter, therefore, the boundary conditions for the sensitivity equation result into $s_T^a = 1$ at the inlet and $s_T^a = 0$ on the walls. Again we perform a finite-difference calculation in order to have a base for a comparison with a difference between the two inlet temperatures of $T_{mod} - T = 0.0001°\text{C}$.

The obtained results are shown in Fig. 10.4. One can see that the fluid is heated from the walls as it flows towards the outlet. Since we can only influence the temperature at the inlet, the sensitivity is also maximal at this point and drops

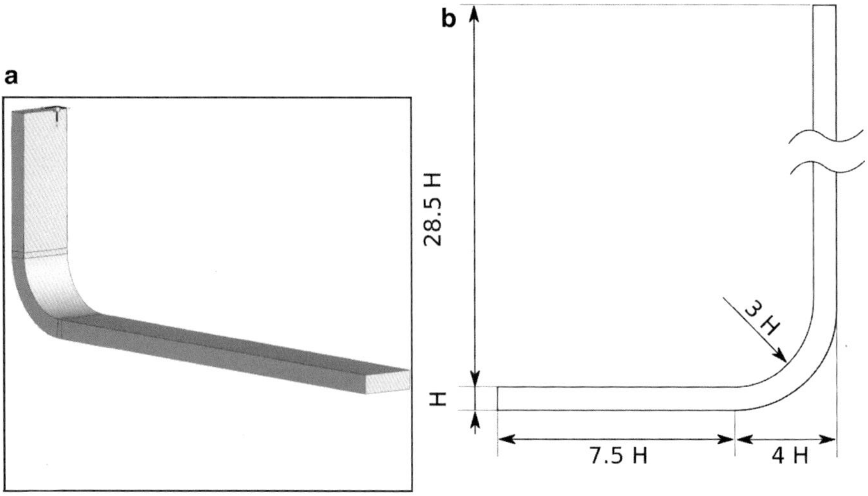

Fig. 10.5 Geometry of the curved duct testcase. (**a**) Geometry, (**b**) Dimensions

towards the outlet. In comparison with the finite-difference method we observe again that there are only minor and randomly distributed differences between the two computations (see Fig. 10.4c). In terms of computing times we again save up to 90% in comparison to the time needed for a second computation of the problem.

10.3.1.3 Verification of Turbulent Sensitivities

The underlying test case for the verification of the turbulence quantities is the well-known flow in a curved duct out of the ERCOFTAC database. The geometry is shown in Fig. 10.5. With a given inlet velocity of $\bar{u} = a$ and $a = 16$ m/s as well as the underlying physical properties the Reynolds number results to $\mathrm{Re} = 2.32 \cdot 10^4$ which is clearly in the turbulent regime. The present turbulent flow is calculated with the methods prescribed in Sect. 10.2.1. Straightforward differentiation of the inlet velocity leads to $s_u^a = 1$ as boundary condition for the sensitivities. For the finite-difference method two calculations are performed with a difference between the two inlet velocities of $\bar{u}_{\text{mod}} - \bar{u} = 0 \cdot 0001$ m/s. The results are shown in Fig. 10.6.

As expected the computations for the velocity and the sensitivities show similar results. Hence the comparison to the finite difference shows some larger differences in the values for the sensitivities. A possible explanation is the fact, that the coupling of the turbulent viscosity and the flow equations as described in Sect. 10.2.2 is different for the flow and the sensitivity equation. This may lead to different requirements regarding the mesh, especially in the critical near wall section, for the two equation system. Both calculations are performed on the same grid. A deeper investigation of this topic is required. However, the results obtained are good enough to verify the basic idea behind the derivation and the correct implementation.

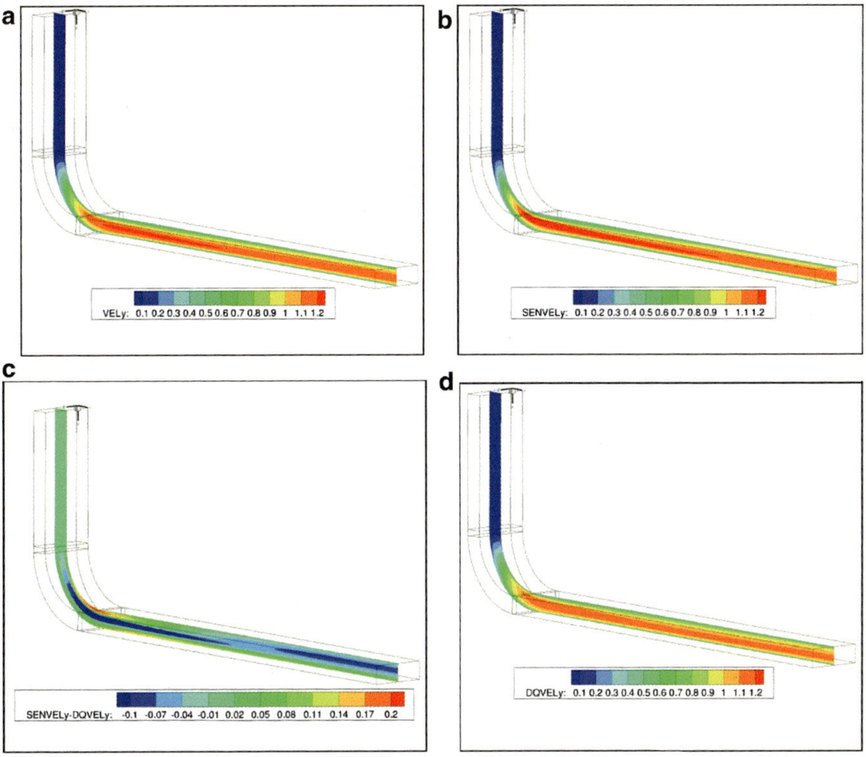

Fig. 10.6 Results for the velocity in a curved duct. (**a**) Velocity in v-direction, (**b**) Sensitivity for the velocity in v-direction, (**c**) Difference between sensitivity and finite-difference, (**d**) Finite-difference for the velocity in v-direction

10.3.2 Flow Over a Two Dimensional Bump

Now we present a test case in which the sensitivities are used in a shape optimization environment. The initial situation is the quasi two-dimensional flow (Re = 1) across a NURBS generated hill to derange the ideal laminar Poiseuille flow (see Fig. 10.7). The Poiseuille flow is set as inlet condition for the channel.

The top and bottom channel faces, including the hill surface, are impermeable walls. The shape of the hill is formed by a NURBS consisting of 6 control points $\mathbf{P}_{s,t}$ ($s = 1, 2;\ t = 1, 2, 3$), see Fig. 10.7b. The four corner points of the hill are fixed. The position of control point $P_{1,2,}$ can be changed along the y-direction by the design parameter a. Control point $P_{2,2}$ moves parallel to $P_{1,2}$. Mathematically this relation is given by the equations

$$\mathbf{P}_{1,2} = a \begin{pmatrix} 0 \\ 1 \\ 0 \end{pmatrix} + \mathbf{P}^{(0)}_{1,2}, \tag{10.39}$$

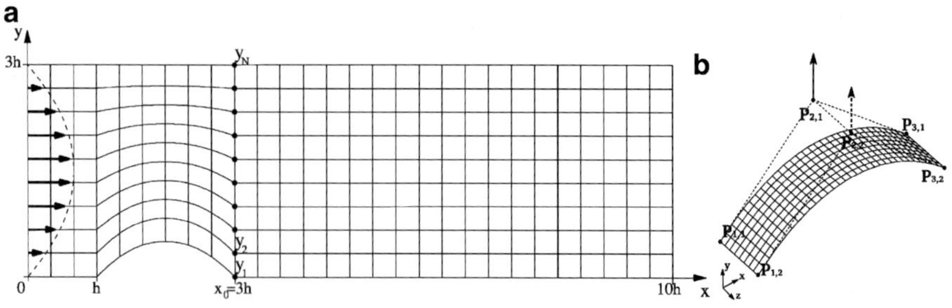

Fig. 10.7 2d hill set up and definition of design variables (*Red arrows*) indicate the direction of movement depending on the movement of the directly by the design variable affected control point (*green arrows*). (**a**) 2d hill problem set up, (**b**) NURBS constructed hill

$$\mathbf{P}_{2,2} = a \begin{pmatrix} 0 \\ 1 \\ 0 \end{pmatrix} + \mathbf{P}^{(0)}_{2,2}, \tag{10.40}$$

where $P^{start}_{s,2}$, s = 1,2 denotes the initial position of the related NURBS control points. The optimization objective is to equalize the flow behind the hill with the flow right at the inlet. Mathematically this leads to the objective functional

$$\mathcal{J}\left(u\left(a_1, x, y\right), u_{x=h}, x_0, y_1, \ldots, y_N\right) = \frac{1}{2}\sum_{i=1}^{N}\left(u\left(a_1, x_0, y_i\right) - u_{x=h}\left(y_i\right)\right)^2 \tag{10.41}$$

where $u_{x=0}$ denotes the velocity in x-direction right at the beginning of the hill and $\partial u\left(a_1,\, x_0,\, y_i\right)/\partial a_1$ denotes the current velocity in x-direction dependent on the shape of the flow domain (controlled by a_1), at $\left(x_o, y_i\right), i = 1, \ldots, N$ (see Fig. 10.7b).

Furthermore the trust region algorithm needs the objective functional value and its derivative to build the quadratic submodel. The derivative is given analytically by

$$\begin{aligned} &\frac{\partial}{\partial a_1}\mathcal{J}\left(u\left(a_1, x, y\right), u_{x=h}, x_0, y_1, \ldots, y_N\right) \\ &= \sum_{i=1}^{N}\left(u\left(a_1, x_0, y_i\right) - u_{x=h}\left(y_i\right)\right)\frac{\partial}{\partial a_1}u\left(a_1, x_0, y_i\right) \end{aligned} \tag{10.42}$$

where $\partial u\left(a_1,\, x_0,\, y_i\right)/\partial a_1$ denotes the sensitivity of the velocity in x-direction, at the position $\left(x_o, y_i\right), i = 1, \ldots, N$, depending on the design parameter a_1 which modifies the height of the hill. The grid deformation within the block above the hill is performed by linear interpolation, while the remaining grid is unchanged. The initial value of the design parameter is set to zero. Therefore the initial position of

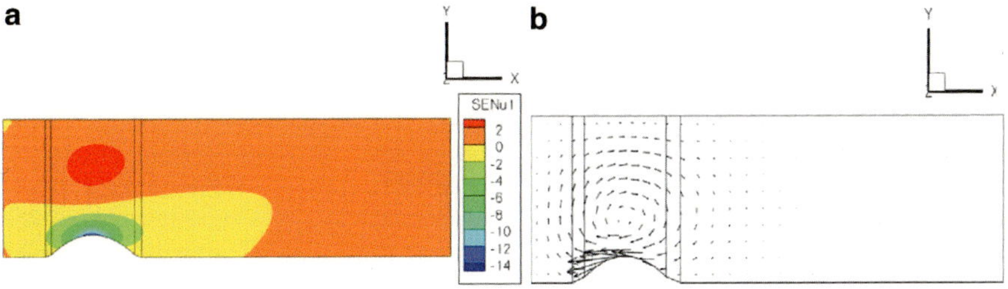

Fig. 10.8 Calculated sensitivities at the beginning of the optimization. (**a**) Distribution of the sensitivities of the u-velocity, (**b**) Distribution of the sensitivities

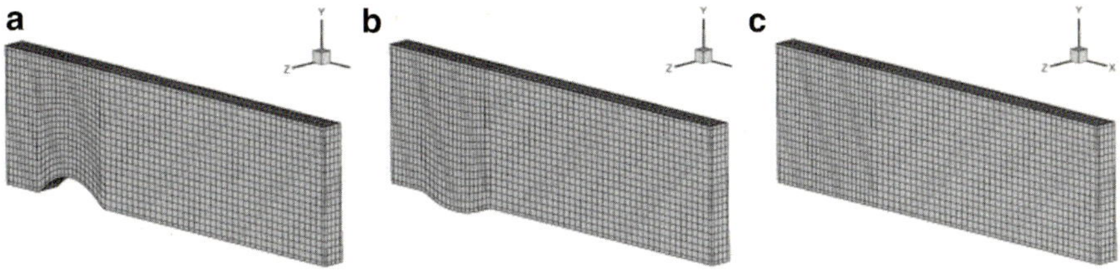

Fig. 10.9 Different shapes during the optimization process. (**a**) Initial shape, (**b**) Shape after 2 optimization steps, (**c**) Final shape

control point $P_{1,2}^{(0)}$ is (0.2 h, 0.1 h, 0), respectively (0.2 h, 0.1 h, 0.5 h) for $P_{2,2}^{(0)}$. A positive a_1 will cause a higher hill peak, while a negative value truncates the hill, or even generates a sink for large negative values. The solution of this simple test case is quite obvious: the design variable is decreased just as much to make the hill vanish, creating the undisturbed Hagen-Poiseuille channel flow. To prevent too large grid deformations and non-physical solutions the design parameter is bounded between −0.15 and 0.1.

In order to calculate the gradient of the objective function the sensitivities have to be computed first. The obtained distribution is shown in Fig. 10.8.

It can be seen, that the influence of the design parameter onto the flow field is strong around the bump with a maximum on its top. With the obtained gradient of the functional the optimizer starts its work. It takes 19 flow evaluations and 8 sensitivity calculations to reach the searched minimum of the cost functional. Figure 10.9 shows the shape of the channel at different points of the optimization process.

10.3.3 Flow Over a Heated Two Dimensional Bump

In a second straightforward example we want to present again the flow over the bump described in Sect. 10.3.2. In this case now the wall of the bump has a given temperature of $T = 600°C$ while the fluid enters the domain with $T = 100°C$ and a

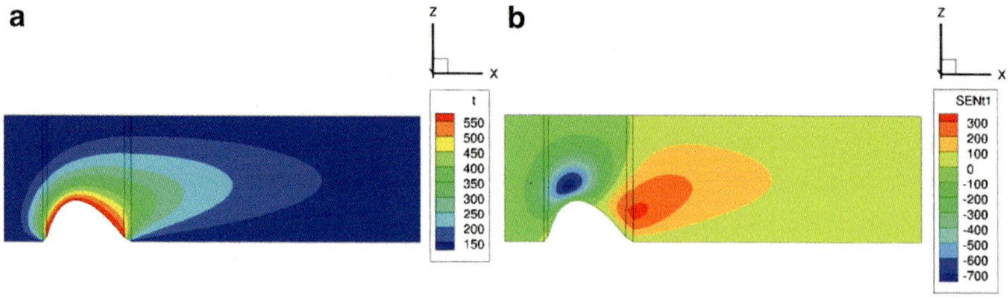

Fig. 10.10 Final shape and temperature distribution of the generic heat exchanger (**a**) Temperature distribution, (**b**) Sensitivity distribution

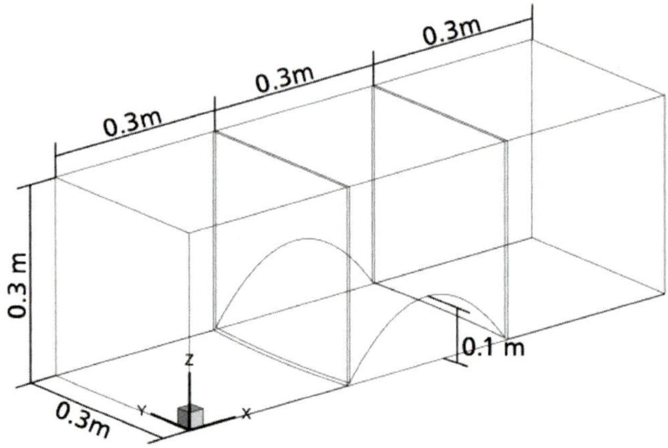

Fig. 10.11 Geometric setup for 3-dimensional flow over NURBS-generated bump

Reynolds number of 1. As we use symmetry boundary conditions before and after the bump in z-direction the case can be seen as an idealized heat exchanger. Goal of the optimization process is to maximize the temperature difference of the fluid between the inlet and the outlet of the domain in dependence of the bump shape. In order to prevent non-feasible solutions we additionally restrict the change in the volume of the bump to 5% based on its original volume. We realize this by the use of a penalty function. The obtained solution after 15 flow evaluations and 6 sensitivity calculations is shown in Fig. 10.10. The temperature difference could be increased by 7%.

10.3.4 Multigrid Analysis

For the verification of the multigrid implementation, a three-dimensional channel flow over a NURBS-generated bump in the middle section is analyzed. The channel geometry is given in Fig. 10.11. Its length is 0.9 m and the cross-section is quadratic

Table 10.1 Number of control volumes on the different grid levels

Grid level	1	2	3	4	5	6
#CV	24	192	1,536	12,288	98,304	786,432

with a height and depth of 0.3 m each. The inlet is located at $x = 0.0$ m and the outlet is situated at $x = 0.9$ m. All the others surrounding channel faces are impermeable walls. The bump ranges from $x = 0.3$ m to $x = 0.6$ m, its peak reaches $z = 0.1$ m. A viscous fluid with $\rho = 1,400$ kg/m^3 and $\mu = 10$ kg/ms is used to assure a Reynolds number in the range of 10 for inlet velocities of about 0.1 m/s.

To generate a block-structured mesh the geometry is divided into three blocks, each block has 64 CVs in each spatial direction. This allows employing 6 grid levels for the computation of the flow variables and the sensitivity values. For each grid-coarsening step we halve the number of nodes in each direction, ending up with only eight CVs per block on the coarsest mesh (see Table 10.1).

From literature (e.g., [1]) it is well known that the computational cost scales approximately linear with the number of CVs when applying a multigrid algorithm. The computation time for solving the sensitivity equation system is measured on a Intel Core2 Duo E8400 (3.00 GHz) with 4,096 MB main memory. The measured time is averaged over several runs to minimize temporal fluctuation.

Another characteristic quantity is the number of necessary fine grid iterations, which usually remain almost constant with grid refinement (e.g., [27]).

Two different control settings are examined to verify this typical multigrid behavior. In the first test case the inlet velocity is set by a design parameter. In the second one the surface of the bump is controlled by a shape variable.

10.3.4.1 Variable Inlet Parameter

We apply the following parabolic x-velocity profile at the inlet

$$u = 0.1\, a \frac{y\,(0.3 - y)}{0.15^2} \frac{z\,(0.3 - z)}{0.15^2} \frac{\text{m}}{\text{s}}, \quad a \geq 0, \tag{10.43}$$

with the control parameter a. The y- and z-component are set to zero.

For the sensitivity computation V-cycles of maximum depth are performed, e.g., on the second grid level a V-cycle over two grids is used, while on the finest grid level a V-cycle over all six meshes is employed. On each grid level we perform 5 SIMPLE iterations and restrict the obtained defect equation to the next coarser grid. Only on the coarsest level 10 SIMPLE iterations are performed, before interpolating the error correction back to the finer grids. On the way back to the finest grid 3 SIMPLE iterations are used on each grid to smooth the high frequent errors resulting from the interpolations between the hierarchical levels. For better

Table 10.2 Multigrid behavior for variable inlet parameter

Scheme (#CVs)	SG1 (24)	MG2 (192)	MG3 (1,536)	MG4 (12,288)	MG5 (98,304)	MG6 (786,432)
Fine grid iterations	132	105	83	94	118	176
Time (sec)	0.05	0.13	0.43	3.50	41.18	510.66

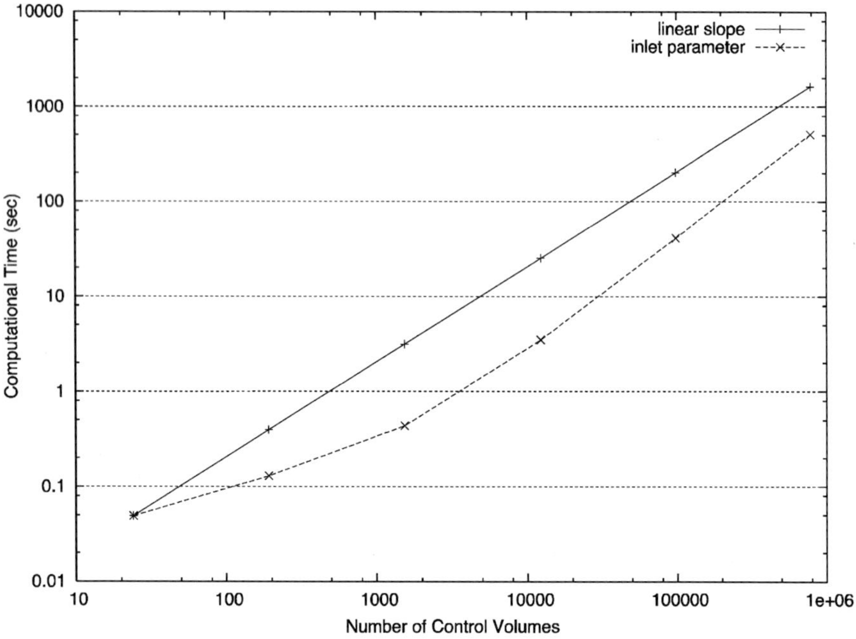

Fig. 10.12 Computational time over the number of control volumes (inlet sensitivity)

comparability of the different sized V-cycles the underrelaxation factors within the SIMPLE algorithm are identically chosen on same grids. The sensitivity calculation is assumed to be converged when the maximum residual value on the finest mesh is smaller then 10^{-6}.

The results are given in Table 10.2. Although the number of fine grid iterations is not perfectly constant, the computational time increases almost linearly with the number of CVs. This correlation is shown in Fig. 10.12 where the computational time is plotted against the number of control volumes in a double logarithmic diagram. The continuous line shows the theoretically expected linear increase of the computational costs.

10.3.4.2 Variable Shape Parameter

In the second test case the design parameter b moves the positions of the control points (CPs) of the NURBS surface, wherein the bump shape can be changed

Table 10.3 Multigrid behavior for controllable bump surface design

Scheme	SG1	MG2	MG3	MG4	MG5	MG6
(#CVs)	(24)	(192)	(1536)	(12288)	(98304)	(786432)
Fine grid iterations	165	100	86	102	126	127
Time (sec)	0.05	0.14	0.48	3.82	43.95	369.15

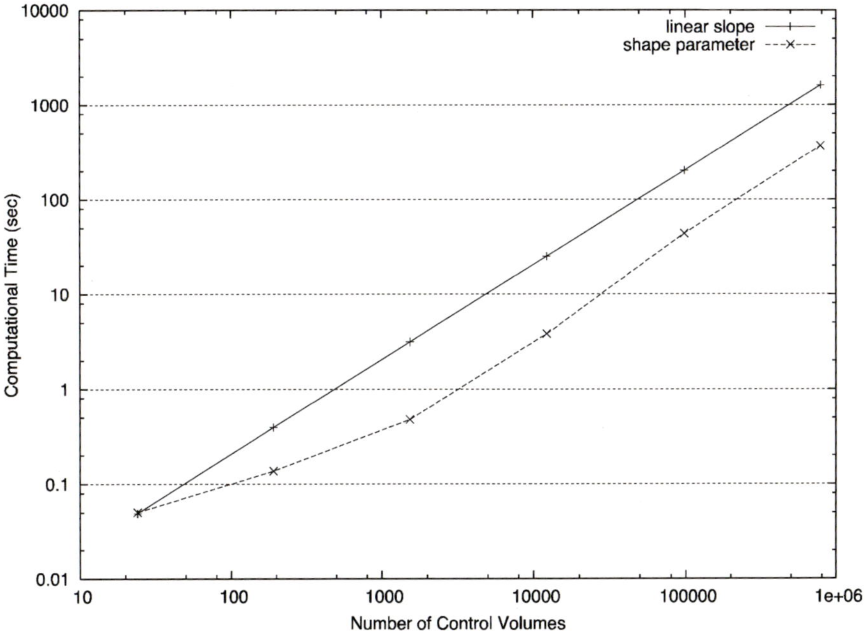

Fig. 10.13 Computational time over the number of control volumes (shape sensitivity)

directly. This test case is well known from literature [4, 34]. 9 CPs are defined to modify the bump. The four corner points are fixed to guarantee geometric continuity at the channel walls, whereas the three CPs at $x = 0.45$ m can be translated in z-direction, i.e.,

$$\mathbf{P}_{i,2}(b) = \mathbf{P}_{i,2}^{(0)} + b \cdot \begin{pmatrix} 0 \\ 0 \\ 1 \end{pmatrix}, \quad i = 1, \dots, 3, \tag{10.44}$$

where $P_{i,2}^{(0)}$ is the initial position of the i-th CP at $x = 0.45$ m. Thus the bump can be raised or lowered by varying b.

The V-cycles setup is the same as used for the inlet sensitivity test case, as well as the underrelaxation parameters and the convergence criterion. We examine as before the behavior of the number of fine grid iterations and the necessary time for the calculations as the main indicators for computational costs. The results can be found in Table 10.3 and Fig. 10.13. The number of fine grid iterations remains

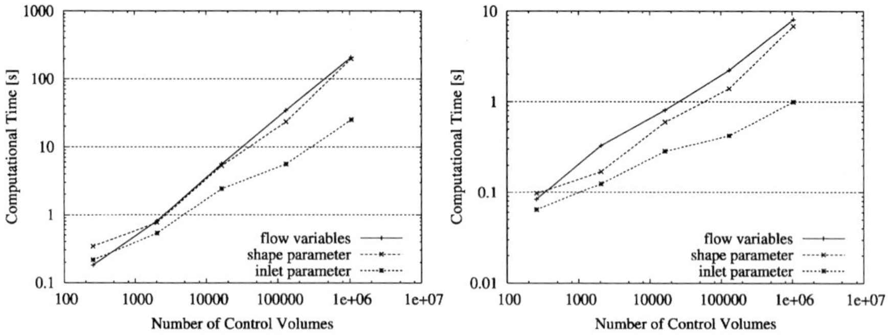

Fig. 10.14 Time required for converged solution of flow and parameter sensitivities performing a FMG of maximum size plotted over the five grid levels. *Left diagram* illustrates computational cost for one processor, *right diagram* for 16 processors

almost constant on the finer grids in agreement with the multigrid theory. Also the (only) linear increase of the computing time with the number of CVs can be seen.

10.3.5 Parallelization Analysis

At first the multigrid behavior is analyzed on one and on 16 processors. We measure the required time for computing the parameter sensitivities, average over several runs and reject outliers to minimize temporal fluctuations. For the sensitivity computation full multigrid (FMG) cycles of maximum depth are performed, e.g., on the second grid level a cycle over two grids is used, while on the finest grid a FMG-cycle over all five available meshes is employed.

The computational cost are plotted in Fig. 10.14. We can verify results from literature that the computational cost scales approximately linear with the number of CVs while applying a FMG algorithm [1], for both, the single processor and the multicore calculation. For estimating the performance of the parallel algorithm we analyze FMG cycles over five grids on a variable number of processors and define the quantities speed-up S_P and efficiency E_P

$$S_p = \frac{T_1}{T_P} \quad \text{and} \quad E_p = \frac{T_1}{PT_P}, \tag{10.45}$$

where T_1 and T_P are the time needed on one processor, and P processors, respectively. The time measurements are again averaged over several runs.

Table 10.4 shows the computational cost for getting a converged state solution as well as for the both control sensitivities. Two aspects are particularly striking: The fast convergence of the non-shape parameter and the gain by switching from one to dual-processor computation. The shape parameter induces high values for the sensitivity boundary constraints along the bump, resulting in a major influence

Table 10.4 Computational cost (in seconds) for performing a full multigrid algorithm over five grid levels on a variable number of processors

Number of processors	State variables	Shape sensitivity	Inlet sensitivity
1	207.4676	196.5054	24.9875
2	55.4051	49.8231	6.5939
4	31.6752	27.3760	3.6872
8	14.7673	12.7305	1.7621
16	8.1280	6.8148	0.9988

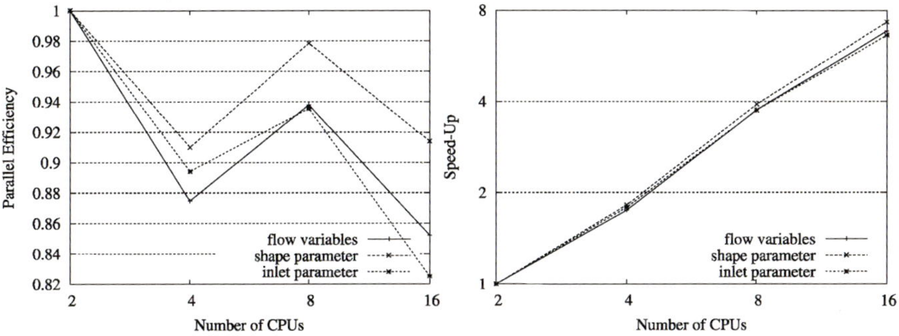

Fig. 10.15 Parallel efficiency (*left*) and speed-up (*right*) of flow and parameter sensitivities plotted over the number of involved processors. Both indicators are standardized to dual-processor computational cost

above the NURBS surface while greater parts of the domain are rather unimpressed by changes in the bump contour.

However, the inlet parameter induces a sensitive inlet value, which permeates the complete computation domain. The used discretization following the state solver, e.g., with the upwind scheme, seems intuitively better suited for this kind of parameters.

The outperforming efficiency gain by switching from a single to a dual processor job might be a consequence from cache level effects or an awkward implementation of serial jobs on the parallel computer. To overcome this blemish we redefine the speed-up and efficiency and standardized them by the dual-processor computational cost:

$$S_P^{(2)} = \frac{T_2}{T_P} \quad \text{and} \quad E_P^{(2)} = \frac{T_2}{(P/2)\,T_P}, \tag{10.46}$$

Both quantities are plotted in Fig. 10.15. On the left side the high efficiency is illustrated, for the shape parameter sensitivity the value is above 90%, even better than for the flow solver. The little bit smaller performance of the inlet parameter sensitivity might occur from the fast convergence in absolute terms, see Table 10.4.

It is obvious that the communication overhead increases relatively for shorter computations. The right diagram shows the good speed-up property of the parallel algorithm with an almost linearly increasing gain for using up to 16 processors.

10.4 Conclusion

A general sensitivity-based approach for the optimization of turbulent flows was presented. We introduced the basic flow describing equations, a general optimization problem and its solution via a trust region algorithm utilizing sensitivities. For geometry variation a NURBS based surface description was employed. In a first step the numerical results of the sensitivity computation could be verified via the comparison with a straightforward finite-difference approach. In a second step the functionality of the used algorithm could be shown with the help of the example flow over a hill. Furthermore the functionality of the NURBS surface to reduce the design parameters for the shape could be shown. Additionally we presented the flow within a generic heat exchanger in order to prove the functionality of the approach within a gas turbine related environment. Furthermore, multigrid and parallelization techniques were introduced for accelerating the computations. It could be shown, that both approaches reduce the computational time in the expected amount. Altogether the combination of a trust-region algorithm using sensitivities and NURBS surfaces seems to be a promising approach in flow and shape optimization within gas turbines.

References

1. Durst, F., Schäfer, M.: A parallel block-structured multigrid method for the prediction of incompressible flows. Int. J. Numer. Methods Fluids **22**, 549–565 (1996)
2. Harth, Z., Sun, H., Schäfer, M.: Comparison of derivative free Newton-based and evolutionary methods for shape optimization of flow problems. Int. J. Numer. Methods Fluids **53**, 753–777 (2007)
3. Hirschen, K., Schäfer, M.: Artifical neural networks for shape optimization in CFD. In: Proceedings of Design Optimization International Conference, Athens (2004)
4. Gunzburger, M.D.: Perspectives in Flow Control and Optimization. Society for Industrial and Applied Mathematics, Philadelphia (2003)
5. Gunzburger, M.: Adjoint equation-based methods for control problems in viscous, incompressible flows. Flow Tubul. Combust. **65**, 249–272 (2000)
6. Borggaard, J., Burns, J.: A PDE sensitivity equation method for optimal aerodynamic design. Int. J. Comput. Phys. **136**, 366–384 (1997)
7. Burkhardt, J., Peterson, J.: Control of steady incompressible 2d channel flows. Flow Control IMA Vol. Math. Appl. **68**, 111–216 (1995)
8. Ferziger, J.H., Perić, M.: Computational Methods for Fluid Dynamics. Springer, Berlin (2002)
9. Launder, B.E., Spalding, D.B.: The numerical computation of turbulent flows. Comput. Methods Appl. Mech. Eng. **3**, 269–289 (1974)
10. Chien, K.H.: Predictions of channel and boundary-layer flows with a low-Reynolds-number turbulence model. AIAA J. **22**, 33–38 (1982)

11. Griewank, A., Corliss, G.F.: Automatic Differentiation of Algorithms: Theory, Implementation, and Application. Defense Technical Information Center OAI-PMH Repository (1998)
12. É, T., Pelletier, D., Borggaard, J.: A continuous sensitivity equation approach to optimal design in mixed convection. Numer. Heat Trans. A Appl. **38**(8), 869–885 (2000)
13. Duvigneau, R., Pelletier, D.: On accurate boundary conditions for a shape sensitivity equation method. Int. J. Numer. Methods Fluids **50**, 147–164 (2006)
14. Siegmann, J., Becker, G., Michaelis, J., Schäfer, M.: A general sensitivity based shape optimization approach for arbitrary surfaces. In: Proceedings of the 8th World Congress on Structural and Multidisciplinary Optimization, WCSMO, Lisbon (2009)
15. Jacoby, S.L.S., Kowalik, J.S., Pizzo, J.T.: Iterative Methods for Nonlinear Optimization Problems. Prentice-Hall, Englewood Cliffs (1972)
16. Mohammadi, B., Pironneau, O.: Applied Shape Optimization for Fluids. Oxford Science, Oxford (2001)
17. Becker, G., Siegmann, J., Michaelis, J., Schäfer, M.: Comparison of a derivative-free and a gradient-based shape optimization method in the context of fluid–structure interaction. In: 8th World Congress on Structural and Multidisciplinary Optimization, WCSMO, Lisbon (2009)
18. Nocedal, J., Wright, S.J.: Numerical Optimization. Springer, New York (1999)
19. Becker, G., Falk, U., Schäfer, M.: Shape optimization with higher-order surfaces in consideration of fluid–structure interaction. In: Fluid–Structure Interaction: Theory, Numerics and Applications. Kassel University Press, Kassel (2009)
20. Schäfer, M., Heck, M., Yigit, S.: An implicit partitoned method for the numerical simulation of fluid–structure interaction. In: Bungartz, H.-J., Schäfer, M. (eds.) Fluid–Structure Interaction: Modelling, Simulation, Optimization, 53, pp. 171–194. Springer, Berlin/Heidelberg (2006)
21. Lehnhäuser, T., Ertem-Müller, S., Schäfer, M., Janicka, J.: Advances in numerical methods for simulating turbulent flows. Prog. Comput. Fluid Dyn. **4**(3–5), 208–228 (2004)
22. Lehnhäuser, T.: Eine effiziente numerische Methode zur Gestaltoptimierung von Strömungsgebieten. TU Darmstadt (2005)
23. Lehnhäuser, T., Schäfer, M.: Improved linear practice for finite-volume schemes on complex grids. Int. J. Numer. Methods Fluids **38**, 625–645 (2002)
24. Gauß, F., Schäfer, M.: A local adaptive grid refinement strategy for block-structured finite-volume solvers. In: Proceedings of the 6th International Conference on Engineering Computational Technology, Athens (2008)
25. Gauß, F., Schäfer, M., Siegmann, J., Sternel, D.: On the influence of boundary discretization schemes on the accuracy of flow simulation with local refinement. In: IV International Conference on Adaptive Modelling and Simulation 2009, Brussels (2009)
26. Gauß, F.: Strategien zur lokalen adaptiven Gitterverfeinerung für Strömungslöser. TU Darmstadt (2005)
27. Schäfer, M.: Computational Engineering – Introduction to Numerical Methods. Springer, Berlin (2006)
28. Briggs, W.L., Henson, V.E., Kaspar, B.: A Multigrid Tutorial. SIAM, Philadelphia (2000)
29. Hackbusch, W.: Multi-Grid Methods and Applications. Springer, Berlin (1985)
30. Trottenberg, U., Oosterlee, C.W., Schüller, A.: Multigrid. Elsevier Academic Press, San Diego (2001)
31. Michaelis, J., Siegmann, J., Becker, G., Schäfer, M.: Efficiency of geometric multigrid methods for solving the sensitivity equatins within gradient based flow optimzazion problems. In Proceedings of the V European Conference on Computational Fluid Dynamics, ECCOMAS CFD 2010, Lisbon (2010)
32. Michaelis, J., Siegmann, J., Schäfer, M.: Parallel efficiency of multigrid methods for solving flow sensitivity equations. In: Proceedings of the CFD and OPTIMIZATION 2011–002, Antalya (2011)
33. Hempel, R.: The MPI standard for message passing. Lect. Notes Comput. Sci. **797**, 247–252 (1994)
34. Stanley, L.G., Stewart, D.L.: Design Sensitivity Analysis. SIAM, Philadelphia (2002)

Chapter 11
Integral Model for Simulating Gas Turbine Combustion Chambers

S. Kneissl, D.C. Sternel, M. Schäfer, P. Pantangi, A. Sadiki, and J. Janicka

Abstract Relying on the software platform FASTEST different sub-models of various complexities designed in the framework of the Collaborative Research Centre 568 were combined to form integral models. As review paper, this contribution summarizes the developments of a Large Eddy Simulation (LES) based integral model for reliably simulating gas turbine combustion chambers. The model was analyzed, validated and evaluated in terms of fuel flexibility, flow, mixing and combustion modeling and predictability. The parallel efficiency for large simulations with the integral model could be improved by using a load balancing strategy based on graph theory.

Keywords LES based integral combustion model • Parallel flow simulation • Block structured grids • Load balancing • Domain decomposition • High performance computing • Efficient algorithms

11.1 Introduction

In gas turbine combustors, various phenomena such as turbulence, heat and mass transfer, radiation, multiphase processes, thermo-acoustics, soot formation, compressor-combustor interaction, combustor-turbine interaction, as well as different areas such as injection or combustion chamber walls, can be distinguished, which in terms of functionality and efficiency are important for the system as a

S. Kneissl • D.C. Sternel • M. Schäfer (✉)
Department of Mechanical Engineering, Institute for Numerical Methods in Mechanical Engineering, Technische Universität Darmstadt, Dolivostr. 15, 64293 Darmstadt, Germany
e-mail: schaefer@fnb.tu-darmstadt.de

P. Pantangi • A. Sadiki • J. Janicka
Department of Mechanical Engineering, Institute for Energy and Power Plant Technology, Technische Universität Darmstadt, Petersentr. 30, 64287 Darmstadt, Germany

J. Janicka et al., *Flow and Combustion in Advanced Gas Turbine Combustors*, Fluid Mechanics and Its Applications 102, DOI 10.1007/978-94-007-5320-4_11,

whole. In particular gas turbine combustion systems are characterized by flows that typically exhibit large scale structures and evolve in a highly unsteady manner, which remains inaccessible to today's RANS (here Reynolds Averaged Approach based Numerical Simulations) methods widely used in many 3D CFD industrial simulation tools [14–25, 54–59].

Focusing on gas turbine combustion systems, it is worth mentioning, that only a few combustion system designs could be studied and experimentally tested in the past as the cost for combustion testing are expensive. Furthermore, the designs have been optimized for years for steady state [14, 15, 24]. Due to inherent unsteady combustion events, like ignition, relight, quenching, blow out, combustion instabilities that may strongly influence the system operation, the optimization process appeared often ineffective.

To significantly reduce development costs and to simultaneously meet specific optimization targets (efficiency and emission reduction, safety, fuel consumption reduction, etc.), a reliable design tool is highly demanded [14–24]. In this respect, Large Eddy Simulation (LES) that has demonstrated its potential in reasonably accounting for inherent unsteady effects in simple combustion systems (laboratory flames) [1, 25, 37] is a valuable candidate as a compromise between Direct Numerical Simulation (DNS) and RANS. Its application to technical combustion systems has now been made possible due to rapid development of computer performance and application-oriented numerical methods, computational and programming techniques. Recently LES applications have been reported in test cases of high complexity [17, 20–24, 38, 39]. Despite these LES successes, its path to become a validated production tool in the industry is still open [24].

Physical and chemical features of combustion LES have been discussed by Janicka and Sadiki [1] and Pitsch [25] with emphasis focused on important aspects of an overall model. Thereby an overall LES model formulation may contain physics-preserving turbulence/mixing closures. These include an appropriate combustion sub-model able to capture finite chemistry effects along with a sub-model for turbulence-chemistry interactions, and possible sub-models accounting for additional phenomena, such as multiphase flow phenomena, radiative heat transfer and soot formation, etc. Regarding chemistry, the details of chemistry are unavoidable if one has to address auto-ignition, flame stabilization, recirculation products which may include intermediate species, and the prediction of some pollutants [26–31]. The reduction and tabulation of chemical species behavior prior to LES remains one of the available options investigated to downsize combustion chemistry [28–31].

To account for the stabilization of lifted flames via partial premixing as occurred in the gas turbine combustor, the flamelet generated manifolds (FGM) [30, 31] or Flamelet prolongated ILDM (FPI) [28] method is introduced. This is achieved by taking into account a transport equation for the progress variable in the CFD in addition to the mixture fraction equation and the classical governing equations for LES.

This paper is divided into three main parts. The first part documents the achievements related to the creation of an integral model. As this integral model covers phenomena on many different temporal and spatial scales the computational requirements are tremendous, especially for realistic test cases (e.g. [12, 21]). These can only be matched with the use of massive parallel high performance computers. The developed parallelization strategy allowing the computations to be resourceful will be presented. The second part focuses on demonstrating the feasibility of the LES based integral model to complex chemically reacting flows in a realistic single sector combustor. This combustion chamber is fuelled with pre-vaporized kerosene fuel using a nozzle fired at different pressures. It features a strong unsteady swirling flow with recirculation and breakdowns of large scales vertical structures, turbulent mixing, combustion, conjugate heat and mass transfer and pollutant formation. These complex interacting processes make predictions of such a system very complicated and challenging even if only part of the phenomena is considered. Soot formation and radiation modeling are not included in this contribution. The last part of the paper is devoted to the conclusions.

11.2 Integral Model and Parallelization Strategy

11.2.1 Integral Modeling Strategy

In modern gas turbines, either in the field of power generation applications or in aircraft jet engines, efforts to meet the environmental and efficiency requirements need to be supported by a profound understanding and an accurate prediction of the complex flow phenomena and interaction mechanisms occurring in the system. The isolated mechanisms are in general described by appropriate sub-models that have to be coupled in an integral model including interaction models to deal with the whole gas turbine combustion chamber. The integral modeling strategy may then consist of a set of compatible well-tested sub-models capable of describing the essential physical and chemical processes that may strongly interact with each other,

- efficient numerical methods,
- a suitable integration of boundary components and involved interactions,
- well-conceived experimental configurations for model validation.

This has been achieved in the RANS context as reported in [54–61] for single phase and two-phase flow systems. Focused on LES, efforts have been made in developing and applying appropriate subgrid-scale (SGS) models for the essential flow, mixing and combustion processes that occur in gas turbine combustion chambers. These models, also including hybrid RANS/LES techniques, have been postulated, analyzed and validated in different configurations of various complexities including simple generic geometries and industrial configurations [1–7, 9, 13]. Particular objectives were to analyze the integral models, sub-models and interaction of sub-models in terms of the meaningfulness for the overall model and to control the quality of the resulting overall model following [63, 64].

All the filtered modeled governing equations have been implemented in the three dimensional CFD code, FASTEST [62]. The code uses geometry-flexible, block-structured, boundary fitted grids. This enables to represent complex geometries. A collocated grid with a cell-centered variable arrangement is used. The flow solver offers fully second order accuracy. Discretization is based on finite volume method. For spatial discretization specialized central-differencing schemes are used. To assure boundedness of the mixture fraction, the convective term in the scalar transport equations has been discretized using non-oscillatory bounded TVD (Total Variation Diminishing) schemes [40]. For the time stepping multiple stage Runge–Kutta schemes with second order accuracy are used. Following a fractional step formulation, in each stage a momentum correction is carried out in order to satisfy the continuity. FASTEST is parallelized by domain decomposition using the MPI message passing library. Further features of the software platform are described in [9–11, 62].

11.2.2 Load Balancing Method

The accurate solving of complex technically relevant flows is restricted by the capacities of the computers. Therefore, the only way to get quantitatively accurate results in an acceptable time is through the usage of efficient numerical solution procedures combined with parallel computing [41].

The usage of block-structured grids is an acceptable compromise for calculating flow in complex geometries by using efficient algorithms. Such a block structured grid can be made up of more than thousands of blocks, each containing different numbers of control volumes. Domain decomposition can be used to parallelize this approach on distributed memory computers. An appropriate mechanism is then required to split and distribute blocks among the processors. The task is to find domain decompositions with evenly divided load among the processing units and to minimize the communication between processors. This leads to a very difficult optimization problem which is in terms of complexity theory NP-hard. For a small number of blocks, it is possible to solve this problem manually. However, if the number of blocks increases, i.e., in complex three-dimensional configurations, it is impractical, if not impossible.

In contrast to unstructured grids where domain decomposition libraries such as METIS [42] and SCOTCH [43] are available, few works have been carried out for block-structured grids. For example, Ahusborde and Glockner [44] propose a geometric technique suitable for two-dimensional domains. In elsA [45], a software for compressible flows around complex geometries, the so called greedy algorithm [46] and recursive edge bisectioning [47] are used. Defining a cost function allows [48] to employ a genetic algorithm. Rantakokk [49, 50] introduces a general framework and compares different strategies.

Here we present an algorithm for accomplishing the domain decomposition task for block-structured grids automatically. We first split the blocks appropriately and develop a graph partitioning representation for the problem of assigning the splitted blocks to the processors. We solve this problem with a recursive spectral bisectioning algorithm and a refinement heuristic. Exemplary, a LES of the pressure combustion chamber as carried out in [3] will be considered to illustrate the functionality of our approach and to discuss its efficiency.

11.2.3 Load Distributing Method

11.2.3.1 Graph Description for the Domain Decomposition Problem

To make the domain decomposition methodologically accessible a graph description with some simplifications is introduced. It allows to transfer the domain decomposition problem to a graph partitioning problem where well known methods from computer science can be applied. The blocks constitute the vertices and the data dependencies between the blocks constitute the edges. We choose the number of control volumes of a block as vertex-weights. Now assigning each block to a processor, subsequently also referred to as mapping, also defines a division into sub-graphs in the graph representation. Each processor corresponds to one sub-graph. A mapping is given by any graph partitioning.

In order to then assess the quality of a domain decomposition created by the mapping, the following assumptions are made:

- Communication time is proportional to the number of blocks that are connected across processors.
- Computational workload for every processors is proportional to the sum of control volumes being assigned to a processor.

Based on these assumptions the communication overhead of a domain decomposition can be minimized by computing a graph partitioning with a minimal number of edges connecting sub-graphs (edge-cut). The load balancing efficiency is optimal when each sub-graph has the same weighted vertices sum.

11.2.3.2 Mapping Procedure

Our mapping algorithm is composed of three main parts, which are described in the following.

The first step includes the decomposition of the geometric block structure (we call that interim block structure (IBS)). In the second step the partitioning of the IBS on the required numbers of processors will be accomplished by using an optimization algorithm. In the third step we merge blocks located on one processor, if possible. The resulting structure defines the parallel block structure (PBS). Figure 11.1 shows that sequence schematically.

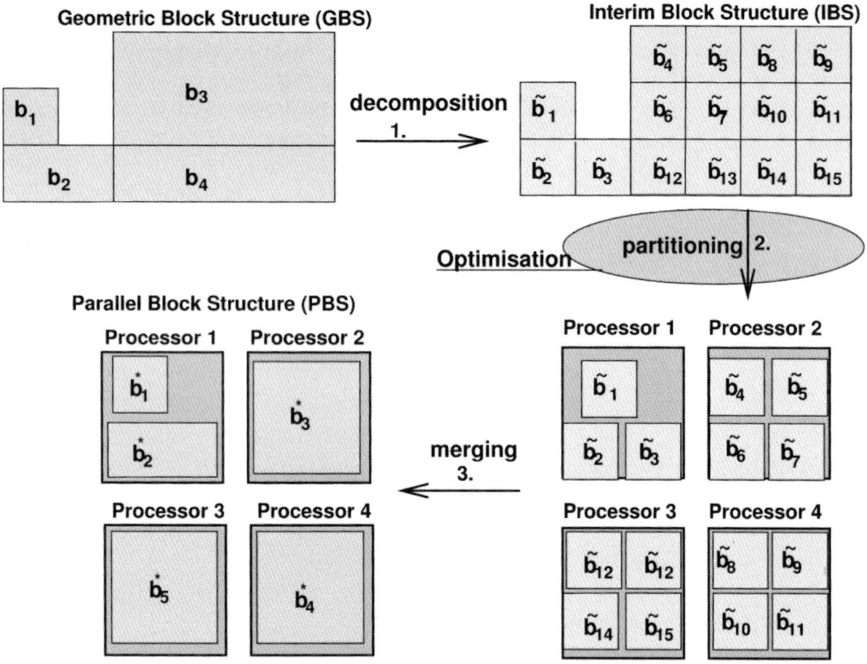

Fig. 11.1 Overview of the complete mapping procedure [13]

11.2.3.3 Pre-processing the Block-Structure

An obvious way to decompose a block-structured grid is to use the existent geometric blocks for the partitioning. That will work well in some special cases but normally a splitting of the geometric blocks is indispensable. This is due to the fact that the capacity of one processor is often smaller than the handling of some large blocks required, or that the geometric blocks could not be distributed onto the processors in a way to get an equal load for each processor. The smaller the block sizes in relation to the overall number of control volumes, the better the domain decomposition can get. The flexibility made possible through small blocks allows to combine an equal distribution of control volumes with a small communication interface between processors.

However, the use of very small blocks adversely affects the solution efficiency on a processor and also requires more memory as the ghost-cell proportion is increased. To counteract this increasing effort, a target size CV_{tar} is set. Only blocks with a size larger than 1.25 CV_{tar} are allowed to be divided. These will be divided perpendicular to the largest dimension by 2 to preserve the geometric multi-grid suitability. This will be repeated until further splitting is not possible. Whether a block can be splitted into further sub-blocks depends on the size of its edges and the requested number of grid levels. The block structure resulting from the splitting algorithm, defines the interim block structure (IBS).

11.2.3.4 Recursive Two Step Bi-sectioning

The graph partitioning problem described previously is complex and computational demanding. It is a combinatorial explosive optimization problem where the computational effort to solve increases with the number of processors n_{proc} and the number of blocks n_{block}. Each block can be assigned to n_{proc} different processors resulting in $(n_{block})\hat{}n_{proc}$ possible mappings. So the first idea to make the graph partitioning feasible is to do n_{proc}-1 bi-sectionings instead of one n_{proc} multiway partitioning at once. This of course shrinks the solution space and the best possible solution might not be obtainable anymore. But as information on communication speed between processors is not included in the graph representation, recursive bi-sectioning is a good way to account for it implicitly. The recursive approach builds a prioritizing hierarchy of minimized communication interfaces corresponding to the hierarchy of communication speeds resulting from the hardware structure. The slowest interfaces are minimized first and are thereby prioritized because the solution space is still large. Let us for example consider two computing nodes connected via slow ethernet. Here the first bi-sectioning already minimizes the communication via the slowest connection in the parallel computation.

As method for bi-secting the graph and all following sub-graphs we use a spectral method [51] to minimize the edge-cut. That gives us two partitions of equal size under the assumption that all vertices have the same size. For practical application that will rarely be possibly to achieve, especially as blocks can only be splitted in half. But if the scatter of block sizes is small after pre-processing, the spectral solution is a good starting point for local refinement heuristics. We have implemented the spectral bi-sectioning method combined with a Kernighan-Lin like local refinement method [52].

At each bi-sectioning step we assign the vertices of the graph according to the Fiedler vector [53] into two parts. The two parts should now have approximately equal workloads. We then improve this assignment by swapping vertices between the two parts using the local refinement method. The refinement is achieved by exploiting gains computed from moved vertices in the graph. The gain calculation considers edge cut and load balance.

If we have a relatively small number of blocks compared to the number of processors, it will be difficult to achieve a balanced workload distribution. To avoid unbalanced solutions in those cases, more splitting in the pre-processing phase is needed.

This method is also suitable for large CFD cases, where a large number of blocks have to be computed on a large number of processors. The computational solution effort scales linearly with the number of processors due to the recursive approach. In the bi-sectioning procedure the most resource consuming part is the spectral method involving the numerical calculation of eigenvectors. But those are affordable as the IBS is unlikely to exceed more than 100,000 blocks.

11.2.3.5 Post-processing the Block-Structure

Often the block sizes in the interim block-structure are larger than required for a sufficient load balance or are small in regions that are not affected by the partitioning. In both cases, if they are assigned to the same processor, it is possible to recombine blocks of the IBS to increase the intra-processor efficiency. The resulting structure is the final parallel block structure (PBS).

11.2.4 Representative Test Case

A representative test case for showing the capabilities of the mapping algorithm should have a large number of blocks with a wide range of numbers of control volumes per block in the geometric block structure.

An actual interesting configuration is the flow field in a gas turbine combustion chamber. For the design of gas turbine combustors with regard to stable and low-pollution combustion, detailed and accurate knowledge of the flow field in the combustion chamber is essential. The complex flow field is turbulent and strongly non-isotropic with strong stream-line curvature. The investigations are performed for the pressure combustion chamber as studied in [3] and [8], which reproduces the characteristics of the flow field in real combustion chambers and for which detailed experimental results are available.

This test case has 2,390,784 control volumes distributed over 154 geometric blocks. It's geometric block-structure is shown in Fig. 11.2.

The smallest block contains 9,472 control volumes, the largest 174,160. Having splitted the blocks with the splitting algorithm and the requirement that the smallest block of the GBS represents the target block size CV_{tar}, the resulting interim block structure, which is shown in Fig. 11.3, contains 386 blocks.

The largest block now only contains 10,885 control volumes. Smaller target block sizes were not considered as the ghost-cell to control volume ratio rapidly increases for smaller blocks as shown in Fig. 11.4.

As the refinement step of each bi-sectioning is restricted to only swap neighbouring blocks the reachable load balance efficiency will be in most cases below the possible maximum. A mapping method just aiming for an optimal load balancing is used as reference. In that reference method all blocks are sequentially assigned to the least loaded processor. To demonstrate the importance of the splitting step the case without splitting is also considered.

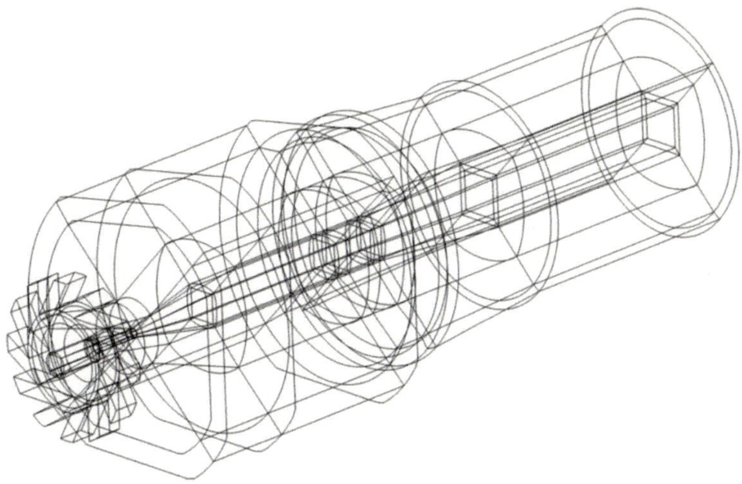

Fig. 11.2 Geometric block-structure of the pressure combustion chamber

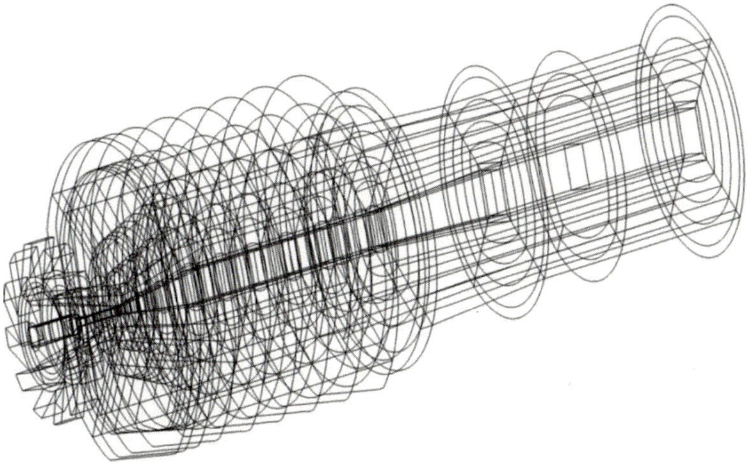

Fig. 11.3 Interim block-structure of the pressure combustion chamber

Table 11.1 shows the load balancing efficiencies that are achieved for different numbers of processors.

By using the splitting algorithm, a good load balancing efficiency can be attained on up to 64 processors. The limitations of the load balancing efficiency for splitted structures depend on the prescribed minimum number of control volumes per block. Without splitting the load balancing efficiency for more than 8 processors drops below an acceptable limit. Concerning load balancing the communication optimized mapping is always slightly worse than the pure load balancing optimized mapping. This disadvantage has to be recovered by a superior communication overhead.

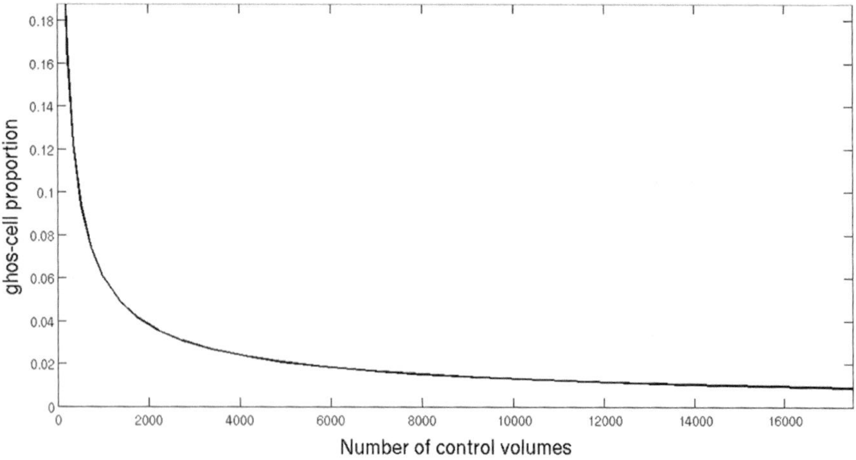

Fig. 11.4 Ghost-cell proportion of the number of control volumes for cubic blocks. For non-cubic blocks the ghost-cell proportion will be even higher

Table 11.1 Attained load balancing efficiency

# Processors	GBS	IBS	IBS opt
2	100	100	100
4	100	100	100
8	87.45	99.20	98.65
16	65.12	98.54	98.26
32	49.78	97.80	96.59
64	12.40	96.17	95.66

GBS: unsplitted geometric block-structure mapped with the reference method. *IBS*: splitted geometric block-structure mapped with the reference method. *IBS opt*: splitted geometric block-structure mapped communication optimized

To demonstrate the importance of minimizing the communication overhead besides optimizing the load balance efficiency, the method presented here is also compared to the reference method just aiming for an optimal load balancing. Figure 11.5 shows the computing time of both methods for different numbers of processors. All measurements correspond to the calculation of 10 time steps with a fixed number of inner iterations of a non reactive LES with Crank Nicholson time marching scheme and were carried out on processors of the IBM p690 system.

For all numbers of processors the communication optimized mapping performs better. The runtime advantage increases the more processors are used because more communication is needed and slower communications paths have to be used. This is a promising result especially as the load balancing efficiency is worse.

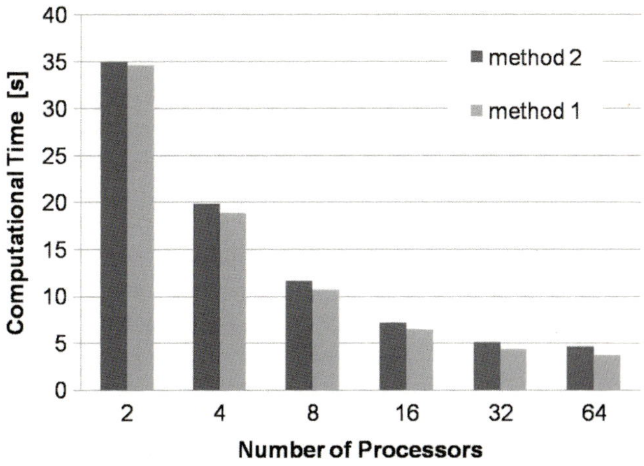

Fig. 11.5 Comparison between the communication overhead optimized mapping (*method 1*) and the just load balance optimized mapping(*method 2*) for different numbers of processors

11.3 LES of Pre-vaporized Kerosene Combustion at High Pressures in a Single Sector Combustor Using the FGM Method

In order to evaluate the capability of the integral model for predicting combustion processes induced by complex real fuels a high pressure single sector combustor (SSC) is considered. This combustion chamber is fuelled with pre-vaporized kerosene fuel and features very complex unsteady swirling flow and partially premixed combustion properties. The validation of the designed tool along with the prediction analysis is carried out in terms of comparison between experimental data and numerical results.

11.3.1 Model Description

A classical approach for LES is used. To separate the large from small-scale structures in LES, filtering operations are applied to the governing equations. According to the FGM methodology, the resulting filtered governing equations are:

- the momentum equation along with the continuity equation used to describe the motion of low Mach number Newtonian fluids,
- the mixture fraction equation to follow the mixing change caused by the turbulent convection and diffusion of a conserved scalar,

- the reaction progress variable (RPV) equation required to track the reaction progress along with the chemical kinetic effects, especially when the combustion takes place in the partially premixed regime.

A Smagorinsky-model with dynamic procedure according to Germano et al. [32] is applied to determine the subgrid scale stresses. In order to stabilize the model, the modification proposed by Sagaut [33] is applied. In addition a clipping approach will reset the negative Germano coefficient C_s to zero to avoid destabilizing values of the model coefficient. No special wall-treatment is included in the sub grid scale model. The dynamic procedure applied here allows to capture the correct asymptotic behavior of the turbulent flow when approaching the wall (see e.g. Wegner et al. [2]). A detailed discussion of this issue was reported by Wegner [3]. To represent the sub-grid scale scalar flux in the mixture fraction and in the RPV equations a gradient ansatz is applied with a constant turbulent Schmidt number of 0.7. Although advanced sub grid scale scalar flux models exist and are known to behave superior in specific flow test cases (Huai et al. [34]), the model combination described above and applied here was chosen due to its simplicity.

The remaining term to be closed, i.e. the chemical reaction term, is modeled following the FGM method [30, 31]. As any flamelet based model, flamelet generated manifolds are based on the idea that a multi-dimensional flame can be represented by a set of one-dimensional flamelets. The method is therefore based on the laminar flamelet equation and includes Intrinsic Low-Dimensional Manifold (ILDM) [35] reduction methodology by solving transport equations for a given number of progress variables. Note that premixed and non-premixed generated manifolds can be constructed, even they can be combined. A comparative study has been reported in [36] for an LES of Sandia flame D and F. Visual observation from experiments show that the flame featuring both lifted and attached behavior exhibiting partially premixed combustion properties. Instead of considering diffusion flamelets, the FGMs used in this work are based on steady 1-D premixed flames to capture the lifted flame behavior. One reaction progress variable as defined as a (linear) combination of the ratios of mass fraction and molar mass of CO_2, H_2O and H_2, respectively, is used. According to this approach a Favre-filtered thermo-chemical quantity is calculated by integrating over the joint PDFs of the mixture fraction and the RPV while accounting for the turbulence-chemistry interaction via a presumed joint PDF ansatz. Here the procedure already presented in Sect. 3.2.3 is followed.

Since kerosene is used as fuel, its chemical composition in the FGM context is represented by a model fuel consisting of 80% n-decane and 20% n-propyl-benzene (by volume), which should represent two different chemical groups: long chained alkanes as the major component and cyclic hydrocarbons as the second large group. This model fuel was developed within the EU project CFD4C [38]. It has been validated against measurements of ignition delay times and burning velocity. The mechanism consists of 180 species and 992 reversible reactions.

This code has been already used to study numerically a series of laboratory classical flames [1, 2, 5, 37, 38] and a generic combustor fueled by methane and n-heptane using FGM method for combustion [3, 5, 6, 7]. Computations are carried

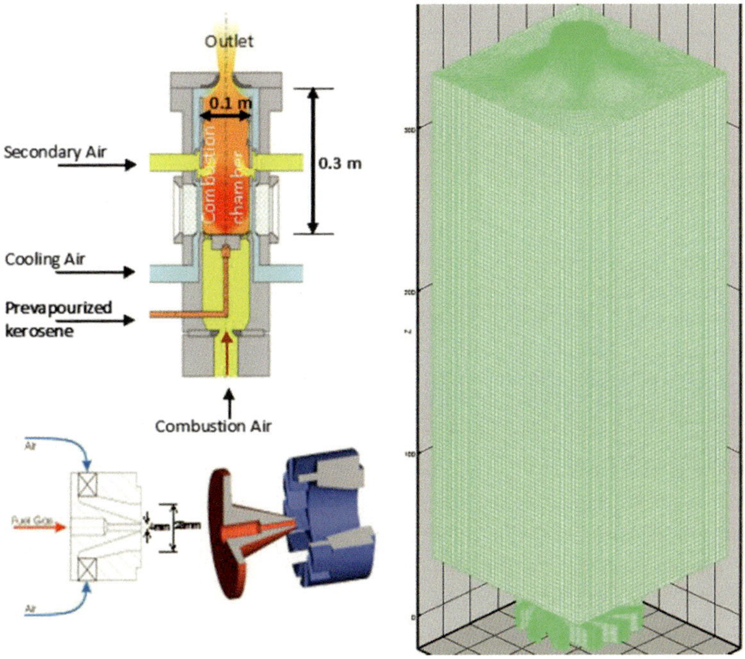

Fig. 11.6 (*left*) Experimental setup (*top*) and swirled nozzle (*bottom*) [6], (*right*) computational domain and grid

out for three flow throughs prior to collecting statistics and four flow throughs for obtained time averaged values.

11.3.2 Configuration

The combustion chamber under study features the DLR (Deutsches Zentrum für Luft- und Raumfahrt) experimental where liquid kerosene was vaporized at a minimum temperature of 673 K in a flowing system. It consists of a squared cross section single sector combustor (SSC) as depicted in Fig. 11.6 (left), where combustion air was supplied through a swirl nozzle. It is optically accessible from three sides in order to allow various modern optical laser diagnostics. The test rig can withstand an operating pressure of up to 2 MPa and can be operated with a combustion air flow up to 1.3 kg/s and cooling air flow rates of up to 3.0 kg/s. The maximum heating temperature of the combustion air is 850 K. Similar to the previous combustion chamber, the fused silica windows of the combustion chamber are cooled by guided cooling air, which is let into the hot exhaust before leaving the combustor through a throttling nozzle. In order to keep further disturbances of the chemical and physical reactions inside the combustion chamber small, no

Table 11.2 Operating conditions for the investigated cases

	Pre-vaporized kerosene: Jet -A1(Experiments)	
Fuel	80% n-decane and 20% n-propylbenzene (FGM)	
Combustor pressure [MPa]	0.4	0.6
Kerosene mass flow rate (g/s)	4.16	6.12
Fuel temperature [K]	673	673
Oxidant	Air	Air
Mass flow of oxidant [g/s]	77	114
Oxidant temperature [K]	623	623
Equivalent ratio	0.9	0.9
Thermal power (kW)	174	250
AFR (air fuel ratio)	18.6	18.6
Nozzle swirl number (geometrical)	1.2	

secondary air was used (Fig. 11.6) in the experiments. Note that the "combustion air" inlet supplies the swirler/injector part, while a "window air" inlet provides fresh gas through films on the front combustion chamber wall. This air entering the main combustion chamber is needed during the experiment to avoid any soot deposition on the windows impacting the optical quality.

In computations the window air was not included. The SSC test rig pressure is controlled by the amount of cooling air let into the system. In the present test case, dilution of hot gases by cold air is not considered for both experimental and numerical studies.

Flow fields and flame stabilization were investigated using state-of-the-art Laser Doppler velocimetry (LDV) and planar laser-induced fluorescence (PLIF) methods. This paper focuses on validation of the model implemented in the in-house CFD code FASTEST and does not give details of the spectroscopic part of the measurements, consisting of kerosene LIF, OH-LIF and chemiluminescence measurements at three different operating points. In fact, the fired nozzle was operated at 0.4, 0.6 and 0.9 MPa, with an air mass flow of 77, 114 and 170 g/s heated up to 623 K, corresponding to a pressure drop across the nozzle of 3.4% for all three experimentally investigated test cases.

The global equivalence ratio was set to $\phi = 0.9$ which is equal to an Air Fuel Ratio (AFR) of 18.6. As a result, a comprehensive database including, the velocity flow fields and the characteristic parameters of the flame derived from the spectroscopic measurements has been provided. This database is suitable for model validation and numerical simulation. For the present investigations, the test cases corresponding to a combustion chamber pressure of 0.4 and 0.6 MPa are considered. Table 11.2 summarizes the corresponding operating conditions.

To simulate the configuration as shown in Fig. 11.6 (left) the computational domain of the combustion chamber and the swirl nozzle are represented by a mesh consisting of 137 grid blocks featuring an O-type structure (Fig. 11.6, right). The total number of grid points is 2.0 millions.

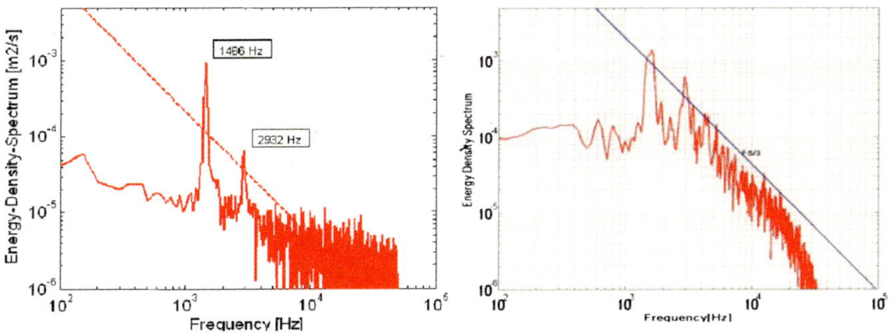

Fig. 11.7 Power spectral density experimental (*left*) numerical (*right*)

As inlet boundary conditions, the mass flows from the experiment were prescribed using laminar unperturbed profiles. A laminar inlet profile used was sufficient for such a simulation as measurements from experiments show that flow field is dominated by the intense recirculation of the swirl flow and not by the inlet turbulence. The resulting mesh is able to resolve more than 85% of total kinetic energy of the flow field in accordance to the so-called Pope-criteria (see in [1]). All simulations were run on sixteen processors.

11.3.3 Results and Discussion

For the 4 bar test case time series were recorded at three radial position at z = 5 mm. Figure 11.7 show (left) power spectral densities deduced from temporal autocorrelations (via FFT) for axial position different axial positions. The peak frequency was approximately at 1,450 Hz. Second peak is recoded at 2,932 Hz. These experimental findings are captured well by the simulations results (Fig. 11.7 (right), both peak frequencies from experimental findings are reproduced by the simulations.

The recirculation zone is typical for highly swirling flows and results from a positive axial pressure gradient that is associated with the vortex breakdown phenomenon. In Fig. 11.8 negative axial velocity is increasing from 5 mm from the exit of the nozzle to the 20 mm. This indicates the presence of a recirculation zone, which is necessary for flame stabilization. Fuel jet penetration is restricting the back flow at 5 mm. All three components of the velocities and the turbulent kinetic energy predicted by the LES are in good agreement with the experimental data. The axial velocity component is becoming strongly negative from x = 5 mm to x = 20 mm away from nozzle exit. The instantaneous temperature and RPV source term on plane passing through centre of nozzle are plotted in Fig. 11.9 RPV source term represents the main reaction zone in the combustion chamber; here main reaction

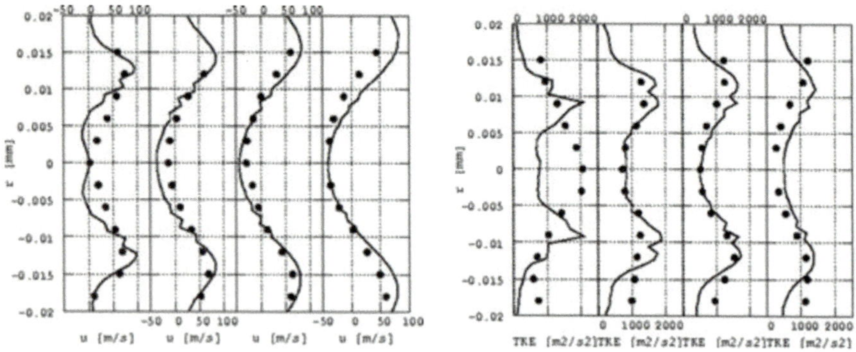

Fig. 11.8 Radial profiles of time averaged axial (u) (*top*) and turbulent kinetic energy (TKE) (*bottom*) at 5, 10, 15 and 20 mm (*left* to *right*) from the exit of the nozzle (_simulated, • Experiment)

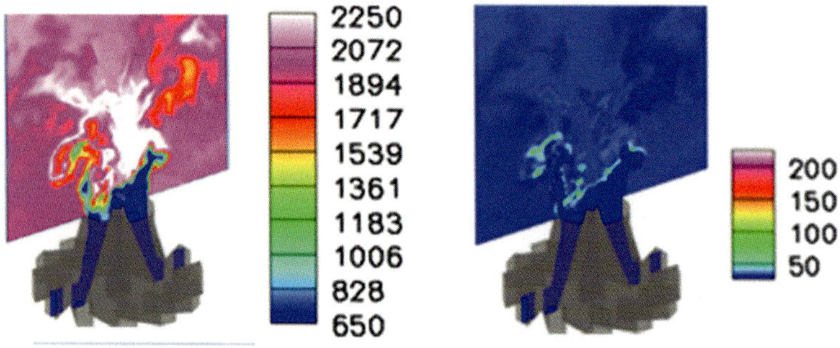

Fig. 11.9 Temperature (K) (*top*) RPV source term (kg/m^3-s) (*bottom*) on a plane passing through the centre of nozzle

zone is lifted, but temperature profile shows that flame is attached to the nozzle. This concluded here flame is exhibiting partially premixed combustion, which was also observed in experiments.

For the 0.6 MPa case the agreement between experimental data and numerical results has also satisfactorily been achieved as shown in Fig. 11.10 in which velocity profiles and turbulent kinetic energies are plotted at different axial positions from the nozzle exit at x = 5, 10, 15, 20 mm. All three components of the velocities and the turbulent kinetic energy predicted by the LES are in good agreement with the experimental data. The axial velocity component is becoming strongly negative from x = 5 mm to x = 20 mm away from nozzle exit. This indicates the presence of a recirculation zone, which is necessary for flame stabilization. In this respect, the instantaneous RPV source term and the temperature plotted in Fig. 11.11 allow to give a first impression of the flame characteristics.

The instantaneous RPV source term in Fig. 11.11 (right) is located in the main reaction zone. The flame seems to stand above the nozzle featuring a lifted flame in agreement with experiments. However it can be observed from the averaged reaction progress variable source term (not plotted) that the flame is attached to the nozzle,

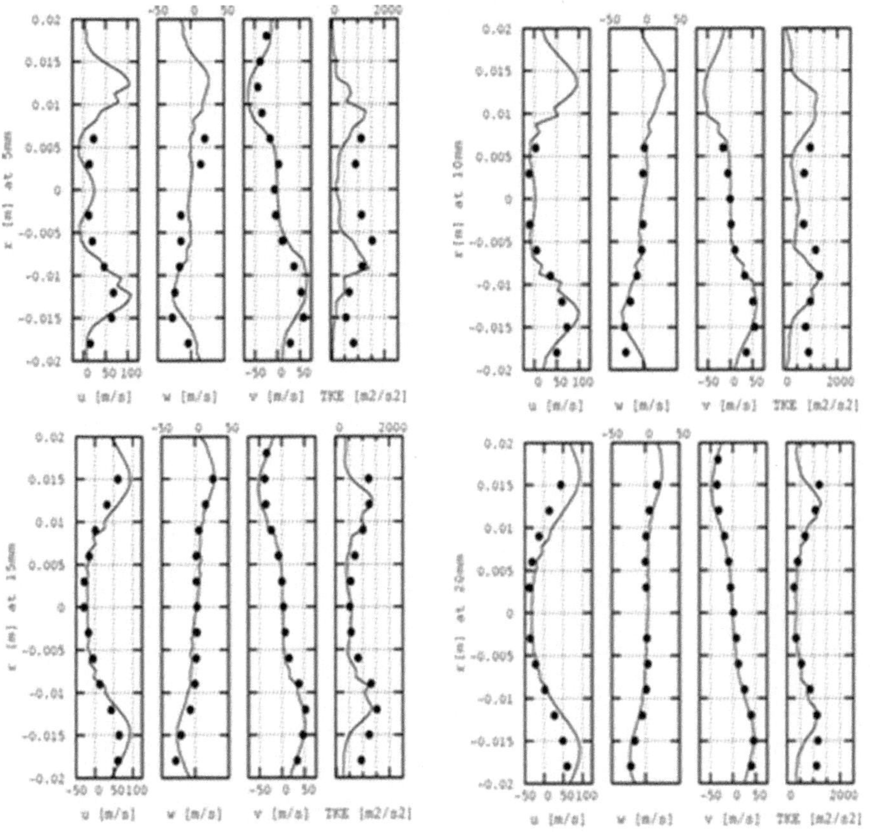

Fig. 11.10 Radial profiles of time averaged axial (u), radial (w) and tangential (v) velocity components and turbulent kinetic energy

though the value of the RPV source term remains very low. This suggests that the flame may be fluctuating between an attached and a lifted regime. In non-premixed swirled combustion as investigated in [3] the flame was found to be lifted while exhibiting a partially premixed nature. The RPV source term variance (not plotted) looks like two thin leafs starting from the swirled nozzle tip. It is worth noting that the reaction progress variable is strongly influenced by the swirl flow. This causes the strong change of the RPV in the mixing layer of the swirled air flow and the fuel.

A high concentration of CO in the reaction zone is observed (not shown). Most of the CO is combusting further downstream to limit the reaction zone within the vicinity of the nozzle. This is strongly influenced by the swirled air. The maximum instantaneous temperature in the reaction zone is found to be 2,250 K (Fig. 11.11, left). This maximum temperature is found at stoichiometric mixture fraction region. Instantaneous temperature contours confirm that the main reaction zone of the flame is lifted. Streamline plots derived from experimental LDV data as well as from numerical simulation show the existence of a recirculation zone responsible for

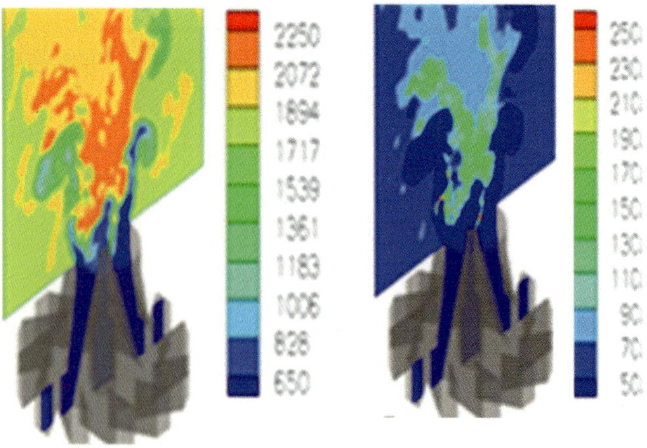

Fig. 11.11 Contour plots of instantaneous. *Left*: Temperature (K). *Right*: RPV source (kg/m^3-s)

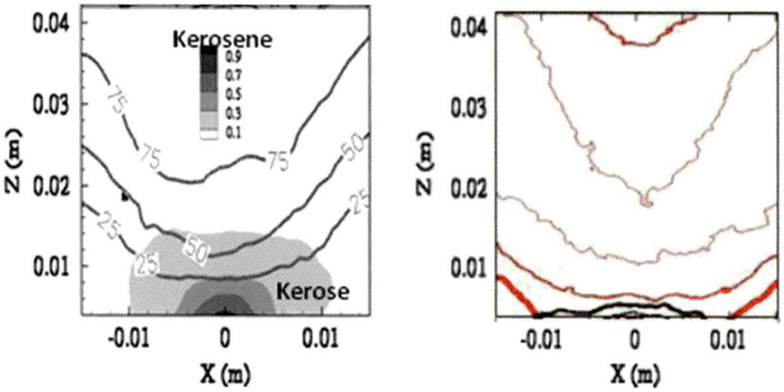

Fig. 11.12 Qualitative representation of OH concentration and kerosene. *Left*: simulation. *Right*: Experiment (bold contour: 25% of maximum; middle contour: 50% of maximum; light contour: 75% of maximum)

stabilization of the flame and a fluctuating stagnation point near the nozzle, causing the flame position to also fluctuate between a lifted to an attached behavior.

A comparison of predicted mass fraction of OH species and kerosene by LES against pixel intensity from experiments is shown in Fig. 11.12 for the main reaction zone region. The different red line contours of Fig. 11.12 (right) show the distributions of averaged OH measured in experiments. Gray line contours of Fig. 11.7 (left), with corresponding percentage of time averaged maximum OH mass fraction are estimated by LES. Experiments show the maximum OH concentration on top the fuel jet. This is also confirmed by LES.

In general, though OH mass fractions from the LES calculations are qualitatively comparable with those from experiments, they differ from each other by a few millimeters in physical space. One of the reasons may be the window air effect that was not included in computations. In particular the greater part of the vaporized kerosene (dark lines in Fig. 11.12 (left)) that is located in the vicinity of the nozzle is captured by the LES (gray scale contours of Fig. 11.12 (right)) with a slight deviation.

Due to the lack of detailed experimental data regarding the other species concentration and the temperature distribution further comparison studies to this respect have not been achieved.

11.4 Conclusions

Based on the software platform FASTEST different sub-models for LES designed for describing essential processes occurring in gas turbine combustion chambers were combined to form integral models.

A procedure for optimizing the distribution of computational workload has been first presented for parallel flow simulations on block-structured grids. Modeling the block-structure as weighted graph, allowed to use well known methods from graph partitioning theory to solve the domain decomposition problem. The considered test case showed that optimizing load balancing efficiency and communication costs simultaneously is superior to just optimizing the load balancing efficiency. Especially for computations with many processors large amounts of computing time can be saved.

The automation of splitting, mapping and recombining blocks enabled us to use a highly efficient block-structured flow solver with large and complex geometries, for which manual domain decomposition would not be feasible. Together with highly efficient numerical algorithms it will further enhance the possibilities to accomplish large, technical relevant simulations of turbulent flows in complex geometries.

Then, a combustion LES based integral model, that has been developed for a reliable description of combustion processes in gas turbine combustion chambers, has been validated and assessed in terms of its prediction capability of flow and combustion characteristics under different operating conditions. The complex flow field properties in the SSC are captured well. However, the turbulent kinetic energy at 5 mm from exit of the nozzle does not match the experimental data. The results further show that the flame is not always attached to nozzle and appears to fluctuate in time in agreement with experimental observations.

Acknowledgments The authors are grateful to the financial support by the German Research Council (DFG).

References

Project-Related Publications

1. Janicka, J., Sadiki, A.: Large eddy simulation of turbulent combustion systems. Proc. Combust. Inst. **30**, 537–547 (2005)
2. Wegner, B., Maltsev, A., Schneider, C., Sadiki, A., Dreizler, A., Janicka, J.: Assessment of unsteady RANS in predicting swirl flow instability based on LES and experiments. Int. J. Heat Fluid Flow **25**, 528–536 (2004)
3. Wegner, B.: A large-eddy simulation technique for the prediction of flow, mixing and combustion in gas turbine combustors. PhD thesis, Technische Universitaet Darmstadt, VDI Verlag GmbH, ISBN 978-3-18-354906-1 (2007)
4. Ketelheun, A., Olbricht, C., Hahn, F., Janicka, F.: NO prediction in turbulent flames using LES/FGM with additional transport equations. Proc. Combust. Inst. **33**(2), 2975–2982 (2011)
5. Olbricht, C., Hahn, F., Sadiki, A., Janicka, J.: Analysis of subgrid scale mixing using a hybrid LES-Monte-Carlo PDF method. Int. J. Heat Fluid Flow **28**(6), 1215–1226 (2007)
6. Pantangi, P., Sadiki, A., Janicka, J., Hage, M., Dreizler, A., van Oijen, J.A., Hassa, C., Heinze, J., Meier, U.: LES of pre-vaporized kerosene combustion at high pressures in a single sector combustor taking advantage of the flamelet generated manifolds method. In: Proceedings of ASME Turbo Expo 2011 (GT2011-45819), Vancouver, Canada, 6–10 June 2011
7. Chrigui, M., Gounder, J., Sadiki, A., Masri, A.R., Janicka, J.: Partially premixed reacting acetone spray using LES and FGM tabulated chemistry. Combust. Flame **159**(8), 2718–2741 (2012)
8. Janus, B., Dreizler, A., Janicka, J.: Experiments on swirl stabilized non-premixed natural gas flames in a model gas turbine combustor. Proc. Combust. Inst. **31**(2), 3091–3098 (2006)
9. Kornhaas, M., Sternel, D.C., Schäfer, M.: Influence of time step size and convergence criteria on large eddy simulations with implicit time discretization. In: Meyers, J., Geurts, B., Sagaut, P. (eds.) Quality and Reliability of Large-Eddy Simulations. ERCOFTAC Series 12, pp. 119–130. Springer, Berlin (2008)
10. Gauß, F., Schäfer, M., Siegmann, J., Sternel, D.: On the influence of boundary discretization schemes on the accuracy of flow simulation with local refinement. In: Diez, P, Bouillard, Ph. (eds.) IV International Conference on Adaptive Modelling and Simulation, pp. 97–100. ECCOMAS, International Center for Numerical Methods in Engineering, May 2009
11. Jakirlic, S., Kadavelil, G., Kornhaas, M., Schäfer, M., Sternel, D.C., Tropea, C.: Numerical and physical aspects in LES and hybrid LES/RANS of turbulent flow separation in a 3-D diffuser. Int. J. Heat Fluid Flow **31**, 820–832 (2010)
12. Sternel, D.C., Kornhaas, M., Schäfer, M.: High-performance computing techniques for coupled fluid, structure and acoustics simulations. In: Competence in High Performance Computing 2010: Proceedings of an International Conference on Competence in High Performance Computing, June 2010. Springer, Germany (2011)
13. Sternel, D.C., Junglas, D., Martin, A., Schäfer, M.: Optimisation of partitioning for parallel flow simulation on block structured grids. Proceedings of the Fourth International Conference on Engineering Computational Technology, Civil-Comp Press (2004)

Other Publications

14. Patel, N., Kirtas, M., Sankaran, V., Menon, S.: Simulation of spray combustion in a lean-direct injection combustor. Proc. Combust. Inst. **31**(2), 2327–2334 (2007)

15. Boudier, G., Gicquel, L., Poinsot, T., Bissieres, D., Berat, C.: Comparison of LES, RANS and experiments in an aeronautical gas turbine combustion chamber. Proc. Combust. Inst. **31**(2), 3075–3982 (2007)
16. Boudier, G., Staffelbach, G., Gicquel, L., Poinsot, T.: Reliability of large eddy simulation of combustion in complex test cases. In: Proceedings of Workshop "Quality and Reliability of LES", Leuven, Belgium, 24–26 October 2007
17. Vervisch, L., Domingo, P., Lodato, G., Veynante, D.: Scalar energy fluctuations in large-eddy simulation of turbulent flames: statistical budgets and mesh quality criterion. Combust. Flame **157**(4) (2010)
18. Wang, H., Stephen, B.P.: Large eddy simulation/probability density function modeling of a turbulent CH4/H2/N2 jet flame. Proc. Combust. Inst. **33**(1), 1319–1330 (2011)
19. Moin, P., Apte, S.V.: Large eddy simulation of realistic gas turbine combustors. AIAA J. **44**(4), 698–708 (2006)
20. Fureby, C., Grinstein, F.F., Li, G., Gutmark, E.J.: An experimental and computational study of a multi-swirl gas turbine combustor. Proc Combust. Inst. **31**, 3107–3114 (2007)
21. Staffelbach, G., Gicquel, L., Poinsot, T.: Highly parallel large eddy simulations of multiburner test cases in industrial gas turbines. In: The Cyprus International Symposium on Complex Effects in LES, Limassol, September 20–24, (2005)
22. Vermorel, O., Richard, S., Colin, O., Benkenida, A., Angelberger, C., Veynante, D.: Multi-cycle LES simulations of flow and combustion in a PFI SI 4-valve production engine. SAE Technical paper series, 2007-01-0151 (2007)
23. Jones, W.P., Lyra, S., Navarro-Martinez, S.: Large eddy simulation of a swirl stabilized spray flame. Proc. Combust. Inst. **33**(2), 2153–2160 (2011)
24. James, S., Zhu, J., Anand, M.S.: Large eddy simulations as a design tool for gas turbine combustion systems. AIAA J. **44**(4), 674–686 (2006)
25. Pitsch, H.: Large eddy simulation of turbulent combustion. Annu. Rev. Fluid Mech. **38**, 453–482 (2006)
26. Warnatz, J., Maas U., Dibble, R.W.: Combustion. Physical and Chemical Fundamentals, Modeling and Simulation, Experiments, Pollutant Formation. 4th Edition, Springer Berlin, ISBN 978-3540677512 (2006)
27. Pope, S.B.: Turbulent Flows. Cambridge University Press, Cambridge (2000)
28. Fiorina, B., Gicquel, O., Vervisch, L., Carpentier, S., Darabiha, N.: Premixed turbulent combustion modeling using tabulated detailed chemistry and PDF. Proc. Combust. Inst. **30**, 867–874 (2005)
29. Vervisch, L., Lodato, D., Domingo, P.: Proceedings of Workshop "Quality and Reliability of LES", Leuven, Belgium, 24–26 October 2007
30. van Oijen, J.A., De Goey, L.P.H.: Modelling of premixed laminar flames using flamelet-generated manifolds. Combust. Sci. Technol. **161**, 113–137 (2000)
31. van Oijen, J.A., de Goey, L.P.H.: A numerical study of confined triple flames using a flamelet-generated manifold. Combust. Theory Model. **8**(1), 141–163 (2004)
32. Germano, M., Piomelli, U., Moin, P., Cabot, W.H.: A dynamic sub-grid scale eddy viscosity model. Phys. Fluids A **3**, 1760–1765 (1991)
33. Sagaut, P.: Large Eddy Simulation for Incompressible Flows. Springer, Berlin (2001)
34. Huai, Y., Sadiki, A.: Analysis and optimization of turbulent mixing with large eddy simulation. In: ASME 2nd Joint U.S.-European Fluids Engineering Summer Meeting, FEDSM 2006–98416, Miami (2006)
35. Maas, U., Pope, S.B.: Simplifying chemical kinetics: intrinsic low-dimensional manifolds in composition space. Combust. Flame **88**, 239–264 (1992)
36. Vreman, A.W., Albrecht, B.A., van Oijen, J.A., de Goey, L.P.H., Bastiaans, R.J.M.: Premixed and non-premixed generated manifolds in Large-eddy simulation of Sandia flame D and F. Combust. Flame **253**, 394–416 (2008)
37. Kempf, A.: Large-eddy simulation of non-premixed turbulent flames. PhD thesis, Darmstadt University of Technology (2003)

38. Final Report, Computational Fluid Dynamics for Combustion (CFD4C), CEC Project No: GRD1-1999-10325 (2003)
39. Final Report, Modeling of Low Emissions Combustors using Large Eddy Simulations, (MOLECULES), EC Project GRD1-2000-25221 (2004)
40. Waterson, N.P., Deconinck, H.: Development of bounded higher-order convection scheme for general industrial applications. Project Report 1994–33, Von Karman Institute (1994)
41. Durst, F., Peric, M., Schäfer, M., Schreck, E.: Parallelization of efficient numerical methods for flows in complex geometries. Notes Numer. Fluid Mech. **38**, 79–92 (1993)
42. Karypis, G., Kumar, V.: Metis – unstructured graph partitioning and sparse matrix ordering system. Technical report, University of Minnesota, Department of Computer Science (1995)
43. Pellegrini, F.: Graph partitioning based methods and tools for scientific computing. Parallel Comput. **23**(12), 153–164 (1997)
44. Ahusborde, E., Glockner, S.: A 2d block-structured mesh partitioner for accurate flow simulations on nonrectangular geometries. Comput. Fluids **43**(1), 2–13 (2011)
45. Gourdain, N., Gicquel, L., Montagnac, M., Vermorel, O., Gazaix, M., Staelbach, G., Garcia, M., Boussuge, J.F., Poinsot, T.: High performance parallel computing of flows in complex geometries: I. methods. Comput. Sci. Discov. **2**(1), art. no. 015003 (2009)
46. Ytterström, A.: A tool for partitioning structured multiblock meshes for parallel computational mechanics. Int. J. High Perform. Comput. Appl. **11**(4), 336–343 (1997)
47. Berger, M.J., Bokhari, S.H.: A partitioning strategy for nonuniform problems on multiprocessors computers. IEEE Trans. Comput **C-36**(5), 570–580 (1987)
48. Quagliarella, D.: Optimal Domain Decomposition for Parallel Multiblock Flow Field Solvers Using Genetic Algorithms, pp. 215–222. North- Holland, Amsterdam (1995)
49. Rantakokko, J.: Comparison of partitioning strategies for pde solvers on multiblock grids. In: Kagström, B. et al. (eds.) Applied Parallel Computing Large Scale Scientic and Industrial Problems, vol. 1541 of Lecture Notes in Computer Science, pp. 468–475. Springer, Berlin/Heidelberg (1998)
50. Rantakokko, J.: Partitioning strategies for structured multiblock grids. Parallel Comput. **26**(12), 1661–1680 (2000)
51. Hagen, L., Kahng, A.B.: New spectral methods for ratio cut partitioning and clustering. IEEE Trans. Comput. Aided Des. Integr. Circuits Syst. **11**(9), 1074–1085 (1992)
52. Kernighan, B.W., Lin, S.: An efficient Heuristic procedure for partitioning graphs. Bell Syst. Technol. J. **49**(1), 291–307 (1970)
53. Chung, F.R.K.: Spectral graph theory. Number 92 in Regional Conference Series in Mathematics. American Mathematical Society, Providence (1997)
54. Maltsev, A.: Towards the development and assessment of complete CFD models for the simulation of stationary Gas turbine combustion process. Dissertation, VDI Verlag GmBH, Duesseldorf, Reihe 6, Nr. 515 (2004)
55. Yun, A.: Development and analysis of advanced explicit algebraic turbulence and scalar flux models for complex engineering configurations, Dissertation, Darmstadt (2005)
56. Schneider, E., Sadiki, A., Janicka, J.: Modeling and 3D-simulation of the kinetic effects in the post-flame region of turbulent premixed flames based on the G-equation approach. Flow Turbul. Combust. **75**, 191–216 (2005)
57. Schneider, E., Sadiki, A., Kurenkov, A., Janicka, J., Oberlack, M.: Modelling of Premixed Combustion Using G-Equation in RANS Context. Modelling of Premixed Combustion Using G-Equation in RANS Context. Advances in Turbulence, ETC., 10, Trondheim (2004)
58. Schneider, E.: Numerische Simulation turbulenter vorgemischter Verbrennungssysteme: Entwicklung und Anwendung eines RANS-basierten Gesamtmodells. Dissertation, Darmstadt (2005)
59. Schneider, E., Maltsev, A., Sadiki, A., Janicka, J.: Numerical study of the precessing vortex core in a confined partially premixed swirl combustion system. GT2007-27272 ASME Turbo Expo (2007)

60. Chrigui, M.: Eulerian–Lagrangian approach for modeling and simulations of turbulent reactive multi-phase flows under gas turbine combustor conditions, Dissertation, TU Darmstadt (2005)
61. Akula, R.K.: Study of the performance of different subgrid scale models and large eddy simulation of premixed combustion. Dissertation. TU Darmstadt (2007)
62. Sternel, D.C. (ed.): FASTEST Manual.: Institute of numerical methods in mechanical engineering, Department of Mechanical Engineering, Technische Universität Darmstadt, Germany (2005)
63. Schlesinger, S.: Terminology for model credibility. Simulation **32**(3), 103–104 (1979)
64. Oberkampf, W.L., Trucano, T.G., Hirsch, C.: Verification, Validation, and Predictive Capability in Computational Engineering and Physics, Report SAND2003-3769, February 2003

Chapter 12
Adaptive Large Eddy Simulation and Reduced-Order Modeling

S. Ullmann, S. Löbig, and J. Lang

Abstract The quality of large eddy simulations can be substantially improved through optimizing the positions of the grid points. LES-specific spatial coordinates are computed using a dynamic mesh moving PDE defined by means of physically motivated design criteria such as equidistributed resolution of turbulent kinetic energy and shear stresses. This moving mesh approach is applied to a three-dimensional flow over periodic hills at $\mathrm{Re} = 10{,}595$ and the numerical results are compared to a highly resolved LES reference solution. Further, the applicability of reduced-order techniques to the context of large eddy simulations is explored. A Galerkin projection of the incompressible Navier–Stokes equations with Smagorinsky sub-grid filtering on a set of reduced basis functions is used to obtain a reduced-order model that contains the dynamics of the LES. As an alternative method, a reduced-order model of the un-filtered equations is calibrated to a set of LES solutions. Both approaches are tested with POD and CVT modes as underlying reduced basis functions.

Keywords Large eddy simulation • Moving mesh method • Reduced-order modeling • Adaptivity

S. Ullmann (✉) • S. Löbig • J. Lang
Department of Mathematics, Technische Universität Darmstadt, Dolivostr. 15, D-64293 Darmstadt, Germany
e-mail: ullmann@mathematik.tu-darmstadt.de; sloebig@mathematik.tu-darmstadt.de; lang@mathematik.tu-darmstadt.de

J. Lang
Center of Smart Interfaces, Technische Universität Darmstadt, Petersenstr. 30, D-64287 Darmstadt, Germany
e-mail: lang@mathematik.tu-darmstadt.de

Graduate School Computational Engineering, Technische Universität Darmstadt, Dolivostr. 15, D-64293 Darmstadt, Germany

J. Janicka et al., *Flow and Combustion in Advanced Gas Turbine Combustors*, Fluid Mechanics and Its Applications 102, DOI 10.1007/978-94-007-5320-4_12,

12.1 Introduction

Locally large solution variations are best resolved by a high concentration of mesh points there and few points in the remaining domain with less solution activity. This is especially true for modeling turbulent flows by large eddy simulations (LES). Here, the characteristic length of turbulent fluctuation varies significantly over the physical domain, which demands for an LES-specific adaptation of the grid size to turbulent length scales.

In the first part of this work we enhance the quality of our LES by applying the moving mesh method developed by Huang and Russell [8]. This r-adaptive method moves the grid points according to a time dependent mesh moving PDE to achieve higher resolution in important areas of the domain while keeping the data structure unchanged, i.e., topology and number of degrees of freedom of the spatial discretization once chosen are kept unchanged. The significance of areas is defined via a so called monitor function, which links in a natural and smooth way the mesh adaptation process to properties of the physical solution. It is commonly designed by solution-dependent quantities of interests (QoI), which are equidistributed over the adaptively chosen grid cells of the physical domain in an integral sense.

The moving mesh method has successfully been used for the computation of a turbulent flow over periodic hills using single physical QoIs [1, 9] and various combinations of physical QoIs [2]. In Lang et al. [3] a moving mesh method for two-dimensional finite element computations using mathematical QoIs has been studied.

The second part of the paper focuses on reduced-order models for incompressible flows. While in many instances, the accuracy and efficiency of a numerical flow simulation can be greatly improved by adaptive meshing, the additional speed-up promised by reduced-order techniques seems attractive especially for many-query and real-time applications. In our context, by reduced-order techniques we mean models that are obtained by a Galerkin projection of a high dimensional problem on a small set of reduced basis functions with global support, that incorporate information about the solution of the high dimensional problem. Typically, the reduced basis functions are created using snapshots which are computed beforehand with a conventional numerical code. Thus, a large 'off-line' computational cost, depending on the degrees of freedom of the numerical code, is accepted in order to obtain a small 'on-line' cost, depending only on the number of basis functions used for the projection.

A detailed introduction to reduced-order modeling based on the proper orthogonal decomposition (POD) is given by Holmes et al. [10]. While the original purpose of the POD method was to identify coherent structures of turbulent flows and to investigate their dynamics [11], slow progress has been made towards actually simulating turbulent flows using reduced-order models. Two reasons that perhaps contribute to that stagnation are these: Firstly, even for simple laminar flows it has been observed that Galerkin reduced-order models can converge towards spurious limit cycles [12, 13]. Secondly, solutions of turbulent flows are much less amenable to data compression than laminar periodic flows, in other words, turbulent flow solutions can not be approximated well by a linear combination of a small number of snapshots. Still, we want to mention some promising results that could be achieved for three-dimensional transitional and turbulent flows: Telib et al. [14] applied a calibrated POD to the flow in a T-mixer at $Re = 300$ and $Re = 400$, Buffoni et al. [15] simulated a three-dimensional flow around a square cylinder at $Re = 300$ and Couplet at al. [16] built a reduced-order model from snapshots of an LES of the flow past a backward facing step at $Re = 7{,}432$. In our own work [4] we study a reduced-order model of the flow around a cylinder at $Re = 3{,}900$, based on Smagorinsky-LES. Wang et al. [17] present a similar approach for the flow around a cylinder at $Re = 1{,}000$.

We compare two different methods to compute basis functions from the snapshot data: proper orthogonal decomposition and centroidal Voronoi tessellation (CVT). An introduction to the CVT method is provided by Du et al. [18]. We assess the resulting POD or CVT Galerkin reduced-order models by their ability to approximate given snapshot data. Different from the comparisons performed by Burkardt et al. [19], we apply both techniques to the laminar vortex-shedding flow around a circular cylinder, and present detailed quantifications of the model and approximation errors with respect to the number of basis functions. We derive and apply a reduced-order model for the pressure in order to extend the comparisons to the drag and lift forces acting on the cylinder. Finally, we explore techniques with which a progression towards low-dimensional modeling of the large-scale structures of turbulent flows can be made.

12.2 Adaptive Moving Meshes for Large Eddy Simulations

12.2.1 Large Eddy Simulation

The spatially filtered incompressible Navier–Stokes equations are given by

$$\frac{\partial \bar{u}_i}{\partial t} + \frac{\partial}{\partial x_j}\left(\bar{u}_i \bar{u}_j\right) = -\frac{\partial \bar{p}}{\partial x_i} + \frac{\partial (2\nu \bar{S}_{ij})}{\partial x_j} - \frac{\partial \tau_{ij}}{\partial x_j} + f_i,$$

$$\frac{\partial \bar{u}_j}{\partial x_j} = 0,$$

where the filtered velocity and pressure are denoted by $\bar{\mathbf{u}} = (\bar{u}_1, \bar{u}_2, \bar{u}_3)^T$ and $\bar{p}$, respectively. The elements of the filtered strain rate tensor $\bar{\mathbf{S}}$ are defined by $\bar{S}_{ij} = (\partial_{x_j}\bar{u}_i + \partial_{x_i}\bar{u}_j)/2$ and ν represents the molecular viscosity. The external force is given by f_i and $\tau_{ij} = \overline{u_i u_j} - \bar{u}_i \bar{u}_j$ constitutes the subgrid-scale tensor.

To model the subgrid stresses we employ the eddy viscosity approach of Smagorinsky [20],

$$\tau_{ij} \approx -2\nu_t \bar{S}_{ij} + \frac{1}{3}\tau_{kk}\delta_{ij} \qquad \text{with} \qquad \nu_t = (C_S \Delta)^2 \left|\bar{\mathbf{S}}\right| .$$

Here, ν_t is the turbulent eddy viscosity, C_S denotes the Smagorinsky constant, Δ is the filter width and the norm of the filtered strain rate tensor is given by $|\bar{\mathbf{S}}| = (2\bar{S}_{ij}\bar{S}_{ij})^{1/2}$.

12.2.2 Moving Mesh Method

The moving mesh method developed by Huang and Russell [8] employs a time-dependent moving mesh partial differential equation (MMPDE) to move the grid points in a way such that the coordinates minimize an adaptation functional. The essential part of the functional is formed by the monitor function which controls the concentration of the mesh via some solution-dependent quantity of interest (QoI).

The MMPDE is defined as a coordinate transformation from a physical domain Ω to a computational domain Ω_c, where the coordinates are denoted by $\mathbf{x} = (x_1, \dots, x_n)^T$ and $\boldsymbol{\xi} = (\xi_1, \dots, \xi_n)^T$, respectively. For actual computations, however, it is more convenient to have the MMPDE in the non-conservative formulation for the inverse mapping $\mathbf{x} = \mathbf{x}(\boldsymbol{\xi}, t)$,

$$\tau \frac{\partial \mathbf{x}}{\partial t} = P \left[\sum_{i,j} \underbrace{\nabla \xi_i^T G^{-1} \nabla \xi_j}_{a_{ij}} \frac{\partial^2 \mathbf{x}}{\partial \xi_i \partial \xi_j} - \sum_i \underbrace{\sum_j \nabla \xi_i^T \frac{\partial G^{-1}}{\partial \xi_j} \nabla \xi_j}_{b_i} \frac{\partial \mathbf{x}}{\partial \xi_i} \right],$$

since the location of the mesh points is then defined explicitly. The parameter τ is used to adjust the time-scale of the mesh movement and is usually held fixed during the computation. For large values of τ, the mesh movement is smoother and therefore the MMPDE is easier to integrate numerically, whereas a smaller value results in faster adaption of the mesh to changes of the monitor function G. The parameter P is used to achieve a spatially balanced MMPDE and is often chosen as some bound on the coefficients, i.e.,

$$P = \frac{1}{\sqrt{\sum_i (a_{ii}^2 + b_i^2)}}.$$

As suitable boundary condition for our MMPDE we move the boundary points by extrapolating their position from interior cells in such a way that we obtain orthogonal cells.

The monitor function G in the MMPDE is a positive definite and symmetric $n \times n$ matrix and also the heart of the moving mesh method. It contains some quantity of interest (QoI) that is responsible for the assignment of important areas since the MMPDE equidistributes the product of the QoI and the cell volume over the domain.

These QoI can be mathematically or physically motivated, like error indicators or the solution's gradient. A common choice is Winslow's monitor function [21] together with an intensity parameter γ

$$G = \omega I, \qquad \omega = \sqrt{1 + \gamma \|\psi\|^2}.$$

We use this as basis for constructing our monitor function for multiple QoIs. Our technique is related to the concept of the balanced monitor function (BMF) by van Dam [22]. The resulting monitor function reads

$$G = \omega I,$$
$$\omega(\boldsymbol{\psi}) = 1 + \gamma \sum_{p=1}^{N_p} g_p \omega_p(\boldsymbol{\psi}) = 1 + \gamma \sum_{p=1}^{N_p} g_p \left[\frac{(1-\beta)\alpha_p + \beta \psi_p}{(1-\beta)\alpha_p + \beta M_p} \right].$$

Here, $\boldsymbol{\psi} = (\boldsymbol{\psi}_1, \ldots, \boldsymbol{\psi}_{N_p})^T$ is the vector of N_p different QoI, α_p and M_p are the corresponding average and maximum values, respectively. We normalize each QoI with its maximum value to ensure that all are at least approximately of the same range. The additional normalization of each QoI by its respective average prevents single large maximum values from dominating all other monitor values on the rest of the domain. Additionally, weights g_i with sum one are introduced that can be assigned to the different QoIs.

The monitor function is usually very non-smooth and therefore we apply a local averaging technique. This leads to smoother meshes and reduces the stiffness of the MMPDE. A description of the smoothing technique can be found in Hertel et al. [2].

12.2.3 Numerical Realization

To solve the three-dimensional momentum and mass conservation on moving grids the Arbitrary Lagrangian–Eulerian (ALE) Method is employed in its filtered form. The governing equations read, see Ferziger [23],

$$\frac{\mathrm{d}}{\mathrm{d}t} \int_{V(t)} \bar{u}_i \mathrm{d}V + \int_{V(t)} \frac{\partial \left[\bar{u}_i (\bar{u}_j - u_{G,j}) \right]}{\partial x_j} \mathrm{d}V$$
$$= - \int_{V(t)} \frac{\partial \bar{p}}{\partial x_i} \mathrm{d}V + \int_{V(t)} \frac{\partial (2\nu \bar{S}_{ij})}{\partial x_j} \mathrm{d}V - \int_{V(t)} \frac{\partial \tau_{ij}}{\partial x_j} \mathrm{d}V + \int_{V(t)} f_i \mathrm{d}V$$

$$\frac{\mathrm{d}}{\mathrm{d}t} \int_{V(t)} \mathrm{d}V + \int_{V(t)} \frac{\partial \left(\bar{u}_j - u_{G,j} \right)}{\partial x_j} \mathrm{d}V = 0.$$

Here, u_G and $V(t)$ are the velocity and time-dependent volume of a computational cell, respectively. When discretizing the governing equations there is a risk of introducing mass sources or sinks in the flow field when the velocity u_G and the change of volume over time are not treated consistently.

For this reason, the Geometric Conservation Law (GCL), see Demirdzic [24],

$$\frac{\mathrm{d}}{\mathrm{d}t}\int_{V(t)} \mathrm{d}V - \int_{V(t)} \frac{\partial u_{G,j}}{\partial x_j}\,\mathrm{d}V = 0$$

needs to be satisfied. The GCL is used to determine the velocity by the given change of volume of the computational cell to ensure consistency. The resulting expressions for the components of u_G are included in the discretized governing equations, so that only the change of volume over time remains, split into separate expressions for each face of the cell [23, 24]. Hence, the GCL is not treated as separate equation here, but is introduced into the discrete form of the first governing equation.

In each time step, the MMPDE has to be integrated first since the described ALE formulation for the governing equations requires the new grid points at the end of the time step to evaluate the change of volume over time for each cell. Central Finite Differences in space together with an implicit time scheme are used for the discretization of the MMPDE. It is sufficient to solve the MMPDE up to moderate accuracy since not the exact position of the grid points but just an appropriate direction for moving the grid points with a suitable velocity is needed and therefore the implicit Euler scheme is adequate. To use the implicit Euler scheme, the coefficients a_{ij} and b_i in the MMPDE are frozen during the time step, i.e., they are only once determined at the starting time. The strongly implicit procedure (SIP) introduced by Stone [25] is used to solve the resulting system of equations. The MMPDE approach has been successfully implemented in the LESOCC2 code [2, 9], which is used for numerical tests in the following.

12.2.4 Turbulent Flows Over Periodic Hills

The turbulent flow through a channel with streamwise periodically arranged constrictions was proposed as a test case for separation from curved surfaces, e.g. in Fröhlich et al. [26]. Figure 12.1 shows the extent of the computational domain with respect to the hill height. The coordinates x, y and z denote streamwise, normal and spanwise direction, respectively, and we have $L_x = 9h$, $L_y = 3.035h$ and $L_z = 4.5h$. The Reynolds number is $Re = 10{,}595$, based on the hill height and the bulk velocity at the crest of the hill. The main features of this flow are a free shear layer above a recirculation area downstream of the hill, reattaching roughly at half the domain length followed by a strong acceleration at the windward side of the next hill. The resolution of the flow near the separation point has a strong impact on the reattachment of the recirculating flow as well as on the characteristics of the shear

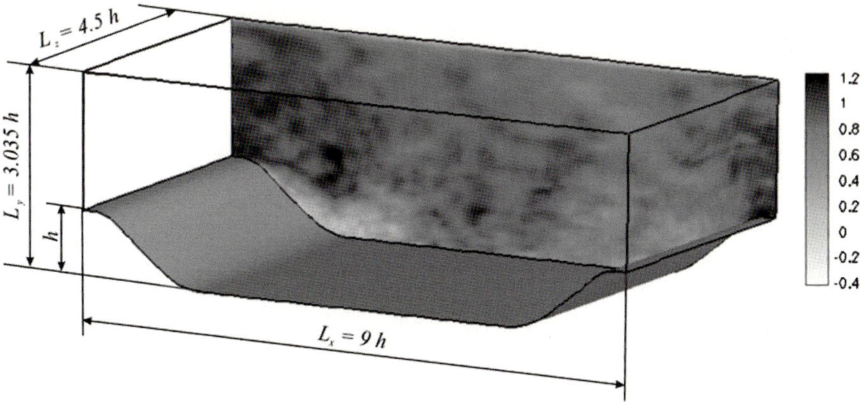

Fig. 12.1 Geometry and computational domain for the flow over periodic hills. The contour plot at the back shows a snapshot of the instantaneous streamwise velocity u

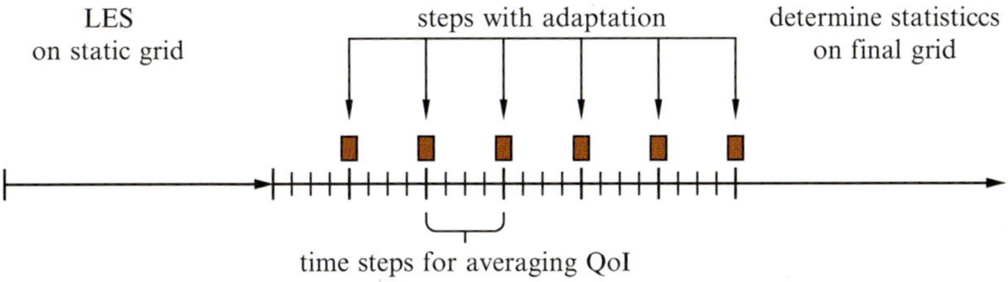

Fig. 12.2 Illustrated outline of the adaptation procedure applied

layer. The domain has periodic boundary conditions in streamwise and spanwise direction and walls on the boundaries in y-direction.

The reference solution for this test case is provided by Fröhlich et al. [26]. This highly resolved LES was carried out on a wall-resolving, nearly orthogonal 196 × 128 × 186 grid. For our computations we use an initial grid with 89 × 33 × 49 points equispaced in x-direction and equidistant grid points along each vertical line. Figure 12.2 illustrates the adaptation process. We begin our computation with a statistically converged LES on the initial grid. Since we want to improve the LES with respect to the statistical flow properties it is not necessary to move the grid in every time step. Instead we adapt the grid every N_{aver} time steps. During these N_{aver} time steps temporal averages (with additional average in homogeneous z-direction) are computed for use in the MMPDE. The adaptation is carried out here only in two dimensions (x, y) because of the statistically homogeneous nature of the flow with respect to the spanwise direction. The grid is adapted N_{adapt} times until it approaches a nearly steady state. Finally, a simulation on the obtained stationary grid is carried out to determine the statistics.

For the computations presented here we chose $N_{\text{aver}} = 225$, i.e., approximately a half flow through time. Tests showed that all adaptive grids in each computation of the flow over periodic hills have reached a nearly steady state after $N_{\text{adapt}} = 300$ adaptations. For the time-scaling and the intensity parameter the combination of $\tau = 1.0$ and $\gamma = 10$ lead to satisfactory results in all cases without risking overshooting grid points. Following van Dam [22], we chose $\beta = 0.3$ in the balanced monitor function.

Various physically motivated criteria for mesh movement as well as combinations of these criteria have been investigated in Hertel et al. [2]. Some of these criteria will be introduced as well as the results of the respective computations will be presented here. Statistically averaged values were used since the focus was on the improvement of the LES with respect to statistical flow properties. In the equations below, $\langle . \rangle$ indicates averaging in time and statistically homogeneous z-direction.

The *gradient of the streamwise velocity* (GU)

$$\psi_{\text{gu}} = |\nabla \langle \bar{u}_1 \rangle|$$

is high along the hill and promises a high mesh concentration near the wall and around the separation point of the recirculation area. The gradient includes all regions of the flow field with at least one large derivative.

Equidistributing the criterion resulting from the *turbulent kinetic energy* (TKE)

$$\psi_{k,\text{tot}} = \frac{\langle k_{\text{sgs}} \rangle}{k_{\text{tot, max}}} \quad \text{with} \quad k_{\text{tot, max}} = \max_{\Omega_c} \left(\langle k_{\text{res}} \rangle + \langle k_{\text{sgs}} \rangle \right),$$

i.e., the ratio between the modeled TKE k_{sgs} of the subgrid-scales and the maximum of the total TKE over the flow field $k_{\text{tot, max}}$, over the domain is motivated by the idea of LES. The unresolved TKE is determined here via the approach by Berselli et al. [27], $k_{\text{sgs}} \approx \left(2^{1/3} - 1\right) 0.5 \left|\bar{\mathbf{u}} - \bar{\bar{\mathbf{u}}}\right|^2$. In order to avoid unphysically high values in regions where k_{res} is small, the maximum of the total TKE had to be used for the QoI instead of the local total amount of TKE.

Also motivated by the idea of LES is the balance of modeled to total *shear stress* (ST)

$$\psi_\tau = \frac{\langle \tau_{12}^{\text{mod}} \rangle}{\left|\langle \tau_{12}^{\text{mod}} \rangle\right| + \left|\langle \bar{u}_1'' \bar{u}_2'' \rangle\right|} \quad \text{with} \quad \tau_{12}^{\text{mod}} = \nu_t \left(\frac{\partial \bar{u}_1}{\partial y} + \frac{\partial \bar{u}_2}{\partial x} \right).$$

The modeled SGS shear stress is available when the Smagorinsky model is used. Refinement is needed when this value is large since then more shear stress is modeled than resolved. To avoid unphysically high values for ψ_τ the absolute values had to be applied in the denominator.

Investigations on moving grids using single criteria have shown that neither TKE nor ST lead to satisfactory results for the turbulent flow over periodic hills [2, 9]. The performance of the gradient of the streamwise velocity however exceeds those of every other physically motivated criterion tested so far. Therefore, it seems natural

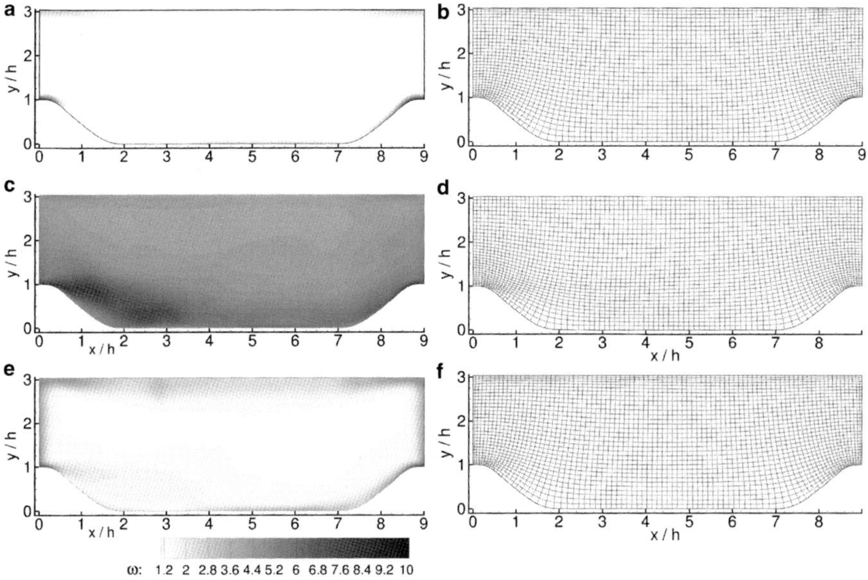

Fig. 12.3 *Left*: monitor function ω. *Right*: final grid after 300 adaptations. QoI: (**a**, **b**) gradient of streamwise velocity; (**c**, **d**) turbulent kinetic energy; (**e**, **f**) shear stress

to try to further improve the achieved quality by combining the gradient with other criteria when adapting the grid.

Figure 12.3 presents pictures of the monitor function and the final grid after 300 adaptations for the gradient of the streamwise velocity ψ_{gu} and the combination of the gradient with the TKE ψ_{gu} and $\psi_{k,\mathrm{tot}}$ and the ST ψ_{gu} and ψ_{τ}.

For the gradient of the streamwise velocity the monitor function is expectedly high near the crest of the hill, especially on the windward side of the hill where a strong acceleration of the flow takes place. Also near the separation point high values of the monitor function can be observed. Moderate values are achieved near the upper wall. In the rest of the domain the monitor function is negligibly small. The grid refinement is according to the monitor values strong near the hill crest and the separation point and achieves nearly wall resolution in those regions, but only in wall normal direction. Combining the gradient of the streamwise velocity with other criteria leads to monitor functions with high values in regions where the gradient or the second criterion proposed large values. In the case of the combination with the TKE, ψ_{gu} and $\psi_{k,\mathrm{tot}}$, this yields high monitor values and therefore grid refinement in the free shear layer after the separation from the hill. Closer examination reveals that the refinement does not happen at the exact separation point but some small distance away from it downstream. The combination of gradient and shear stress leads to some refinement near the upper wall in addition to the crest of the hill. In the other regions of the domain the monitor function for this combination stays negligibly small.

12.2.5 *Results*

We are interested in an improvement of the statistical quality of LES. Therefore we compare our results with the reference data and our initial grid concerning the averaged streamwise velocity $\langle \bar{u}_1 \rangle$ and the Reynolds shear stress $\langle \bar{u''}_1 \bar{u''}_2 \rangle$. Figure 12.4 shows the mean profiles for these quantities at positions $x/h = 0.5, 4.0, 8.0$.

Significant improvement compared to the initial grid can be observed for every QoI. Although the average streamwise velocity is well predicted in position and amplitude of the maximum, the gradient and its combination with the shear stress have small problems in capturing the velocity near the bottom of the domain at $x/h = 4.0$. Greater differences in the predictions can be observed when it comes to the Reynolds shear stress. Here, the gradient of the streamwise velocity captures the amplitude of the maxima overall best despite its weakness again at $x/h = 4.0$. For $x/h = 0.5$ the combination of gradient and TKE predicts the position of the maximum Reynolds shear stress well but fails to capture the right amplitude.

The comparison of the predicted separation and reattachment points for the different QoI, see Table 12.1, shows how strong the gradient of the streamwise velocity already is on its own. The combination with other criteria corrupts the prediction of the separation point in both cases. Only the reattachment point is better predicted by the combination of the gradient of the streamwise velocity with the shear stress. Despite the inferior performance of the combined criteria, here each QoI improves the prediction of the initial grid considerably.

12.3 Reduced-Order Modelling

The construction of the reduced-order models is based on the following matrix form of a spatial semi-discretization of the Smagorinsky LES equations:

$$\begin{aligned} \mathcal{M}\dot{\mathbf{U}} &= \mathcal{B}(\mathbf{U})\mathbf{U} + \mathcal{C}P + \mathcal{D}(\mathbf{U})\mathbf{U}, \\ 0 &= \mathcal{C}^T\mathbf{U}. \end{aligned}$$

Here, $\mathcal{M}$ is a mass matrix and $\mathcal{C}$ is a discretized gradient operator. Later on, the $\mathcal{M}$ inner product and norm,

$$(\mathbf{V}, \mathbf{W})_{\mathcal{M}} = \mathbf{V}^T\mathcal{M}\mathbf{W}, \qquad \|\mathbf{V}\|_{\mathcal{M}} = (\mathbf{V}^T\mathcal{M}\mathbf{V})^{1/2},$$

respectively, will be used, which are discretizations of the L^2 inner product and norm. The advection matrix $\mathcal{B}(\mathbf{U})$ depends linearly on $\mathbf{U}$ and the viscosity matrix $\mathcal{D}(\mathbf{U})$ depends non-linearly on $\mathbf{U}$ via the turbulent viscosity. In case of a direct numerical simulation (DNS), the viscosity matrix becomes independent of the velocity.

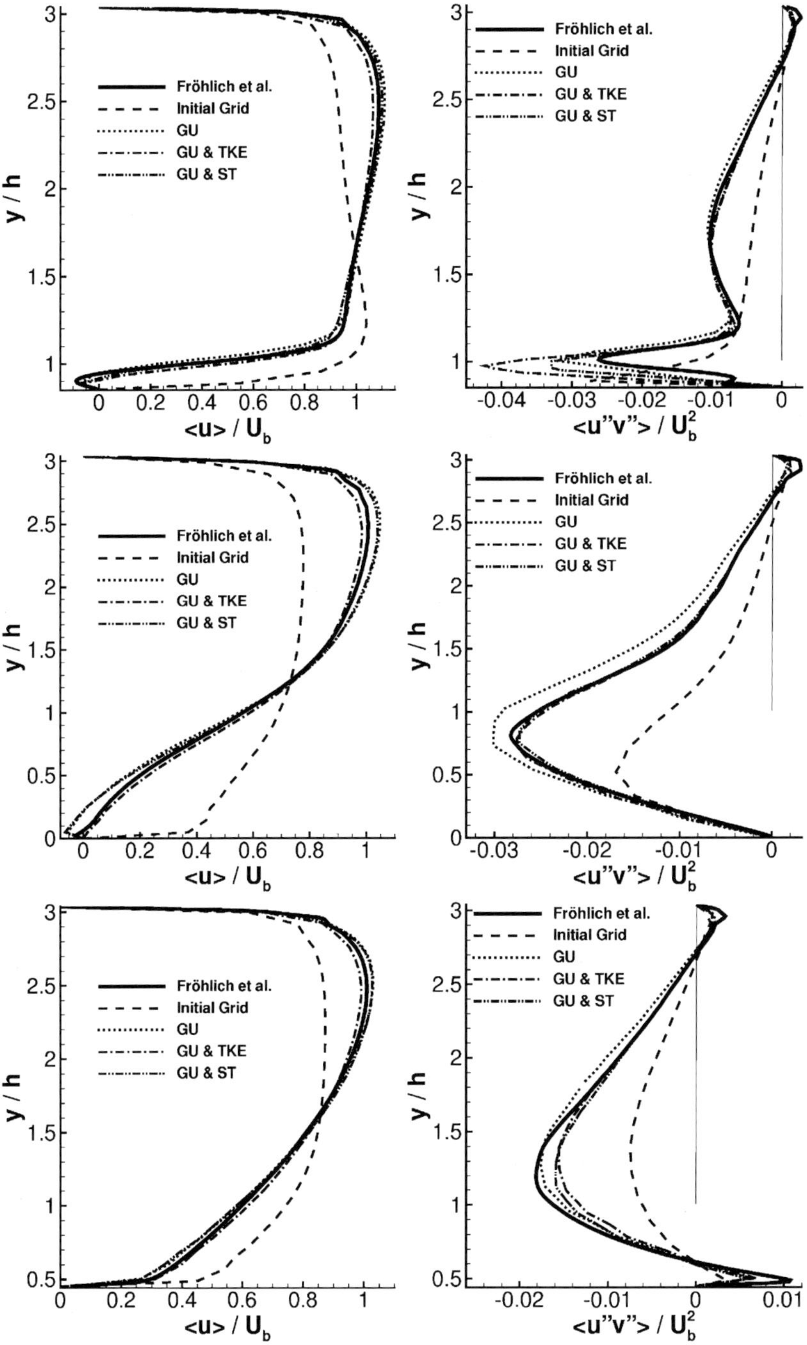

Fig. 12.4 Averaged streamwise velocity $\langle \bar{u}_1 \rangle$ $(= \langle \bar{u} \rangle)$ and Reynolds shear stress $\langle \bar{u''}_1 \bar{u''}_2 \rangle$ $(= \langle \bar{u}'' \bar{v}'' \rangle)$ at positions $x/h = 0.5$ (*top*), $x/h = 4.0$ (*middle*) and $x/h = 8.0$ (*bottom*) in comparison with data from Fröhlich et al. [26] for adaptation using the following criteria: ψ_{gu}, ψ_{gu} and $\psi_{k,\mathrm{tot}}$ and ψ_{gu} and ψ_τ

Table 12.1 Separation and reattachment points for different QoIs. Values obtained with final adapted grids compared to the reference data

	Reference	Initial	ψ_{gu}	ψ_{gu} and $\psi_{k,\text{tot}}$	ψ_{gu} and ψ_τ
x_{sep}/h	0.20	0.53	0.29	0.40	0.37
x_{rea}/h	4.56	3.08	4.48	3.90	4.53

12.3.1 Proper Orthogonal Decomposition

Let $\mathbf{U}_1, \ldots, \mathbf{U}_N$ be a set of discrete velocity snapshot vectors that are numerically obtained from a DNS or LES at time steps $t_1, \ldots, t_N$. The elements of each snapshot vector are the solution values at the mesh nodes. We define the snapshot fluctuations $\mathbf{U}'_n = \mathbf{U}_n - \overline{\mathbf{U}}$ for $n = 1, \ldots, N$ with respect to some discrete reference velocity vector $\overline{\mathbf{U}}$. In the following we present the proper orthogonal decomposition in this setting, by applying the method of snapshots [11] to the snapshot fluctuations $\mathbf{U}'_1, \ldots, \mathbf{U}'_N$. The theory is derived in Holmes et al. [10] for a more general context.

POD modes of a snapshot fluctuation matrix $\mathcal{S} = (\mathbf{U}'_1, \ldots, \mathbf{U}'_N)$ are a set of vectors $\mathbf{\Phi}_1, \ldots, \mathbf{\Phi}_R$ that satisfy $(\mathbf{\Phi}_i, \mathbf{\Phi}_j)_{\mathcal{M}} = \delta_{ij}$ for all combinations of $i, j = 1, \ldots, R$ and that solve

$$\min_{\{\mathbf{\Phi}_r\}} \sum_{n=1}^{N} \left\| \mathbf{U}'_n - \sum_{r=1}^{R} (\mathbf{U}'_n, \mathbf{\Phi}_r)_{\mathcal{M}} \mathbf{\Phi}_r \right\|_{\mathcal{M}}^2 \tag{12.1}$$

for a given R that is not larger than the rank of $\mathcal{S}$. A solution of the minimization problem can be obtained by the following procedure:

- Compute the eigendecomposition $\mathcal{S}^T \mathcal{M} \mathcal{S} = \mathcal{V} \Sigma^T \Sigma \mathcal{V}^T$, where $\Sigma \in \mathbb{R}^{N \times N}$ is a diagonal matrix that contains the square roots of the (non-negative) eigenvalues $\lambda_1, \ldots, \lambda_N$ ordered non-increasingly on the diagonal and $\mathcal{V} \in \mathbb{R}^{N \times N}$ contains the corresponding eigenvectors.
- Choose an approximation rank R, not larger than the number of non-zero eigenvalues. Define $\mathcal{V}_R \in \mathbb{R}^{N \times R}$ as the matrix that contains the left R columns of $\mathcal{V}$ and define $\Sigma_R \in \mathbb{R}^{R \times R}$ as the upper left portion of Σ.
- Compute the matrix $\mathcal{U}_R = (\mathbf{\Phi}_1, \ldots, \mathbf{\Phi}_R)$ by $\mathcal{U}_R = \mathcal{S} \mathcal{V}_R \Sigma_R^{-1}$.

The POD can also be defined using singular value decomposition,

$$\mathcal{M}^{1/2} \mathcal{S} = \widehat{\mathcal{U}} \Sigma \mathcal{V}^T, \qquad \mathcal{M}^{1/2} \mathcal{U} = \widehat{\mathcal{U}}.$$

In this case $\mathcal{U}_R$ is formed by the left R columns of $\mathcal{U}$ and the matrices $\mathcal{V}_R$ and Σ_R are defined as above. This approach requires a factorization $\mathcal{M} = (\mathcal{M}^{1/2})^T \mathcal{M}^{1/2}$, so for non-diagonal mass matrices the eigendecomposition of $\mathcal{S}^T \mathcal{M} \mathcal{S}$ is the preferred method.

For the derivation of the reduced-order model it will be useful that the POD modes are a linear combination of the snapshot fluctuation vectors,

$$\mathbf{\Phi}_r = \sum_{n=1}^{N} \frac{(\mathbf{U}'_n, \mathbf{\Phi}_r)_{\mathcal{M}}}{\lambda_r} \mathbf{U}'_n, \tag{12.2}$$

which means that some properties of the snapshot fluctuations are carried over to the POD modes.

12.3.2 Centroidal Voronoi Tessellation

The Centroidal Voronoi tessellation is a method that groups the set of discrete snapshot fluctuation vectors $\mathbf{U}'_1, \ldots, \mathbf{U}'_R$ in clusters, such that on average the distances between each cluster's center and its members are small. More precisely, the CVT is defined as a set of clusters $V_1, \ldots, V_R$ and a set of modes $\mathbf{\Phi}_1, \ldots \mathbf{\Phi}_R$ that solve the minimization problem

$$\min_{\{\mathbf{\Phi}_r\},\{V_r\}} \sum_{r=1}^{R} \sum_{\mathbf{U}'_n \in V_r} \|\mathbf{U}'_n - \mathbf{\Phi}_r\|_{\mathcal{M}}^2. \tag{12.3}$$

The CVT of a collection of snapshot fluctuation vectors can be computed using a variant of Lloyd's method [28]:

1. Assign each vector to some cluster.
2. Handle empty clusters.
3. Compute all cluster mid-points.
4. Assign each vector to the cluster whose mid-point is closest.
5. If in step 4 any vector was shifted, then go to step 2, otherwise quit.

The CVT modes are then given by the cluster mid-points,

$$\mathbf{\Phi}_r = \frac{1}{|V_r|} \sum_{\mathbf{U}'_n \in V_r} \mathbf{U}'_n, \tag{12.4}$$

where $|V_r|$ is the number of snapshots contained in the cluster V_r. Thus, for the CVT it holds that each mode is a linear combination of snapshot fluctuation vectors.

While the POD modes are orthonormal with respect to the $\mathcal{M}$-inner product, the CVT modes are not $\mathcal{M}$-orthonormal in general. It is even possible that they are linearly dependent, in which case they do not form a linear basis. In the special case that the CVT modes are computed from a set of snapshot fluctuations with respect to the snapshot mean, the CVT modes are always linearly dependent. To prove the linear dependence we need to find a set of coefficients $a'_1, \dots, a'_R$, not all equal to zero, for which $a'_1\mathbf{\Phi}_1 + \dots + a'_R\mathbf{\Phi}_R = 0$. Choosing $a'_r = |V_r|$ for $r = 1, \dots, R$, we can write for any reference flow

$$\sum_{r=1}^{R} a'_r \mathbf{\Phi}_r = \sum_{r=1}^{R} \frac{a'_r}{|V_r|} \sum_{\mathbf{U}'_n \in V_r} \mathbf{U}'_n = \sum_{n=1}^{N} (\mathbf{U}_n - \overline{\mathbf{U}}) = \left(\sum_{n=1}^{N} \mathbf{U}_n \right) - N\overline{\mathbf{U}}.$$

If the reference flow is the snapshot mean, it follows that

$$\left(\sum_{n=1}^{N} \mathbf{U}_n \right) - N\overline{\mathbf{U}} = \sum_{n=1}^{N} \mathbf{U}_n - N \left(\frac{1}{N} \sum_{n=1}^{N} \mathbf{U}_n \right) = 0.$$

As a consequence, a reference flow different from the snapshot mean must be used. If a stationary solution of the flow problem was available, we could use this stationary solution as a reference, as done in Burkardt et al. [19]. Otherwise we could also select a snapshot from the available solution data of the start-up phase of the simulation. Taking some snapshot $\mathbf{U}_n$ as a reference flow is not recommended, because this will lead to $\mathbf{U}'_n = 0$, which results in a zero CVT mode if $\mathbf{U}'_n$ happens to be the only member of its cluster. In any case, it is recommended to check the output of Lloyd's method for linear independence.

12.3.3 Reduced-Order Velocity Model

The derivation of the Galerkin reduced-order model for the direct numerical simulation is based on the reduced basis approximation $\mathbf{U}^R$ of some discrete velocity vector $\mathbf{U}$,

$$\mathbf{U}^R = \overline{\mathbf{U}} + \sum_{r=1}^{R} a_r \mathbf{\Phi}_r \tag{12.5}$$

where $\mathbf{\Phi}_1, \dots, \mathbf{\Phi}_R$ can be POD or CVT basis functions. We assume that the discrete velocity snapshots $\mathbf{U}_1, \dots, \mathbf{U}_N$ fulfill the discrete continuity equation $C^T\mathbf{U} = 0$ sufficiently well. Using the properties (12.2) and (12.4) of the POD and CVT basis functions, one can verify that divergence-free snapshots imply divergence-free basis functions, $C^T\mathbf{\Phi}_r = 0$ for $r = 1, \dots, R$, which imply divergence-free reduced basis approximations, $C^T\mathbf{U}^R = 0$. Therefore, we concentrate solely on the discrete momentum equation

$$\mathcal{M}\dot{\mathbf{U}} = \mathcal{B}(\mathbf{U})\mathbf{U} + \mathcal{D}\mathbf{U} + \mathcal{C}P.$$

Projecting this system of equations on the reduced basis functions $\mathbf{\Phi}_1, \ldots, \mathbf{\Phi}_R$ results in

$$\mathbf{\Phi}_r^T \mathcal{M} \dot{\mathbf{U}} = \mathbf{\Phi}_r^T \mathcal{B}(\mathbf{U})\mathbf{U} + \mathbf{\Phi}_r^T \mathcal{D}\mathbf{U} + \mathbf{\Phi}_r^T \mathcal{C} P, \qquad r = 1, \ldots, R. \tag{12.6}$$

Assuming that the reduced basis functions are discretely divergence-free, it holds that $\mathbf{\Phi}_r^T \mathcal{C} P = (\mathcal{C}^T \mathbf{\Phi}_r)^T P = 0$ for any $r = 1, \ldots, R$. Substituting (12.5) into (12.6) leads to

$$\sum_{i=1}^{R} A_{ri} \dot{a}_i = \sum_{i,j=1}^{R} B_{rij} a_i a_j + \sum_{i=1}^{R} C_{ri} a_i + D_r \qquad r = 1, \ldots, R. \tag{12.7}$$

with

$$\begin{aligned}
A_{ri} &= \mathbf{\Phi}_r^T \mathcal{M} \mathbf{\Phi}_i, & r, i &= 1, \ldots, R, \\
B_{rij} &= \mathbf{\Phi}_r^T \mathcal{B}(\mathbf{\Phi}_j) \mathbf{\Phi}_i, & r, i, j &= 1, \ldots, R, \\
C_{ri} &= \mathbf{\Phi}_r^T \mathcal{B}(\overline{\mathbf{U}}) \mathbf{\Phi}_i + \mathbf{\Phi}_r^T \mathcal{B}(\mathbf{\Phi}_i) \overline{\mathbf{U}} + \mathbf{\Phi}_r^T \mathcal{D} \mathbf{\Phi}_i, & r, i &= 1, \ldots, R, \\
D_r &= \mathbf{\Phi}_r^T \mathcal{B}(\overline{\mathbf{U}}) \overline{\mathbf{U}} + \mathbf{\Phi}_r^T \mathcal{D} \overline{\mathbf{U}}. & r &= 1, \ldots, R.
\end{aligned}$$

Suitable initial conditions can be derived by projecting the initial conditions of the spatial semi-discretization on the reduced basis functions. The computing time of the reduced-order system of ordinary differential equations (12.7) is only dependent on the number of basis functions R, but not on the number of mesh nodes any more.

12.3.4 Reduced-Order Pressure Model

The pressure has been eliminated from the reduced-order model described in the last section. For the computation of the drag and lift coefficients, however, the pressure field is needed. In the following, we describe a way to compute the pressure from the solution of reduced-order velocity models. The method is based on the ideas of Rempfer [29] and Noack et al. [30], who substituted the reduced-order approximation of the velocity field in the right-hand side of a continuous pressure Poisson equation. We extend this method to the case of a discrete pressure Poisson equation. For details on continuous and discrete pressure Poisson equations, see Gresho and Sani [31].

The discrete pressure Poisson equation is derived by substituting the discrete momentum equations in the time derivative $\mathcal{C}^T \dot{\mathbf{U}} = 0$ of the discrete continuity equation, which yields

$$-\mathcal{C}^T \mathcal{M}^{-1} \mathcal{C} P = \mathcal{C}^T \mathcal{M}^{-1} (\mathcal{B}(\mathbf{U}) + \mathcal{D}) \mathbf{U}. \tag{12.8}$$

With this equation we can compute the pressure associated with a discrete velocity field $\mathbf{U}$. If we substitute a reduced basis approximation $\mathbf{U}^R$ of $\mathbf{U}$, see (12.5), and denote the resulting pressure by P^R, we can rewrite the discrete pressure Poisson equation (12.8) as

$$\begin{aligned}-\mathcal{C}^T\mathcal{M}^{-1}\mathcal{C}P^R &= \sum_{i,j=1}^{R}\mathcal{C}^T\mathcal{M}^{-1}\mathcal{B}(\mathbf{\Phi}_i)\mathbf{\Phi}_j a_i a_j \\ &+ \sum_{i=1}^{R}\mathcal{C}^T\mathcal{M}^{-1}(\mathcal{B}(\overline{\mathbf{U}})\mathbf{\Phi}_i + \mathcal{B}(\mathbf{\Phi}_i)\overline{\mathbf{U}} + \mathcal{D}\mathbf{\Phi}_i)a_i \\ &+ \mathcal{C}^T\mathcal{M}^{-1}(\mathcal{B}(\overline{\mathbf{U}})\overline{\mathbf{U}} + \mathcal{D}\overline{\mathbf{U}})\end{aligned}$$

From this equation we can deduce that P^R can be decomposed as a linear combination of partial pressures,

$$P^R = \sum_{i,j=1}^{R} P_{ij}^R a_i a_j + \sum_{i=1}^{R} P_i^R a_i + P_0^R, \tag{12.9}$$

where the partial pressures satisfy

$$\begin{aligned}-\mathcal{C}^T\mathcal{M}^{-1}\mathcal{C}P_{ij}^R &= \mathcal{C}^T\mathcal{M}^{-1}\mathcal{B}(\mathbf{\Phi}_i)\mathbf{\Phi}_j, \\ -\mathcal{C}^T\mathcal{M}^{-1}\mathcal{C}P_i^R &= \mathcal{C}^T\mathcal{M}^{-1}(\mathcal{B}(\overline{\mathbf{U}})\mathbf{\Phi}_i + \mathcal{B}(\mathbf{\Phi}_i)\overline{\mathbf{U}} + \mathcal{D}\mathbf{\Phi}_i), \\ -\mathcal{C}^T\mathcal{M}^{-1}\mathcal{C}P_0^R &= \mathcal{C}^T\mathcal{M}^{-1}(\mathcal{B}(\overline{\mathbf{U}})\overline{\mathbf{U}} + \mathcal{D}\overline{\mathbf{U}}).\end{aligned}$$

This means, we have to compute the solution of these $R^2 + R + 1$ systems of linear algebraic equations once, store the resulting partial pressures, and then we can compute the pressure fields associated with the reduced velocities cheaply using (12.9).

12.3.5 Application to LES Using Updated Coefficients

In this section we provide a reduced-order model that is based on the governing equations of a large eddy simulation instead of the equations of a direct numerical simulation. The method is not a reduced-order model in the strong sense, however, because its on-line computation time is dependent on the number of unknowns of the original simulation, the reason being the non-linearity in the eddy viscosity model.

We perform a Galerkin projection of the spatially semi-discretized weak form of the LES momentum equations on the POD or CVT modes, respectively. The resulting system of ordinary differential equations is given by

$$\sum_{i=1}^{R} A_{ri}\dot{a}_i = \sum_{i,j=1}^{R} B_{rij}a_i a_j + \sum_{i=1}^{R} C^{\mathrm{t}}_{ri}(\mathbf{U}^R)a_i + D^{\mathrm{t}}_r(\mathbf{U}^R), \quad r = 1, \ldots, R,$$

where the solution dependent model coefficients are given as

$$C^{\mathrm{t}}_{ri}(\mathbf{U}^R) = \mathbf{\Phi}_r^T \mathcal{B}(\overline{\mathbf{U}})\mathbf{\Phi}_i + \mathbf{\Phi}_r^T \mathcal{B}(\mathbf{\Phi}_i)\overline{\mathbf{U}} + \mathbf{\Phi}_r^T \mathcal{D}(\mathbf{U}^R)\mathbf{\Phi}_i, \quad r, i = 1, \ldots, R$$

$$D^{\mathrm{t}}_r(\mathbf{U}^R) = \mathbf{\Phi}_r^T \mathcal{B}(\overline{\mathbf{U}})\overline{\mathbf{U}} + \mathbf{\Phi}_r^T \mathcal{D}(\mathbf{U}^R)\overline{\mathbf{U}}, \quad r = 1, \ldots, R.$$

For linear finite elements the turbulent viscosity ν_{t} is constant within each grid cell. In this case, the expressions for the solution dependent model coefficients can be reformulated, so that the computational cost for computing them reduces to the time needed for constructing ν_{t} from the current solution of the reduced-order model plus $R^2 + R$ inner products of vectors whose length is equal to the number of grid cells. To save more computing time, we can extrapolate the solution-dependent coefficients over a short time period before updating them. To this end, we choose a sequence of update times. During the time integration, at the update times we compute and store the coefficients, and between the update times we reconstruct the coefficients by quadratic extrapolation using previously stored values. For the first time steps, where we do not have enough data for a quadratic extrapolation, we use a full computation of the coefficients for each evaluation of the right-hand side [4].

12.3.6 Application to LES Using Calibration

In this section we formulate a reduced-order model for LES that has the same on-line computational cost as the standard reduced-order model based on a direct simulation of the incompressible Navier–Stokes equations. The method uses a calibration of the model coefficients of the constant and linear terms of the reduced-order model, so that it mimics the behavior of the LES, even if it does not contain the non-linear dynamics of the eddy viscosity model.

Let $a_{r,n} = (\mathbf{U}_n - \overline{\mathbf{U}}, \mathbf{\Phi}_r)_{\mathcal{M}}$ and let $\hat{a}_{r,n}$ be the solution of a reduced-order model at t_n, for $r = 1, \ldots, R$ and $n = 1, \ldots, N$. Further, let $a_{r,\mathrm{s}}$ and $\hat{a}_{r,\mathrm{s}}$ be respective cubic spline interpolations for $r = 1, \ldots, R$, so that

$$\mathbf{U}^R_{\mathrm{s}} = \overline{\mathbf{U}} + \sum_{r=1}^{R} a_{r,\mathrm{s}}\mathbf{\Phi}_r, \qquad \hat{\mathbf{U}}^R_{\mathrm{s}} = \overline{\mathbf{U}} + \sum_{r=1}^{R} \hat{a}_{r,\mathrm{s}}\mathbf{\Phi}_r.$$

A cost functional that measures the difference between the projected solution and the model solution is given as

$$\mathcal{J} = \int_0^T \left\| \mathbf{U}^R_{\mathrm{s}} - \hat{\mathbf{U}}^R_{\mathrm{s}} \right\|^2_{\mathcal{M}} \mathrm{d}t = \int_0^T \sum_{i,r=1}^{R} (a_{i,\mathrm{s}} - \hat{a}_{i,\mathrm{s}})A_{ir}(a_{r,\mathrm{s}} - \hat{a}_{r,\mathrm{s}})\, \mathrm{d}t.$$

We want to let a gradient-based optimization routine find the model coefficients C_{kl} and D_k, for $k, l = 1, \ldots, R$, that minimize the cost functional. In each step of the optimization routine we have to provide $\mathcal{J}$ and its derivatives with respect to the model coefficients. To compute the derivatives we apply adjoint techniques [32] to our problem. One can derive the expressions

$$\frac{D\mathcal{J}}{DC_{kl}} = \int_0^T \hat{b}_k \hat{a}_l \, \mathrm{d}t, \qquad \frac{D\mathcal{J}}{DD_k} = \int_0^T \hat{b}_k \, \mathrm{d}t,$$

where $\hat{b}_1, \ldots, \hat{b}_R$ are the solutions of the adjoint reduced-order model

$$-\sum_{r=1}^{R} \dot{\hat{b}}_r A_{ri} = \sum_{r,j=1}^{R} \hat{b}_r (B_{rij} + B_{rji}) \hat{a}_j + \sum_{r=1}^{R} (\hat{b}_r C_{ri} + 2A_{ir}(a_{r,s} - \hat{a}_{r,s})),$$

$$\hat{b}_i(T) = 0, \qquad i = 1, \ldots, R.$$

While in theory $\hat{a}_1, \ldots, \hat{a}_R$ and $\hat{b}_1, \ldots, \hat{b}_R$ are exact solutions of the reduced-order models, in practice they can be replaced by dense output or spline interpolations of a numerical solution.

In each iteration of the optimization routine we are given a set of model coefficients, for which we perform the following steps:

1. Solve the reduced-order model forward in time.
2. Solve the adjoint reduced-order model backward in time.
3. Compute the values of the functional and its derivatives.

By using the adjoint of the reduced-order model, the calibration can be done much faster than by using a finite difference approximation of the gradients of the cost functional, as the computational cost in each step of the optimization is mainly one forward and one backward solution.

12.3.7 Results for Laminar Vortex-Shedding Flow

The accuracy of the reduced-order models was studied for a two-dimensional simulation of a flow around a circular cylinder at a Reynolds number of $Re = 100$. The domain with boundary conditions is sketched in Fig. 12.5, where Γ_{N}^i denotes Neumann conditions and Γ_{D}^i denotes Dirichlet conditions for u^i.

We used the geometric parameters $D_{\mathrm{cyl}} = 0.1$ m, $L = 3$ m and $H = 2$ m, the inflow velocity was set to $u_{\mathrm{in}} = 1\,\mathrm{m/s}$ and the kinematic viscosity was set to $\nu = 0.001\,\mathrm{m/s^2}$. The simulation was performed with the finite element solver Kardos [5], using stabilized linear finite elements [6] and the ROS3P time integration method [7]. For the spatial discretization we used a triangular mesh with 79,723 mesh nodes, locally refined near the cylinder. The time was discretized with

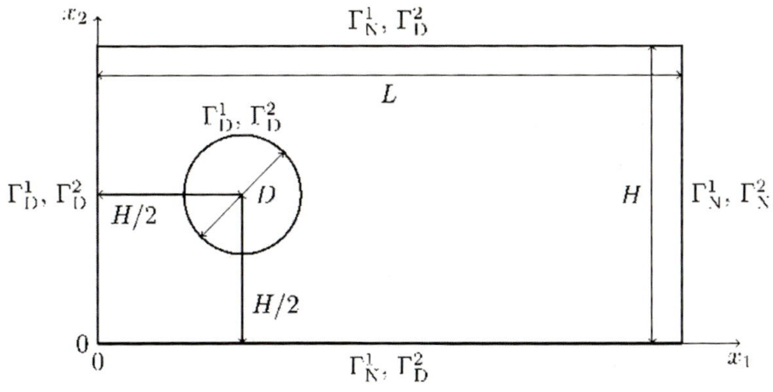

Fig. 12.5 Sketch of the geometry (not to scale) and boundary conditions of the two-dimensional flow around a circular cylinder

Fig. 12.6 First snapshot of the absolute velocity field of the two-dimensional numerical solution

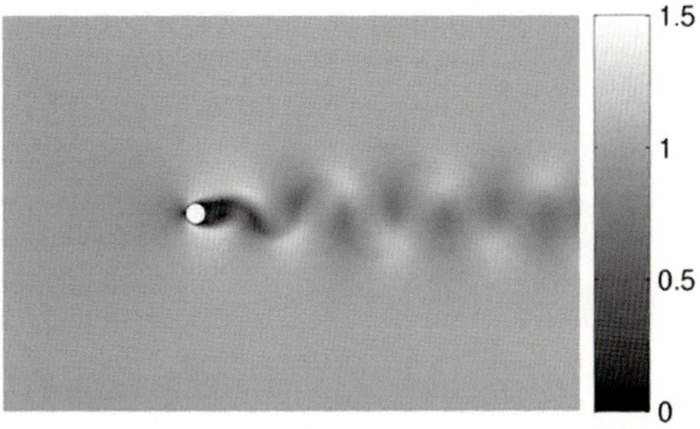

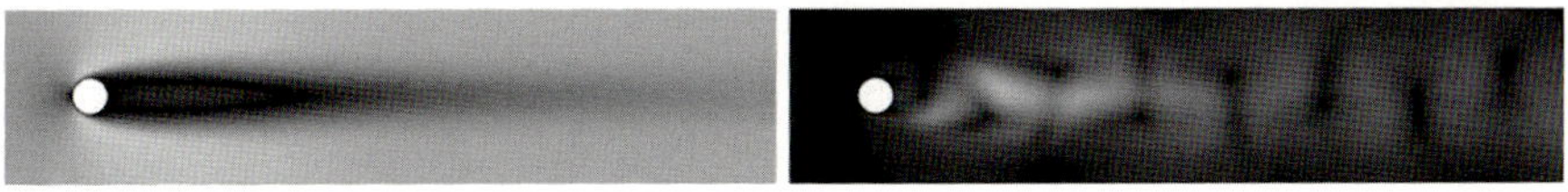

Fig. 12.7 Central part of the absolute velocity field of the reference flow, sampled at t = 3.89 s (*left*). Snapshot fluctuation at t = 20 s with respect to the reference flow (*right*). The same color scale as in Fig. 12.6 is used for both images

a constant step size of $\Delta t = 0.001$ s, but only every tenth solution was stored. The simulation was started with a fluid at rest. After a transient simulation time of less than 20 s the solution became periodic. A snapshot of the solution at $t = 20$ s is presented in Fig. 12.6, which demonstrates the appearance of a regularly shaped von Kármán vortex street.

We chose the snapshot at $t = 3.89$ s as a reference flow $\overline{\mathbf{U}}$, which is shown in Fig. 12.7 together with the snapshot fluctuation at $t = 20$ s with respect to the reference flow. Note that only a cut-out of the domain is plotted in Fig. 12.7 and in the following figures. Using the snapshot fluctuations at times $t = 20.00$ s, 20.01 s, $\ldots$, 20.58 s, corresponding to one shedding cycle, we created

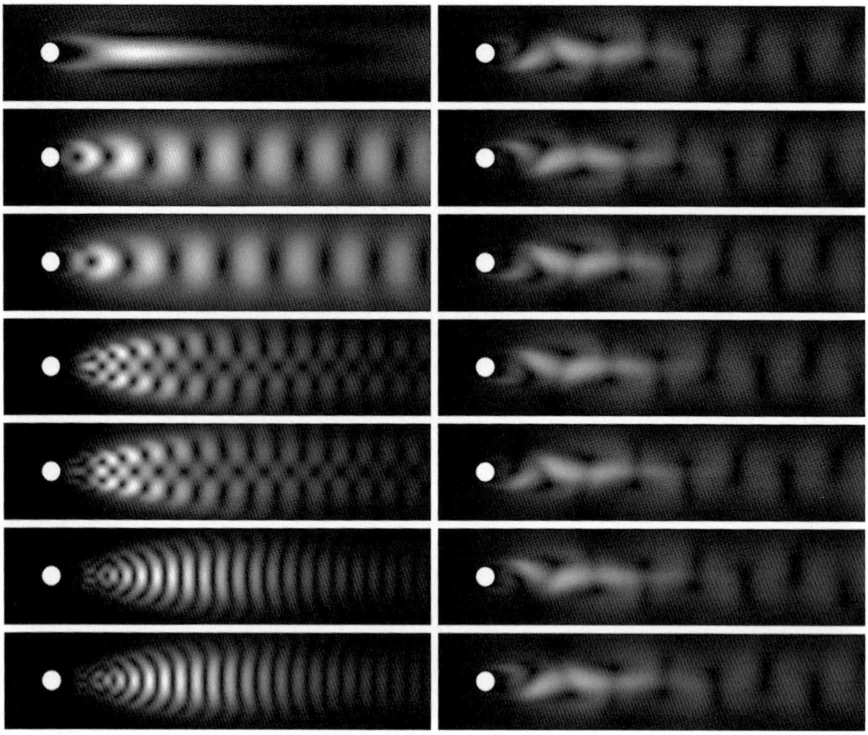

Fig. 12.8 Modes of a POD (*left*) of rank 7 and a CVT (*right*) of rank 7. The plots show the absolute values of the modal vector fields

sets of POD and CVT reduced basis functions. Figure 12.8 displays the reduced basis functions of a 7 mode POD and a 7 mode CVT.

To compare the snapshots with their reduced-order representations a few definitions are required: Let a set of POD or CVT reduced basis functions $\mathbf{\Phi}_1, \ldots, \mathbf{\Phi}_R$ be given for some fixed rank R. Let $\mathbf{U}_1, \ldots, \mathbf{U}_N$ and $P_1, \ldots, P_N$ be velocity and pressure solution vectors, respectively, sampled at $t_1, \ldots, t_N$. These solution vectors may be different from the snapshots with which the reduced basis functions were created. Using the reduced-order coefficients

$$a_{r,n} = (\mathbf{U}_n - \overline{\mathbf{U}}, \mathbf{\Phi}_r)_{\mathcal{M}}, \quad n = 1, \ldots, N, \quad r = 1, \ldots, R,$$

we define the reduced-order approximations, based on (12.5) and (12.9), as

$$\mathbf{U}_n^R = \overline{\mathbf{U}} + \sum_{r=1}^{R} a_{r,n} \mathbf{\Phi}_r, \quad n = 1, \ldots, N,$$

$$P_n^R = \sum_{i,j=1}^{R} P_{ij}^R a_{i,n} a_{j,n} + \sum_{i=1}^{R} P_i^R a_{i,n} + P_0^R, \quad n = 1, \ldots, N.$$

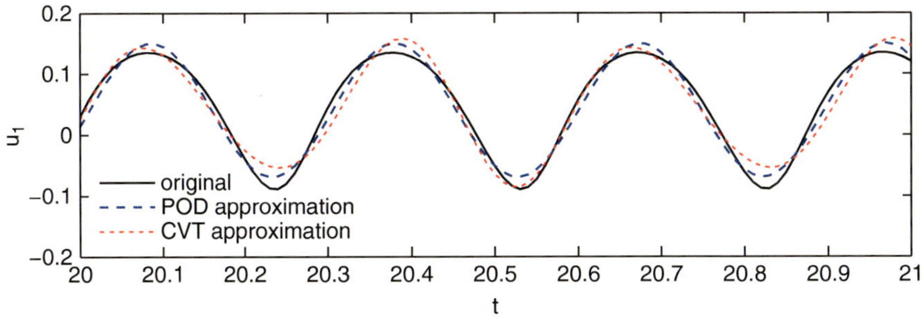

Fig. 12.9 Comparison of the velocity components u_1 at the monitoring point $\mathbf{x} = (1.2, 1)$ for the original simulation, the POD approximation of rank 7 and the CVT approximation of rank 7

We used the reference flow shown in Fig. 12.7 and the reduced basis functions shown in Fig. 12.8 to approximate the velocity field from time $t = 20$ s to time $t = 21$ s. We chose a monitoring point $\mathbf{x} = (1.2, 1)$ in the wake of the flow. In Fig. 12.9 the time dependent streamwise velocity component at the monitoring point is compared to the reduced-order approximations of rank 7. The POD and CVT basis functions have lead to different approximations at the monitoring point, but the approximation quality was similar.

To compare the solutions of the reduced-order model with the reduced-order approximations, we denote the numerical solutions of the reduced-order model at times $t_1, \ldots, t_N$ as $\hat{a}_{r,n}$ for $r = 1, \ldots, R$ and $n = 1, \ldots, N$. We define the reduced-order modeled velocity and pressure fields, based on (12.5) and (12.9) as

$$\hat{\mathbf{U}}_n^R = \overline{\mathbf{U}} + \sum_{r=1}^{R} \hat{a}_{r,n} \mathbf{\Phi}_r, \quad n = 1, \ldots, N,$$

$$\hat{P}_n^R = \sum_{i,j=1}^{R} P_{ij}^R \hat{a}_{i,n} \hat{a}_{j,n} + \sum_{i=1}^{R} P_i^R \hat{a}_{i,n} + P_0^R, \quad n = 1, \ldots, N.$$

Now we define, for $n = 1, \ldots, N$, the L^2 model, approximation and total errors, respectively, as

$$E_{n,R}^{\mathbf{U},\mathrm{m}} = \left\| \hat{\mathbf{U}}_n^R - \mathbf{U}_n^R \right\|_{\mathcal{M}},$$

$$E_{n,R}^{\mathbf{U},\mathrm{a}} = \left\| \mathbf{U}_n^R - \mathbf{U}_n \right\|_{\mathcal{M}},$$

$$E_{n,R}^{\mathbf{U},\mathrm{t}} = \left\| \hat{\mathbf{U}}_n^R - \mathbf{U}_n \right\|_{\mathcal{M}}.$$

We also measure the errors in the drag force F_{D} and the lift force F_{L} acting on the cylinder. These forces are given as

$$F_{\mathrm{D}}(p, \mathbf{u}) = \int_S (p n_1 - \nu (\nabla u_1 + \partial_{x_1} \mathbf{u}) \cdot \mathbf{n}) \, \mathrm{d}S,$$

$$F_{\mathrm{L}}(p, \mathbf{u}) = \int_S (p n_2 - \nu (\nabla u_2 + \partial_{x_2} \mathbf{u}) \cdot \mathbf{n}) \, \mathrm{d}S,$$

where S is the surface of the cylinder and $\mathbf{n}$ points outside of the domain. By using the finite element approximations of p and $\mathbf{u}$ we can write the drag and lift forces as functions of the nodal vectors, $F_{\mathrm{D}}(P, \mathbf{U})$ and $F_{\mathrm{L}}(P, \mathbf{U})$, respectively. Now we define the drag model, approximation and total errors, respectively, as

$$E_{n,R}^{\mathrm{D,m}} = F_{\mathrm{D}}(\hat{P}_n^R, \hat{\mathbf{U}}_n^R) - F_{\mathrm{D}}(P_n^R, \mathbf{U}_n^R),$$

$$E_{n,R}^{\mathrm{D,a}} = F_{\mathrm{D}}(P_n^R, \mathbf{U}_n^R) - F_{\mathrm{D}}(P_n, \mathbf{U}_n),$$

$$E_{n,R}^{\mathrm{D,t}} = F_{\mathrm{D}}(\hat{P}_n^R, \hat{\mathbf{U}}_n^R) - F_{\mathrm{D}}(P_n, \mathbf{U}_n).$$

The lift errors are defined in an analogous way. To measure the errors over all time steps we use the error norm

$$\|E_R\| = \sqrt{\sum_{n=1}^{N} E_{n,R}^2},$$

where $E_{n,R}$ can be any of the L^2, drag or lift errors defined above.

We created POD and CVT reduced-order models of the cylinder flow simulation, using the snapshots at $t = 20.00$ s, 20.01 s, $\ldots, 20.58$ s and the reference flow at $t = 3.89$ s. The ranks R of the reduced bases were varied from 1 to 25. Reduced-order approximations and solutions of the reduced-order model were computed within the larger time interval $t \in [20, 30]$.

The L^2 velocity errors and the drag and lift errors are presented in Fig. 12.10. It turns out that the POD and CVT models behave qualitatively similar.

In the L^2 case the modeling error is relatively close to the approximation error, which means that the reduced-order model is nearly optimal. Around $R = 25$, however, the modeling error stops to decrease any further. By comparing different mesh and time step sizes of the original simulations it could be verified that this accuracy limit of the reduced-order model is determined by the numerical errors of the snapshots. For the drag and lift computations, the approximation error first drops rapidly and then stays nearly constant for $R \geq 10$. This behavior can be explained by the fact that the reduced-order approximation of the pressure field was not obtained by a projection, but by solving a set of pressure Poisson equations. The computation of the snapshots, however, did not involve a pressure Poisson equation, so even the reduced-order approximations of the drag and lift forces contain some modeling error.

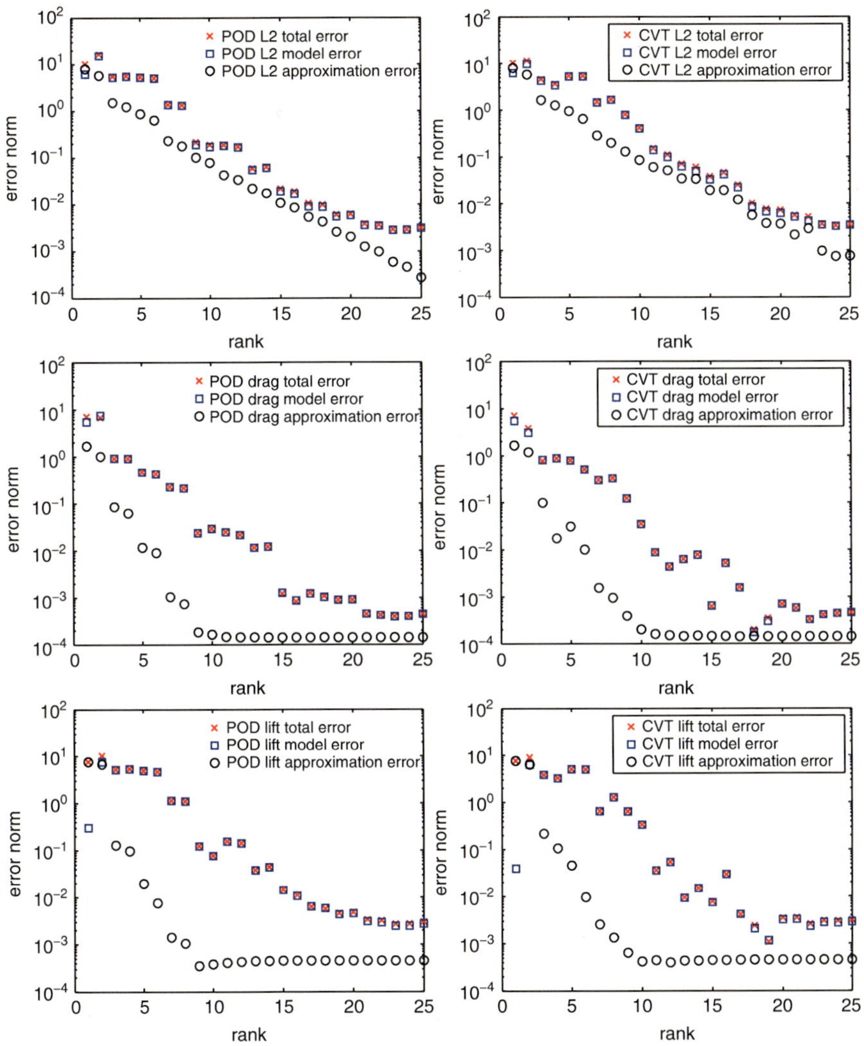

Fig. 12.10 Errors norms, depending on the rank of the reduced bases, using POD (*left*) and CVT (*right*) basis functions. From *top* to *bottom*: L^2 velocity error, drag error, lift error

12.3.8 Results for Turbulent Flow

In this section we study the accuracy of the reduced-order models for a three-dimensional large eddy simulation performed with the finite element software Kardos [5]. The domain was obtained by extruding the two-dimensional geometry of the preceding section in the third dimension by a depth of 0.3 m. We chose an

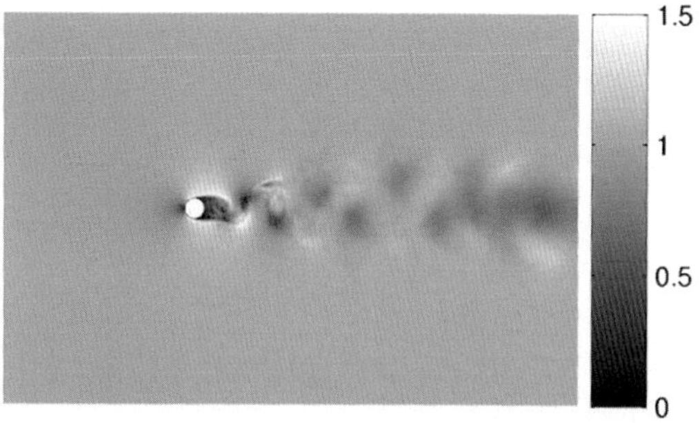

Fig. 12.11 First snapshot of the absolute Smagorinsky-filtered velocity field of a three-dimensional large eddy simulation

Fig. 12.12 Central part of the absolute velocity field of the reference flow (*left*). Central part of the absolute velocity field of the fluctuation based on the reference flow and the first snapshot (*right*)

inflow velocity of $u_{\text{in}} = 1$ m/s and a kinematic viscosity of 1/39,000 m/s^2, so that the Reynolds number became $Re = 3{,}900$. For the modeling of the turbulent stress tensor we used a Smagorinsky constant of $C_S = 0.15$. The tetrahedral mesh was manually refined near the cylinder and in the near-cylinder wake region, resulting in a number of 123,553 mesh nodes. For the time discretization a constant step size of 0.002 s was chosen. The initial velocity field was taken from a simulation that was performed with an adaptive time stepping, starting from a fluid at rest and running through the transient flow phase until a developed turbulent flow was reached. A snapshot of the initial velocity field is shown in Fig. 12.11.

After computing the solution of the LES within a time interval of 4 s, we selected the 1,000 available LES solutions from $t_1 = 0$ s to $t_{1000} = 1.998$ s as snapshots and picked a solution from the transient start-up simulation as a reference flow. In Fig. 12.12 the reference velocity field and a velocity fluctuation around the reference are shown. The reason for not including all available numerical solutions up to $t = 4$ s in the snapshot data base was to enable the assessment of the reduced-order models with respect to reproducing 'known' snapshots compared to reproducing 'unknown' snapshots. While the first case is a necessary premise for the success of the methods, the latter case will be more significant for real-life applications, where the reduced model is to be used as a surrogate for the original model.

The snapshot fluctuations around the reference flow were used to generate POD and CVT reduced basis functions. In Fig. 12.13 the reduced basis functions of a 7 mode POD are compared with a 7 mode CVT.

As a visual criterion for the comparison of the different reduced-order models for large eddy simulations we took the time evolution of the streamwise velocity component at the monitoring point $(1.2, 1, 0.15)$ from $t = 0$ s up to $t = 4$ s. In Fig. 12.14 the different reduced-order solutions of rank 64 are compared to the original simulation results.

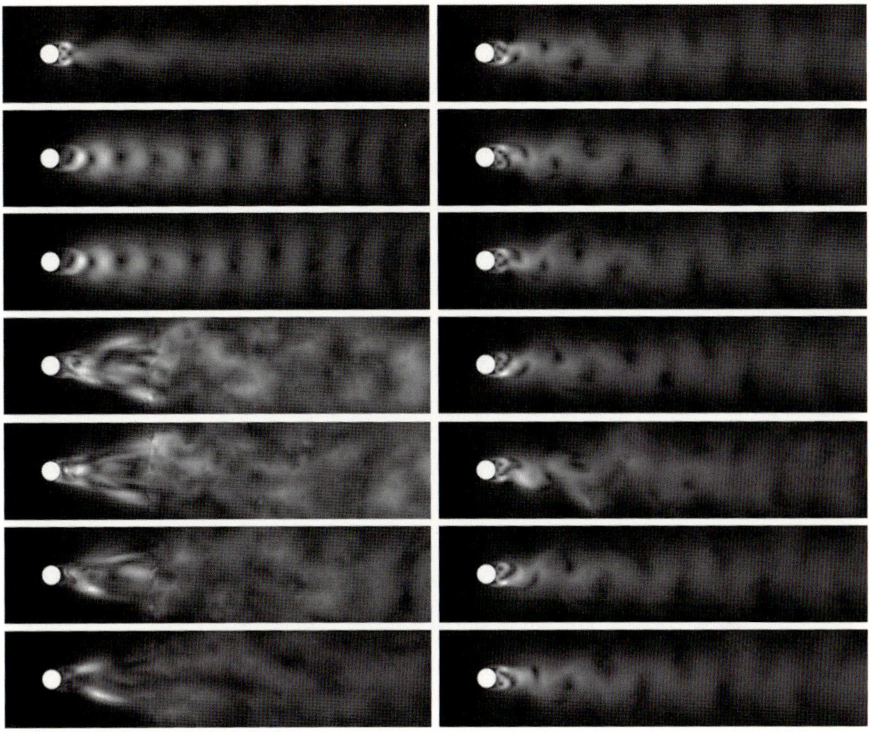

Fig. 12.13 Modes of a velocity POD (*left*) of rank 7 and a velocity CVT (*right*) of rank 7. The plots show the absolute values of the vector fields

In the first plot of Fig. 12.14, the projection of the solution on the reduced-basis functions is shown, which serves as a reference for the other models, as the projection is the L^2 optimal approximation of the original velocity field by a linear combination of the chosen reduced-basis functions. It can be observed that the first half of the snapshots, which formed the snapshot data base for building the reduced basis functions, are approximated well by the projected snapshots, while the other half is approximated poorly. The difference between the projection on the POD and on the CVT basis functions is relatively small over the whole time interval.

The second plot of Fig. 12.14 shows the solution of the DNS reduced-order models applied to the LES test case. The model solutions quickly diverge from the original simulation results and exhibit strong oscillations. The results of the POD and CVT are different, but qualitatively similar.

The solution of the models with updated coefficients, based on the governing equations of the LES, is shown in the third plot of Fig. 12.14. The model coefficients were updated at every 10th snapshot time. While the models are able to capture the dynamics of the snapshot phase well, they fail to approximate the dynamics of the unknown phase, due to the inability of the reduced basis functions to approximate these solutions. The accuracies of the POD and CVT models are similar.

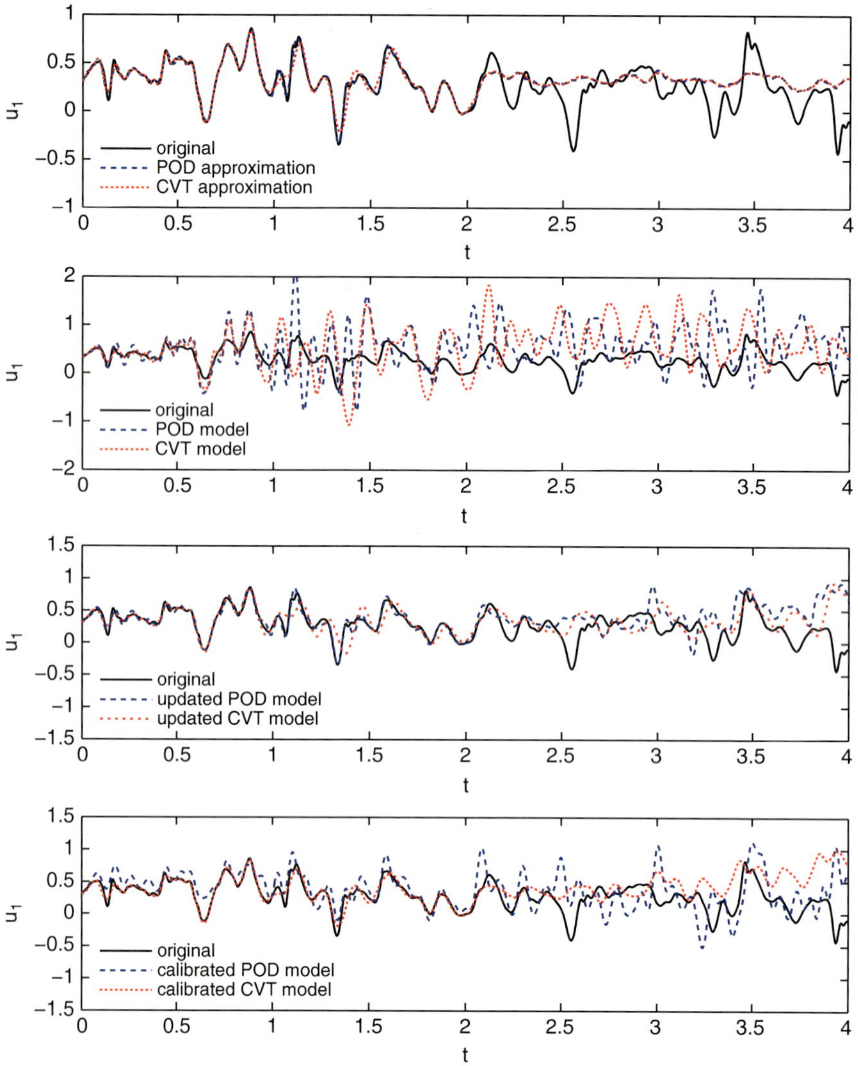

Fig. 12.14 Comparison of the streamwise velocity component u_1 of the original simulation with the POD and CVT reduced-order solutions of rank 64 at the monitoring point $\mathbf{x} = (1.2, 1, 0.15)$. For $t \geq 2$, the solutions are not contained in the snapshot data base used for constructing the reduced basis function. From *top* to *bottom*: projection of the velocity on the reduced basis functions, solution of reduced DNS model, solution of the reduced LES model, solution of the reduced DNS model calibrated to the snapshot data base

In the last plot of Fig. 12.14 we show the results of the calibrated models. The models are based on the equations of a direct numerical simulation, but the coefficients were calibrated with respect to the snapshots that were used to generate the basis functions. Within the first half of the time interval the performance is similar to the LES reduced-order model, in the second half the solutions are less accurate. In the calibrated case, the CVT model is more accurate than the POD model.

Finally, we compare the computation times of the updated and the calibrated models. Creating the updated POD and CVT models of rank 64 from the snapshot set took 1,712 s and 3,863 s of wall-clock time, respectively, while running the models took 5,335 s and 4,621 s. Creating the calibrated models took 15,301 s and 32,588 s for POD and CVT, respectively, but their time integrations took merely 21 s and 26 s. The time integration of the finite element model over the same time interval took almost 5 weeks. All timings were obtained using one computational core of a 3.0 GHz AMD Opteron processor.

12.4 Summary and Conclusions

In the first part of this paper we have presented a mesh adaptation method based on LES-specific design criteria in order to improve the quality of statistical flow properties. The heart of the method is an iterative moving mesh strategy that balances statistically averaged values over the whole spatial grid. As a test problem the flow over periodic hills at $Re = 10{,}595$ was considered. Inspired by the remarkably good results for the mean streamwise velocity reported in Hertel et al. [2], we have combined this quantity with the statistically averaged turbulent kinetic energy and shear stresses within the framework of balanced monitor functions, following the approach developed by van Dam [22]. The key idea is to improve the local resolution of these important LES quantities through a higher concentration of grid points in areas where they are insufficiently resolved.

Based on our numerical experiments for the periodic hill flow we can draw the following main conclusions: (i) Significant improvement compared to the initial grid approximation can be observed for all combinations of LES-specific quantities. (ii) Even though the additional consideration of the turbulent kinetic energy and part of the shear stress to steer the moving mesh approach can locally improve the flow resolution with respect to specific quantities such as the reattachment point, the overall and well balanced quality of the streamwise velocity cannot be beaten. Further studies on different flows are planned for future research.

In the second part of this work, reduced-order models were studied for a laminar direct numerical simulation and a turbulent large eddy simulation of the flow around a cylinder. The reduced-order models for the velocity fields were obtained by a Galerkin projection of the semi-discretized flow problems onto relatively small sets of global basis functions. The reduced-order models for the pressure fields were derived by using a discrete pressure Poisson equation. Two different methods were used to generate basis functions for the projection, namely the proper orthogonal decomposition and the centroidal Voronoi tessellation. The solutions of the resulting reduced-order models were compared by their accuracy with respect to the solutions of the underlying finite element simulations.

For the DNS test case, the errors in the reduced-order solutions could be diminished to the order of magnitude of the numerical error of the original finite element solution. These results were possible with using about 25 POD or CVT modes, which validated the applicability and correctness of the models. The POD and CVT results did not differ much with respect to the error in the velocity field and in the time-dependent drag and lift forces. The findings indicate that both methods are similarly suitable for the construction of reduced-order models.

For the LES test case, besides the different types of basis functions, we compared two different approaches for modeling the dynamics: a model based on the LES equations and a calibrated DNS model. Both models succeeded to capture the 'known' solutions that were present in the snapshot database used for the creation of the reduced basis functions, but they failed to capture 'unknown' solutions that were not present in the snapshot database. As in the laminar case, the differences between the POD and CVT based models turned out to be rather small. Comparing the computational times, running the calibrated models took much less time than running the updated models, but the setup phase of the calibrated models was longer.

The fact that the turbulent reduced-order models could not reproduce 'unknown' snapshots highlights an important issue of reduced-order modeling in a turbulent context. While a lot can be done to improve the dynamics of the POD and CVT models, as shown in this work, the output of the models can, by definition, only be a linear combination of snapshots. Therefore a sufficiently large number of snapshots and, consequently, a large number of reduced basis functions must be provided to enable accurate computations of turbulent flows.

Acknowledgements The authors acknowledge the financial support from the German Research Council (DFG) through the SFB568. We would also like to thank Jochen Fröhlich and Claudia Hertel (TU Dresden) for making the turbulent flow solver LESOCC2 available to us and for their kind programming support during our numerical studies.

References

Project-Related Publications

1. Löbig, S., Dörnbrack, A., Fröhlich, J., Hertel, C., Kühnlein, C., Lang, J.: Towards large eddy simulation on moving grids. Proc. Appl. Math. Mech. **9**, 445–446 (2009)
2. Hertel, C., Schümichen, M., Löbig, S., Fröhlich, J., Lang, J.: Adaptive large eddy simulation with moving grids. Preprint Technische Universität Dresden, accepted for publication in Theoretical and Computational Fluid Dynamics (2012)
3. Lang, J., Cao, W., Huang, W., Russell, R.D.: A two-dimensional moving finite element method with local refinement based on a posteriori error estimates. Appl. Numer. Math. **46**, 75–94 (2003)
4. Ullmann, S., Lang, J.: A POD-Galerkin reduced model with updated coefficients for Smagorinsky LES. In: Pereira, J.C.F., Sequeira, A., Pereira, J.M.C. (eds) Proceedings of the V European Conference on Computational Fluid Dynamics ECCOMAS CFD 2010, Lisbon, Portugal (2010)

5. Erdmann, B., Lang, J., Roitzsch, R.: Kardos user's guide. ZIB-Report 02–42, ZIB (2002)
6. Lang, J.: Adaptive incompressible flow computations with linearly implicit time discretization and stabilized finite elements. In: Papailiou, K., Tsahalis, D., Periaux, J., Hirsch, C., Pandolfi, M. (eds.) Computational Fluid Dynamics '98. Chichester, New York (1998)
7. Lang, J., Verwer, J.: ROS3P—an accurate third-order Rosenbrock solver designed for parabolic problems. BIT Numer. Math. **41**, 730–737 (2001)

Other Publications

8. Huang, W., Russell, R.D.: Adaptive Moving Mesh Methods, 1st edn. Springer, New York (2011)
9. Hertel, C., Fröhlich, J.: Error reduction in LES via adaptive moving grids, QLES II, Pisa, Italien. In: M.-V. Salvetti et al. (Hsg.) Proceedings: Quality and Reliability of Large-Eddy Simulations II, Springer, 9–11 September 2009
10. Holmes, P., Lumley, J., Berkooz, G.: Turbulence, Coherent Structures, Dynamical Systems and Symmetry. Cambridge University Press, Cambridge (1996)
11. Sirovich, L.: Turbulence and the dynamics of coherent structures, parts I, II and III. Q. Appl. Math. **45**, 561–571 (1987)
12. Rempfer, D.: On low-dimensional Galerkin models for fluid flow. Theor. Comput. Fluid Dyn. **14**(2), 75–88 (2000)
13. Bergmann, M., Bruneau, C., Iollo, A.: Enablers for robust POD models. J. Comput. Phys. **228**(2), 516–538 (2009)
14. Telib, H., Manhart, M., Iollo, A.: Analysis and low-order modeling of the inhomogeneous transitional flow inside a T-mixer. Phys. Fluids **16**, 2717–2731 (2004)
15. Buffoni, M., Camarri, S., Iollo, A., Salvetti, M.: Low-dimensional modelling of a confined three-dimensional wake flow. J. Fluid Mech. **569**, 141–150 (2006)
16. Couplet, M., Basdevant, C., Sagaut, P.: Calibrated reduced-order POD-Galerkin system for fluid flow modeling. J. Comput. Phys. **207**, 192–220 (2005)
17. Wang, Z., Akhtar, I., Borggaard, J., Iliescu, T.: Two-level discretizations of nonlinear closure models for proper orthogonal decomposition. J. Comput. Phys. **230**, 126–146 (2011)
18. Du, Q., Faber, V., Gunzburger, M.: Centroidal Voronoi tessellations: applications and algorithms. SIAM Rev. **41**(4), 637–676 (1999)
19. Burkardt, J., Gunzburger, M., Lee, H.C.: POD and CVT-based reduced-order modeling of Navier–Stokes flows. Comput. Methods Appl. Mech. Eng. **196**(1–3), 337–355 (2006)
20. Smagorinsky, J.: General circulation experiments with the primitive equations, I, The basic experiment. Mon. Weather Rev. **91**, 99–164 (1963)
21. Winslow, A.M.: Numerical solution of the quasilinear poisson equation in a nonuniform triangle mesh. J. Comput. Phys. **2**, 149–172 (1967)
22. van Dam, A.: Go with the flow. In: Ph.D. Thesis, Utrecht University (2009)
23. Ferziger, J.H., Peric, M.: Computational Methods for Fluid Dynamics, 3rd edn. Springer, Berlin (2002)
24. Demirdzic, I., Peric, M.: Space conservation law in finite volume calculations of fluid flow. Int. J. Numer. Methods Fluids **8**, 1037–1050 (1988)
25. Stone, H.L.: Iterative solution of implicit approximation of multidimensional partial differential equations. SIAM J. Numer. Anal. **5**, 530–558 (1968)
26. Fröhlich, J., Mellen, C.P., Rodi, W., Temmermann, L., Leschziner, M.A.: Highly resolved large-eddy simulation of separated flow in a channel with streamwise periodic constrictions. J. Fluid Mech. **526**, 19–66 (2005)
27. Berselli, L., Iliescu, T., Layton, M.: Mathematics of Large Eddy Simulation of Turbulent Flows, 1st edn. Springer, Heidelberg/Berlin (2006)

28. Lloyd, S.P.: Least squares quantization in PCM. IEEE Trans. Inform. Theory **28**(2), 129–137 (1982)
29. Rempfer, D.: Investigations of boundary layer transition via Galerkin projection on empirical eigenfunctions. Phys. Fluids **8**(1), 175–188 (1996)
30. Noack, B., Papas, P., Monkewitz, P.A.: The need for a pressure-term representation in empirical Galerkin models of incompressible shear flows. J. Fluid Mech. **523**, 339–365 (2005)
31. Gresho, P.M., Sani, R.L.: Incompressible Flow and the Finite Element Method. Wiley, New York (2000)
32. Gunzburger, M.: Perspectives in Flow Control and Optimization. SIAM, Philadelphia (2003)

Chapter 13
Efficient Numerical Multilevel Methods for the Optimization of Gas Turbine Combustion Chambers

S. Ulbrich and R. Roth

Abstract In this paper we present an approach for the optimization of turbulent flows. To accomplish such a complex task, the general strategy has to be carefully designed. On the optimization side, we incorporate multilevel optimization algorithms. With this kind of algorithms, different levels describing a problem can be efficiently used for the optimization. Typical examples are discretization levels or models of different physical fidelity. Many optimization algorithms rely on gradient information, which is generally not available for complex problems described by partial differential equations (PDEs). Nevertheless, gradient information can be obtained from computer programs by the use of Automatic Differentiation (AD) techniques. We present a discrete adjoint approach, which was applied to the flow solver FASTEST. The numerical results show the efficiency of the adjoint mode and the optimization algorithms. They include shape optimization and boundary control examples for the Navier-Stokes Equations (NSE), Large Eddy Simulation (LES) and Reynolds Averaged Navier-Stokes (RANS) Equations.

Keywords Discrete Adjoint • Optimal Control • Multilevel Optimization • Shape Optimization

13.1 Introduction

Optimization problems governed by flows typically have a large number of design variables, such that derivative-free optimization algorithm are not efficient. In contrast, gradient based optimization algorithms have shown to be able to complete

S. Ulbrich • R. Roth (✉)
Nonlinear Optimization, Department of Mathematics, Technische Universität Darmstadt, Dolivostr. 15, 64293 Darmstadt, Germany
e-mail: ulbrich@mathematik.tu-darmstadt.de; rroth@mathematik.tu-darmstadt.de

J. Janicka et al., *Flow and Combustion in Advanced Gas Turbine Combustors*, Fluid Mechanics and Its Applications 102, DOI 10.1007/978-94-007-5320-4_13,

the optimization task in an effort, that can be estimated within several flow solves, independent of the number of design variables. However, this was only achieved for model problems and stationary flows.

In the last few years, multilevel optimization techniques have been proven to be able to handle large scale optimization problems [3]. These techniques exploit a hierarchical structure describing a problem and the different levels are used for the optimization procedure. In the context of flow control, there are many opportunities for the application of this kind of methods. Efficient solvers almost always use multigrid techniques for the solution of the nonlinear systems. Therefore a hierarchy involving grid levels is directly at hand for the optimization. A completely different multilevel structure arises from different mathematical models describing the same underlying problem. In computational fluid dynamics (CFD) for example, turbulence modeling provides models of varying physical fidelity and computational effort. These models are often used in surrogate modeling [1] approaches. Closely related is the work of subproject D4 with the use of POD models. These models try to build a low dimensional model from snapshots of the flow field.

A special multilevel optimization approach is motivated by the Full Approximation Scheme (see [4, p.98 ff]), known from nonlinear multigrid methods. These optimization algorithms solely work with function and gradient evaluations and therefore are very versatile, in contrast to other methods, that explicitly use the structure of the underlying PDE. Based on different discretization levels or different models, first order accurate models are constructed and used for the optimization. Two well known algorithms are MG/OPT by Nash [26] and Lewis and Nash [23] and the RMTR by Gratton et al. [12–14]. We will concentrate on the RMTR, since it provides a natural treatment of bound constraints. Most of the developments are also applicable to MG/OPT.

Many optimization algorithms, as well as the RMTR, rely on gradient information. Usually this is not directly at hand for complex problems described by PDEs. For the evaluation of a target function for a certain design a highly specialized solver is used. Often the computational code of the solver is already at hand, but lacks gradient information needed for the optimization. This gradient information can be obtained with the *sensitivity* or the *adjoint approach*.

For the sensitivity approach, linearized state equations have to be solved. The total number of additional equations equals the number of design variables. The method performs well on problems with a small number of design variables, but the effort increases linearly with the number of design variables. For the adjoint approach in contrast, only one linear equation for the whole gradient has to be solved. Therefore it always outperforms the sensitivity approach for large scale optimization problems.

The general approach, where sensitivity or adjoint equations are derived on the discrete level is called *discretize-then-optimize* approach. The gradient information can then be obtained with AD techniques [17]. There is also a second possibility, called *optimize-then-discretize* approach. For optimize-then-discretize, we first derive the optimality system on the continuous level and then discretizes the PDEs, see [20] for an introduction. A major drawback of this approach is that it leads to inconsistent gradient information if no care is taken and one has to use special algorithms to control the error, see Ziems and Ulbrich [37]. Both approaches have advantages and disadvantages. For a comparison see Griesse and Walther [15].

Within the SFB 568, the authors had access to the source code of the flow solver FASTEST [9]. In close cooperation with subproject D2, gradients have been implemented via an adjoint mode, described in Sect. 13.3. We therefore have access to gradient information for various turbulence models and discretization levels. To make full use of the information available, a versatile optimization algorithm is needed. We therefore considered the $RMTR_\infty$ algorithm, presented in Sect. 13.2.

We are interested in optimization problems for unsteady flows governed by the incompressible Navier-Stokes equations involving possibly a turbulence model. An example for the distributed control of the Smagorinsky or Germano LES model by a volume force is given by

$$\min_{\mathbf{y},\mathbf{u}} \quad f(\mathbf{y},\mathbf{u}) \quad \text{subject to}$$

$$\begin{aligned}
&\mathbf{v}_t + (\mathbf{v}\cdot\nabla)\,\mathbf{v} - \nu\Delta\mathbf{v} - \nabla\cdot R\left(\nabla^s\mathbf{v}\right) + \tfrac{1}{\rho}\nabla p = \tfrac{1}{\rho}B\left(\mathbf{u}\right) && \text{on } \Omega\times I,\\
&\nabla\cdot\mathbf{v} = 0 && \text{on } \Omega\times I,\\
&\mathbf{v} = \mathbf{v}_D && \text{on } \Gamma_D\times I,\\
&\tfrac{\partial\mathbf{v}}{\partial\mathbf{n}} = 0 && \text{on } \Gamma_N\times I,\\
&\mathbf{v}\left(x,0\right) = \mathbf{v}_0(x) && \text{on } \Omega,\\
&\mathbf{u}\in U_{ad}.
\end{aligned}$$

Here, $\mathbf{u}$ is the control, $\mathbf{y} = (\mathbf{v}, p)$ the state, ρ the density, $R\left(\nabla\mathbf{v} + \nabla\mathbf{v}^T\right)$ the model for the Reynolds stress tensor with the symmetrized gradient $\nabla^s\mathbf{v} = \frac{1}{2}\left(\nabla\mathbf{v} + \nabla\mathbf{v}^T\right)$, ν the kinematic viscosity and $B(\mathbf{u})$ the control operator. Moreover, f is the objective function (e.g. drag, distance to a desired state, etc.) and U_{ad} is the set of admissible controls. After discretization by using a flow solver, we obtain a problem of the form

$$\min_{\mathbf{y},\mathbf{u}} \quad f(\mathbf{y},\mathbf{u}) \quad \text{subject to} \quad \mathbf{C}(\mathbf{y},\mathbf{u}) = 0, \quad \mathbf{u}\in U_{ad},$$

where $\mathbf{y}$, $\mathbf{u}$ are the discrete state and control, respectively, $f : \mathbb{R}^l \times \mathbb{R}^d \mapsto \mathbb{R}$ is the discrete objective function and $\mathbf{C} : \mathbb{R}^l \times \mathbb{R}^d \mapsto \mathbb{R}^l$ is the discretized PDE constraint resulting from the flow solver.

Under suitable assumptions discussed in Sect. 13.3, this problem can be equivalently rewritten as

$$\min_{\mathbf{u}} \hat{f}(\mathbf{u}) \quad \text{s.t.} \quad \mathbf{u}\in U_{ad}$$

and now can be handled with suitable optimization algorithms.

The outline of this paper is as follows. In Sect. 13.2 a presentation of the $RMTR_\infty$ multilevel optimization algorithm is given, along with some modifications.

The algorithm needs gradient information and therefore in Sect. 13.3 we will present Automatic Differentiation (AD) techniques and a discrete adjoint approach for optimization problems with PDE constraints. Section 13.4 is devoted to the implementation and some further details of the methods. The numerical results in Sect. 13.5 will show the efficiency and general applicability of our methods. In the final Sect. 13.6 we draw conclusions.

13.2 Multilevel Optimization

We will now motivate the *RMTR*$_\infty$ algorithm along with our modifications. The algorithm relies on gradient information and the efficient implementation of the gradient for a complex flow solver is discussed in Sect. 13.3.

The *RMTR*$_\infty$ algorithm of Gratton et al. [12, 13] for bound constrained optimization problems was developed from the original *RMTR* algorithm by Gratton, Sartenaer and Toint [14] for unconstrained problems. In many applications upper and lower bounds are of great interest and therefore we will only study the *RMTR*$_\infty$ method. Closely related is the MG/OPT algorithm of Nash and Lewis [23, 26], that incorporates a line-search procedure for unconstrained optimization problems.

We consider $L+1$ functions $f_0, \ldots, f_L$, where $f_l : \mathbb{R}^{d_l} \to \mathbb{R}$, $d_l \in \mathbb{N}^+$, $l = 0, \ldots, L$. As already motivated, these functions may represent an evaluation of (the same) cost function on different discretization levels or with different underlying models. Actually for the convergence of the algorithm no link between the functions is needed at all, but is the main motivation for the efficiency. Therefore one usually assumes the lower level functions are cheaper to minimize in terms of computational cost than the higher levels functions. Also the underlying problem should be represented best on the highest level. Often the decision which model to use on which level is not trivial.

We are interested in the solution of the optimization problem on the highest level:

$$\min_{\mathbf{u}\in\mathbb{R}^{d_L}} f_L(\mathbf{u})$$

$$s.t.\ \mathbf{b} \le \mathbf{u} \le \mathbf{t}. \tag{13.1}$$

Here $\mathbf{b}, \mathbf{t} \in \mathbb{R}^{d_L}$ are lower and upper bounds for the optimization variable $\mathbf{u}$.

The *RMTR*$_\infty$ algorithm uses a trust-region methodology to solve this optimization problem. Trust-region methods approximate the function to be minimized (f_L in our case) by a model that is in some sense easier to optimize.

The usual choice is a quadratic model (13.5), but we will additionally make use of the functions $f_0, \ldots, f_{L-1}$.

All the models we will use are only approximations to the upper level functions. Therefore we introduce a suitable neighborhood, called *trust-region*, where a model is considered accurate enough. We refer to Conn, Gould and Toint [8] for an

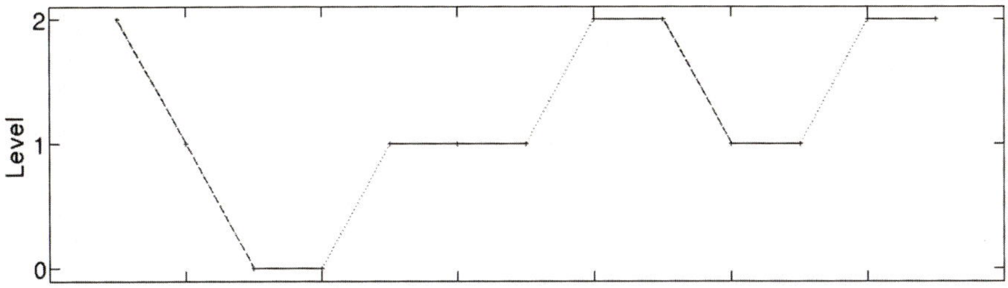

Fig. 13.1 Recursions of the multilevel optimization algorithm

introduction to trust-region methods. A specific difficulty for the multilevel structure arises from the bounds and the trust-region. A consistent setup is therefore discussed in Sect. 13.2.2.

Figure 13.1 depicts possible iterations for the multilevel optimization with three levels. If were are in some iteration on a current level, the possibilities are to use a coarser level model for the optimization (depicted with the dashed lines) or to stay on the current level and calculate a step with the quadratic model (depicted with the solid lines).

One of the assumptions for the convergence of the (basic) trust-region algorithm is that the coarse model gradient equals the gradient on the fine level. Due to the possibly different dimensions of the models in our case, this is not directly applicable to our problem structure. Instead of this, we will assume that the restriction of the gradient on the finer level equals the gradient on the coarser level. This will be motivated in the following.

13.2.1 Construction of the Models

Assume we already have constructed some model function h_{l+1} for the level $l+1$. Note that this will involve the function f_{l+1}. In iteration k of the optimization algorithm on that level we have arrived at some intermediate iterate $\mathbf{u}_{l+1,k}$. We set the starting or reference point on the coarser level l to $\mathbf{u}_{l,0} = R^{\mathbf{u}}_{l+1}\mathbf{u}_{l+1,k}$, where $R^{\mathbf{u}}_{l+1} \in \mathbb{R}^{d_{l+1}\times d_l}$ is a restriction operator for the iterates. The model h_l for the coarser level is now constructed around this reference point with the help of the function f_l such that

$$\nabla h_l\left(\mathbf{u}_{l,0}\right) = R_{l+1}\nabla h_{l+1}\left(\mathbf{u}_{l+1,k}\right). \tag{13.2}$$

Here $R_{l+1} \in \mathbb{R}^{d_{l+1}\times d_l}$ is a full rank restriction operator for the gradient, see remark 1 for further explanations about the two restriction operators.

In general the models shall satisfy the following assumption.

Assumption 1 For iterations k, where the model h_l, $l = 0, \ldots, L-1$ *is used, it holds that*

1. $h_l \in C^2(\mathcal{F}_l)$,
2. $\nabla h_l(\mathbf{u}_{l,0}) (= \mathbf{g}_{l,0}) = R_{l+1}\mathbf{g}_{l+1,k}$ *for* $l \neq L$,
3. $\|\nabla h_l(\mathbf{u}_l)\|_\infty \leq \kappa_g, \kappa_g \geq 1$ *for all* $\mathbf{u}_l \in \mathcal{F}_l$,
4. $\left\|\nabla^2 h_l(\mathbf{u}_l)\right\|_\infty \leq \kappa_H - 1, \quad \kappa_H \geq 1$ *for all* $\mathbf{u}_l \in \mathcal{F}_l$.

Here, we have used the notation $\mathbf{g}_{l,k} := \nabla h_l(\mathbf{u}_{l,k})$. On the highest level L we define $h_L := f_L$ and the set $\mathcal{F}_l$ is defined as

$$\mathcal{F}_l := \left\{\mathbf{u}_l \in \mathbb{R}^{d_l} \,|\mathbf{b}_l \leq \mathbf{u}_l \leq \mathbf{t}_l\right\},$$

where the vectors $\mathbf{t}_l, \mathbf{b}_l$ are upper and lower bounds introduced in Sect. 13.2.2.

A common choice to construct first order consistent models h_l from the functions f_l is the additive correction

$$h_l\left(\mathbf{u}_{l,0} + \mathbf{s}_l\right) = f_l\left(\mathbf{u}_{l,0} + \mathbf{s}_l\right) + \mathbf{v}_l^T\mathbf{s}_l$$

with

$$\mathbf{v}_l = R_{l+1}\nabla h_{l+1}\left(\mathbf{u}_{l+1,k}\right) - \nabla f_l\left(\mathbf{u}_{l,0}\right).$$

We see that the gradient on the coarse level equals the restricted fine level gradient since

$$\nabla h_l\left(\mathbf{u}_{l,0}\right) = \nabla f_l\left(\mathbf{u}_{l,0} + \mathbf{s}_l\right) + \mathbf{v}_l = R_{l+1}\nabla h_{l+1}\left(\mathbf{u}_{l+1,k}\right).$$

This approach is similar to the ones used in nonlinear multigrid, see [4, p.98 ff] and was first proposed by Nash and Lewis [23, 26] in the optimization context.

The optimization of h_l on level l will produce a step $\mathbf{s}_l$ that will then be prolongated back to the finer level by

$$\mathbf{s}_{l+1} := P_{l+1}\mathbf{s}_l,$$

where we assume $\sigma_{l+1}P_{l+1} = R_{l+1}^T \; (\sigma_{l+1} > 0)$, from which follows that

$$\begin{aligned}\mathbf{s}_{l+1}^T\nabla h_{l+1}\left(\mathbf{u}_{l+1,k}\right) &= (P_{l+1}\mathbf{s}_l)^T\nabla h_l\left(\mathbf{u}_{l+1,k}\right)\\ &= \frac{1}{\sigma_l}\mathbf{s}_l^T R_{l+1}\nabla h_l\left(\mathbf{u}_{l+1,k}\right) = \frac{1}{\sigma_l}\mathbf{s}_l^T\nabla h_l\left(\mathbf{u}_{l,0}\right).\end{aligned}$$

If $\mathbf{s}_l$ is a direction of descent for the coarse model then the latter value is less than zero and $\mathbf{s}_{l+1}$ is a direction of descent for the finer level model.

Remark 1 (On the Restriction Operators). In [23] and [14] it is assumed that $R_l^{\mathbf{u}} = R_l,\ l = 1, \ldots, L$. In practical applications this may not be a proper choice, because the gradient may have a different scaling than the iterates. In fact the restriction operator for the iterates may be chosen independently from the restriction for the gradients, since the linear consistency condition (2) will hold regardless of the reference point $\mathbf{u}_{l,0}$. The convergence proofs are independent of the assumption $R_l^{\mathbf{u}} = R_l$ and remain valid for differing restrictions. This is not surprising at all, since for nonlinear multigrid methods, where the *RMTR* is motivated from, these generalizations are common.

After having discussed the appropriate choice of the models on the different levels, we will deal with the bound constraints of the optimization problem (1).

13.2.2 Treatment of the Bounds

We introduce a multilevel structure for the bounds and define $\mathbf{b}_L := \mathbf{b}$ and $\mathbf{t}_L := \mathbf{t}$ along with the already mentioned set

$$\mathcal{F}_l := \left\{ \mathbf{u}_l \in \mathbb{R}^{d_l} \,\middle|\, \mathbf{b}_l \leq \mathbf{u}_l \leq \mathbf{t}_l \right\}, \tag{13.3}$$

where the vectors $\mathbf{b}_l, \mathbf{t}_l$ for $l \neq L$ will be defined in the following.

The algorithm will keep the iterates in the feasible domain. Therefore we need to make sure that if $\mathbf{u}_{l,k}$ is in the feasible region, then the same holds for the next iterate $\mathbf{u}_{l,k+1} = \mathbf{u}_{l,k} + \mathbf{s}_l = \mathbf{u}_{l,k} + P_l \mathbf{s}_{l-1}$, where $\mathbf{s}_{l-1}$ is a step obtained from the lower level model. In [13] the bounds on the coarser level are derived from the higher level in the following manner. We set

$$(\mathbf{b}_{l-1})_j := (\mathbf{u}_{l-1,0})_j + \frac{1}{\|P_l\|_\infty} \max_{i=1,\ldots,d_l} \begin{cases} (\mathbf{b}_l - \mathbf{u}_{l,k})_i & \text{if } (P_l)_{ij} > 0, \\ (\mathbf{u}_{l,k} - \mathbf{t}_l)_i & \text{if } (P_l)_{ij} < 0 \end{cases}$$

and

$$(\mathbf{t}_{l-1})_j := (\mathbf{u}_{l-1,0})_j + \frac{1}{\|P_l\|_\infty} \min_{i=1,\ldots,d_l} \begin{cases} (\mathbf{t}_l - \mathbf{u}_{l,k})_i & \text{if } (P_l)_{ij} > 0, \\ (\mathbf{u}_{l,k} - \mathbf{b}_l)_i & \text{if } (P_l)_{ij} < 0, \end{cases}$$

where $\mathbf{u}_{l-1,0}$ is the starting point for the optimization on the coarser level and $\mathbf{u}_{l,k}$ is the current iterate on level l. In the case where the restriction and prolongation have solely positive entries this equals the approach of Gelman and Mandel from [10].

This definition enforces

$$\mathbf{u}_{l,k} + P_l(\mathbf{u}_{l-1} - \mathbf{u}_{l-1}) \in \mathcal{F}_l \text{ for all } \mathbf{u}_{l-1} \in \mathcal{F}_{l-1},$$

see [13, Lemma 4.3]. As we can see regardless of the coarse level iterate, the prolongated step will produce a feasible iterate for the higher level. This is true as long as it fulfills the bound constraints on that level.

Now we know how to transfer the bound constraints to the lower level and next we discuss the treatment of the trust-region bounds. The trust-region (i.e. $\|\mathbf{s}\|_\infty \leq \Delta$) of the algorithm on each level has to be brought to the lower level. In general we could use the same kind of set up as before, but in [13] a different approach is used. The trust-region bounds from the upper level can be seen as soft constraints in the sense that they may be violated up to some factor and convergence can still be achieved.

The representation of these trust-region bounds uses two vectors $\mathbf{v}_{l-1}, \mathbf{w}_{l-1} \in \mathbb{R}^{d_{l-1}}$ and we define

$$\mathcal{A}_{l-1} := \left\{\mathbf{u}_{l-1} \in \mathbb{R}^{d_{l-1}} \,|\mathbf{v}_{l-1} \leq \mathbf{u}_{l-1} \leq \mathbf{w}_{l-1}\right\}$$

with

$$\begin{aligned}(\mathbf{v}_{l-1})_j := & \sum_{i,\left(R_l^{\mathbf{u}}\right)_{ji}>0} \left(R_l^{\mathbf{u}}\right)_{ji} (\max(\mathbf{v}_l, \mathbf{u}_{l,k} - \Delta_{l,k}\mathbf{e}))_i \\ & + \sum_{i,\left(R_l^{\mathbf{u}}\right)_{ji}<0} \left(R_l^{\mathbf{u}}\right)_{ji} (\min(\mathbf{w}_l, \mathbf{u}_{l,k} - \Delta_{l,k}\mathbf{e}))_i\end{aligned}$$

and

$$\begin{aligned}(\mathbf{w}_{l-1})_j := & \sum_{i,\left(R_l^{\mathbf{u}}\right)_{ji}>0} \left(R_l^{\mathbf{u}}\right)_{ji} (\min(\mathbf{w}_l, \mathbf{u}_{l,k} - \Delta_{l,k}\mathbf{e}))_i \\ & + \sum_{i,\left(R_l^{\mathbf{u}}\right)_{ji}<0} \left(R_l^{\mathbf{u}}\right)_{ji} (\max(\mathbf{v}_l, \mathbf{u}_{l,k} - \Delta_{l,k}\mathbf{e}))_i.\end{aligned}$$

Here $\mathbf{e}$ is a vector of ones and $\Delta_{l,k}$ is the current trust-region radius on level l in iteration k. For a consistent setting we define on the highest level $\mathbf{w}_L := +\infty$ and $\mathbf{v}_L := -\infty$.

The lower level model seeks for the minimum within the set

$$\mathcal{L}_{l-1} := \mathcal{F}_{l-1} \cap \mathcal{A}_{l-1}.$$

In addition to the restricted trust-region, on the level $l-1$ will also be a trust-region. This trust-region in iteration k is defined there by the set

$$\mathcal{B}_{l-1,k} := \left\{\mathbf{u}_{l-1,k} + \mathbf{s}_{l-1} \in \mathbb{R}^{d_{l-1}} \,\middle|\, \|\mathbf{s}_{l-1}\|_\infty \leq \Delta_{l-1,k}\right\}.$$

In every iteration, the optimization algorithm for the lower level model will try to find a step within the bounds

$$\mathcal{W}_{l-1,k} := \mathcal{F}_{l-1} \cap \mathcal{A}_{l-1} \cap \mathcal{B}_{l-1,k}.$$

13.2.3 Stopping Criteria

The algorithm relies on a stopping criterion that measures the progress towards the solution of the optimization problem. The check for convergence in an iteration k on a level l is done with the use of a *criticality measure* $\chi(\mathbf{u}_{l,k}) = \chi_{i,k} \in \mathbb{R}^+$.

In difference to [13] we are using the criticality measure

$$\chi_{i,k} = \left\| \mathcal{P}_{U_l,\mathcal{L}_l} \left(\mathbf{u}_{l,k} - M_l^{-1} \mathbf{g}_{l,k} \right) - \mathbf{u}_{l,k} \right\|_{U_l}, \tag{13.4}$$

where $\left\| M_l^{-1} \right\|_\infty \leq \kappa_M$ is a symmetric positive definite matrix, $\| \cdot \|_{U_l}$ is a level dependent (finite dimensional) norm defined by $\|\mathbf{u}_l\|_{U_l}^2 = \mathbf{u}_l^T M_l \mathbf{u}_l$ and $\mathcal{P}_{U_l,\mathcal{L}_l}$ is the projection onto the box $\mathcal{L}_l$ in the U_l norm.

This is a common setting for discretized control problems, where $\| \cdot \|_{U_l}$ is a discrete variant of the norm of the control space and $M_l^{-1} g_{l,k}$ corresponds to the Riesz representation of the gradient.

In the thesis of Mouffe [25] it was shown that the algorithm will convergence if the criticality measure meets the requirements

$$|\chi(\mathbf{u}) - \chi(\mathbf{v})| \leq \kappa_L \|\mathbf{u} - \mathbf{v}\|_\infty, \qquad \chi(\mathbf{u}_{l-1,0}) \leq 2\kappa_g \Delta_{l,k}$$

for a constant $\kappa_L > 0$. This can be shown for the measure defined in (13.4).

An important feature of the RMTR$_\infty$ is a check, whether it is useful to invoke the lower level model. If the criticality measure on the coarser level in the initial iterate is too small compared to the measure in the current iterate on the current level, we will not use the coarser level model. More specifically, we will only use the coarser level if

$$\chi_{l-1,0} \geq \kappa_\chi \chi_{l,k} \text{ for a } \kappa_\chi \in (0, 1).$$

Note that we are (usually) able to calculate $\chi_{l-1,0}$ without function evaluations on the coarser level. If the use of the coarser level makes sense, we will optimize until in some iteration m it holds that

$$\chi_{l-1,m} < \kappa_\chi \chi_{l,k} := \varepsilon_{l-1}.$$

In fact we may even terminate earlier. It is enough to make at least one successful step on the coarser level, see [13]. This gives the algorithm a great degree of freedom.

If the coarser level is not useful, we will have to use a quadratic model

$$m_{l,k}\left(\mathbf{u}_{l,k}+\mathbf{s}_l\right)=h_l\left(\mathbf{u}_{l,k}\right)+\mathbf{g}_{l,k}^T\mathbf{s}_l+\frac{1}{2}\mathbf{s}_l^T H_{l,k}\mathbf{s}_l, \tag{13.5}$$

called *Taylor model*, with

$$1+\|H_{l,k}\|_\infty =: \beta_{l,k}\le \kappa_H \quad \text{for } l=0,\ \ldots,\ L \ \text{ for all } k.$$

The step $\mathbf{s}_{l,k}$ that is produced by this model is required to satisfy the condition

$$m_{l,k}\left(\mathbf{u}_{l,k}\right)-m_{l,k}\left(\mathbf{u}_{l,k}+\mathbf{s}_{l,k}\right)\ge \kappa_{red}\,\chi_{l,k}\ \min\left(1,\ \frac{\chi_{l,k}}{\beta_{l,k}},\ \Delta_{l,k}\right), \tag{13.6}$$

with $\kappa_{red}\in(0,0.5)$. This *generalized fraction of Cauchy decrease* condition is a standard assumption and holds for some efficient algorithms (see [12]) and for the actual minimizer of the model.

Now we have motivated all the important steps of the algorithm.

13.2.4 The Algorithm

The RMTR$_\infty$ algorithm can be described as follows. We choose constants $0<\eta_1\le\eta_2<1$, $0<\gamma_1\le\gamma_2<1$, $\kappa_\chi\in(0,\ 1)$ and initial trust-region radii $\Delta_l^s\in\mathbb{R}^+$ (compare the basic trust-region algorithm). For the initial iterate $\mathbf{u}_{L,0}\in\mathbb{R}^{d_L}$ we calculate $\mathbf{g}_{L,0},\ \chi_{L,0},\ \mathcal{F}_L,\ \mathcal{A}_L$ and to drive the criticality measure lower than ε_L, we call

$$\mathbf{u}_{L,*}=\mathrm{RMTR}_\infty\left(L,\ \mathbf{u}_{L,0},\ g_{L,0},\ \chi_{L,0},\mathcal{F}_L,\ \mathcal{A}_L,\varepsilon_L\right).$$

The description of the function RMTR$_\infty$ (...) is given with algorithm 1.

Our modifications do not change the convergence result and we have the following theorem (see Mouffe [25]):

Theorem *Assume we start algorithm 1 with* $\varepsilon_L=\varepsilon_L^j>0$ *from a sequence* $\{\varepsilon_L^j\}_{j=1}^\infty$ *that is monotonically converging to zero. We denote the solution as* $\mathbf{u}_{L,j}$ *with criticality measure* $\chi_{L,j}$. *Then it holds that*

$$\lim_{j\to\infty}\chi_{L,j}=0.$$

Remark 2 For large scale optimization problems, step 3 of the algorithm needs efficient solvers. Choices like *Sequential Coordinate Minimization* and the *Projected*

Truncated Conjugate Gradient algorithm are discussed in [13]. The applications we aim for are very complex, but medium scale. Problems like shape optimization or boundary control have at most as many design variables as there are grid points on the two dimensional surface they are defined on. Also, usually practically relevant distributed control problem do not have all grid points as design variables. Therefore we are able to fully solve the quadratic problem with a standard solver.

Algorithm 1 $\mathrm{RMTR}_\infty\,(l,\ \mathbf{u}_{l,0},\ \mathbf{g}_{l,0},\ \chi_{l,0},\ \mathcal{F}_l,\ \mathcal{A}_l, \varepsilon_l)$

0. (Initialization) Construct a model h_l according to Assumption 1. Set $k = 0,\ \Delta_{l,0} = \Delta_l^s$ and initialize the constraints with

$$\mathcal{L}_l = \mathcal{F}_l \cap \mathcal{A}_l,\quad \mathcal{W}_{l,0} = \mathcal{L}_l \cap \left\{\mathbf{u} \middle| \, \|\mathbf{u} - \mathbf{u}_{l,0}\|_\infty \leq \Delta_{l,0}\right\}.$$

1. (Model choice) If $l = 0$ goto step 3, else compute $\mathcal{F}_{l-1},\ \mathcal{A}_{l-1},\ \mathcal{L}_{l-1}$ according to Sect. 13.2.2 and then $\chi_{l-1,0}$. If $\chi_{l-1,0} < \kappa_\chi \chi_{l,k}$ goto step 3, else choose between step 2 and 3.
2. (Recursive step) Call

$$\mathrm{RMTR}_\infty \left(l-1,\, R_l^{\mathbf{u}}\mathbf{u}_{l,k},\, R_l\mathbf{g}_{l,k},\, \chi_{l-1,0},\, \mathcal{F}_{l-1},\, \mathcal{A}_{l-1}\kappa_\chi\chi_{l,k}\right)$$

with result $\mathbf{u}_{l-1,*}$. Set

$$\mathbf{s}_{l,k} = P_l\left(\mathbf{u}_{l-1,*} - R_l^{\mathbf{u}}\mathbf{u}_{l,k}\right),\quad \delta_{l,k} = \frac{1}{\sigma_l}\left(h_{l-1}\left(R_l^{\mathbf{u}}\mathbf{u}_{l,k}\right) - h_{l-1}\left(\mathbf{u}_{l-1,*}\right)\right)$$

and go to step 4.
3. (Taylor step) Choose $H_{l,k}$ and compute a step $\mathbf{s}_{l,k}$ that reduces the model $m_{l,k}$ given by (5) such that the sufficient decrease condition (6) holds and $\mathbf{u}_{l,k} + \mathbf{s}_{l,k} \in \mathcal{W}_{l,k}$. Set $\delta_{l,k} = m_{l,k}\left(\mathbf{u}_{l,k}\right) - m_{l,k}\left(\mathbf{u}_{l,k} + \mathbf{s}_{l,k}\right)$.
4. (Step acceptance) Compute $h_l\left(\mathbf{u}_{l,k} + \mathbf{s}_{l,k}\right)$ and set

$$\rho_{l,k} = \frac{1}{\delta_{l,k}}\left(h_l\left(\mathbf{u}_{l,k}\right) - h_l\left(\mathbf{u}_{l,k} + \mathbf{s}_{l,k}\right)\right).$$

If $\rho_{l,k} \geq \eta_1$ set $\mathbf{u}_{k+1,l} = \mathbf{u}_{l,k} + \mathbf{s}_{l,k}$, else $\mathbf{u}_{k+1,l} = \mathbf{u}_{l,k}$.
5. (Termination) Compute $\mathbf{g}_{l,k+1},\ \chi_{l,k+1}$. If $\chi_{l,k+1} \leq \varepsilon_l$ or $\mathbf{u}_{l,k+1} \notin \mathcal{L}_l$ then return with $\mathbf{u}_{l,*} = \mathbf{u}_{l,k+1}$.
6. (Trust-region update) Set

$$\Delta_{l,k+1} \in \begin{cases} [\Delta_{l,k}, \infty) & \text{if } \rho_{l,k} \geq \eta_2, \\ [\gamma_2\Delta_{l,k},\ \Delta_{l,k}] & \text{if } \eta_1 \leq \rho_{l,k} < \eta_2, \\ [\gamma_1\Delta_{l,k},\ \gamma_2\Delta_{l,k}] & \text{if } \rho_{l,k} < \eta_1 \end{cases}$$

and $\mathcal{W}_{l,k+1} = \mathcal{L}_l \cap \left\{\mathbf{u} \middle| \, \|\mathbf{u} - \mathbf{u}_{l,k+1}\|_\infty \leq \Delta_{l,k+1}\right\}$. Set $k = k + 1$ and go to step 1.

13.2.5 Additive and Multiplicative Correction

The usual correction method for the multilevel optimization algorithms is the additive correction

$$h_{l-1}(\mathbf{u}_{l-1,0}+\mathbf{s}_{l-1}) := f_{l-1}(\mathbf{u}_{l-1,0}+\mathbf{s}_{l-1}) + (R_l \nabla h_l(\mathbf{u}_{l,k}) - \nabla f_{l-1}(\mathbf{u}_{l-1,0}))^T \mathbf{s}_l.$$

Another popular correction method frequently used in surrogate modeling is the *multiplicative correction* (or *β-correlation*, due to [5]) method. Assume that $f_l, h_l > 0$, $\sigma_l = 1$ and set

$$\beta(\mathbf{u}_{l-1,0}+\mathbf{s}_{l-1}) = \frac{h_l(\mathbf{u}_{l,k}+P_l\mathbf{s}_{l-1})}{f_{l-1}(\mathbf{u}_{l-1,0}+\mathbf{s}_{l-1})}.$$

By construction it holds that

$$h_l(\mathbf{u}_{l,k}+P_l\mathbf{s}_{l-1}) = \beta(\mathbf{u}_{l-1,0}+\mathbf{s}_{l-1}) f_{l-1}(\mathbf{u}_{l-1,0}+\mathbf{s}_{l-1}).$$

As we can see, β is a scaling factor, but knowing β is the same as knowing the whole function h_l. For the multiplicative correction we now use a first order expansion of β and use this as a scaling for h_{l-1}. Consequently

$$\beta(\mathbf{u}_{l-1,0}+\mathbf{s}_{l-1}) \approx \beta(\mathbf{u}_{l-1,0}) + \nabla\beta(\mathbf{u}_{l-1,0})^T \mathbf{s}_{l-1} := \beta_c(\mathbf{u}_{l-1,0}+\mathbf{s}_{l-1})$$

and define the lower level model as

$$h_{l-1}(\mathbf{u}_{l-1,0}+\mathbf{s}_{l-1}) := \beta_c(\mathbf{u}_{l-1,0}+\mathbf{s}_{l-1}) f_{l-1}(\mathbf{u}_{l-1,0}+\mathbf{s}_{l-1}).$$

Note that

$$\nabla\beta(\mathbf{u}_{l-1,0}) = \frac{R_l \nabla h_l(\mathbf{u}_{l,k}) f_{l-1}(\mathbf{u}_{l-1,0}) - h_l(\mathbf{u}_{l,k}) \nabla f_{l-1}(\mathbf{u}_{l-1,0})}{f_{l-1}(\mathbf{u}_{l-1,0})^2}$$

and therefore the linear consistency holds

$$\nabla h_{l-1}(\mathbf{u}_{l-1,0}) = R_l \nabla h_l(\mathbf{u}_{l,k}).$$

We now come back to the assumption $f_l, h_l > 0$, which is needed for the correction to be well defined. The positivity of the objective functions f_l is often true in practical situation, e.g. if it results from a tracking-type cost functional or if it is the drag of an object. The positivity of h_l can not be assured and may be violated in practice. If on some level $h_l(\mathbf{u}_{l,k}) < 0$ then the coarser level function f_{l-1} will be multiplied with a negative term. In the context of the tracking-type functional, we would then try to maximize the difference between state and reference state, which is against the physical interpretation. Still the linear consistency holds, but the model will possibly only be accurate in a small neighborhood.

We propose the following adjustment of the method which will give models that fulfill assumption 1, if the functions f_l are uniformly bounded from below by a positive constant. Consider the levels $l+1$ and l in iterations m and k, with $h_{l+1}(\mathbf{u}_{l+1,m}) > 0$ and $h_l(\mathbf{u}_{l+k}) < 0$. Note that $h_l(\mathbf{u}_{l,0}) > 0$ by construction (see also below). The fact that h_l turns negative is not an indicator that the model is very bad, but in this case the definition of the coarser level model h_{l-1} is a problem. We therefore take a closer look at h_l.

$$\begin{aligned}
h_l(\mathbf{u}_{l,k}) &= h_l(\mathbf{u}_{l,0} + s_l) \\
&= \left(\beta(\mathbf{u}_{l,0}) + \nabla\beta(\mathbf{u}_{l,0})^T s_l\right) f_l(\mathbf{u}_{l,0} + \mathbf{s}_l) \\
&= \beta(\mathbf{u}_{l,0}) f_1(\mathbf{u}_{l,0} + s_l) + \left(\nabla\beta(\mathbf{u}_{l,0})^T \mathbf{s}_l\right) f_l(\mathbf{u}_{l,0} + \mathbf{s}_l) \\
&= \frac{h_{l+1}(\mathbf{u}_{l+1,m})}{f_l(\mathbf{u}_{l,0})} f_l(\mathbf{u}_{l,0} + \mathbf{s}_l) + \left(\nabla\beta(\mathbf{u}_{l,0})^T \mathbf{s}_l\right) f_l(\mathbf{u}_{l,0} + \mathbf{s}_l)
\end{aligned}$$

Since $h_{l+1}(\mathbf{u}_{l+1,m}) > 0$ and $f_l > 0$, we see that for the first term

$$\frac{h_{l+1}(\mathbf{u}_{l+1,m})}{f_l(\mathbf{u}_{l,0})} f_l(\mathbf{u}_{l,0} + \mathbf{s}_l) > 0$$

holds. Obviously the model may only turn negative because

$$\left(\nabla\beta(\mathbf{u}_{l,0})^T \mathbf{s}_l\right) f_l(\mathbf{u}_{l,0} + \mathbf{s}_l) < 0.$$

Therefore we propose the following correction method. We treat

$$\frac{h_{l+1}(\mathbf{u}_{l+1,m})}{f_l(\mathbf{u}_{l,0})} f_1(\mathbf{u}_{l,0} + \mathbf{s}_l) =: \tilde{h}_l^1(\mathbf{u}_{l,0} + \mathbf{s}_l)$$

and

$$\left(\nabla\beta(\mathbf{u}_{l,0})^T \mathbf{s}_l\right) f_l(\mathbf{u}_{l,0} + \mathbf{s}_l) =: \tilde{h}_l^2(\mathbf{u}_{l,0} + \mathbf{s}_l)$$

as two functions. A multiplicative correction for f_{l-1} will be carried out with $\tilde{h}_l^1$ and $\tilde{h}_l^2$ will be used in a linear correction with the 0 function. In the general case now $h_l < 0$ may be true, but on every level a meaningful multiplicative correction can be used.

13.3 A Discrete Adjoint Approach

In this section we present how gradients that are consistent on the discrete level can be obtained with the *discrete adjoint approach.* We will give guidelines on how to apply the method to complex problems in an efficient manner. Additionally we will give an brief overview of AD methods.

We consider the following discretized optimal control problem with *design variables* $\mathbf{u}$, *state variables* $\mathbf{y}$, *state equation* $\mathbf{C}$ and U_{ad} is some admissible set, e.g. upper and lower bounds for $\mathbf{u}$.

Consequently we want to solve

$$\min_{\mathbf{y},\mathbf{u}} f(\mathbf{y},\mathbf{u}) \quad \text{s.t.} \quad \mathbf{C}(\mathbf{y},\mathbf{u}) = 0, \quad \mathbf{u} \in U_{ad}.$$

Here $f : \mathbb{R}^l \times \mathbb{R}^d \mapsto \mathbb{R}$, $\mathbf{C} : \mathbb{R}^l \times \mathbb{R}^d \mapsto \mathbb{R}^l$. We assume that for all $\mathbf{u} \in U_{ad} \subset \mathbb{R}^d$ the equation $\mathbf{C}(\mathbf{y},\mathbf{u}) = 0$ has a unique solution and so we are able to find for every $\mathbf{u} \in U_{ad}$ a unique state $\mathbf{y} = \mathbf{y}(\mathbf{u})$. Additionally we assume that $\mathbf{C}$ fulfills the assumptions for the implicit function theorem i.e. $\mathbf{C}$ is continuously differentiable and $C_{\mathbf{y}}(\mathbf{y}, \mathbf{u})$ is invertible. Now we are able to recast the optimization problem in its reduced form

$$\min_{\mathbf{u}} \hat{f}(\mathbf{u}) \quad \text{s.t.} \quad \mathbf{u} \in U_{ad}.$$

with the reduced objective function

$$\hat{f}(\mathbf{u}) = f(\mathbf{y}(\mathbf{u}), \mathbf{u}).$$

By the implicit function theorem $\mathbf{u} \mapsto \mathbf{y}(\mathbf{u})$ is continuously differentiable and this can be used to calculate the reduced derivative, given by

$$\frac{d\hat{f}(\mathbf{u})}{d\mathbf{u}} = f_{\mathbf{y}}(\mathbf{y}(\mathbf{u}), \mathbf{u}) \frac{d\mathbf{y}(\mathbf{u})}{d\mathbf{u}} + f_{\mathbf{u}}(\mathbf{y}(\mathbf{u}), \mathbf{u}). \tag{13.7}$$

Similarly we obtain the expression

$$\frac{d\mathbf{C}(\mathbf{y}(\mathbf{u}), \mathbf{u})}{d\mathbf{u}} = C_{\mathbf{y}}(\mathbf{y}(\mathbf{u}), \mathbf{u}) \frac{d\mathbf{y}(\mathbf{u})}{d\mathbf{u}} + \mathbf{C}_{\mathbf{u}}(\mathbf{y}(\mathbf{u}), \mathbf{u}) = 0. \tag{13.8}$$

Due to the assumptions, $\mathbf{C}_{\mathbf{y}}$ is invertible and consequently $\mathbf{C}_{\mathbf{y}}^T$. The adjoint $\boldsymbol{\lambda}$ may therefore be defined as the unique solution to the linear system

$$\mathbf{C}_{\mathbf{y}}^T(\mathbf{y}(\mathbf{u}), \mathbf{u}) \boldsymbol{\lambda} = -f_{\mathbf{y}}^T(\mathbf{y}(\mathbf{u}), \mathbf{u}) \quad \text{(adjoint equation)}. \tag{13.9}$$

Combining this with (13.7) and (13.8), we get the following representation of the gradient

$$\begin{aligned}\frac{d\hat{f}(\mathbf{u})}{d\mathbf{u}} &= f_{\mathbf{y}}(\mathbf{y}(\mathbf{u}), \mathbf{u})\frac{d\mathbf{y}(\mathbf{u})}{d\mathbf{u}} + f_{\mathbf{u}}(\mathbf{y}(\mathbf{u}), \mathbf{u}) \\ &= -\boldsymbol{\lambda}^T \mathbf{C}_{\mathbf{y}}(\mathbf{y}(\mathbf{u}), \mathbf{u})\frac{d\mathbf{y}(\mathbf{u})}{d\mathbf{u}} + f_{\mathbf{u}}(\mathbf{y}(\mathbf{u}), \mathbf{u}) \\ &= \boldsymbol{\lambda}^T \mathbf{C}_{\mathbf{u}}(\mathbf{y}(\mathbf{u}), \mathbf{u}) + f_{\mathbf{u}}(\mathbf{y}(\mathbf{u}), \mathbf{u}).\end{aligned}$$

These calculations will be the basis of our approach to the discrete gradient in Sect. 13.3.2. But before we will show other ways to obtain the discrete gradient.

13.3.1 Automatic Differentiation

In the following, we will give a short description of the forward and reverse mode of AD. These are the commonly used techniques to obtain reliable derivative information from complex computer programs. We refer to the book of Griewank and Walther [17] for an extensive introduction.

The forward mode of AD amounts to the application of the chain rule to the underlying computer program. The computational cost of obtaining the whole gradient of the function $\hat{f}$ is then linearly depends on the number of design variables $\mathbf{u}$. The forward mode has proven to allow for a quick implementation and is efficient for a small number of design variables.

For a large number of design variables, the forward mode is not efficient anymore and the reverse mode is the method of choice. It may be seen as an reverse application of the chain rule and has a computational cost that is independent of the number of design variables. Unfortunately the method relies on the storage of all intermediate results of the computational code. Although the storage cost is linear in the size of the code, it is prohibitive in practical situation.

There is a way to circumvent the storage problem in the reverse mode called *checkpointing* (see Griewank [16] or Walther [35]). These algorithms use a clever rule for the recomputation of values instead of storing them all. The result is a logarithmic increase in memory accompanied by a logarithmic increase of runtime because of the recomputation of values.

We would like to mention here that there exists a variant of the reverse mode in cases where the solution of a (nonlinear) system is found by a fixed point iteration. In these cases a fixed point algorithm for the forward and reverse mode may be derived, see Christianson [7]. It eventually has the same convergence speed as the original iterations and the storage requirements are significantly less than for the usual reverse mode.

With the use of complexity models, the runtime of the forward and reverse mode can be estimated [17]. For a function $\mathbf{g} : \mathbb{R}^d \mapsto \mathbb{R}^m$ the cost of the reverse mode for the computation of its derivative is less or equal than 1.5 + 2.5 m times the cost of a function evaluation. The cost of the forward mode in this case is bounded by 1 + 1.5d times the cost for a function evaluation.

13.3.1.1 Implementation of AD

In principle one could implement the ideas of AD by hand, going through the code and changing it line by line. But as the transformations follow specific rules, this can be done by a program. Generally now there are two ways to implement AD. One possibility is *source transformation*. Here the original source code of a program is transformed into a code that additionally calculates the derivatives. Such a code has to be able to handle the full language standard of the underlying programming language. Easier to implement is the *operator overloading* approach. In object oriented programming languages operator overloading allows operators to have different implementations depending on their arguments. For the implementation of a forward mode, one would introduce a new data type, that carries the original function value and its gradient. For each operation now the corresponding operation for the gradient is implemented by overloading the operator. This allows an implementation in a very reasonable time. We again refer to the book of Griewank and Walther [17] for additional information.

13.3.2 A Discrete Adjoint Approach

As mentioned before, we pursue the idea of solving

$$\mathbf{C}_y^T(\mathbf{y}, \mathbf{u})\,\boldsymbol{\lambda} = -f_{\mathbf{y}}(\mathbf{y}, \mathbf{u})$$

directly. In the following we will address some points concerning the applicability and efficiency of the method. First we will show the application to time dependent problems, followed by a part about how to obtain the derivative $\mathbf{C}_{\mathbf{y}}$. Finally we will motivate methods to solve the equation.

13.3.2.1 Application to Time Dependent Problems

For a time dependent problem with K time steps, solved by a time stepping algorithm, $\mathbf{C}$ looks like $\mathbf{C}(\mathbf{y}, \mathbf{u}) = \left[\mathbf{C}_1^T(\mathbf{y}, \mathbf{u}), \ldots, \mathbf{C}_K^T(\mathbf{y}, \mathbf{u})\right]^T$. The equations $\mathbf{C}_i$ depend only on a limited number of $\mathbf{y}_i$, $\mathbf{u}_i$, $i = 1, \ldots, K$. We will assume that there are two time step solutions involved. Examples are the implicit Euler or Crank-Nicolson time stepping schemes. The idea easily applies to schemes involving more time steps. Therefore

$$\mathbf{C}(\mathbf{y},\mathbf{u}) = \begin{pmatrix} \mathbf{C}_1(\mathbf{y}_0,\mathbf{u}_0,\mathbf{y}_1,\mathbf{u}_1) \\ \vdots \\ \mathbf{C}_N(\mathbf{y}_{K-1},\mathbf{u}_{K-1},\mathbf{y}_K,\mathbf{u}_K) \end{pmatrix}.$$

Due to the local dependencies $\mathbf{C}_\mathbf{y}$ is an extremely sparse, large, block triangular matrix

$$\mathbf{C}_\mathbf{y} = \begin{pmatrix} (\mathbf{C}_1)_{y_1} & & & \\ (\mathbf{C}_1)_{y_1} & (\mathbf{C}_1)_{\mathbf{y}_2} & & \\ & \cdots & \cdots & \\ & & (\mathbf{C}_K)_{\mathbf{y}_{K-1}} & (\mathbf{C}_K)_{\mathbf{y}_K} \end{pmatrix}.$$

Therefore we see that the adjoint equation

$$\mathbf{C}_\mathbf{y}^T \boldsymbol{\lambda} = -f_\mathbf{y}^T$$

decomposes into equations for different time steps

$$\begin{aligned} (\mathbf{C}_K)_{\mathbf{y}_K}^T \boldsymbol{\lambda}_K &= -f_{\mathbf{y}_K}^T, \\ (\mathbf{C}_{K-1})_{\mathbf{y}_{K-1}}^T \boldsymbol{\lambda}_{K-1} &= -f_{\mathbf{y}_{K-1}}^T - (\mathbf{C}_K)_{\mathbf{y}_K}^T \boldsymbol{\lambda}_K, \\ &\vdots \\ (\mathbf{C}_1)_{\mathbf{y}_1}^T \boldsymbol{\lambda}_1 &= -f_{\mathbf{y}_1}^T - (\mathbf{C}_2)_{\mathbf{y}_1}^T \boldsymbol{\lambda}_2. \end{aligned}$$

This is a time stepping scheme of the adjoint.

For the application of the method to time dependent problems, one solely has to save the time steps $\mathbf{y}_i$. This effort is less than taping the computational procedure for the time step. As we will see in the numerical results in Sect. 13.5.1, one may not need to use checkpointing for the solution of the adjoint, but we have the theory and algorithms at hand in the case we need it.

13.3.2.2 Assembling the Linear System

In Sect. 13.3.1 we have mentioned the two main approaches of AD. These may be used for obtaining the matrix $\mathbf{C}_\mathbf{y}$ (as well as $\mathbf{C}_\mathbf{u}$, $f_\mathbf{y}$, $f_\mathbf{u}$). This will be efficient, i.e. the cost will be linear in the number of grid points used to discretize the PDE because of the following reasons. Standard discretization schemes like Finite Element, Finite Volume (FV) or Finite Difference schemes will produce a sparse (linear) system, that is solved by some method. The number of nonzero elements is proportional to the number of grid points times the stencil used. The FV scheme we used for the numerical results uses seven elements for the stencil.

The forward and the reverse mode will be efficient because $\mathbf{C} : \mathbb{R}^{l \times d} \mapsto \mathbb{R}^l$. In general it is crucial that the AD method employed makes use of the sparsity of the problem. For our numerical results, we implemented the sparse forward mode of M. Ulbrich and S. Ulbrich [34].

For the implementation we need to have a routine at hand, that evaluates the residual $((\mathbf{C}(\mathbf{y}^*, \mathbf{u}) \approx 0))$ for a design variable $\mathbf{u}$. Here $\mathbf{y}^*$ is the approximate solution from the flow solver. An application of AD to the routine gives the derivatives and a quick implementation can be realized with operator overloading.

13.3.2.3 Employing Multigrid Techniques

Multigrid techniques are used by most solvers for PDEs. The structure can be used for the construction of an efficient adjoint multigrid solver. The matrices for the different levels can be obtained by differentiation of the residual equation on the different grid levels. Only the solution on the finest grid has to be stored and is then restricted to the coarser grids with the available operators. Now a standard linear multigrid scheme can be utilized to make use of the levels, see for example Briggs, Henson and McCormick [4] for an introduction to multigrid. The adjoint multigrid procedure needs the transposed restriction operator of the original scheme as prolongation and vice versa. This yields no problem since a sparse representation of the operators can be obtained by the application of sparse AD to the restriction/prolongation routines.

13.3.2.4 Parallelism

If the state solver runs in parallel, the adjoint solver may be parallelized with the help of the available routines. Parallel solvers often use a *domain decomposition* approach [28] along with the Message Passing Interface (MPI) [24]. Now the parallel exchange routines have to be rewritten, such that they are able to communicate the variables of the forward or reverse mode. For the calculation of the residual no solver is involved and therefore the forward mode will run very well in parallel in case of a domain decomposition.

For the solution of the linear systems in parallel one may incorporate ideas from the forward solver and use the structures. For a quick implementation one can use parallel linear solver packages like PETSc [2]. We will give some additional information about our implementation in Sect. 13.4.2.

13.4 Software and Implementation

The optimization algorithm presented in Sect. 13.2 was implemented in MATLAB ® and the discrete adjoint mode from Sect. 13.3 was implemented in Fortran 95, based on the flow solver FASTEST [9]. FASTEST is a parallel, block structured Finite Volume (FV) code. The solver uses a geometric, nonlinear multigrid method and for the solution of the linear systems a SIMPLE scheme is employed. Several LES and RANS turbulence models are implemented and a grid deformation routine allows for changes of the computational domain. Therefore we are also able to realize shape optimization. The code is fully three dimensional. For the parallelization a distribution of the blocks to different cores and MPI is used.

We will give details on the implementation of the optimization algorithm and the discrete adjoint mode in the following sections.

The coupling between the two codes is achieved via a file exchange. The adjoint mode reads in the design variables, calculates the gradient and writes it out along with the function value. The optimization code reads them in, calculates a new design and writes it out again. As already mentioned, the optimization algorithm takes only a fraction of the time we need for the evaluation of a function value or the gradient. Therefore this approach is efficient and very flexible.

13.4.1 Implementation of the Recursive Multilevel Trust-Region Algorithm

In Gratton et al. [12] and Nash and Lewis [27] some guidelines for the implementation of multilevel algorithms are presented. These are usually suitable for very large problems, where there are as many design variables as grid points. Our applications are smaller, but more complex.

The implementation of the RMTR is therefore tailored to small and medium sized problems, for which the function and gradient evaluations are expensive. These problems at most have several thousand design variables and the Taylor models can be solved by the standard MATLAB QP solver up to full accuracy. For the problems the time to evaluate the function or the gradient is much higher than the solution time for the bound constrained QP.

In the publications of the RMTR and MG/OPT, second order accurate models are used. We do not have second order information available and because of this we use proper Quasi-Newton updates. According to Heinkenschloss and Vicente [19] for example the BFGS update in function space is given by

$$H_{k+1} = H_k + \frac{M_l^{-1}\mathbf{y}_k\mathbf{y}_k^T}{\mathbf{y}_k^T\mathbf{s}_k} - \frac{H_k\mathbf{s}_k(M_l H_k\mathbf{s}_k)^T}{\mathbf{s}_k^T M_l H_k\mathbf{s}_k},$$

where $\mathbf{y}_k = \nabla f_l(\mathbf{u}_{k+1}) - \nabla f_l(\mathbf{u}_k)$, $\mathbf{s}_k = \mathbf{u}_{k+1} - \mathbf{u}_k$. For $M_l = I$ one recovers the common BFGS formula. For the numerical results, we choose M_l such that $\mathbf{u}^T M_l \mathbf{u}$ approximates the square of the L^2 norm.

The update of the trust-region radius is done as follows. If the step is not successful ($\rho_{l,k} < \eta_1$), we shrink the radius by a factor of 2, if the step is very successful ($\rho_{l,k} < \eta_2$), the radius is enlarged by a factor of 2 and not changed otherwise.

13.4.2 Implementation of the Adjoint Mode

The implementation breaks down into the implementation of AD for the construction of the matrix $\mathbf{C_y}$(and $\mathbf{C_u}$, $f_\mathbf{u}$, $f_\mathbf{y}$) and the linear solvers that use multigrid methods.

13.4.2.1 Implementation of AD

We have implemented our own sparse forward mode via operator overloading in Fortran 95. The standard of the language allows for allocatable arrays in derived types, i.e. we may dynamically allocate memory for new nonzero elements. Due to the compatibility of the Fortran standards, the mix with the FORTRAN-77 code yields no problems. The implementation of the sparse forward mode follows the algorithm presented by Ulbrich and Ulbrich in [34]. We mention the main points incorporated by us.

The data structure for the AD variables stores the original intermediate result along with a sparse representation of the gradient. An integer variable stores the current number of nonzeros in the gradient. By setting this variable to zero, one may easily zero out the whole gradient. For a fast multiplication with constants, a prefactor is stored. Therefore a multiplication with the gradient is only of cost one and not related to the number of nonzero elements in the gradient. The main part of the information is stored in two allocatable arrays, one for the nonzero gradient entries and one for their position in the gradient. As mentioned one may add new elements by reallocation of the arrays. We call the prefactor α, the number of nonzeros nnz, the gradient entries (without prefactor) $\tilde{\mathbf{g}}$ and the position array $\mathbf{c}$. Then $\tilde{\mathbf{g}}$ and $\mathbf{c}$ are arrays of length at least nnz and the gradient is zero except for the positions $\mathbf{c}(i)$, where the entry is $\alpha\tilde{\mathbf{g}}(i)$, $i = 1, \ldots, nnz$. For the addition of two sparse vectors, we use a merge sort.

The parallelization of the sparse forward mode is straightforward, as one only has to augment the original MPI routines to transport the gradient information. We observed that the parallel efficiency is better than for the state and adjoint solver.

The FV code uses a cell-centered arrangement, i.e. the pressure and the velocities are defined at the cell centers. The mass fluxes are defined at the centers of the cell faces. Systems for the velocities and pressure correction are solved and then the mass fluxes are updated in an explicit manner, but iteratively. For a consistent discrete adjoint one therefore needs to define the mass fluxes as independent variables and add the defining equations to the adjoint linear system.

It was observed, that the definition of dependent variables as independent variables can give a huge computational speed up. The general size of the linear system to be solved then rises, but has possibly much less entries. The computational effort for the solution of the linear systems and the AD depends directly on the number of nonzero elements in $\mathbf{C_y}$. For the turbulence models, we define the turbulent viscosity as an independent variable. This reduces the number of matrix elements by factor 1.1 (RANS) and 4 (LES). The same procedure was done for a scaling factor for the mass fluxes at the outflow, that caused a huge fill-in when not used as independent variable.

Finally the number of matrix elements for the adjoint system is linear in the number of grid points. This is one of the main prerequisites for the efficient solution of the adjoint equation.

13.4.2.2 Implementation of Linear Solvers

For the solution of the linear systems a standard linear multigrid solver was implemented, see e.g. [4]. The smoothing and solving is carried out by preconditioned parallel Krylov subspace methods. Therefore a parallel GMRES algorithm [30] and the Improved BiCGStab(2) algorithm from [18] are implemented. For the preconditioning of the velocity and pressure variables, the SIMPLE and SIMPLER preconditioner ([29]) are used. The SIMPLE preconditioner is employed for unsteady problems and SIMPLER for steady problems. An underrelaxation of variables did not have a significant effect, therefore we do not use it. The remaining flow variables are then preconditioned block-wise with an ILU or Jacobi preconditioner. The package DLAP [31] carries out the ILU decompositions.

The SIMPLE-type preconditioners involve approximate solves. We use the same kind of approach as the state solver, namely an iterative refinement procedure, where an ILU is used for the approximate solution. The pressure-correction equation uses $5-10$ iterations, the remaining equations use only one iteration.

We use a processor-wise preconditioning for the linear systems for the parallel computations.

13.5 Numerical Results

The correctness of the implementation of the adjoint mode was tested for the available models in steady and unsteady calculations. Available turbulence models are the Smagorinsky [32] and Germano [11] LES models, the standard $k-\epsilon$ RANS model [21] with and without the low Reynolds modification of Chien [6] and the Renormalization Group Theory $k-\epsilon$ model [36].

The relative error versus central differences with properly chosen step lengths varies between $5{\cdot}10^{-9}$ and $5{\cdot}10^{-11}$, depending on the model and the kind of optimization variables. The limiting factor is the accuracy of the divided differences. For these testing the state and the adjoint solver were fully converged.

13.5.1 Accuracy and Efficiency of Adjoint Mode

We will demonstrate the accuracy and efficiency of the adjoint mode with the following example that uses one design variable, motivated by [33].

Depending on the direction, a spinning object in a viscous fluid will create a lift or a down force. This is commonly known as the Magnus effect. We will consider the flow around a rotating cylinder. By the control of the rate of rotation, the drag of the cylinder may be influenced. A corresponding optimization problem would be the search for the optimal rate of rotation for a given flow condition. Nevertheless we are only interested in the validation of the adjoint mode, such that only one gradient computation is carried out. The computations were done on a machine with 16 Dual-Core AMD Opteron 8384 processors with 2.7 GHz and 320 GB RAM.

The general setup is as follows. The inflow boundary is located at $x = -1$, the outflow at $x = 2$. Periodic boundary conditions are used in z-direction at $z = -0.15$ and $z = 0.15$. We use Dirichlet boundary conditions of 1 for the inflow boundary and the walls at $y = 1$ and $y = -1$. The center of the cylinder is located at $(x,y) = (0,0)$ and the cylinder has a diameter of $D = 0.1$. We are using a density of one and so the Reynolds number can be set directly with ν as $Re_D = \frac{0.1}{\nu}$. The design variable is the circumferential velocity of the cylinder u_d and in our test cases we use $\frac{u_d}{u_\infty} = 0.1$, where u_∞ is the free-stream velocity.

We will show the efficiency of the adjoint mode with two large scale simulations.

13.5.1.1 Smagorinsky Model, Re = 1,000

For this test case we set $\mathbf{v} = 10^{-4}$ resulting in $Re = 1{,}000$ and we are using the Smagorinsky LES model with Smagoringsky constant $Cs = 0.1$. The grid has 143,616 CVs and is distributed to 11 cores. For the spatial discretization we use a second order accurate scheme with a 5% blending of an upwind scheme. The temporal discretization is done with the Crank-Nicolson scheme with 5,000 time steps of $2 \cdot 10^{-3}$ s, such that the CFL number is approximately 1.

Figure 13.2 illustrates the several interesting features of the flow. After a transitional phase of about 1,000 time steps the flow gets instationary, but is two dimensional. Then the two dimensional flow breaks up and the flow gets three dimensional. This can be seen even better in the evolution of the drag in Fig. 13.3. The general appearance clearly changes between the 2,200th and 3,000th time step. The flow is never periodic.

The target function is computed by taking the average of the drag of the last 2,000 time steps, yielding a value of ~ 1.12. Note that for the adjoint calculation still the whole 5,000 time steps are needed. The flow equations are solved up to an error of 10^{-7}, while every time step of the adjoint is solved up to an error of 10^{-8}.

Central differences with a step size of 10^{-6} yield an approximate gradient $9.8595 \cdot 10^{-2}$, the adjoint calculation gives $9.8518 \cdot 10^{-2}$, resulting in a relative error of $7.8 \cdot 10^{-4}$. This shows the capability of the adjoint mode to calculate very accurate gradients over a longer time interval.

The original flow simulation takes about 11.1 h of CPU time, while the adjoint calculation takes 40.1 h. This results in a runtime ratio of 3.6.

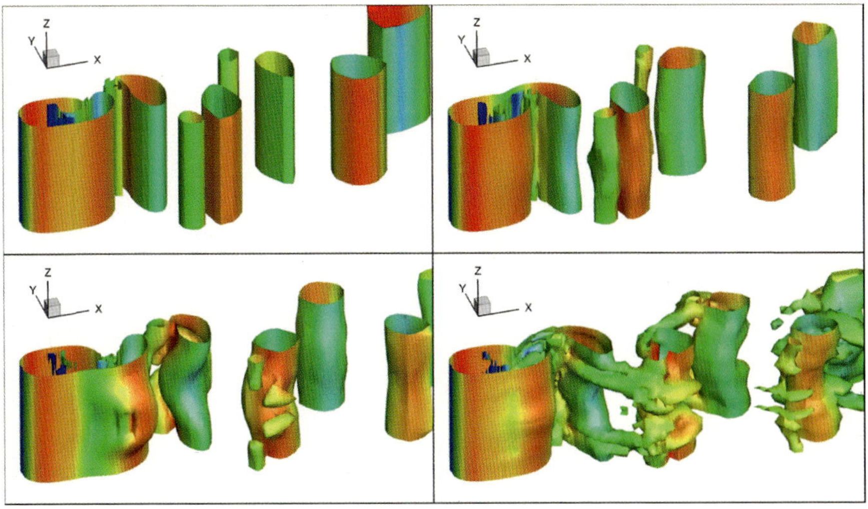

Fig. 13.2 Instantaneous velocity snapshots for $Re = 1{,}000$

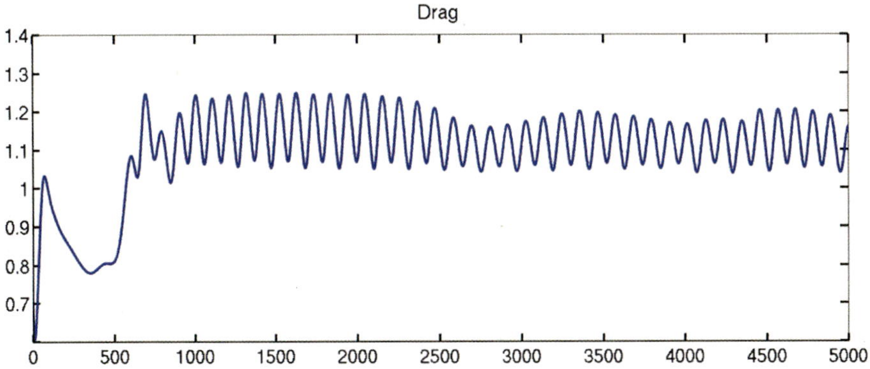

Fig. 13.3 Evolution of the drag for $Re = 1{,}000$

One time step needs 12.41 MB of disc space, resulting in a total requirement of 60.6 GB. The time for the storage of one step is ~0.03 s on our system and is included in the forward solve.

13.5.1.2 Germano Model, Re = 5,000

For this test case we set $\nu = 2{\cdot}10^{-5}$ resulting in $Re = 5{,}000$ and we are using the Germano LES model. The grid has 1,148,928 CVs and is distributed to 22 cores.

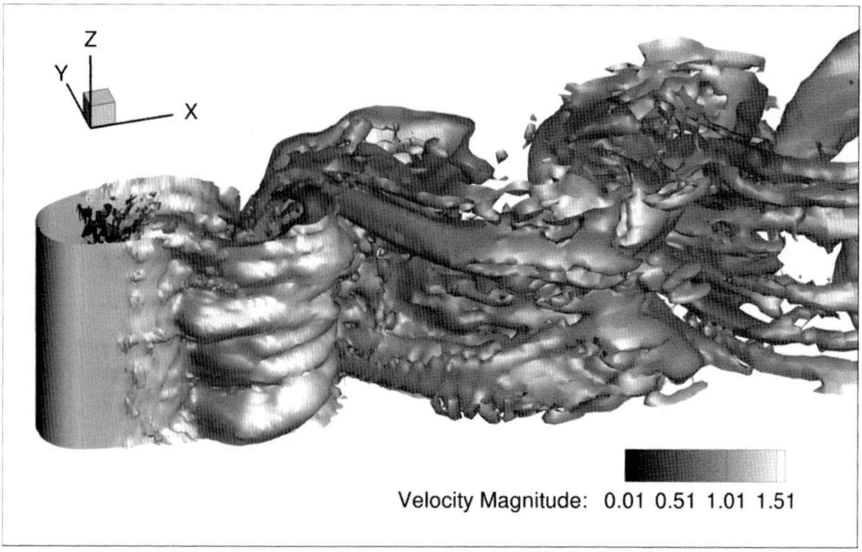

Fig. 13.4 Instantaneous velocity snapshot for $Re = 5{,}000$

For the spatial discretization we use a second order accurate scheme with a 5% blending of an upwind scheme. The temporal discretization is done with the Crank-Nicolson scheme with 1,000 time steps of 10^{-3} s, yielding again a CFL number close to 1.

The simulation takes a snapshot from a fully developed flow as initial velocity. As one can see in Fig. 13.4, the flow is fully turbulent and three dimensional.

The target function is computed by taking the average of the drag of the last 999 time steps, yielding a value of $\sim$1.06. The flow equations are solved up to an error of 10^{-5}, while every time step of the adjoint is solved up to an error of 10^{-6}.

Central differences with a step size of 10^{-6} yield a gradient approximation of $-8.53 \cdot 10^{-2}$, the adjoint calculation gives $-8.73 \cdot 10^{-2}$, resulting in a relative error of $2.3 \cdot 10^{-2}$. This is a very good result, considering the comparatively low accuracy of the solves and the Reynolds number.

The original flow simulation takes about 7.3 h of CPU time, while the adjoint calculation takes 60.9 h. This results in a runtime ratio of 8.3. The increase in runtime ratio versus the simulation with the Smagorinsky model comes from the less accurate solve of the state equation and a huge fill in caused by the Germano model.

One time step needs 79.33 MB of disc space, resulting in a total requirement of 77.5 GB. The time for the storage of one step is $\sim$0.23 s on our system and is included in the forward solve.

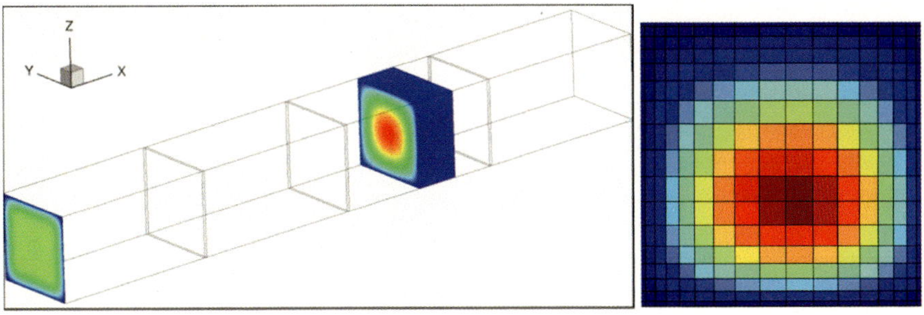

Fig. 13.5 Wind tunnel setup (*left*) and velocity on inflow boundary (*right*)

13.5.2 Optimization

13.5.2.1 Boundary Control

We will first consider a stationary model problem, that has a favorable structure for the RMTR. For this example, the goal is to match a certain velocity profile in a tunnel and we need to find the corresponding inflow velocity. The situation is depicted on the left side in Fig. 13.5.

To exploit a maximal degree of freedom, we are able to prescribe the inflow velocity for every grid point on the inflow boundary. This situation can be seen on the right side in Fig. 13.5, which also depicts the initial velocity for the optimization.

The optimization problem reads

$$\min_{u_{in} \in U_{ad}} \frac{1}{2} \|u - u_m\|^2_{L^2(\Omega')} + \frac{\alpha}{2} \|u_{in}\|^2_{H_0^2(\Gamma_{in})}$$

$$\begin{aligned} s.t.\ (\mathbf{v} \cdot \nabla)\mathbf{v} - \nu \Delta \mathbf{v} + \nabla p &= 0 && \text{in } \Omega, \\ \nabla \cdot \mathbf{v} &= 0 && \text{in } \Omega, \\ \mathbf{v} &= 0 && \text{on } \Gamma_D, \\ (u, v, w) &= (u_{in}, 0, 0) && \text{on } \Gamma_{in}, \\ \tfrac{\partial \mathbf{v}}{\partial \mathbf{n}} &= 0 && \text{on } \Gamma_N. \end{aligned}$$

The computational domain is $\Omega = [0, 8] \times [0, 1] \times [0, 1]$ and the domain where the velocity is matched is

$$\Omega' = [5, 5.5] \times [0, 1] \times [0, 1].$$

Fig. 13.6 Multilevel optimization results

level	#CVs	#DVs	eqv. f	eqv. g
1	256	16	5	5
2	2048	64	5.38	5.38
3	16384	256	5.63	5.63
4	131072	1024	5.83	5.83
5	1048576	4096	5.71	5.71

A no-slip condition is imposed on the wall Γ_D. On the inflow boundary Γ_{in} the velocity in x-direction u_{in} can be adjusted. The velocity to match is

$$u_m(x, y, z) = 0.8 \left(\frac{y(1-y) \cdot z(1-z)}{0.5^4} \right).$$

The viscosity is chosen as $\nu = 0.01$ and the regularization parameter α is $2 \cdot 10^{-6}$. The admissible set is

$$U_{ad} = \left\{ (u, v, w) = \mathbf{v} \in W_0^2(\Gamma_{in}) \, | 0 \le u \le 0.5 \right\}.$$

All grid points on the inflow boundary are design variables. We are using five grid levels, such that the number of control volumes varies from 256 to 1,048,576 and the number of design variables from 16 to 4,096.

The restriction and prolongation operators for the RMTR$_\infty$ were taken from the code of the flow solver.

They are used by the adjoint multigrid procedure anyway and we wrote out the relevant part for the inflow boundary. The other parameters of the RMTR are set as follows: $\eta_1 = 0.01$, $\eta_2 = 0.7$, $\kappa_\chi = 0.1$, the initial trust-region radii are chosen larger than the bound constraints.

For the recursions, we use free-form cycles with one post optimization step, i.e. we always do a recursion if allowed, then followed by a step of the quadratic model. For the quadratic model the SR1 update is used.

The optimization algorithm stops, if the criticality measure is below 10^{-8}. This is near the maximal accuracy that can be reached for this example. The results of the optimization can be seen in Fig. 13.6. The effort for the whole optimization task is expressed in function and gradient evaluations on the finest level. So *eqv. f* and *eqv. g* include the costs of the lower level function and gradient evaluations, converted to the costs of fine level evaluations.

As we can see, the cost stays nearly constant on the all levels, as expected from a multilevel algorithm. The number of function and gradient evaluations are always the same, which means that no steps are rejected.

13.5.3 Shape Optimization

Next we will show two multilevel optimization examples, based on RANS and LES calculations. Basically, the idea is to modify the shape of a wind tunnel, such that

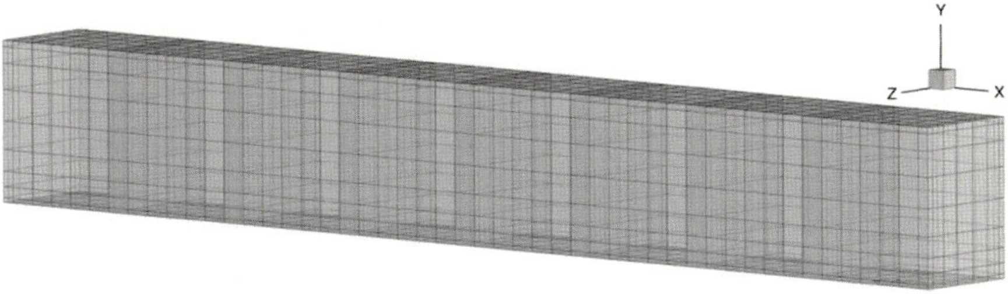

Fig. 13.7 Unperturbed wind tunnel

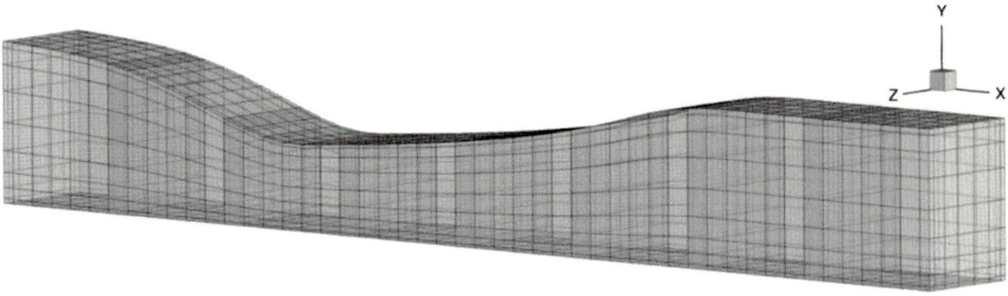

Fig. 13.8 Perturbed wind tunnel

a certain pressure gradient is obtained in a section of the tunnel. This optimization test case originated from the Institute for Fluid Mechanics and Aerodynamics at TU Darmstadt and the diploma thesis [22].

In practice, the design procedure involves the construction of many different displacement bodies and the use of proper optimization techniques may speed up the process significantly.

We will present an example, where the RMTR is used with a steady RANS model and an example, where a successive refinement approach is used with an instationary LES model.

The optimization starts from the unperturbed domain Ω, that describes the wind tunnel geometry. It is defined as

$$\Omega = [0, 2] \times [-0.27, 0] \times [0, 0.2].$$

See Fig. 13.7 to get an impression. Admissible designs change the ceiling of the tunnel, see Fig. 13.8.

The fluid is air, so we set the dynamic viscosity μ to $1.806 \cdot 10^{-5}$ and the density ρ to 1.225. We set Dirichlet boundary conditions of 6(m/s) for the block inflow and no-slip boundary conditions on the four walls.

13.5.3.1 Test Case I

We are searching for a function $\mathbf{u} : [0, 2] \to [-0.27, 0]$ with $\mathbf{u}(0) = \mathbf{u}(2) = 0$ that defines the ceiling. The changes are kept constant in z-direction. The computational

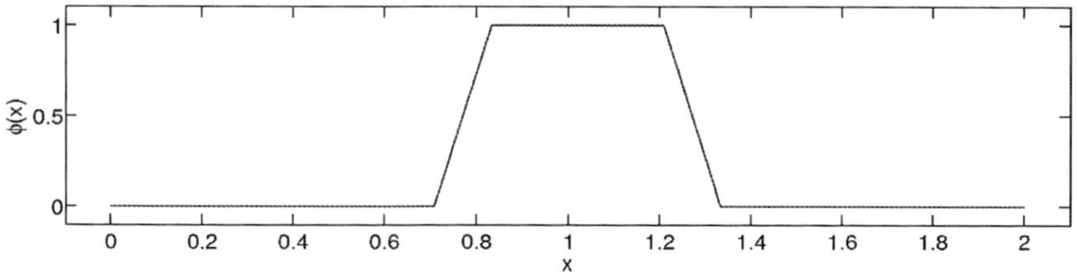

Fig. 13.9 Weighting function

domain is then defined by

$$\Omega(\mathbf{u}) = \{\mathbf{x} = (x, y, z) \in \Omega \,|\, y \le \mathbf{u}(x)\}.$$

The restriction on the range of the function **u** is later implemented by point-wise constraints. The optimization problem uses a RANS model and is given by

$$\min_{\mathbf{u} \in U_{ad}} \frac{1}{2} \int_{\Omega(\mathbf{u})} \phi(x) \left(\frac{\partial p}{\partial x} - 67 \right)^2 d\mathbf{x} + \frac{\alpha}{2} \|\mathbf{u}\|_{L^2([0,2])} + \frac{\beta}{2} \|\mathbf{u}\|_{H^2([0,2])}$$

$s.t.$ RANS model of Chien

$$\mathbf{v} = \mathbf{v}_D \qquad \text{on } \partial\Gamma_D(\mathbf{u}),$$

$$\frac{\partial \mathbf{v}}{\partial \mathbf{n}} = 0 \qquad \text{on } \partial\Gamma_N(\mathbf{u}).$$

A pressure gradient in x-direction of 67 shall be matched between $x = 0.83$ and $x = 1.23$. This is implemented with the weighting function $\phi(x)$, that realizes a smooth weighting of the cost function, depicted in Fig. 13.9.

With α, we can penalize the L^2 norm of **u**. This is to achieve the use of less material for the construction for the displacement body. With β, we can penalize the H^2 norm of **u**, to get a smooth shape of the ceiling. We set $\alpha = 400, \; \beta = 0.2$.

Additionally, we impose point wise constraints via

$$U_{ad} = \left\{ \mathbf{u} \in H_0^2([0,2]) \,|\, {-0.26} \le \mathbf{u} \le 0 \right\}.$$

The upper bound is due to the fact, that we may only add new material to the ceiling, the lower bound is because we may not take away material from the bottom of the tunnel. For meaningful shapes of the tunnel, the lower bounds will never be active and they also stabilize the optimization algorithm.

We are using three grid levels with 2,304, 18,432 and 147,456 CVs, that also define the discretization of **u**. Accordingly the number of design variables from lowest to highest grid level is 47, 95 and 191. For the restriction of the design variables, we use an injection. The prolongation is

Fig. 13.10 Number of function/gradient evaluations for the optimization

Algorithm\Level	0	1	2
RMTR	89/70	76.3/59.9	104.0/70.3
Successive Refinement	89/70	95.4/69.0	203.0/149.6

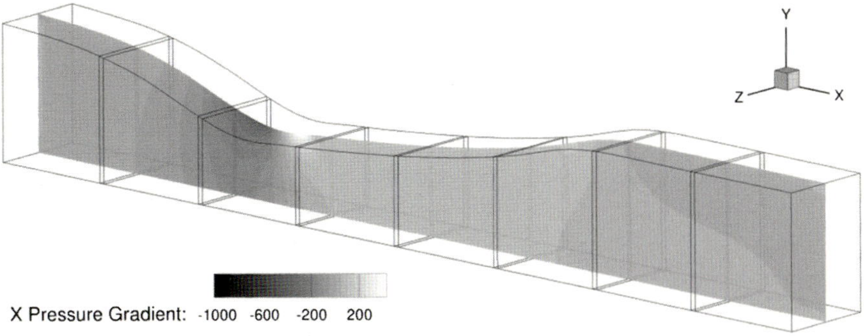

Fig. 13.11 Optimal shape

$$\frac{1}{4}\begin{bmatrix} 1 & 2 & 1 & & & \\ & 2 & 2 & & & \\ & 1 & 2 & 1 & & \\ & & 2 & 2 & & \\ & & 1 & 2 & 1 & \end{bmatrix}.$$

The optimization results can be found in Fig. 13.10, where the costs of the optimization are expressed as function and gradient evaluations on the currently highest grid. Here we used $(\varepsilon_0, \varepsilon_1, \varepsilon_2) = (0.01, 0.1, 0.05)$ and the criticality measure starts at a value of $\sim$5. The optimal shape is shown in Fig. 13.11.

For the highest level the recursive algorithm is about twice as fast as the successive refinement approach, where the prolongated optimum of the lower level was used as initial value and on the finest level only Taylor iterations were performed. We expect the recursive algorithm to perform even better with more design variables.

For the first level we used $\kappa_x = 0.1$ and for the second level $\kappa_x = 0.2$. With the choice of κ_x we can steer to what extent the lower levels are used. For this example it takes the recursive algorithm few steps to get close to the optimum, but then needs additional iterations to build up second order information and finally fulfill the convergence criterion. Note that we do not have second order information available.

The runtimes for the evaluation timing of the function/gradient are 1 s/1 s on the first level, 14 s/14 s on the second level and 110 s/125 s on the finest level. The state and adjoint solver were converged up to the same accuracy of 10^{-8}, 10^{-7}, 10^{-6} from coarsest to finest level. We used 8 cores of the AMD Opteron 8384 machine mentioned before.

The runtimes for the evaluation timing of the function/gradient are 1 s/1 s on the first level, 14 s/14 s on the second level and 110 s/125 s on the finest level. The state and adjoint solver were converged up to the same accuracy of, 10^{-8}, 10^{-7}, 10^{-6} from coarsest to finest level. We used 8 cores of the AMD Opteron 8384 machine mentioned before.

13.5.3.2 Test Case II

For this optimization case, we use the instationary Smagorinsky turbulence model. The setup is a bit different than before, as the ceiling is parameterized by cubical B-Splines with 10 design variables. It may be changed between 0 and 1.45 in x-direction. Again the constraints $\mathbf{u} \leq 0$ are used.

The problem accordingly reads:

$$\min_{\mathbf{u}} \frac{1}{2} \int_{t_0}^{T} \int_{\Omega_x(\mathbf{u})} \left(\frac{\partial p}{\partial x} - 67 \right)^2 dx\, dt$$

$$s.t.\ \rho \mathbf{v}_t + \rho(\mathbf{v} \cdot \nabla)\mathbf{v}$$

$$\begin{aligned}
&-\nabla \cdot \left(\left(\mu + \rho (C_s \Delta)^2 \left(2 \left\| \nabla^s \mathbf{v} \right\|_F^2 \right)^{\frac{1}{2}} \right) \nabla^s \mathbf{v} \right) + \nabla p = 0 && \text{in } \Omega(\mathbf{u}) \times I, \\
&\nabla \cdot \mathbf{v} = 0 && \text{in } \Omega(\mathbf{u}) \times I, \\
&\mathbf{v} = \mathbf{v}_D && \text{on } \Gamma_D(\mathbf{u}) \times I, \\
&\tfrac{\partial \mathbf{v}}{\partial \mathbf{n}} = 0 && \text{on } \Gamma_N(\mathbf{u}) \times I, \\
&\mathbf{v}(x,0) = \mathbf{v}_0(x) = 0 && \text{in } \Omega(\mathbf{u}).
\end{aligned}$$

Here $\|\cdot\|\ F$ is the Frobenius norm and we use the symmetric part of the gradient

$$\nabla^s \mathbf{v} = \frac{1}{2} \left(\nabla \mathbf{v} + \nabla^T \mathbf{v} \right).$$

We set the Smagoringsky constant C_s to 0.1 and the local filter width is defined in relation to the local grid widths Δx, Δy, Δz as

$$\Delta = \left(\frac{1}{3} \left((\Delta x)^2 + (\Delta y)^2 + (\Delta z)^2 \right) \right)^{\frac{1}{2}}.$$

The time interval is $I = (0,T)$. For our numerical tests, we use t_0=0.4 and $T = 1.6$. The pressure gradient is matched in the following part of the wind tunnel

$$\Omega_x(\mathbf{u}) = \Omega(\mathbf{u}) \cap \{(x, y, z) \in \Omega(\mathbf{u})\, |x \in [0.83, 1.23]\}.$$

For the time discretization the Crank-Nicholson scheme is used with 800 time steps of $2 \cdot 10^{-3}$s and we use a second order accurate scheme in space.

Level	Control Volumes	Objective	Iterations	Time (min)
1	12800	0.33106	23	780
2	24192	0.56990	11	995
3	49152	0.64311	5	706
4	102400	0.89729	17	7083

Fig. 13.12 Convergence history

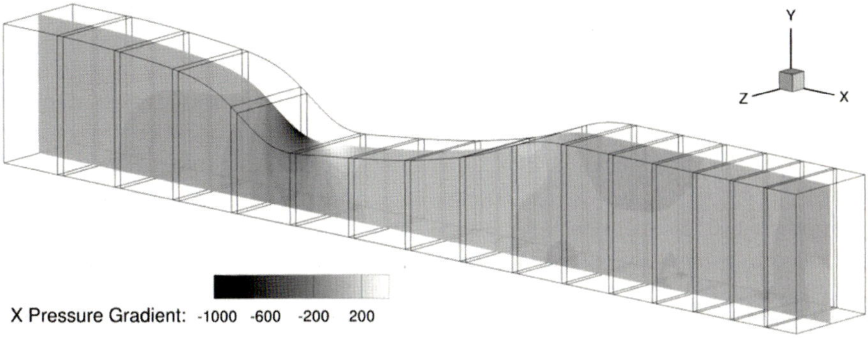

Fig. 13.13 Optimal shape

The optimization is done with a successive refinement approach. The optimization problem is solved on the coarsest grid with a SQP-solver that was developed in our group. Then the solution is used as a starting point for the next iteration. Also the approximation of the Hessian and dual variables are reused. The optimization stops, if the reduced gradient has decreased by a factor of 10^{-3} from the initial value.

In total we have used four different grid levels and the optimization results can be found in Fig. 13.12 and the optimal shape is shown in Fig. 13.13.

The flow is unsteady on the final grid level and steady on the coarser levels because it is underresolved there. This is also reflected in the number of optimization steps taken on each level. The most number of iterations of the optimization algorithm are needed on the coarsest level and decrease for the next two levels. On the fourth level the problem changes significantly and correspondingly the number of iterations rises. Still we have possibly saved computational time on the finest grid, which are very costly compared to steps on the coarser grids.

The acquired shapes are very similar to those obtained and verified in the diploma thesis [22]. The slope of the wind tunnel very similar in the RANS and the LES configurations.

For the adjoint method, we observe a factor of 5.1 in computational time on the finest grid compared to the flow solution. The time spent for obtaining all the derivatives $f_{\mathbf{u}}$,$f_{\mathbf{y}}$,$C_{\mathbf{u}}$,$C_{\mathbf{y}}$ takes about 42% of the total time for the adjoint. The forward solver was converged to 10^{-5} and the adjoint solver to 10^{-6}. On the highest level we used 16 cores of a machine with 8 Dual-Core AMD Opteron 8220 processors with 2.8 GHz and 128 GB RAM.

13.6 Conclusions

In this contribution we have discussed an approach to the optimization of (turbulent) flows. The first part was devoted to an efficient multilevel optimization algorithm. Here, different models may be used to speed up the optimization process. Some modifications of the algorithm have been given to adapt it to the setting of discretized optimal control problems. In the second part we have introduced a discrete adjoint approach that is suitable for large scale steady and unsteady flow computations. For the successful application, there are several important points. We use a sparse forward mode of AD to obtain a linear system for the adjoint. This system is then solved with the use of multigrid techniques and Krylov subspace methods with appropriate preconditioners.

Combining the optimization and adjoint techniques, we are able to handle a wide variety of optimization problems. The numerical results show the application of the adjoint mode and the optimization algorithm to complex problems. These problems include stationary RANS and instationary LES calculations for high Reynolds number flows.

Acknowledgements The authors gratefully acknowledge the support of the Sonderforschungbereich 568 funded by the German Research Foundation (DFG). Moreover, the first author was supported by the Graduate School Computational Engineering and the Center of Smart Interfaces at TU Darmstadt, which are both funded by the German Research Foundation (DFG).

References

1. Alexandrov, N., Lewis, R.: First-order frameworks for managing models in engineering optimization. In: 1st International Workshop on Surrogate Modelling and Space Mapping for Engineering Optimisation, vol. 11, pp. 16–19 (2000)
2. Balay, S., Brown, J., Buschelman, K., Eijkhout, V., Gropp, W., Kaushik, D., Knepley, M., McInnes, L., Smith, B., Zhang, H.: Petsc users manual revision 3.2 (2011)
3. Borzı, A., Schulz, V.: Multigrid methods for PDE optimization. SIAM Rev. **51**(2), 361–395 (2009)
4. Briggs, W., Henson, V., McCormick, S.: A Multigrid Tutorial. Society for Industrial Mathematics, Philadelphia (2000)
5. Chang, K.J., Haftka, R.T., Giles, G.L., Kao, P.J.: Sensitivity-based scaling for approximating structural response. J. Aircr. **30**, 283–288 (1993)
6. Chien, K.Y.: Predictions of channel and boundary-layer flows with a low-Reynolds-number turbulence model. AIAA J. **20**(1), 33–38 (1982)
7. Christianson, B.: Reverse accumulation and attractive fixed points. Optim. Methods Softw. **3**(4), 311–326 (1994)
8. Conn, A., Gould, N., Toint, P.: Trust-Region Methods, vol. 1. Society for Industrial Mathematics, Philadelphia (2000)
9. FASTEST User Manual.: Fachgebiet Numerische Berechnungsverfahren im Maschinenbau, Technische Universität Darmstadt (2005)
10. Gelman, E., Mandel, J.: On multilevel iterative methods for optimization problems. Math. Programming **48**(1), 1–17 (1990)

11. Germano, M.: Turbulence: the filtering approach. J. Fluid Mech. **238**, 325–336 (1992)
12. Gratton, S., Mouffe, M., Sartenaer, A., Toint, P., Tomanos, D.: Numerical experience with a recursive trust-region method for multilevel nonlinear bound-constrained optimization. Optim. Methods Softw. **25**(3), 359–386 (2010)
13. Gratton, S., Mouffe, M., Toint, P., Weber-Mendonça, M.: A recursive trust-region method for bound-constrained nonlinear optimization. IMA J. Nume. Anal. **28**(4), 827–861 (2008)
14. Gratton, S., Sartenaer, A., Toint, P.: Recursive trust-region methods for multiscale nonlinear optimization. SIAM J. Optim. **19**(1), 414–444 (2008)
15. Griesse, R., Walther, A.: Evaluating gradients in optimal control: continuous adjoints versus automatic differentiation. J. Optim. Theory Appl. **122**(1), 63–86 (2004)
16. Griewank, A.: Achieving logarithmic growth of temporal and spatial complexity in reverse automatic differentiation. Optim. Methods Softw. **1**, 35–54 (1992)
17. Griewank, A., Walther, A.: Evaluating Derivatives, 2nd edn. SIAM, Philadephia (2008)
18. Gu, T., Zuo, X., Liu, X., Li, P.: An improved parallel hybrid bi-conjugate gradient method suitable for distributed parallel computing. J. Comput. Appl. Math. **226**(1), 55–65 (2009)
19. Heinkenschloss, M., Vicente, L.N.: An interface between optimization and application for the numerical solution of optimal control problems. ACM Trans. Math. Softw. **25**, 157–190 (1998)
20. Hinze, M., Pinnau, R., Ulbrich, M., Ulbrich, S.: Optimization with PDE Constraints, Mathematical Modelling, vol. 23. Springer, Dordrecht (2009)
21. Jones, W., Launder, B.: The prediction of laminarization with a two-equation model of turbulence. Int. J. Heat Mass Trans. **15**, 301–314 (1972)
22. Kretz, F.: Auslegung, konstruktion und erprobung einer druckgradienten-messstecke. Technische Universität Darmstadt, Diplomarbeit (2006)
23. Lewis, R., Nash, S.: A multigrid approach to the optimization of systems governed by differential equations. In: 8-th AIAA/USAF/ISSMO Symposium Multidisciplinary Analysis and Optimization (2000)
24. MPI: A message-passing interface standard, Version 2.2. Message Passing Interface Forum (2009)
25. Mouffe, M.: Multilevel optimization in infinity norm and associated stopping criteria. In: Ph.D. thesis, Institut National Polytechnique de Toulouse (2009)
26. Nash, S.: A multigrid approach to discretized optimization problems. Optim. Methods Softw. **14**(1/2), 99–116 (2000)
27. Nash, S., Lewis, R.: Assessing the performance of an optimization-based multilevel method. Optim. Methods Softw. **26**(4–5), 693–717 (2011)
28. Quarteroni, A., Valli, A.: Domain Decomposition Methods for Partial Differential Equations. Clarendon Press, New York (1999)
29. ur Rehman, M., Vuik, C., Segal, G.: Preconditioners for the steady incompressible Navier-Stokes problem. Int. J. Appl. Math. **38**, 223–232 (2008)
30. Saad, Y., Schultz, M.: Gmres: a generalized minimal residual algorithm for solving nonsymmetric linear systems. SIAM J. Sci. Stat. Comput. **7**(3), 856–869 (1986)
31. Segar, M.K.: A SLAP for the Masses, pp. 135–155. Wiley, New York (1989)
32. Smagorinsky, J.: General circulation experiments with the primitive equation. Mon. Weather Rev. **91**(3), 99–164 (1963)
33. Tuncer, I. (ed.): Automatic generation of discrete adjoints for unsteady optimal flow control. ECCOMAS CFD & Optimization, Paper No. 2011-043 (2011)
34. Ulbrich, M., Ulbrich, S.: Automatic differentiation: a structure-exploiting forward mode with almost optimal complexity for kantorovic trees. Appl. Math. Parallel Comput. 327–357 (1996)
35. Walther, A.: Program reversal schedules for single- and multi-processor machines. In: Ph.D. thesis, Institute of Scientific Computing, Technical University Dresden, Germany (1999)
36. Yakhot, V., Orszag, S.A., Thangam, S., Gatski, T.B., Speziale, C.G.: Development of turbulence models for shear flows by a double expansion technique. Phys. Fluids **4**, 1510–1520 (1992)
37. Ziems, J.C., Ulbrich, S.: Adaptive multilevel inexact SQP methods for pde-constrained optimization. SIAM J. Optim. **21**(1), 1–40 (2011)

Part V
Transfer Projects

The last major part of the book assembles a number of research topics that have been formulated and carried out in a partnership between industry and university to demonstrate the transferability of the models, methods and techniques developed to other application fields dealing with flow and combustion. Chapter 14 summarizes the results of assessing the state of the art model capabilities of the two-phase LES as existing in FLUENT/ANSYS code. From various analyses and parameter studies weak points have been identified and efforts have been achieved in order to increase the predictive ability of the LES tool in a wide range of applications.

Droplet size measurements are also crucial for characterizing practical sprays in automotive applications. Besides phase Doppler (PD) techniques that allow for highly accurate point-wise measurements of single droplets diameters and velocities there is some need for planar drop sizing methods in engineering applications that call for much faster spray characterizations. Simultaneous Mie scattering and planar laser-induced fluorescence (Mie/PLIF) is one choice that provides access to Sauter-mean-diameters (SMD). Chapter 15 reports on the adaptation of the Mie/PLIF technique to real-world requirements as conducted in cooperation with BMW (F1).

In the final chapter the transfer of laser-based measurement techniques is highlighted, which were used and improved in the context of stationary gas turbine combustion to the investigation of intermittent processes in internal combustion (IC) engines. The focus was on capturing cycle-to-cycle fluctuations as they appear in recent direct injection IC engines. Thereby it is demonstrated in cooperation with the Robert Bosch GmbH how high-speed measurement techniques are applied to investigate the temporal evolution of the in-cylinder flow, fuel distribution and flame propagation. In addition charge motion is investigated by Particle Image Velocimetry (PIV) and spray by imaging of Mie-scattering. Mixture distribution is captured qualitatively by means of Planar Laser-Induced Fluorescence (PLIF) of a fluorescing fuel while OH-PLIF is used to investigate the development of the early flame kernel and turbulent flame propagation.

Chapter 14
Large Eddy Simulation of Dispersed Two-Phase Flows and Premixed Combustion in IC-Engines

D. Dimitrova, M. Braun, J. Janicka, and A. Sadiki

Abstract An accurate prediction of particle dispersion is an essential issue for reactive two-phase flows as they occur in IC-engines. It is also a challenging application for Large Eddy Simulation (LES) based Eulerian–Lagrangian methods. The main objective of this work is to assess the state-of-the-art model capabilities of the LES based Eulerian–Lagrangian method as implemented into the commercial CFD code, FLUENT/ANSYS. This is achieved by carrying out various parameter studies that may enable a deeper understanding of the interactions between the numerics and modeling involved, and thus an increasing of the predictive ability and the reliability of transfer of findings from one configuration to others. In this report, special attention is paid to the prediction of the particle preferential accumulation, because of its importance for simulations of mixing and combustion in turbulent reacting two-phase flows. The combustion itself is not considered. The conclusions are based on a systematic variation of relevant flow parameters, such as the Reynolds number and the particle Stokes number, so that a wide range of applications is covered. Therefore, several particle–laden flow configurations, such as two plane channel flows, a free jet and an evaporating spray at low temperature, have been investigated. The results presented in this report are especially for the two plane channel flows characterized by low and high Reynolds numbers, respectively. It was observed that the maximum preferential accumulation occurs at a constant Stokes number and that this number does not depend on the Reynolds number. The magnitude of the accumulation, however, depends on the Reynolds number of the

D. Dimitrova (✉) • M. Braun
Fluent/ANSYS Deutschland GmbH, Birkenweg 14A, 64295 Darmstadt, Germany
e-mail: desislava.dimitrova@ansys.com; markus.braun@ansys.com

J. Janicka • A. Sadiki
Department of Mechanical and Processing Engineering, Institute for Energy and Powerplant Technology, Technische Universität Darmstadt, Petersenstr. 30, 64287 Darmstadt, Germany
e-mail: janicka@ekt.tu-darmstadt.de; sadiki@ekt.tu-darmstadt.de

J. Janicka et al., *Flow and Combustion in Advanced Gas Turbine Combustors*, Fluid Mechanics and Its Applications 102, DOI 10.1007/978-94-007-5320-4_14,

flow. The effect of a sub-grid dispersion model on the particle accumulation was found to be less pronounced for particles with characteristic time scales in the order of the Kolmogorov scale.

Keywords Two-phase flow • Preferential accumulation • Particle dispersion modeling • Eulerian–Lagrangian method • LES

14.1 Introduction

In the last two decades Large Eddy Simulation (LES) technique for single phase flow simulations has denoted a significant progress in its feasibility and reliability [30, 31, 39, 41, 55, 64, 81, 83]. This could be made possible thanks to the impressively fast development of computational resources. So, the method has been successfully extended for the simulations of reacting single phase flows, as reported by Pitsch [68], Janicka and Sadiki [40] and others (e.g. [68]). Recently special care has being taken of quality assessment of the method (e.g. [13, 58]).

LES for prediction of two-phase flows is still at its beginning stage [22, 34, 59, 110, 112]. It pursues two main development streams. Vance et al. [104], Kuerten and Vreman [43], Vreman et al. [107, 108] and Fede and Simonin [25] conducted research work on fundamental understanding of phenomena in dilute two-phase flows and on the ability of LES to provide reliable results (see also Pozorski et al. [70] and therein quoted references). The authors highlight deficits of the standard Eulerian–Lagrangian method with focus on the turbulent dispersion of small particles. This topic has a vital relevance for aircraft and IC engine technology, since the method can be applied to simulate such combustion systems fired by liquid or solid fuels. Mixing processes are still an issue like in single phase LES based simulations. The additional uncertainties introduced by the consideration of chemical reactions can further violate the reliability of the simulation (e.g. [14, 33, 63]). A second direction for the development of two-phase LES consists in studying the applicability of the method to practical configurations. It is outlined by Riber et al. [76], Apte et al. [3], Boileau et al. [10] and others, e.g. [14, 33, 63, 81]. The configurations mostly investigated exhibit complex flow patterns, involving heat and mass transfer, and include swirling or recirculating flows as well as spray combustion (e.g. [14, 21, 33, 66, 73, 74, 78]). Reasonable achievements in terms of statistical properties of the velocity fields for both phases and for the local particle diameter distribution are generally reported.

Comprehensive experimental data available beyond velocity fields is rather very rare [1, 9, 12, 19, 27, 28, 54, 57, 65, 72, 96]. The LES based Eulerian–Lagrangian approach, that has proven to successfully simulate complex flow configurations, still suffers from a lack of information concerning local and unsteady phenomena and interacting mechanisms between the phases despite the ability of LES to deliver impressive qualitative results. Due to the limited experimental data [45, 46, 81, 92–96], many investigations are restricted to reference data from DNS, [11, 61, 77], which are performed for flows at very low Reynolds number. In contrast to the methods developed for quality assessment of LES in single phase flows [13, 58], the development of controlling mechanisms for quality estimation of LES in the multiphase context is still at the very beginning stage.

The present contribution utilizes the Eulerian–Lagrangian approach to describe dilute dispersed two-phase flows in the context of LES. It aims at assessing the state-of-the-art model capabilities of the LES based Eulerian–Lagrangian method as implemented into the commercial CFD code, FLUENT/ANSYS [2]. For this purpose, a systematic variation of relevant flow parameters, such as the Reynolds number and the particle Stokes number, has been carried out in several particle–laden flow configurations including two plane channel flows, a free jet and an evaporating spray at low temperature. In this paper, attention is especially paid to the prediction of the particle preferential accumulation in the plane channel flow configurations because of the determinant influence of this phenomenon on mixing and combustion in turbulent reacting two-phase flows.

The following section introduces fundamentals of the governing equations and modelling used. Section 14.3 outlines the main characteristic features of the CFD code applied in this work. This will be followed in Sect. 14.4 by the description of the configurations under investigation. The next section (Sect. 14.5) summarizes and discusses the main results, whilst the last section is devoted to conclusions.

14.2 Governing Equations and Modelling

Following the Eulerian–Lagrangian method the turbulent fluid phase is described according to the classical LES approach. The governing equations (continuity equation, Navier–Stokes equations, etc.) are filtered to separate the large-scale and small-scale turbulence. The unsteady, general form of the filtered transport equations appears as:

$$\frac{\partial \left(\bar{\rho}\,\tilde{\phi}\right)}{\partial t} + \frac{\partial \left(\bar{\rho}\tilde{u}_i\,\tilde{\phi}\right)}{\partial x_i} - \frac{\partial}{\partial x_i}\left(\bar{\Gamma}\,\frac{\partial \tilde{\phi}}{\partial x_i} - \Lambda_i^{sgs}\right) = \bar{S}_\phi + \bar{S}_{\phi,p} \tag{14.1}$$

in which ϕ may represent the mass density, velocity components u_i, scalar transported quantity or subgrid scale (SGS) turbulent kinetic energy, etc., respectively. The quantity Γ stands for an effective diffusion coefficient, while $\bar{S}_\phi$ expresses turbulence source terms well known in single phase flows. The additional source term $\bar{S}_{\phi,p}$ in Eq. (14.1) represents the volume averaged rate of exchange of momentum, scalar or SGS turbulent transported quantities between the gas and the particulate phase, respectively. It expresses the interaction between the two phases and account for the two-way coupling of phases without phase transition. Equation (14.1) govern the evolution of the large, energy-carrying, scales of flow field. Thereby $\bar{(\cdot)}$ and $\tilde{(\cdot)}$denote the spatial filter and the mass-weighted spatial filter, respectively. The filter length can be defined as the average box length of the numerical grid as proposed by Schuman [87]. The effect of the small scales appears through the SGS stress tensor (and SGS scalar flux in the case of passive or reactive scalar transported quantity, [14, 21, 33, 63]).

Different authors postulated different requirements for the resolution of an LES. Pope [69] suggested that a reliable LES should resolve at least 80% of the kinetic energy. Consequently and in contrast to the Reynolds stress tensor [75, 111], the sub-grid stress term draws a significantly smaller amount of energy compared to the total kinetic energy in the flow. However, the first model concepts proposed for the sub-grid stress rely on previous methods used for the development of RANS models. Advanced SGS models are presented and discussed in [31, 83]. In the present work the commonly applied sub-grid stress models, namely the Smagorinsky model [90] with a constant model coefficient and its dynamic version following Germano et al. [32] and Lilly [49] are used.

To track the evolution of the dispersed phase a Lagrangian approach is applied [101]. This requires, first, the solution of equations for the particle position and the particle velocity along the trajectory of each computational particle in the carrier flow field. From the derivation of forces acting on a spherical particle [3–5, 15, 16, 62, 63, 80, 82, 84, 88] it is obvious, that depending on the particle to fluid density ratio or the particle diameter, some forces can be up to several orders of magnitude, higher than others. Hjelmfelt and Mockros [37] and Elghobashi and Truesdell [23] performed an analysis to quantify these forces. According to these authors drag, buoyancy and Basset force have the strongest contribution to the particle motion in case of the particle to fluid density ratio greater than one. Furthermore, the Basset force is at least one order of magnitude lower than the drag force. For this reason it is a common practice to consider solely drag and buoyancy effects for the analysis of solid and liquid particle–gas flows.

In this investigation, heat and mass transfer effects are not considered [24, 89]. The particle position and particle motion equations then reduce to:

$$\frac{\mathrm{dx_{pi}}}{\mathrm{dt}} = \mathrm{u_{pi}} \tag{14.2}$$

$$\frac{du_{p,i}}{dt} = \frac{3}{4}\frac{C_D}{D_p}\frac{\rho}{\rho_p}\left|\vec{u} - \vec{u_p}\right|\left(u_i - u_{p,i}\right) + \frac{\left(\rho_p - \rho\right)}{\rho_p} g_i = \frac{1}{m_p}\sum_j F_j \tag{14.3}$$

The drag coefficient C_D used to model the complex dependency between the droplet and the flow is not constant but depends on relative velocities, viscosity of the disperse phase and carrier phase, the particle rand shape and the roughness of the particle's surface [16, 17, 79, 86] . The drag coefficient used within this work is determined for a spherical, not deformable, particle according to Schiller and Naumann [85]:

$$C_D = \frac{24}{\mathrm{Re}_p}\left(1 + \frac{1}{6}\mathrm{Re}_p^{2/3}\right) \quad \mathrm{Re}_p \leq 1{,}000 \tag{14.4}$$

$$C_D = 0.44 \quad \mathrm{Re}_p \gtrsim 1{,}000 \tag{14.5}$$

where Re_p denotes the particle Reynolds number calculated using

$$\mathrm{Re}_p = \frac{D_p \left| u - u_p \right|}{\nu}. \tag{14.6}$$

A more exact approximation of the C_D-curve progression is the polynomial function of Morsi and Alexander [60]. In Eq. (14.3) the first term includes the particle-relaxation time, τ_d, that is given by Ref. [91]

$$\tau_d = \frac{4 D_d \rho_d}{3 C_D \, \bar{\rho} \left| u_i - v_i \right|} \tag{14.7}$$

where D_d is the particle diameter, ρ_d the density of a particle group and ν the kinematic viscosity of the fluid [18, 105, 106].

The effect of turbulence on disperse phase, that is apparent in the fluid-particle relative velocity in Eq. (14.3), has been investigated by numerous experimental studies [15, 79, 100, 114]. According to them, turbulence can significantly affect the particle motion. Nevertheless, often due to the significant discrepancy between the quantitative experimental data, up to date in many cases the turbulence effect is neglected when performing numerical integration of the equation of particle motion. To quantify the influence of the turbulence many authors postulate and pursue the correlation between a characteristic turbulent quantity and the drag coefficient, assuming that the drag force is the most relevant force for the particle motion. The general tendency from the experiments is as follows:

- Increase of the relative turbulence intensity causes an increase of the drag coefficient [102, 114].
- The higher the relative turbulence intensity, the lower the critical particle Reynolds number [15, 100] which indicates the transition from a laminar to a turbulent boundary layer around the particle.
- The integral turbulent length scale is another suitable turbulent quantity for this purpose [16] along with the ratio of this quantity to the particle diameter. Nevertheless, due to the broad scale spectrum in turbulent flows, the phase interaction at different length or time scales addresses different physical phenomena. Such are the flow of particles much smaller than the Kolmogorov length scale [42, 36, 71] or the evolution of the particle boundary layer for particles bigger than the Kolmogorov. The interaction mechanisms are still not fully understood and for this reason the numerical simulation of such flows is increasingly applied for the last two decades [5, 25, 97]. According to the investigations in homogeneous isotropic turbulence together with a Lagrangian particle tracking technique, the turbulent dispersion is mainly driven by the interaction of the particles with energy-containing eddies [29]. Deutsch and Simonin [20] quantify the particle dispersion in homogeneous turbulence with the particle kinetic energy and the particle Lagrangian integral time scale.

An additional phenomenon observed in dilute two-phase flows is the preferential accumulation, which mainly denotes a deviation from the random distribution of particles in the flow [28]. Preferential accumulation can be detected in form of settling of particles in regions with low vorticity or increased particle concentration in near wall regions [9]. In contrast to the turbulent dispersion of particles, where in general the energy-containing eddies are responsible for the distribution, the accumulation does not seem to correlate with a specific flow scale.

In homogeneous turbulence field Fessler et al. [28] detected the highest order of nonrandomness for particles with particle relaxation time similar to the Kolmogorov time scale. Fede [25] again confirmed this findings using Direct Numerical Simulation of homogeneous turbulence field and colliding heavy particles.

Marchioli and Soldati [51] report of near-wall particle accumulation due to interaction with the streaky structures in the near-wall region. In addition the non-homogeneous velocity fluctuations, typical for near-wall turbulence, force the particle accumulation toward the wall.

Squires and Eaton [97] identified regions with high strain rate, typical for the regions between macroscopic turbulence structures, where particles retain longer. Longmire and Eaton [50] were also able to visualize particle preferential accumulation in a field of a periodically enforced round jet, where particles concentrate in the periphery of the macrostructures, generated by the forcing of the flow field.

The dispersion model adopted in the present work has its origin in the eddy crossing/eddy life time concept. The fundamental idea is that, during the computation of the drag force, a random component is added to the fluid velocity seen by the particle. Using this modified velocity the particle equation of motion is integrated. In the so called Discrete RandomWalk model, this fluctuating component is assumed to be piecewise constant in time. The value is kept constant over an interval of time given by the characteristic lifetime of an eddy, modeled by a turbulence model. The prediction of the particle dispersion makes use of the concept of the integral time scale, which describes the time spent in turbulent motion along the particle path [92, 93].

14.3 Numerical Procedure and Code Validation

Before proceeding, some of the characteristic features of the CFD tool used in this work will be listed. FLUENT is a general purpose CFD simulation environment, commercially distributed by ANSYS Inc., Southpointe [103]. The underlying flow simulation algorithm is based on the Finite Volume Method, formulated for arbitrary unstructured non-orthogonal grids [38, 56]. The core solver is efficiently parallelized based on the domain decomposition concept utilizing a variety of communication libraries, like MPIR or PVMR. Next to advanced dynamic mesh capabilities, heavily used for the simulation of internal combustion engines, involving mesh motion, dynamic re-meshing and others, modules for the simulation of turbulence, multiphase flows, compressible flows, reacting flows, radiating heat transfer, solidification and melting and others are included.

The FLUENT solver includes two fundamentally different flow solution algorithms for solving the mass and momentum conservation equations. The difference is in the choice of an independent variable for the mass continuity equation, density on the one hand, pressure on the other hand [26, 48, 98]. Both allow for the assembly of a coupled matrix consisting of the linearized mass and momentum equations. In the case of the density based solver, since its main design target is the simulation of high speed compressible external aerodynamic flows, the energy equation is also included in this matrix, delivering the highest possible degree of implicit equation coupling. In the pressure based solver, in addition to that, segregated approaches like SIMPLE, SIMPLEC, PISO or fractional step methods are implemented allowing for faster time marching in transient simulations [6–8].

The majority of the simulations presented in this work share the same basic numerical setup. A second order implicit time-advancement scheme is applied to the Navier–Stokes equations using the finite-volume discretization. Within the scope of this work, a time efficient non-iterative Fractional Step procedure is applied for the pressure–velocity coupling of the carrier fluid governing equations. The coupling of the method with two-way coupled particle tracking simulations is validated successfully on configurations of isothermal flows (channel flows) using experimental and DNS data. Because this paper mainly concentrates on the model predictability, the details of these studies are not provided here. More details and all the results obtained are reported in [21].

After a necessary validation of the basic prediction ability of the CFD code for single and two-phase flows when employing LES in terms of basic numerical setup, pressure–velocity coupling algorithm and unsteady inflow boundary conditions according to the Lagrangian vortex method, an assessment of the two-phase flow LES have been achieved. Thereby the particle tracking in conjunction with LES has been validated along with the pressure–velocity coupling algorithm in two-way coupling and the number of numerical versus real particles in the LES context.

14.4 Configurations Under Study

This section gives an overview of the configurations investigated and provides the numerical setup applied to the simulations. Several particle–laden flow configurations, such as two plane channel flows [28, 46, 67], a free jet [35] and an evaporating spray at low temperature [93], have been studied. The reader can find the detailed results in [21]. In this report the results presented are especially for the two plane channel flows characterized by low and high Reynolds number, respectively.

14.4.1 Low Reynolds Number Particle Laden Channel Flow

The flow in this configuration is rendered between two parallel plane walls. The extension of the wall boundaries is supposed to be infinite. Therefore, flow inhomogeneities develop only in the wall-normal direction. The carrier fluid is contaminated with small heavy particles.

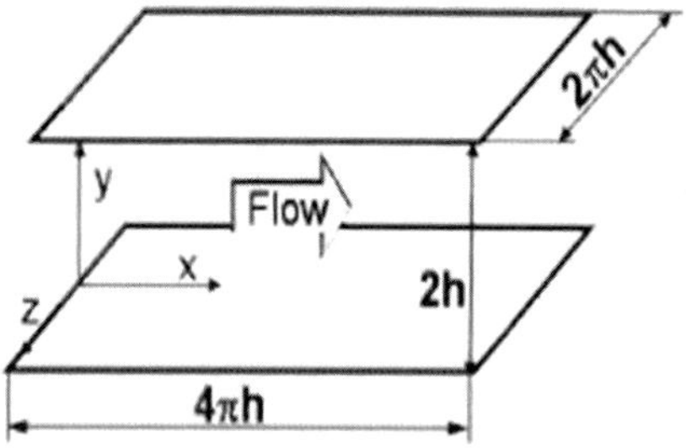

Fig. 14.1 Sketch of the channel flow domain for the DNS

Table 14.1 General configuration and flow properties

Channel half height, h	0.02	m
Friction velocity, u_τ	0.11775	m/s
Re_τ	150	–
Re_{bulk}	2,110	–
Fluid density, ρ_f	1.3	kg/m^3
Fluid viscosity, υ_f	1.5710^{-5}	m^2/s
Particle density, ρ_P	770 ρ_f	kg/m^3

Table 14.2 Particle properties

$St = \tau_p^+$	τ_P (s)	d_p^+	d_P (μm)	St_η	St_ι
1	$1.13 \cdot 10^{-3}$	0.15	20.4	0.08	0.012
5	$5.66 \cdot 10^{-3}$	0.34	45.6	0.4	0.06
25	$2.83 \cdot 10^{-2}$	0.77	102.0	1.9	0.3
125	$1.42 \cdot 10^{-1}$	1.71	228.0	9.6	1.5

The Reynolds number, based on the wall friction velocity and the channel half height, is 150. The intention of this simulation is to investigate the main properties of the dispersed phase, such as particle velocities, particle dispersion and the propensity to accumulate in the near-wall region, depending on particle diameter. Beyond this, the LES results are compared with reference DNS from Picciotto et al. [67] who applied DNS to the carrier phase whilst the motion of the particles is computed from the Newton equation of motion. It should be noted that the DNS is part of a benchmark test of a particle-laden turbulent channel flow, as reported in Marchioli et al. [53]. The fluid is assumed to be incompressible and Newtonian. A sketch of the computational domain, with channel half height h as a scaling parameter, is presented in Fig. 14.1. The flow and particle properties are listed in Tables 14.1 and 14.2, respectively.

The Navier–Stokes equations are solved using a pseudo-spectral method. Details of the numerical procedure can be found in Lam and Banerjee [47]. The domain for the DNS is discretized with $128 \times 128 \times 128$ nodes in all three directions.

In stream- and spanwise directions periodic boundary conditions are imposed on the fluid velocity field, together with a pressure gradient in streamwise direction to retain the flow. A no-slip boundary condition is applied at the walls.

The concentration of the particles is assumed to be low enough to consider one-way coupling conditions. Particle-particle interactions are also neglected.

Table 14.3 Mesh resolution for the large-eddy simulation

Grid label	Grid points	Δx^+	Δy_1^+	Δy_c^+	Δz^+	Grid points below $y^+ = 10$	Time step, *s*
Coarse	$58 \times 60 \times 58$	16.2	0.6	11.7	8.0	9	$5e-3$
Medium	$72 \times 74 \times 72$	13.0	0.5	10.7	6.5	15	$1e-3$
Fine	$88 \times 90 \times 88$	11.0	0.6	7.7	5.4	15	$5e-4$

The motion of the ensemble of 10^5 rigid spherical particles is resolved with a set of ordinary differential equations (Eqs. (14.2) and (14.3)). The only significant forces are drag force and buoyancy [23]. Gravity is also neglected in this simulation. The drag force includes a standard correction for particle Reynolds numbers lower than 1,000, as shown in Eq. (14.4).

Similar to the fluid boundary conditions, a periodic type of boundary is applied to the particle velocity and position when a particle leaves the domain in a stream- or spanwise direction. Following the standard Lagrangian approximation, particles of finite size are replaced by mass points. The boundary condition for the particle approaching a wall is defined as follows: When the distance between particle position and the wall becomes smaller than the physical radius of the particle, the particle rebound elastically from the wall. The particle-laden flow is computed for approximately 400 times the mean flow residence time, defined by the domain length in streamwise direction and the mean centerline velocity. During the last 20,000 flow time steps, which corresponds to approximately 17 times the mean flow residence time, particle velocity and position data are collected every 50 time steps to build up the database for the particle properties. Finally, from the extracted particle data, one-point statistics for velocity and particle number density are used to analyze the preferential particle accumulation.

The computational domain for the large-eddy simulation is reduced to a quarter of the original domain of the DNS by considering the half length and width of the primary domain. At the relatively low Reynolds number the correlation coefficient does not drop to zero. Nevertheless, the domain extension is sufficient to reproduce the fluid and particle one-point statistics, as the comparison with the DNS confirms. The domain axes are as follows: x-in streamwise , y-in wall-normal and z-axis in spanwise direction. The spatial resolution in stream- and spanwise directions is equidistant and in the wall-normal direction, the mesh is stretched using a Poisson function, with growth factor of 1.05.

Table 14.3 contains the mesh size normalized by wall variables. Δy_c^+–denotes the grid size in the centerline plane of the channel, and Δy_1^+– the size of the wall next cell. In addition to the two-phase flow simulation, for the purpose of validating, the simulation of the single phase flow is carried out on three different numerical meshes.

The two-phase LES is performed on the medium grid. The current two-phase flow simulation setup computes 15 times more particles than those for the reference DNS. The number of the particles does not change the two-phase flow behavior since inter-particle collisions are neglected and only one-way coupling is

Table 14.4 Summary DNS vs. LES setup

Case	Physical domain	Numerical method	STOKES number, *St*	Particles number	Simulated time	Sampled time
DNS	$4\pi h \times 2h \times 2\pi h$	pseudo-spectral	1, 5, 25, 125	10^5	$400 \cdot t_R$	$17 \cdot t_R$
LES	$2\pi h \times 2h \times \pi h$	finite-volume	1, 5, 25	$15 \cdot 10^5$	$400 \cdot t_R$	$xx \cdot t_R$

considered. Therefore, the simulated time and thus the time for sampling particle properties is shorter. The full details concerning the differences of the DNS vs. the LES are summarized in Table 14.4. Thereby t_R denotes the flow residence time.

14.4.1.1 Boundary and Initial Conditions

Periodic boundary conditions are applied in streamwise and spanwise directions for both the continuous and the dispersed phases. Particles are reflected elastically when the particle center reaches the wall. Pressure gradient in x-direction is applied to retain the flow. The initial field of the air flow is taken from a steady state k-Epsilon simulation. The averaged velocity field has been superimposed with random fluctuations to accelerate the development of the turbulent field in the large-eddy simulation. The two-phase flow simulation starts from the fully developed turbulent single phase flow. The particles are initially homogeneously distributed within the domain.

Initial particle velocity is the fluid velocity interpolated at the particle position. Since only one-way coupling is considered for the LES, similar to the reference DNS, the number of particles does not have an influence on the particle distribution and accumulation, whereas larger number of particles accelerate the convergence of the statistical particle properties, such as mean and root mean square (rms) velocity or particle number density. Hence, significantly larger number of particles compared with the reference DNS setup are simulated (Table 14.4).

The sub-grid stress tensor, within the scope of the LES, is modeled using the Smagorinsky model with dynamic estimation of the model coefficient. Particle tracking is initialized after the single phase flow has reached a statistically steady state. Particle motion is controlled by the drag force, assuming that the Rep $< 1{,}000$. A second order interpolation scheme is used to determine the fluid velocity at particle position. The integration in time of the particle equation of motion is evaluated applying an adaptive scheme, which switches between a 1st order Euler implicit and 2nd order trapezoidal scheme. Details of the numerical methods for particle tracing can be found in Ref. [21].

14.4.2 High Reynolds Number Particle Laden Channel Flow

The experimental setup for this case has been developed and used for numerous measurements of particle-laden turbulent channel flows using various parameters,

Table 14.5 Numerical grid parameters

Case	Grid points	Δx^+	Δy_1^+	Δy_c^+	Δz^+	Grid point below $y^+ = 10$	Time step, s
Wall-resolved	$129 \times 100 \times 129$	31	0.6	33	15	9	$5.e-5$
Wall function	$129 \times 83 \times 129$	31	15	15	15	–	$5.e-5$

such as particle diameter, particle mass loading and channel wall roughness. The Reynolds, number based on the wall friction velocity and the channel half height, is 644. The experimental investigations in a vertical channel gas-solids flow by Kulick et al. [46] and Fessler et al. [28] are used as a reference case. The experiment focuses on particle dispersion by examination of different Stokes numbers and on turbulence modification due to Stokes number variation combined with mass loading variation.

The high aspect ratio of the real channel cross section allows approximation of the real flow such as the assumption of a two-dimensional flow between two endless flat planes, where its behavior in spanwise and streamwise direction is supposed to be periodic. This assumption allows for an artificial but significant reduction of the volume reproduced in the numerical simulation.

The domain extensions are $2\pi h \times 2h \times \pi h$ where h = 0.02 m represents the channel half-height. The domain axes are as follows: x-in streamwise, y-in wall-normal and z-axis in spanwise direction. Two discretizations in the wall-normal direction aim to investigate the effect of the resolution of the wall boundary layer on the particle properties. The first grid allows for a standard resolution (in LES context) of the boundary layer, where the wall next point is below $y^+ = 1$.

The second grid has an equidistant cell distribution in wall normal direction, so that the wall next point is at $y^+ = 15$ and additional modeling of the wall generated turbulence is required. Details for the two numerical grids are summarized in Table 14.5.

The physical properties for air and particles are kept as defined for the experiment, except for the assumption that each particle class is mono-dispersed. The particle size is set to the mean diameter of the particular particle set.

14.4.2.1 Boundary and Initial Conditions

As for the low Reynolds number case, periodic boundary conditions are applied in streamwise and spanwise directions for both the continuous and dispersed phases. For the periodic boundaries on the fluid side, the chosen domain extensions are supposed to be large enough that the two-point velocity correlation becomes zero within the domain. The pressure gradient in streamwise direction retains the flow parallel to the gravity vector. Gravity also acts on the dispersed phase, together with the drag force induced by the carrier fluid. The influence of the dispersed phase on the fluid properties, i.e. two-way coupling, is considered by utilizing a source term in the momentum Eq. (14.1).

Table 14.6 Topic map of the simulated cases

Topic	Relevant configuration
Particle preferential concentration	$d28\phi03$, $d50\phi02$, $d70\phi02$, $d90\phi02$
Turbulence modification	$d50\phi02$, $d50\phi20$, $d70\phi02$, $d50\phi20$, $d150\phi20$

Table 14.7 Simulation parameters

Case label	Simulated time	Sampled time	Parcel number $*10^3$	Number in Parcels
Single phase	160 t_R	80 t_R	–	–
$d28\phi03$	160 t_R	80 t_R	1362.8	?
$d50\phi02$	160 t_R	80 t_R	516	?
$d70\phi02$	160 t_R	80 t_R	516	?
$d70\phi20$	160 t_R	80 t_R	516	?
$d70\phi20 - nip1$	160 t_R	80 t_R	47.06	1
$d70\phi20 - e0.5$	160 t_R	80 t_R	47.06	1
$d90\phi02$	100 t_R	50 t_R	516	?
$d150\phi02$	100 t_R	50 t_R	250	?
$d70\phi20 - simple$	132 t_R	80 t_R	47.06	1

No additional treatment is applied to model the influence on the subgrid-scale tensor. A no-slip condition is applied at the wall for the air flow. Particles reflect elastically when the particle center reaches the wall. The effect of inelastic collision is investigated in the case of copper particles by setting the restitution coefficient in wall-normal direction to 0.5. It is important to note that, the value of the restitution coefficient chosen here is a very rough estimation because of missing data for the colliding material pair, glass – copper. For comparison, the estimated restitution coefficient for the pair copper-copper is approximately 0.22 and for the pair glass-glass it is about 0.94. To retain as much similarity with the experiment as possible, the gravitational force is considered for both the air flow and the dispersed phase.

The choice of the simulated setups allows the investigation of the potential of large-eddy simulation coupled with Lagrangian particle tracking in two key aspects:

Particle preferential concentration and particle influence on fluid turbulence characteristics. The results from the simulations are split into two groups, as shown in Table 14.6, based on their fundamental interaction phenomenon.

One group focusses on the preferential concentration of particles as induced by the turbulent flow. The second targets the turbulence modification by the particulate flow. Only the results of the preferential concentration of particles are presented and discussed here. Detailed information about the simulation conditions for the single and two-phase flows are provided in Table 14.7, see also [21].

As well as the configurations performed in the experiment due to variation in particle diameter, material or mass loading, a number numerical variations are also performed. These are listed in Table 14.7 as well. The first variation considers the number of real particles per numerical particle (also denoted as a parcel).:

Every numerical particle represents only a small fraction of a real particle, ($<$10% depending on material and mass loading). The following injection procedure is applied: Particles are released from a plane normal to the main flow direction, with 129×100 homogeneously distributed injection positions. The particles are injected at equidistant time intervals, so that the final particle distribution throughout the domain is nearly homogeneous. The duration of the injection is governed by the desired number of numerical particles. In this case, the limit is set to 516,000 numerical particles.

One numerical particle represents exactly one real particle. To achieve this, particles are released from the same cross-sectional plane as above but in accordance with the mass flow and the additional restriction that, every parcel contains the mass of one real particle. The duration of the injection is limited by the averaged volume fraction, estimated by the particle and mass loading case. Particle positions are randomized for every injection event using a normal distribution to achieve a nearly homogeneous particle distribution throughout the domain.

The effect of the number of numerical particles on flow characteristics has been investigated for the configuration of copper particles at 20% mass loading.

The second numerical variation considers the choice for pressure–velocity coupling algorithm.

The case $d70\phi20$ has been computed using the Fractional Step method and the standard SIMPLE method, referred to as $d70\phi20$-*simple*, and compared for fluid and particle properties. The choice of the configuration is based on the expectation that higher particle mass loading can affect the stability and convergence of a simulation in a negative manner when considering the two-way coupling. This allowed to prove that the Fractional Step method, which includes various numerical simplifications and thus is computationally much more efficient, is nevertheless capable of dealing with the higher mass loadings.

14.5 Results and Discussion

Particle distribution due to turbulent structures in the carrier fluid has several physical phenomena. Turbulence augments particle dispersion. Another effect is the preferential accumulation of particles in low vorticity or low turbulence regions, which depends on the ratio of the characteristic scales of the fluid and the dispersed phase. Another well-known phenomenon is the accumulation of particles in the near wall region. To quantify the prediction of these physical phenomena using the LES method, the discussion is based largely on the channel flow configurations, namely the low and high Reynolds number channel flows introduced above. Since two of the particle classes used in both configurations have similar time constants, it is useful to compare the overlapping findings on particle behavior in flows having different Reynolds numbers.

In the following, several perspectives on the available results are presented.

The particle velocity field is investigated and compared with reference data.

Instantaneous particle data are applied to demonstrate typical concentration patterns for particles with different time constants. With the aid of particle number distributions, the amount of preferential accumulation is quantified.

Particle accumulation in the near wall region is analyzed for concentration profiles in the wall-normal direction.

14.5.1 Particle Dispersion and Preferential Accumulation

In the following, the particle concentration in a turbulent flow is investigated in terms of averaged and instantaneous quantities. In addition, instantaneous concentration plots from flow regions with nearly homogeneous and isotropic turbulence are compared to those obtained in a strongly inhomogeneous flow field. The findings from the LES are put in the context of those from previous DNS and experimental studies.

14.5.1.1 Local Particle Accumulation

The Influence of the Stokes and Fluid Reynolds Number

Typical particle behavior in a turbulent flow is demonstrated in Fig. 14.2. The plots represent particle positions in an xz-plane along the centerline in a channel flow (the midplane). The different views correspond to particle classes with different Stokes numbers St_η. It should be emphasized that the results presented correspond to simulations on channel flows with two different Reynolds numbers: the low Reynolds number flow case with $\mathrm{Re_{bulk}} = 2{,}110$ and the high Reynolds number flow case with $\mathrm{Re_{bulk}} = 13{,}800$. The range of $\mathrm{St_n}$ considered here varies from 0.04 up to 19. It is further evident that particle classes with $St_\eta \approx 1$ demonstrate the highest propensity to accumulate. In addition, the particle class with $St_\eta = 0.74$, which is tracked in the high Reynolds number flow, fits into the given order well, i.e. according to their visual degree of accumulation. This qualitative observation is in agreement with the investigations of Wang and Maxey [109]. They showed that the maximum preferential accumulation is observed for particles with $St_\eta \approx 1$ and that the degree of accumulation scales with the Kolmogorov time scale. Nevertheless, their findings are based on Lagrangian particle tracking in homogeneous isotropic turbulence (DNS), which is unambiguously related to the fact that the turbulent structures with the highest intensity are in the order of the smallest scales [80].

Since this particular observations on the results from the present work are restricted to the channel midplane (i.e. far away from the wall), comparison with findings of Wang and Maxey [109] and similar works is suitable.

To investigate the effect of the flow Reynolds number on the behavior of particles with similar Stokes numbers, results from both channel flow configurations are considered and analyzed in parallel. An overview of these, together with

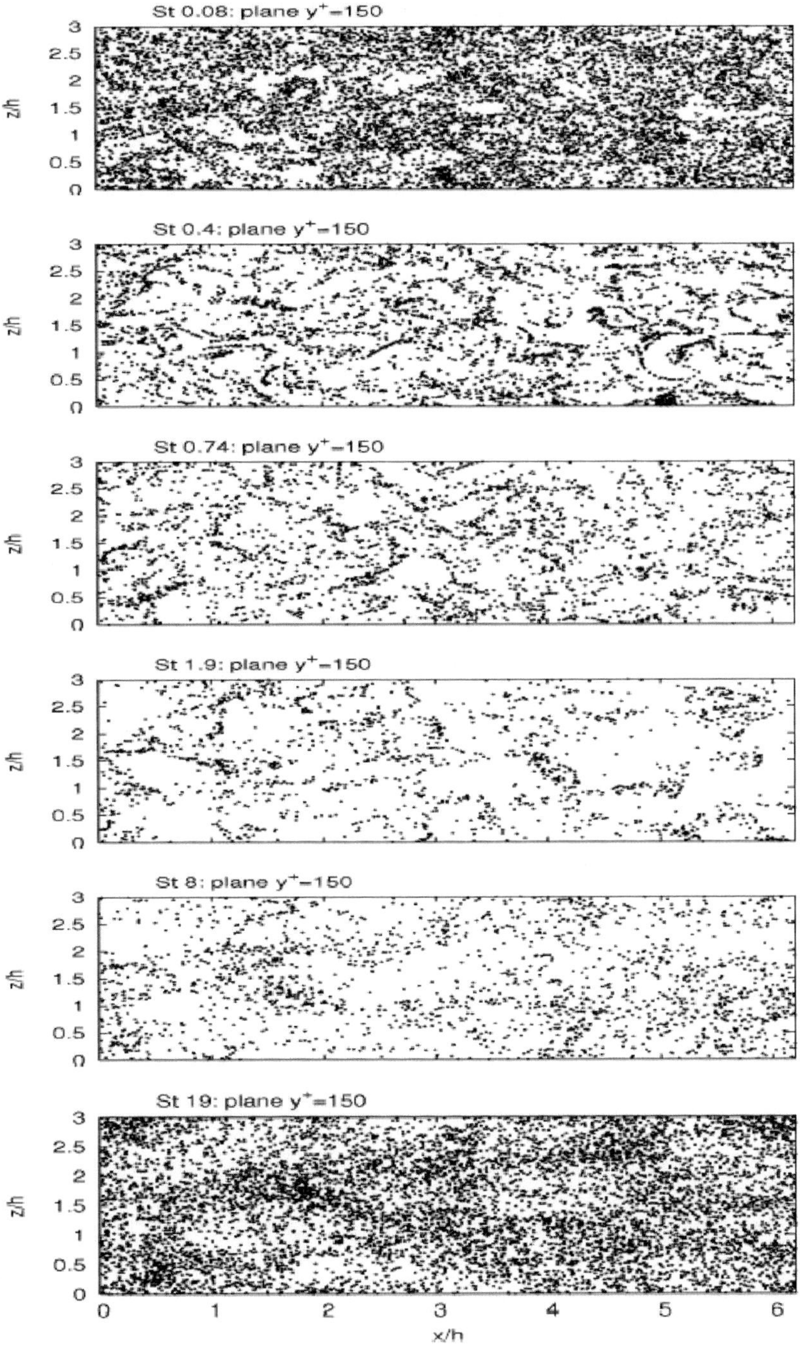

Fig. 14.2 Channel flow $Re_T = 150$ and $Re_T = 644$: Instantaneous particle positions at $y^+ = 150$ (channel midplane). Particles with $St_n = 0.08, 0.4, 1{,}9$ belong to the low Reynolds number channel flow simulation. Particles with $St_n = 0.74, 8, 19$ belong to the high Reynolds number channel flow simulation

Table 14.8 Configurations used for investigation of local preferential accumulation of particles

Reference configuration	Case	St_η
Channel flow, $Re_\tau = 150$	St1	0.08
	St5	0.4
	St25	1.9
Channel flow, $Re_\tau = 644$	$d28\phi03$	0.74
	$d50\phi02$	8
	$d90\phi02$	19

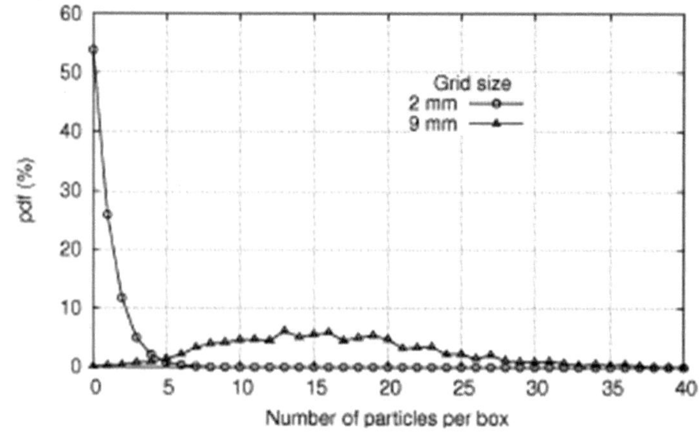

Fig. 14.3 PDF of the particle number density for case d28Phi03 (channel midplane) estimated for different grids

the corresponding particle classes in terms of Stokes numbers, is provided in Table 14.8. The data from the high Reynolds number channel flow simulation address the configurations with particle mass loading of 2%, except for the case with Lycopodium particles, where the loading is 3%. The effect of the dispersed phase on the fluid turbulence is assumed to be negligible.

More quantitative information, compared to the instantaneous plots in Fig. 14.2, permits the investigation of probability density functions (PDF) for the particle number density. For this purpose, a procedure for the PDF evaluation, similar to that used in Fessler et al. [28], is used. The reliability of the simulations in predicting of preferential accumulation is assessed in terms of information extracted from the PDFs of both the LES and experimental data from Fessler et al. [28]. To retain similarity with the experiment, a 1 mm thick sheet in the midplane of the channel is considered. The plane is discretized with regular mesh of equidistant square cells (boxes). This mesh is independent from the numerical grid used for the simulations. All particles with spatial coordinates within a particular box are assigned to it and the associated number density is computed. The final PDF is computed from the number densities. The PDF has a different form depending on the grid size. In Fig. 14.3, an example of PDFs for the case $d28\phi03$ obtained on grids with cell length of 0.5 and 2 mm is shown. The evaluation of the degree of preferential accumulation is based on the comparison of the calculated PDF with a PDF obtained for the same number of particles, assuming purely random distribution.

For randomly distributed particles, the probability P(n) of finding n particles in a particular box is a Poisson distributed quantity. According to this, the random distribution is given by equation

$$P(n) = \frac{e^{-\mu}\mu^n}{n!} \tag{14.8}$$

μ denotes here the mean number of particles per box. The normalized difference D_{dev} is used to quantify the deviation of the simulated distribution from the Poisson distribution. It is defined as the standard deviation σ of the computed distribution and that of a Poisson distribution, $\sigma_{Poisson}$ having the same mean value μ.

$$D_{dev} = \frac{\sigma - \sigma_{Poisson}}{\mu} \tag{14.9}$$

D_{dev} is normalized by the mean particle number per box. Positive values for D_{dev} correspond to situations with preferential accumulation, since a large σ value indicates that there are boxes with a very high number of particles and such a with very low number. In a uniform particle distribution, the standard deviation is zero and D_{dev} becomes negative. $D_{dev} = 0$ describe randomly distributed particles.

As already mentioned, depending on the grid size applied for the PDF evaluation, its form can vary greatly because the typical length scale of accumulation varies. Therefore, the PDF with maximum deviation from the Poisson distribution, according to (9), represents the degree of particle accumulation most accurately, since it matches the scale present best. The limited a priori information about the optimal grid size implies an iterative process for its estimation.

In the present context, the PDFs of the particle number density are computed for a grid size ranging between 0.5 mm and 10 mm. The estimated degree of non-randomness in terms of D_{dev} is shown in Fig. 14.4 as a function of the grid size used for the PDF computation.

The grid size is normalized with the channel half height, h. The deviation from the Poisson distribution exhibits a local extremum (maximum) for all particle classes. This appears for particles with $St_\eta \leq 1$ for a similar grid resolution of approximately (≈ 0.1). For particles with higher Stokes number, though, this maximum seems to be shifted toward larger box sizes.

This can be interpreted as further information related to the scales at which the accumulation occurs. According to conclusions made by Wang and Maxey [109], particles with $\tau_p / \tau_\eta \approx 1$ experience the highest degree of accumulation. Particles with $\tau_p / \tau_\eta \approx 2$ have almost 40% lower values of Ddev. Squires and Eaton [97] confirm maximum accumulation at $\tau_p / \tau_\eta \approx 1$ as well by performing DNS of isotropic turbulence with heavy particles. In contrast to this, the highest degree of non-randomness, according to the LES here, is detected for particles with $\tau_p / \tau_\eta =$

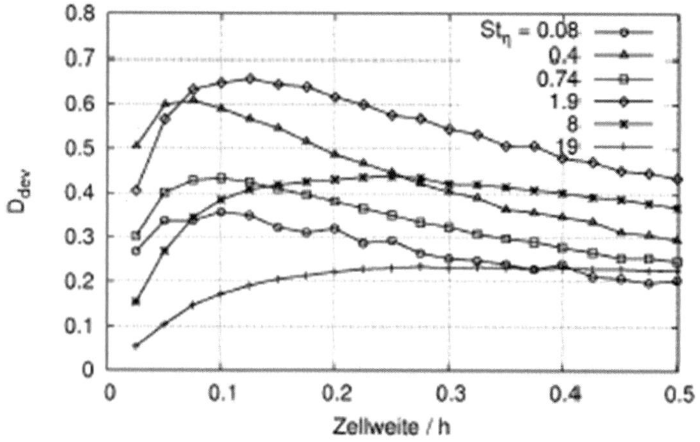

Fig. 14.4 D_{dev} as defined in Eq. (14.9) as function of the grid size for different particle Stokes numbers. Particle classes corresponding to Table 14.8

$St_\eta = 1.9$. The investigations of Fessler et al. [28] on the particle dispersion in a channel flow turbulence have shown that the preferential accumulation is most pronounced for particles with $\tau_p / \tau_\eta = 2.2$. However, this behavior was claimed to be caused by the wider range of scales in the experiment ($t_0 / \tau_\eta = 22$). In contrast to this, the simulation of Squires and Eaton [97] considers flow with ($t_0 / \tau_\eta = 10$).

Another possible explanation, given by Fessler et al. [28], was that the PDFs obtained from DNS are based on the computational grid and their dependency on the grid size is not further evaluated. The two-phase LES considered here is consistent with the experiment. The second, and to the authors best knowledge, it is a new observation that the findings of Fessler et al. [28] in a high Reynolds number flow are herewith reported for very low Reynolds number flow. The estimated scale separation for this flow, estimated at $t_0 / \tau_\eta = 6$, is lower than that used in the flow simulation of Squires and Eaton [97].

The dependency of the maximum value estimated for D_{dev} on the particle Stokes number is presented in Fig. 14.5a. All particle classes are considered in the plot. The maximum value for D_{dev} is the same as estimated in Fig. 14.4, i.e. independent from the grid size for which was detected. There is a significant decrease of D_{dev} for the particle class Stokes number 0.74.

An alternative plot of the same data is presented in Fig. 14.5b, which seems to resolve the contradictory behavior of D_{dev}. Separate lines explicitly highlight the data corresponding to the different channel flows. The tendency of particle accumulation is confirmed in both channel flows. Nevertheless, the magnitude of the deviation from the random state shows significant discrepancies, depending on the flow Reynolds number. In addition, the magnitude of D_{dev} is given, as detected for the particles with St = 0.74 and St = 2:2 in the experiment of Fessler et al. [28]. The second class denotes maximum degree of accumulation. This particle class is not included in the present simulation set.

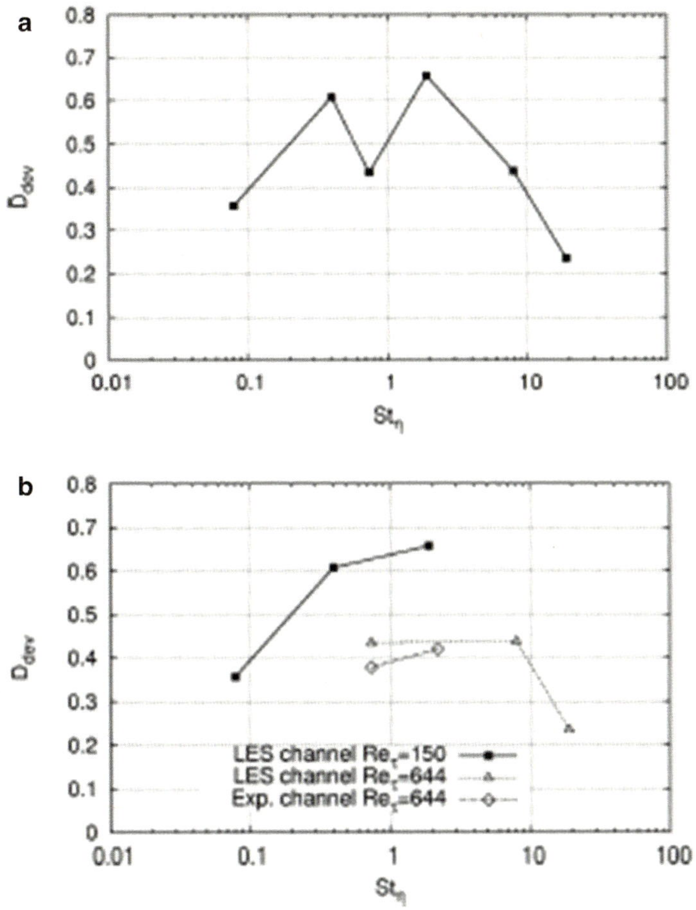

Fig. 14.5 Maximum deviation D_{dev} from the Poisson distribution (Eq. (14.9)) as a function of the particle Stokes number (**a**) for different channel flows (**b**)

It was not possible to obtain a reliable estimation of the local accumulation in the channel from the available database for the dispersed phase in the low Reynolds number channel flow [52], since the number of particles used by the authors is too low for that purpose. Marchioli and Soldati [52] computed 10^5 particles, which was sufficient to obtain reliable statistical information in the context of their objectives. The two-phase LES performed here has 15×10^5 particles.

In conclusion, predicting local accumulation of particles according to the results follows the findings from previous works. However, particularly anomalous behavior of D_{dev} (commented in the context of Fig. 14.5a) leads to the assumption that the influence of the flow Reynolds number leads to varying accumulation behavior.

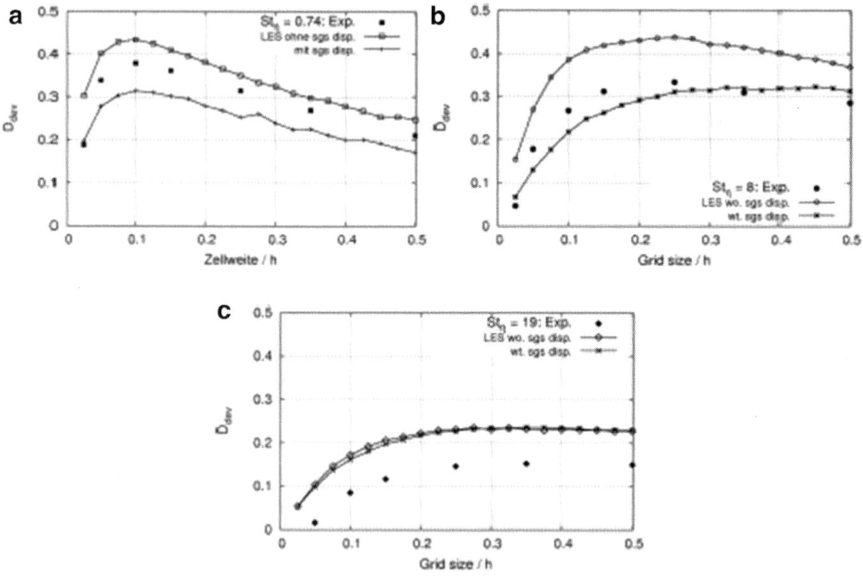

Fig. 14.6 D_{dev} as defined in Eq. (14.9), for $St_n = 0.74$, 8, 19 (case d28Phi03, d50Phi02): Comparison with experimental data from Fessler et al. [28] in (**a**), (**b**) and (**c**), respectively

Influence of the Sub-grid Turbulent Dispersion

As discussed in [21], the consideration of a model for turbulent particle dispersion that accounts for the effect of sub-grid velocity fluctuations does not affect the particle velocity statistics, according to the simulations here. Nevertheless, it has been shown [25, 44, 70] that unresolved scales contribute to the evolution of the local and the mean particle concentration. In the following, this aspect is investigated in the context of the high Reynolds number channel flow with low particle mass loading of 2%. Figure 14.6 presents D_{dev}, (Eq. (14.9)), as estimated for the midplane of the channel as a function of the grid size, used to compute the particle number density.

Deviation from the Poisson distribution is evaluated from particle data corresponding to simulations with and without an additional model for the sub-grid stress turbulent dispersion. In Fig. 14.6 a, D_{dev}, estimated for both simulations together with the experimental data from Fessler et al. [28], is shown as a function of the grid size. The particle class addressed here has $St = 0.74$ (configuration 28Phi03). In general, two issues stand out from this figure. Both simulations predict the evolution of D_{dev} through the different grid sizes estimated from the experiment. Secondly, the increased particle dispersion due to the additional model introduces a large correction in the stochastic particle motion with a tendency to an artificial randomization of the particle positions. This is confirmed by the fact that the experimental data predict a systematically higher preferential accumulation.

Contrariwise, a rough estimate of the relative error between the experiment and the standard simulation, without dispersion model, computed as $(D_{dev}^{less} - D_{dev}^{\exp})/D_{dev}^{\exp}$ leads to an averaged value of 15%. Considering the main mechanism for particle accumulation, it is clear that small particles will accumulate if they are heavy enough to be exposed to the centrifugal force due to the eddy rotation, but in the same time, have a response time shorter than the time for crossing the eddy, or the time for the eddy dissipation. Since particle accumulation evidently occurs, it is clear that the large eddy simulation properly resolves the turbulent structures relevant for the accumulation.

From the perspective of the standard Lagrangian particle tracking method, the particle accumulation in a certain volume is not restricted to the particle packing limit. Without an additional control mechanism from the particle tracking point of view accounting for "permitted" and "not permitted" particle positions, depending on the allowed particle volume fraction in the computational cell, the interaction of the particles with the fluid is the only mechanism for their redistribution from regions with high particle concentration. Nevertheless, this mechanism plays an effective role only for particles with sufficiently small residence time and diameter. Otherwise, the particle collision probability in these regions is much higher compared to the collision probability of homogeneously distributed particles at the same mass loading. It turns out that particle–particle collision is the more natural mechanism for particle redistribution. This fact is especially important when transferring conclusions from here to the large-eddy simulation of liquid sprays. The collision there will lead eventually to the coalescence (depending on the relative droplet velocity and the surface tension coefficients) rather than to further randomization of the droplet position. It is therefore crucial to differentiate between turbulent and collision dispersion observed in solid particle flows.

Despite the alternative of applying a particle collision model, a possible correction of the final particle PDF is to perform a posteriori particle redistribution (different conditions can be specified) or clip the number to the packing limit for the purpose of data evaluation only. The problem with such an assumption is, however, that it is limited to the instantaneous snapshots of the particle positions. These are taken every 200 simulation time steps. Therefore, the natural dispersion due to collisions cannot be captured. Such a method is expected to introduce only a small overall correction and is not applied on the data sets discussed here.

The identical solution from both simulations for the case $\mathrm{St_n} = 19$ shows the limited area of influence of the dispersion model. The intermediate particle class exhibits the largest deviation between the simulations with and without the model. Furthermore, if the sub-grid stress fluctuations are considered, the discrepancy between experimental and simulation result becomes negligible. However, the comparison with the experimental data from Fessler et al. [28] should be considered with caution, since these correspond to a measurement of flow with mass loading of 40%. According to the experiment of Kulick et al. [46], the particles do not influence the fluid statistics even at higher mass loading, which means that the particles are exposed to the same interaction with the fluid as those in the low mass loading case. Of course, a higher mass loading certainly leads to an increased particle–particle collision probability, which leads to effective dispersion.

Observations of the dispersion model and its effect on both low-Stokes-number particle classes, suggest that the model will certainly lead to the desired results if direct comparison with an experimental data is possible, however, the appropriate model constant should be estimated for every application and particle type.

Other authors (e.g. Oefelein [63]) have shown, that this type of models based on eddy–life and particle–eddy interaction time can be successfully extended for LES applications. Segura [88] used several values for the constant in the model proposed by Oefelein [63] in order to facilitate the optimal for each particular configuration. The model constant for all simulations reported here is set to 0.15.

14.5.2 Mean Particle Concentration Profiles

A typical phenomenon in two-phase wall–bounded flows is the accumulation of particles in the near wall region. In the present context, both channel flow configurations exhibit this behavior for most of the particle classes. It is possible to compare the DNS results for the particle concentration in wall-normal direction [67] with those obtained from the LES.

The comparison is presented in Fig. 14.7a. In a similar way, the mean concentration profiles of the high Reynolds number channel flow are shown (Fig. 14.7b). Since concentration data from the experiments of Fessler et al. [28] and Kulick et al. [46] are not available, only results obtained from the two-phase LES are discussed. The wall-normal coordinate is normalized with wall units. The particle number density is normalized by the value corresponding to a homogeneous particle distribution in the same direction. Both configurations demonstrate, that the particle concentration in the wall region is strongly affected by the particle STOKES number, St_{-}. The main discrepancy between the DNS [67] and the LES is localized in the region below $y^+ = 1$ (Fig. 14.7a). One significant difference between both simulations is the position for which the particle-wall collision is computed, despite the numerical technique and implementation. In Picciotto et al. [67], the particle reflection at the wall is applied at a distance of $Dp = 2$ from the wall. In contrast to this particle-wall collision in the present simulations occurs when the particle center encounters the wall. Therefore it is reasonable to suggest that the particle concentration estimated by the LES overestimates the one from the DNS, especially for the larger particles with $St = 1{:}9$. Otherwise, from a practical engineering point of view, the wall region considered is less important, since the viscous sublayer is not resolved for most practical applications. Moreover, the prediction of the mean particle concentration in the near wall region seems to be an issue also for the DNS technique, as reported in Marchioli et al. [53]. Data obtained by independent working groups using DNS, expose significant discrepancies in the prediction of the particle concentration. Marchioli et al. [53] conclude that the combinations of numerical schemes employed by the various codes provide different errors. It is further not clear, which data should be used as a reference. From the concentration profiles corresponding to the high Reynolds number case, it is evident that particle preferential accumulation near the wall can be observed up to $St_n = 19$.

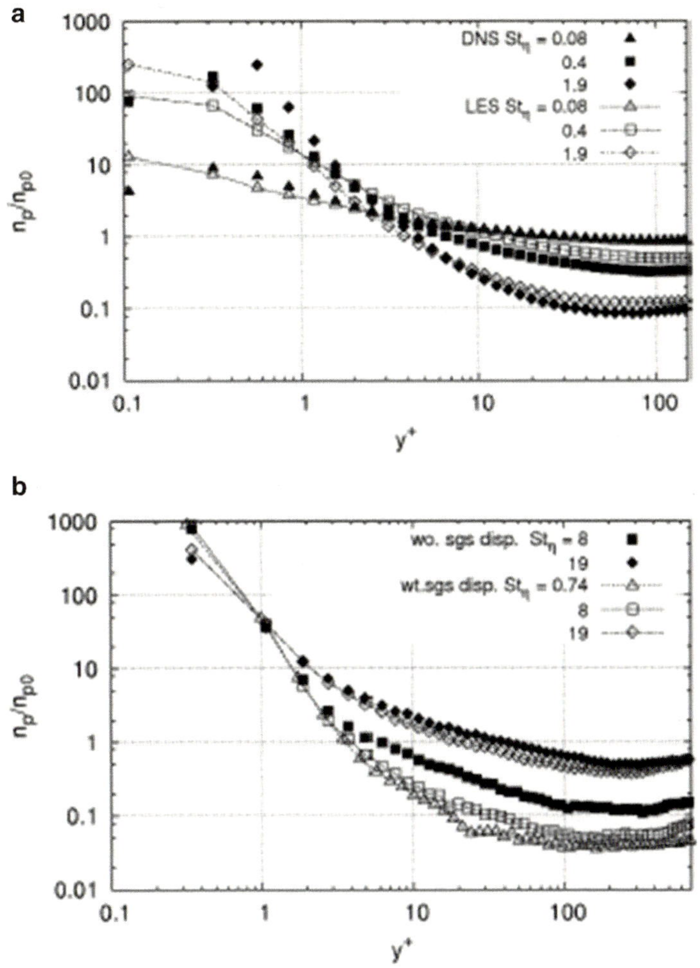

Fig. 14.7 Channel flow $Re_T = 150$, (**a**), and $Re_T = 644$, (**b**): Mean particle concentration profiles in wall-normal direction for particles with the STOKES numbers $St_n = 0{:}08;\ 0{:}4;\ 1{:}9$ (**a**) and $St_n = 0{:}74;\ 8;\ 19;\ 41$ (**b**) (DNS data in (**a**) are taken from Picciotto et al. [67])

Concerning the influence of the sub-grid stress dispersion model (the hollow symbols in Fig. 14.7b), it can be concluded that the particles with $St = 19$ are not affected by the additional fluid velocity fluctuations, which is an expected result according to the work of Young and Leeming [113]. Otherwise, the intermediate particle class ($St = 8$) seems to be sensitive to the sub-grid fluctuations. Nevertheless, it has a negligible effect on the maximum near wall concentration. The influence of the model remains mainly beyond the region of $y^+ < 10$. The sub-grid stress velocity fluctuations are computed dynamically, in a similar way to the sub-grid stress tensor. The fluid sub-grid stress velocity fluctuations are naturally dumped if the dynamic Smagorinsky model is used.

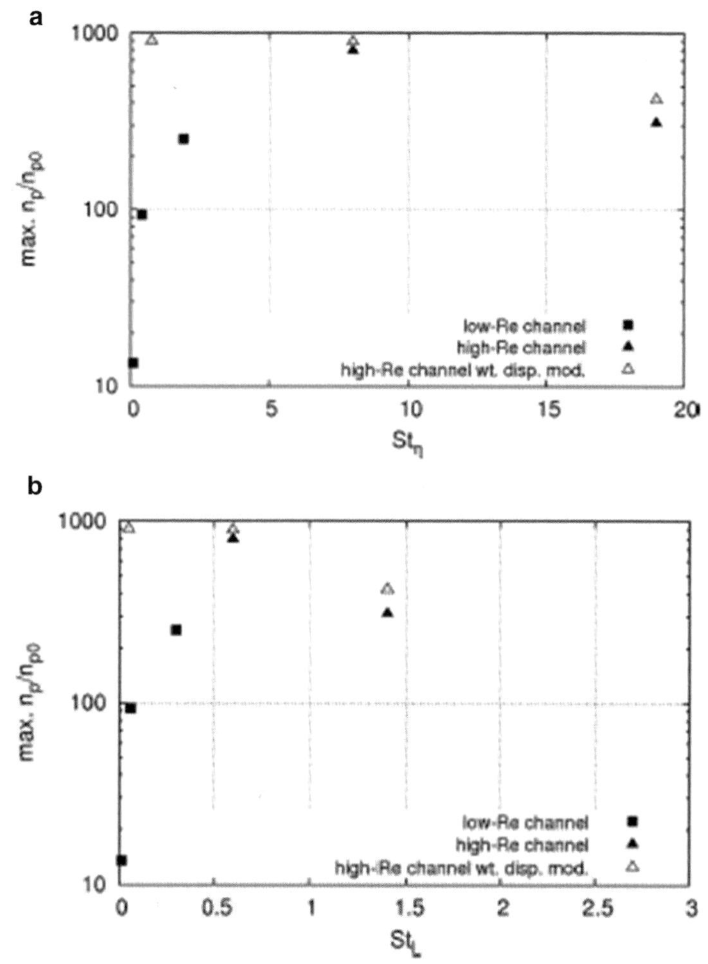

Fig. 14.8 Maximum particle concentration in the near wall region as a function of St_n (**a**) and St_L (**b**) in a turbulent channel flow. Results from the two-phase flow LES on the high and low Reynolds number channel flow, respectively

The relation between particle Stokes number and the maximum concentration is further analyzed in Fig. 14.8. Two fluid scales are used to define the Stokes number – the Kolmogorov and the integral scale of the flow, which leads to St_n and St_L, respectively. The particle concentration is normalized by the corresponding initial uniform concentration value. The available data for the mean particle concentration are extracted from the case with Lycopodium particles. It is evident that the near wall concentration denotes a maximum. The representation of the concentration as a function of the Stokes number based on the integral scale is useful (Fig. 14.8b). According to this figure, the maximum occurs for $0.5 < St_L < 1$.

From this information, it is possible to draw parallels to the work of Marchioli and Soldati [51] and Tang et al. [99]. Both groups consider the large streaky structures near the wall as responsible for the particles to reside in these regions.

14.6 Conclusions

This contribution focused on assessing the capability of the LES based Eulerian–Lagrangian method as implemented into the commercial CFD codes, FLUENT, in reliably predicting particle preferential accumulation and turbulent dispersion. This aspect arises from the essential influence of these phenomena on the prediction of mixing and chemical reaction processes as occurring in IC engine spray. Two alternative generation mechanisms are investigated: the direct influence on turbulent statistics and turbulence modification induced by overall momentum change in the fluid due to particles. The results for local particle accumulation and dispersion investigated in the context of channel flow configurations with both low and high Reynolds number can be summarized as follows:

Using LES to describe the carrier fluid motion explicitly implies that these phenomena become partially resolved. The difference in the standard deviations of the computed number density PDFs and a Poisson distribution presents a parameter for characterization of the preferential accumulation.

The findings for the high Reynolds number flow confirmed the ability of LES to predict the experimental results from Fessler et al. [28] that particles with $\mathrm{St}=2$, where the Stokes number is based on the Kolmogorov time scale, exhibit maximal accumulation. Moreover, this behavior is observed in the low Reynolds number flow as well. The Stokes number exhibiting maximal accumulation contradicts previous findings of particle tracking in the DNS context [97, 109]). The authors, however, did not consider different length scales at which preferential accumulation might possibly occur. In the present work it is evident that the scale of maximum accumulation increases proportional to the St. An additional dependency of the magnitude of accumulation on the flow Reynolds number is demonstrated by comparing both channel flows. A lower turbulence intensity induces higher levels of particle preferential accumulation.

A stochastic model accounting for sub-grid induced particle dispersion that was investigated in the context of the high Reynolds number channel flow based on the eddy–life/eddy–crossing time hypothesis leads to a significant decrease in particle accumulation. The main issue detected here is that the model constant seems to be case dependent, since the amount of the unresolved kinetic energy cannot be fixed, as in the case of turbulent models based on statistical averaging.

Following the resulting local particle number density field, it is further apparent that the need for a collisional model in LES context is explicit, even for very low particle mass loading. Considering particle collision will further increase the particle dispersion in a natural way. For the results presented here chemical reactions were not considered.

Using the code FLUENT/ANSYS, it was found for the type of pressure–velocity coupling, that the Fractional Step Method (FSM) for transient problems preserves the required second order accuracy of the time advancement scheme. The method abstains by definition from the global iterations between the continuity and the momentum equation within a single time step. Consequently, the solution could accelerate up to several times compared to commonly used iterative time advancement methods, such as SIMPLE, SIMPLEC or PISO.

Acknowledgment The authors are grateful to the financial support by the German Research Council (DFG).

References

1. AGARD: A selection of test cases for the validation of large-Eddy simulations of turbulent flows. Agard Advisory Report 345, Neuilly-Sur-Seine, France (1998)
2. ANSYS Fluent 12.0 Documentation. ANSYS Inc. (2009)
3. Apte, S.V., Mahesh, K., Moin, P., Oefelein, J.C.: Large-eddy simulation of swirling particle-laden flows in a coaxial-jet combustor. Int. J. Multiphase Flow **29**, 1311–1331 (2003)
4. Apte, S.V., Mahesh, K., Moin, P.: Large-eddy simulation of evaporating spray in a coaxial combustor. Proc. Combust. Inst. **32**, 2247–2256 (2009)
5. Armenio, V., Piomelli, U., Fiorotto, V.: Effect of the subgrid scales on particle motion. Phys. Fluids **11**, 3030–3042 (1999)
6. Armsfield, S., Street, R.: The fractional step method for the Navier–Stokes equations on staggered grids: accuracy of three variations. J. Comput. Phys. **153**, 660–665 (1999)
7. Axerio, J., Iaccarino, G.: Flow asymmetry and vortical structures behind a rotating tire. Bull. Am. Phys. Soc. **54** (2009)
8. Barth, T.J., Jespersen, D.: The design and application of upwind schemes on unstructured meshes. Technical Report AIAA-89-0366, AIAA 27th Aerospace Science Meeting, Reno, Nevada (1989)
9. Benson, M.J., Eaton, J.K.: The effects of wall roughness on the particle velocity field in fully developed channel flow. Report No. TSD-150. Thermosciences Division, Stanford University (2003)
10. Boileau, M., Pascaud, S., Riber, E., Cuenot, B., Gicquel, L.Y.M., Poinsot, T.J., Cazalens, M.: Investigation of two-fluid method for large eddy simulation of spray combustion in gas turbines. Flow Turb. Combust. **80**, 291–321 (2008)
11. Burton, T.M., Eaton, J.K.: Fully resolved simulations of particle-turbulence interaction. J. Fluid Mech. **545**, 67–111 (2005)
12. Caraman, N., Boree, J., Simonin, O.: Effect of collisions on the dispersed phase fluctuations in a dilute tube flow: experimental and theoretical analysis. Phys. Fluids **15**, 3602–3612 (2003)
13. Celik, I., Klein, M., Freitag, M., Janicka, J.: Assessment measures for urans/des/les: an overview with applications. J. Turbul. **7**(48), 48 (2006)
14. Chrigui, M., Gounder, J., Sadiki, A., Masri, A.R., Janicka, J.: Partially premixed reacting acetone spray using LES and FGM tabulated chemistry. Combust. Flame **159**(8), 2718–2741 (2012)
15. Clamen, A., Gauvin, W.H.: Effects of turbulence on the drag coefficients of spheres in a supercritical flow regime. AIChE J. **15**, 184 (1961)
16. Crowe, C., Sommerfeld, M., Tsuji, Y.: Multiphase Flows with Droplets and Particles. CRC Press, Boca Raton (1998)

17. Crowe, T.: Modeling turbulence in multiphase flows. Eng. Turbul. Model. Exp. **2**, 899–913 (1993)
18. Daubert, T.E., Danner, R.D.: Data compilation tables of properties of pure compounds. Technical report, Design Institute for Physical Property Data. AIChE, New York (1987)
19. Dennis, S.C.R., Singh, S.N., Ingham, D.B.: The steady flow due to a rotating sphere at low and moderate Reynolds numbers. J. Fluid Mech. **101**, 257–279 (1980)
20. Deutsch, E., Simonin, O.: Large eddy simulation applied to the motion of particles in stationary homogeneous fluid turbulence. In: Turbulence Modification in Multiphase Flows, vol. 110, pp. 35–42. ASME, Portland (1991)
21. Dimitrova, D.: On the reliability of large-eddy simulation for dispersed two-phase flows. PhD thesis, TU-Darmstadt, Germany (2010)
22. Drew, D.A., Passman, S.L.: Theory of Multicomponent Fluids. Springer, New York (1999)
23. Elghobashi, S., Truesdell, G.C.: Direct simulation of particle dispersion in a decaying isotropic turbulence. J. Fluid Mech. **242**, 655–700 (1992)
24. Faeth, G.M.: Evaporation and combustion of sprays. Prog. Energy Combust. Sci. **9**, 1–76 (1998)
25. Fede, P., Simonin, O.: Numerical study of the subgrid turbulence effects on the statistics of heavy colliding particles. Phys. Fluids **18**, 045103 (2006)
26. Ferziger, J.H., Peric, M.: Computational Methods for Fluid Mechanics, 3rd edn. Springer, Berlin/New York (2002)
27. Fessler, J.R., Eaton, J.K.: Turbulence modification by particles in a backward-facing step flow. J. Fluid Mech. **394**, 97–117 (1999)
28. Fessler, J.R., Kulick, J.D., Eaton, J.K.: Preferential concentration of heavy particles in a turbulent channel flow. Phys. Fluids **6**, 3742–3749 (1994)
29. Fevrier, P., Simonin, O., Squires, K.D.: Partitioning of particle velocities in gas-solid turbulent flows into a continuous field and a spatially uncorrelated random distribution: theoretical formalism and numerical study. J. Fluid Mech. **533**, 1–46 (2005)
30. Vittoria, M.S., Geurts, B.J., Meyers, J., Sagaut, P.: Quality and reliability of large-eddy simulations. ERCOFTAC Series; vol. 16, ISBN/EAN: 978-94-007-0230-1, Springer (2010)
31. Fröhlich, J.: Large Eddy Simulation Turbulenter Strömungen. B. G. Teubner Verlag, Wiesbaden (2006)
32. Germano, M., Piomelli, U., Moin, P., Cabot, W.H.: Dynamic subgrid-scale eddy viscosity model. Phys. Fluids A **3**(7), 1760–1765 (1991)
33. Goryntsev, D., Sadiki, A., Klein, M., Janicka, J.: Analysis of cycle variation of liquid fuel-air mixing processes in a realistic DISI IC-engine using large eddy simulation. Int. J. Heat Fluid Flow **31**, 845–849 (2010)
34. Hahn, F., Olbricht, C., Janicka, J.: Large eddy simulation of an evaporating spray based on an Eulerian–Lagrangian approach. In: Proceedings of ILASS, Como Lake, Italy, pp. ILASS08-2-9, 8–10 Sept 2008
35. Hardalupas, Y., Taylor, A.M.K.P., Whitelaw, J.H.: Velocity and particle flux characteristics of turbulent particle-laden jets. Proc. R. Soc. Lond. A **426**, 31–78 (1989)
36. Hinze, J.O.: Turbulence, 2nd edn. McGraw-Hill, New York (1975)
37. Hjelmfelt Jr., T., Mockros, L.F.: Motion of discrete particles in a turbulent fluid. Appl. Sci. Res. **16**, 149–161 (1966)
38. Holmes, D.G., Conell, S.D.: Solution of the 2d Navier–Stokes equations on unstructured adaptive grids. In: AIAA 9th Computational Fluid Dynamics Conference, Buffalo (1989)
39. James, S., Zhu, J., Anand, M.S.: Large eddy simulations as a design tool for gas turbine combustion systems. AIAA J. **44**, 674–686 (2006)
40. Janicka, J., Sadiki, A.: Large eddy simulation for turbulent combustion systems. Proc. Combust. Inst. **30**, 573–574 (2005)
41. Kim, S.-E., Makarov, B.: In: 17th AIAA Computational Fluid Dynamics Conference, Toronto Ontario, AIAA Paper 2005–5253, 6–9 June 2005
42. Kolmogorov, N.: Local structure of turbulence in incompressible viscous fluid for very large Reynolds number. Dokl. Akad. Nauk SSSR **30**, 299–303 (1941)

43. Kuerten, H., Vreman, A.W.: Can turbophoresis be predicted by large-eddy simulation? Phys. Fluids **17**, 011701 (2005)
44. Kuerten, J.G.M.: Subgrid modeling in particle-laden channel flow. Phys. Fluids **18**, 025108 (2006)
45. Kulick, J.D., Fessler, J.R., Eaton, J.K.: On the interactions between particles and turbulence in a fully-developed channel flow in air. Mechanical Engineering Report MD-66, Stanford University (1993)
46. Kulick, J.D., Fessler, J.R., Eaton, J.K.: Particle response and turbulence modification in fully developed channel flow. J. Fluid Mech. **277**, 109–134 (1994)
47. Lam, K., Banerjee, S.: On the condition of streak formation in bounded flows. Phys. Fluids A **4**, 306–320 (1992)
48. Leonard, B.P.: The ultimate conservative difference scheme applied to unsteady one dimensional advection. Comput. Methods Appl. Mech. Eng. **88**, 17–74 (1991)
49. Lilly, D.K.: A proposed modification of the germano subgrid-scale closure model. Phys. Fluids **4**, 633–635 (1992)
50. Longmire, E.K., Eaton, J.K.: Structure of a particle-laden round jet. J. Fluid Mech. **236**, 217–257 (1992)
51. Marchioli, C., Soldati, A.: Mechanisms for particle transfer and segregation in a turbulent boundary layer. J. Fluid Mech. **468**, 283–315 (2002)
52. Marchioli, C., Soldati, A.: Dns of particle-laden turbulent channel flow. In: Proceedings of 11th Workshop on Two-Phase Flow Predictions, Merseburg, Germany (2005)
53. Marchioli, C., Soldati, A., Kuerten, J.G.M., Arcen, B., Tanière, A., Goldensoph, G., Squires, K.D., Cargnelutti, M.F., Portela, L.M.: Statistics of particle dispersion in direct numerical simulations of wall-bounded turbulence: results of an international collaborative benchmark test. Int. J. Multiphase Flow **34**, 879–893 (2008)
54. Masoudi, M., Sirignano, W.A.: Nonlinear capillary waves on swirling, axisymmetric free liquid films. Int. J. Multiphase Flow **27**, 1707–1734 (2001)
55. Mathey, F., Cokljat, D.: Assessment of the vortex method for large-eddy simulation inlet conditions. Prog. Comput. Fluid Dyn. **6**, 58–67 (2006)
56. Mathur, S.R., Murthy, J.Y.: A pressure-based method for unstructured meshes. Num. Heat Trans. **31**, 195–215 (1997)
57. Mei, R.: An approximate expression for the shear lift force on a spherical particle at finite Reynolds number. Int. J. Multiphase Flow **18**, 145–147 (1992)
58. Meyers, J., Geurts, B., Sagaut, P.: Quality Reliability of Large Eddy Simulations. Ercoftac Series, vol. 13. Springer, Berlin (2008)
59. Moin, P., Apte, S.V.: Large eddy simulation of realistic gas turbine combustor. AIAA J. **44**, 698–708 (2006)
60. Morsi, S.A., Alexander, A.J.: An investigation of particle trajectories in two-phase flow systems. J. Fluid Mech. **55**, 193–208 (1972)
61. Moser, R.D., Kim, J., Mansour, N.N.: Direct numerical simulation of turbulent channel flow up to $re_{\_} = 590$. Phys. Fluids **11**, 943–945 (1999)
62. Odar, F., Hamilton, W.S.: Forces on a sphere accelerating in a viscous fluid. J. Fluid Mech. **18**, 302–314 (1964)
63. Oefelein, J.C.: Simulation and analysis of turbulent multiphase combustion at high pressures. PhD thesis, The Pennsylvania State University (1997)
64. Olbricht, C.: Numerische Verbrennung technischer Verbrennungssysteme. PhD thesis, TU Darmstadt (2009)
65. Paris, D.: Turbulence attenuation in a particle-laden channel flow. PhD thesis, Stanford University (2001)
66. Patankar, S.V.: Numerical Heat Transfer and Fluid Flow. Hemisphere, Washington, DC (1980)
67. Picciotto, M., Marchioli, C., Reeks, M.W., Soldati, A.: Statistics of velocity and preferential accumulation of micro-particles in boundary layer turbulence. Nucl. Eng. Des. **235**, 1239–1249 (2005)

68. Pitsch, H.: Large-eddy simulation of turbulent combustion. Annu. Rev. Fluid Mech. **38**, 453–482 (2006)
69. Pope, S.B.: Turbulent Flows. Cambridge University Press, Cambridge (2001)
70. Pozorski, J., Apte, S.V.: Filtered particle tracking in isotropic turbulence and stochastic modeling of subgrid–dispersion. Int. J. Multiphase Flow **35**, 118–128 (2009)
71. Prandtl, L.: über die ausgeprägte turbulenz. Z. Angew. Math. Mech. **5**, 136–139 (1925)
72. Qiu, H.-H., Sommerfeld, M.: A reliable method for determinig the measurement volume size and particle mass fluxes using phase-Doppler anemometry. Exp. Fluids **13**, 393–404 (1992)
73. Ranz, W.E., Marshall Jr., W.R.: Evaporation from drops, part i. Chem. Eng. Prog. **48**(3), 141–146 (1952)
74. Ranz, W.E., Marshall Jr., W.R.: Evaporation from drops, part ii. Chem. Eng. Prog. **48**(4), 173–180 (1952)
75. Rhie, C.M., Chow, W.L.: Numerical study of the turbulent flow part an airfoil with trailing edge separation. AIAA J. **21**, 1525–1532 (1983)
76. Riber, E., Moureau, V., Garcia, M., Poinsot, T., Simonin, O.: Evaluation of numerical strategies for large eddy simulation of particulate two-phase recirculating flows. J. Comput. Phys.. doi:10.1016/j.jcp.2008.10.001, 2008
77. Richardson, L.F.: Weather Prediction by Numerical Process. Cambridge University Press, Cambridge (1922)
78. Rubinow, S.I., Keller, J.B.: The transverse force on spinning sphere moving in a viscous fluid. J. Fluid Mech. **11**, 447–459 (1961)
79. Rudoff, R.R., Bachalo, W.D.: Measurements of droplet drag coefficients in polydispersed turbulent flow field. AIAA, pp. 80–0235 (1988)
80. Ruetsch, G.R., Maxey, M.R.: The evolution of small–scale structures in homogeneous turbulence. Phys. Fluids A **4**, 2747 (1992)
81. Rutland, C.J.: Large eddy simulation for internal combustion engine: a review. Int. J. Eng. Resour. **12**(5), 421–445 (2011)
82. Saffman, G.G.: The lift on a small sphere in a slow shear flow. J. Fluid Mech. **22**, 385–400 (1965)
83. Sagaut, P.: Large Eddy Simulation of Incompressible Flows. Springer, Berlin/New York (2001)
84. Sawatzki, O.: Strömungsfeld um eine rotierende kugel. Acta Mech. **9**, 159–214 (1970)
85. Schiller, L., Naumann, Z.: A drag coefficient correlation. VDI Zeitschrift **77**, 318 (1935)
86. Schlichting, H., Gersten, K.: Grenzschicht-Theorie, 9th edn. Springer, Berlin/Heidelberg (1997)
87. Schuman, U.: Subgrid scale model for finite difference simulations of turbulent flows in plane channels and annuli. J. Comput. Phys. **18**, 376–404 (1975)
88. Segura, J.C.: Predictive capabilities of particle-laden large eddy simulation. PhD thesis, Department of Mechanical Engineering, Stanford University (2004)
89. Sirignano, W.A.: Fluid Dynamics and Transport of Droplets and Sprays. Applied Mathematical Sciences 135. Cambridge University Press, New York (2005)
90. Smagorinsky, J.: General circulation experiments with the primitive equations. i. The basic experiment. Mon. Weather Rev. **91**, 99–164 (1963)
91. Snyder, W.H., Lumley, J.L.: Some measurements of particle velocity autocorrelation functions in a turbulent flow. J. Fluid Mech. **48**, 41–71 (1971)
92. Sommerfeld, M.: Modellierung und numerische Berechnung von partikelbeladenen turbulenten Strömungen mit Hilfe des Euler/Lagrange-Verfahrens (1998)
93. Sommerfeld, M.: Theoretical and experimental modelling of particulate flows. Technical Report Lecture Series 2000–06, von Karman Institute for Fluid Dynamics (2000)
94. Sommerfeld, M., Qiu, H.H.: Particle concentration measurements by phase-Doppler anemometry in complex dispersed two-phase flows. Exp. Fluids **18**, 187–198 (1995)
95. Sommerfeld, M., Qiu, H.H.: Spray evaporation in turbulent flow URL http://www-mvt.iw.uni-halle.de/index.php?spray_evaporation. Data base (1998)

96. Sommerfeld, M., Qiu, H.H.: Experimental studies of spray evaporation in turbulent flow. Int. J. Heat Fluid Flow **19**, 10–22 (1998)
97. Squires, K.D., Eaton, J.K.: Preferential concentration of particles by turbulence. Phys. Fluids A **3**, 1169–1178 (1991)
98. Stokes, G.G.: On the effect of the inertial friction of fluids on the motion of pendulums. Trans. Camb. Phil. Soc. **9**, 8–106 (1851)
99. Tang, L., Wen, F., Yang, Y., Crowe, C.T., Chung, J.N., Troutt, T.R.: Self-organizing particle dispersion mechanism in a plane wake. Phys. Fluids A **4**, 2244–2251 (1992)
100. Torobin, L.B., Gauvin, W.H.: The drag coefficient of single sphere moving in steady and accelerated motion in a turbulent fluid. AIChE J. **7**, 615–619 (1961)
101. Truesdell, C., Toupin, R.: The Classical Field Theories, volume III of Handbuch der Physik, chapter 1 Part. Springer, Berlin (1960)
102. Uhlherr, P.H.T., Sinclair, C.G.: The effect of free stream turbulence on the drag coefficient of spheres. In: Proceedings of Chemca '70, 1:1 (1970)
103. URL http://www.eacc.fluent.com/papers.html (2007)
104. Vance, M., Squires, K.W., Simonin, O.: Properties of the particle velocity field in gas-solid turbulent channel flow. Phys. Fluids **18**, 063302 (2006)
105. Vargaftik, N.B.: Handbook of Physical Properties of Liquids and Gases, 2nd edn. Taylor & Francis Inc, Washington, DC (1983)
106. VDI-Gesellschaft Verfahrenstechnik und Chemieingenieurwesen (Hrsg.): VDI-Wärmeatlas: Berechnungsblätter für den Wärmeübergang. Springer, Berlin (2002)
107. Vreman, W.: Turbulence characteristics of particle-laden pipe flow. J. Fluid Mech. **584**, 235–279 (2007)
108. Vreman, A.W., Geurts, B.J., Deen, N.G., Kuipers, J.A.M.: Large-eddy simulation of a particle-laden turbulent channel flow. In: Friedrich, R., et al. (eds.) Direct and Large-Eddy Simulation, vol. 5, pp. 271–278, Kluwer Academic Publishers (2004)
109. Wang, L.P., Maxey, M.R.: Settling velocity and concentration distribution of heavy particles in homogeneous isotropic turbulence. J. Fluid Mech. **256**, 27–68 (1993)
110. Wang, Q., Squires, K.D.: Large-eddy simulation of particle-laden turbulent channel flow. Phys. Fluids **8**, 1207–1223 (1996)
111. Wilcox, D.C.: Turbulence Modeling for CFD. D C W Industries (2000)
112. Yamamoto, Y., Potthoff, M., Tanaka, T., Kajishima, T., Tsuji, Y.: Large-eddy simulation of turbulent gas-particle flow in a vertical channel: effect of considering inter-particle collisions. J. Fluid Mech. **442**, 303–334 (2001)
113. Young, J., Leeming, A.: A theory of particle deposition in turbulent pipe flow. J. Fluid Mech. **340**, 129–159 (1997)
114. Zarin, N.A., Nicholls, J.A.: Sphere drag in solid rockets – non continuum and turbulence effects. Combust. Sci. Technol. **3**, 273 (1971)

Chapter 15
Planar Droplet Sizing for Characterization of Automotive Sprays in Port Fuel Injection Applications Using Commercial Fuel

S. Bareiss, N. Fuhrmann, A. Dreizler, H. Bacher, J. Höffner, R. Weishäupl, and D. Kügler

Abstract Drop size measurements are crucial for characterizing practical sprays in automotive applications. Phase-Doppler (PD) techniques allow for highly accurate point-wise measurements of single droplets diameters and velocities. Statistical parameters such as mean diameters are recorded at single locations that are composed of thousands of individual droplet measurements. Therefore the assessment of an entire spray by the phase Doppler technique is time-consuming. For this reason there is some need for planar drop sizing methods in engineering applications that call for much faster spray characterizations. Simultaneous Mie scattering and planar laser-induced fluorescence (Mie/PLIF) is one choice that provides access to Sauter-mean-diameters (SMD). The advantage in reduced measuring times comes along with various challenging issues, such as accuracy limited by multiple-scattering, absorption effects, and calibration procedures. Its application especially to practically relevant sprays claims for real fuel composed of a mixture of various hydrocarbons with intense absorption bands in the ultraviolet spectral range. The purpose of this project is the adaptation of the Mie/PLIF technique to real-world requirements and is conducted in cooperation with BMW F1. For a multi-hole injector operated with commercial gasoline fuel and port-fuel injection conditions

S. Bareiss • N. Fuhrmann • A. Dreizler (✉)
Reactive Flows and Diagnostics, Center of Smart Interfaces, Mechanical Engineering,
Technische Universität Darmstadt, Petersenstr. 30, D-64287 Darmstadt, Germany
e-mail: dreizler@csi.tu-darmstadt.de

H. Bacher • J. Höffner
BMW AG EA-833, 80788 München, Germany

R. Weishäupl
BMW AG EA-433, 80788 München, Germany

D. Kügler
BMW AG UX-EA-4, 80788 München, Germany

J. Janicka et al., *Flow and Combustion in Advanced Gas Turbine Combustors*,
Fluid Mechanics and Its Applications 102, DOI 10.1007/978-94-007-5320-4_15,

typical for motorsport engines it is shown that deviations between SMDs measured by phase Doppler and Mie/PLIF is below 20%. Savings of time are up to 80% justifying the limitation to measuring the SMD only.

Keywords PDS • LSD • PDA • Sauter-mean-diameter • Mie • LIF • Gasoline

15.1 Introduction

Knowledge of the droplet size distribution is a key aspect for spray characterization. In case of automotive applications the performance of a gasoline injector significantly affects the combustion process and therefore the engine performance and emissions in general. Hence providing methods to non-intrusively characterize and assess injectors by their spray properties significantly supports improvements of engine and emission performance.

For measurements of droplet sizes in two-phase flows laser based techniques are widely used due to their non-intrusiveness. The most common and established method for measuring droplet sizes is the phase-Doppler technique (PD). This technique provides information on single-droplet size and velocity without any need for calibration. Droplet size distributions at specific locations in the spray are deduced from a multitude of individual droplet measurements. A major drawback of PD is its point-like probe volume. Measuring the spatial distribution of droplet sizes causes a huge effort since the spray needs to be scanned at least in one plane. Both, its point-wise character and the large number of individual droplet measurements required for deducing reliable statistical properties, cause a significant time need for comprehensive spray characterization.

Automotive industry requires a speed-up of accurate optical methods for more time-efficient characterization of large numbers of different injectors focusing on quality assessment and performance analysis. Other emerging areas requiring fast spray measurements are non-stationary engine operation and – due to growing demands in emission legislation – start-stop-situations.

A speed-up of spray measurements can be achieved by two-dimensional laser diagnostics. Intensity based methods are promising for planar drop sizing that are supposed to be applicable in dense sprays as encountered in practical applications, e.g. internal combustion engines. Intensity ratio methods are based on the calculation of the ratio of two separate signals from the same probe location measured at the same instant. Following a calibration procedure the ratio serves as a scalar reference value for a statistical mean drop size. Depending on the specific light-matter interaction used in the process, the respective statistical mean diameter may vary, with D_{21} and D_{32} being the most prominent ones. Different combinations of planar spectroscopic or Mie-based methods were reported. Malarski et al. [19] used the ratio calculated from planar spontaneous Raman and Mie scattering. In another approach Hofeldt [13] took advantage of s- and p-polarization directions of Mie-scattered light that show different drop size dependencies. The combination of droplet fluorescence (laser-induced fluorescence – LIF) and Mie scattering signal is discussed in the following in some more detail as it forms the basis of the present contribution.

15.2 Background and Preparative Considerations

The combined LIF/Mie technique was investigated by different research groups and was termed as Planar Drop Sizing (PDS) (Domann et al. [9]) or Laser Sheet Drop Sizing (LSD) (Le Gal et al. [17]). For the combined LIF/Mie technique the ratio of both the fluorescence and the elastic Mie scattering signal yields the SMD (D_{32}):

$$\frac{I_{LIF}}{I_{Mie}} = \frac{C_{LIF} D^3}{C_{Mie} D^2} = C \frac{D^3}{D^2}. \tag{15.1}$$

C_{LIF} and C_{Mie} are constants of proportionality to be determined by appropriate calibration. Applied to a spray and when Mie and LIF signals are detected simultaneously by two CCD cameras where individual droplets are no longer resolved, Eq. (15.1) needs to be extended. Depending on the respective spatial resolution the probe volume contains then a multitude of droplets following a case-specific but *a priori* unknown size distribution. Hence the number of droplets (n) in each probe volume (i,j) for all frames (k) needs to be considered for calculating the ensemble-averaged intensity ratio:

$$\left(\frac{I_{LIF}}{I_{Mie}}\right)_{i,j} = \frac{C_{LIF} \sum_k n_{i,j,k} (D_{i,j,k})^3}{C_{Mie} \sum_k n_{i,j,k} (D_{i,j,k})^2} = C \cdot (D_{32})_{i,j} \tag{15.2}$$

Following a calibration determining the constant C, the local Sauter-Mean-Diameter $(D_{32})_{i,j}$ is obtained. So far, Eq. (15.2) does not account for effects such as multiple scattering or molecular absorption of laser radiation within the droplets.

The combination of droplet fluorescence and the elastic Mie scattering signal as an approach for intensity-based planar droplet sizing dates back to Kamimoto et al. [15]. Kamimoto measured the SMD of a non-evaporating spray. The liquid was a blend of hydrocarbons. Tetramethyl-phenylene-diamine (TMPD) was dissolved as fluorescence tracer. The multi-component fuel was injected into an optical accessible chamber of a rapid compression engine. Le Gal et al. [17] and Sankar et al. [21] subsequently applied this technique to a pressure swirl atomizer. Le Gal et al. carefully assessed the influence of the tracer concentration on the droplet fluorescence. Due to molecular absorption on the one hand the exponent for the fluorescence in Eq. (15.1) can vary between two and three. On the other hand, in case of optical resonances causing lasing of droplets, exponents larger than three are possible. Sankar et al. theoretically investigated elastic scattering by means of Lorenz-Mie calculations. A major finding was that interferences in different scattering orders are smoothed even for rather low molecular absorption. Le Gal as well as Sankar et al. recorded Mie and LIF signals in their experiments sequentially and averaged subsequently over several images to achieve a reliable SMD. In both studies PD was used for calibration purposes. Jermy et al. [14] compared Mie/PLIF-measurements to PD-data in a poly-disperse water spray. These authors conducted an error analysis but neglected any effects of multiple scattering. For their conditions they specified the calibration based on PD-data as the dominant error source (7%). Domann et al. [9] compared detailed Lorenz-Mie calculations to measurements obtained from a mono-dispersed droplet stream. These authors investigated the influence of scattering angle, refractive index, droplet-size and absorption effects (by varying dye concentrations in the liquid) on the exponents (compare Eq. (15.2)) in the Mie and fluorescence signals, respectively. As a result an observation angle of 90° is favored since the influence of any absorption on the scattering signal is then minimized. In a subsequent publication Domann et al. [10] discussed a droplet-size dependency of the calibration constant C in Eq. (15.2) and proposed an iterative calibration procedure.

Park et al. [20] first demonstrated the possibility to use "natural" fluorescence of unleaded gasoline for the combined LIF/Mie technique. These authors investigated a spray generated by a GDI injector. The fluorescence was excited at 355 nm and both, Mie- and LIF-signals, were imaged simultaneously on one intensified CCD camera. A mono-dispersed droplet stream was used to verify the quadratic and volumetric dependency of the Mie and LIF signals shown in Eq. (15.2).

Planar drop sizing by combined LIF/Mie was not only applied to investigate sprays with relevance in automotive applications but also to characterize injectors operating continuously, see Zimmer et al. [25] and Zaller et al. [24]. Zaller et al. applied the LIF/Mie method to a spray in a gas turbine combustor operated with liquid jet fuel. Its natural fluorescence was excited by laser radiation at 266 nm but the approach significantly lacked from molecular absorption caused by the multi-component hydrocarbon fuel.

Düwel et al. [11] compared acetone, toluene and coumarine dye as fluorescence tracers for the combined LIF/Mie-technique. These tracers differ in the respective boiling points affecting their application in a spray flame. It was stated that systematic errors can occur due to enrichment and depletion of fluorescence tracers for evaporating conditions. A careful choice of tracers with similar boiling points as the solvent was therefore strongly recommended.

Another important issue is multiple-scattering that gains consideration especially for dense sprays. Berrocal et al. [6] described spatially structured illumination of sprays, termed Structured Laser Illumination Planar Imaging (SLIPI), in order to minimize systematic errors due to multiple scattering. This theoretical work was extended by scattering experiments (Berrocal et al. [7]) using a hollow cone spray. Significant improvements were achieved regarding the suppression of multiple-scattering and enhancement of the image contrast.

The elastically scattering of light off a spherical droplet is theoretically well described and can be calculated by the Lorenz-Mie-Theory (Bohren and

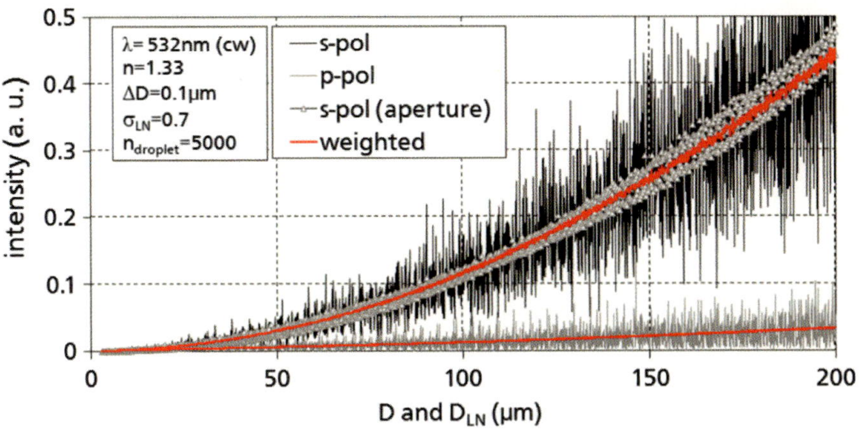

Fig. 15.1 Scattered light intensity for water droplets illuminated by laser light at 532 nm

Huffman [8]). In the present study Lorentz-Mie-calculations were performed for single droplets with a code described by Albrecht et al. [1]. Exemplarily Fig. 15.1 shows a calculation for water droplets illuminated with light of 532 nm and detected in a 90° angle.

Mie-scattered light intensity is shown as a function of the droplet diameter with a droplet size increment of 0.1 μm. The dark line represents the scattered light for vertically polarized (s-pol) light, the bright line for parallel polarized (p-pol) light. Interferences from different scattering orders cause strong oscillations even for small changes in droplet size [4]. For intensity-based planar drop sizing techniques the intensity of the Mie-signal is intended to be linked directly to a mean droplet diameter. Therefore a monotone relation between intensity and droplet diameter is required without any ambiguousness. Damping these oscillations is consequently one key aspect using Mie-scattering light for droplet sizing. Suppression of oscillations can be achieved by different approaches such as ultra-short laser pulses [4], absorbing dye [1], broadband light source [22], and spatial averaging.

To theoretically investigate the effect of spatial averaging the code for the scattering calculation [2] was extended. Based on single droplet calculations a model was created simulating a user-defined number of droplets randomly distributed in size but following a log-normal-distribution. For the spray under investigation the log-normal-distribution was found to be suitable to describe the spray properties. As input for the size distribution the standard deviation (σ_{LN}) and the characteristic diameter (D_{LN}) are required. By presuming a droplet number of 50 in 100 consecutive images and for a standard deviation of 0.7, the calculations shown in Fig. 15.1 exhibit significantly smoothed curves (termed "weighted" in the figure caption) for both the vertically and parallel polarized light, respectively. Hence it can be stated that for the experiments described below a relatively coarse spatial resolution is required to guarantee a sufficient large number of droplets to be contained in each probe volume in order to smooth the scattering oscillations.

Both polarization directions are related differently to the droplet diameter which was already described by Hofeldt [13]. For an observation angle of 90° the vertically polarized light follows approximately a D^2-proportionality whereas the parallel polarized light scales closely to D. These dependencies on their own also can be employed for planar droplet sizing by taking the ratio of vertically and parallel polarized Mie-scattered light. Lettierie et al. [18] first used resonances in the polarization ratio to determine the size of individual levitated droplets. Later the polarization-ratio-technique was investigated first by Hofeldt [13] in more detail. He discovered among other effects a strong dependency of the sensitivity on the refractive index. However, a more detailed discussion of the polarization-ratio-technique is beyond the scope of this report but actually a very promising alternative to the LIF/Mie-technique [5].

Spontaneous emission following a resonant absorption induced by laser light is termed LIF. Fluorescence of large hydrocarbon species is spectrally broad-band and red-shifted relative to the respective absorption band. Therefore fluorescence light can be easily separated spectrally from Mie scattering signals that are observed at the same wavelength as the laser radiation used for excitation. When molecular absorption of the exciting laser radiation is negligible the fluorescence signal is proportional to the droplet volume (compare Eq. (15.2)).

For real-world applications the focus is on commercial fuels and production-type injectors. For this reason the LIF-excitation wavelength has to be chosen to fulfill the requirement of negligible absorption but maintaining a sufficiently high signal-to-noise. Absorption and emission spectra were measured to identify the best compromise in terms of photo-physical properties of the fuel and availability of laser emission wavelengths. The analysis of the commercial fuel was performed by PetroLab, Speyer.

As expected the absorption increases for decreasing excitation wavelengths. Considering common laser emission lines the third harmonic of a pulsed Nd:YAG laser emitting at 355 nm apparently is a reasonable compromise for moderate absorption and reliable laser operation. This choice is supported by numerical simulations of Frackowiak and Tropea [12]. The fluorescence spectrum following upon 355 nm excitation is broadband and reaches up to the green spectral range. A band-pass absorption filter in the blue spectral range is best suited to suppress spurious light from the environment (compare Fig. 15.3).

Another crucial aspect for LIF-experiments is the linearity of its response for varied excitation energies. By means of a variable beam attenuator the linearity was checked carefully. For a light sheet 25 mm in height no significant saturation was observed for laser pulse energies up to 4.5 mJ at 355 nm.

Equation (15.2) presumes that the fluorescence signal scales with the droplet volume. To check this assumption for the present conditions (commercial gasoline, 355 nm excitation), supplementary measurements were carried out using a mono-dispersed droplet stream. The droplet stream was generated by a piezo-electric-droplet-generator, SBG2000 (Palas). The droplets were illuminated by a 355 nm laser sheet of about 25 mm height which was focused onto the droplet stream. The laser sheet thickness in any case was kept larger than the largest droplet diameters.

Hence a homogenous illumination was ensured and the complete droplet volume contributed to the LIF-signal. In this case droplets were detected by a long distance microscope (Infinity K2-S) in conjunction with an intensified CCD camera (see Sect. 15.3). The pixel resolution was approximately 5.3 μm/pix. For resolving even the smallest droplets (20 μm) the intensifier gain was set to 80%. Droplet sizes were varied by using different diameters of the orifice plate, by varying the operating pressure, and the excitation frequency for the piezo crystal.

15.3 Experimental Setup

All measurements were carried out using an experimental test rig at the BMW F1 Division. The rig comprises an optically accessible pressure chamber and the infrastructure to repetitively operate a gasoline spray at 5 Hz. The pressure chamber is sketched in Fig. 15.2. For optical access the chamber is equipped with three quartz windows. The arrangement allows for PD-measurements and planar drop sizing at identical injector positions and identical operation conditions.

Operational conditions are listed in Table 15.1. Apart from the inlet manifold vacuum the conditions of the intake-manifold for port fuel injection in F1 engines can be matched. The injector is a multi-hole injector (7 holes) with symmetric cone angles (60°) in circumferential direction. The spray angle is also symmetric for all except the middle cone that injects the fuel along the center axis of the injector.

Fig. 15.2 Sketch of the pressure vessel

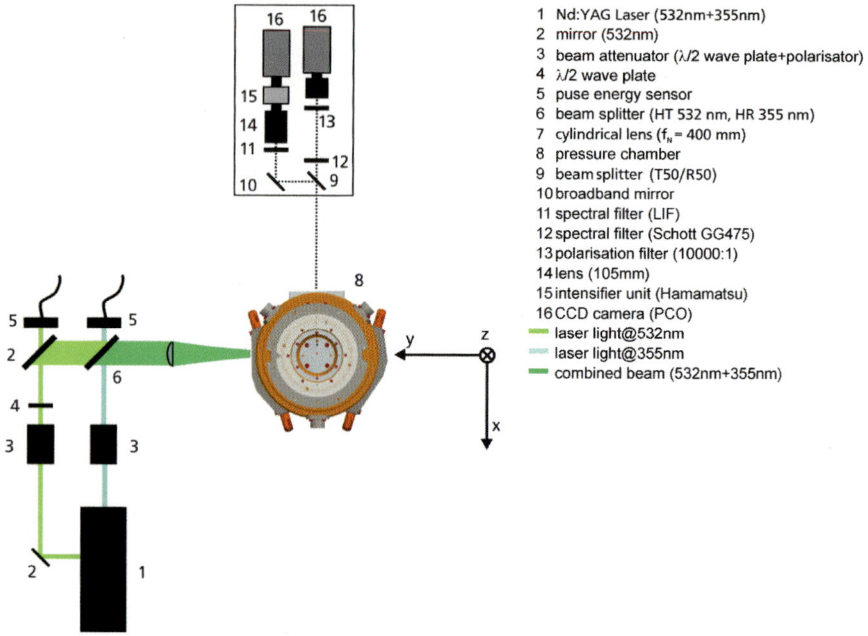

Fig. 15.3 Schematic of the optical setup for combined Mie/LIF drop sizing measurements

Table 15.1 Conditions for the pressure chamber

Chamber temperature (T_c)	Amb. air–100°C
Chamber gauge pressure (p_c)	0.1–4 bar
Injection pressure (p_i)	0–400 bar
Injector position (φ)	0–360°
Liquid for injection	Engine fuel
Chamber atmosphere	N_2

The injector can be rotated via a rotating mount by an angle (φ) of 360° in increments of 0.1°. Unless stated otherwise the injector was operated with a back pressure of 100 bar. The chamber was heated to 80°C and was pressurized to 2 bar.

As discussed in the context of Eq. (15.2) intensity-based planar drop sizing requires appropriate calibration measurements. For this purpose the phase-Doppler technique was used (High Dense PDA System, Dantec Dynamics). Transmitting and receiving PDA-optics were mounted on the same traverse system. This ensured a very high repositioning accuracy better than 100 μm. Any drift in the positioning was corrected for by realigning the foci of the laser beams to match with the center axis of the injector.

All relevant parameters for the transmitting and receiving optics are listed Tables 15.2 and 15.3, respectively. Key parameters such as high-voltage for the photomultipliers, laser-power, signal-to-noise-ratio where chosen following the

Table 15.2 Parameters for the PDA transmitting optics

Name [unit]	Value for PDA channel
Wavelength [nm]	514.5
Focal length [mm]	310
Beam diameter [mm]	2.2
Number of fringes	23
Fringe spacing [μm]	2.03
Beam half-angle [°]	7.28
Probe volume d_X [mm]	0.047
Probe volume d_Y [mm]	0.04662
Probe volume d_Z [mm]	0.3679
Expander ratio [−]	1.98
Beam spacing [mm]	40

Table 15.3 Parameters for the PDA receiving optic

Name [unit]	Value for PDA channel
Receiver type	112 mm fiber PDA
Off-axis angle (scattering angle) θ [°]	70
Receiver focal length [mm]	310
Receiver expander ratio	1
Scattering mode	Refraction (first order)
Aperture mask	Mask B
Spatial filter (slit) [mm]	0.050

Table 15.4 Optical parameters for the BSA-flow-software

Parameter [unit]	Value for PDA channel
Signal gain [dB]	26
SNR [dB]	0
Validation ratio	4
Receiver expander ratio	1
PM1 high voltage [V]	800
PM 2,3 high voltage [V]	800 (balance on PM1)

procedure outlined by Kapulla and Najera [16]. Table 15.4 specifies all user-defined parameters found to be most appropriate for the present configuration.

The optical setup for the planar measurements is sketched in Fig. 15.3. One cavity of a Quantel-Brilliant-B laser system was used to simultaneously generate 532 and 355 nm, respectively, by means of second and third harmonic generation.

Respective laser energies were controlled for both wavelengths independently by variable beam attenuators consisting of λ/2-wave-plate and polarizer. Via periscopes and telescopes in each light path the laser sheet positions in the pressure chamber and the beam divergence for each wavelength were aligned for best match. For 532 nm additionally another λ/2-wave-plate was placed in the laser light path in order to control the polarization direction within the measurement volume (s-pol). Following a telescope (combination of a concave lens $f_N = -40$ mm and a convex lens $f_N = 200$ mm) the beam diameters of 532 and 355 nm radiation were approximately 30 and 25 mm, respectively. The beams were combined by a beam splitter reflecting 355 nm and transmitting 532 nm. The combined beam was directed through a common cylindrical lens ($f_N = 400$ mm) in order to form laser sheets for both wavelengths with a thickness of about 400 μm at the center axis of the injector. For a best match of the laser sheets the individual telescopes in the light paths of 532 nm and 355 nm were crucial. Spatial matching was confirmed using the knife edge method proposed by Araújo et al. [3].

Direction of detection was 90° relative to the light sheet plane. As obvious from Fig. 15.3 two cameras were mounted parallel to each other. Fluorescence and Mie scattered light was divided into two detection paths using a 50/50 beam splitter. Fluorescence-signals were recorded by an image intensifier (Hamamatsu) lens-coupled to a CCD camera (PCO-sensicam qe, 12 bit, 1,376 × 1,040 pixels, pixel size 6.45 μm × 6.45 μm). For imaging a Nikon lens ($f_N = 105$ mm, f-number 2.8) was used. In order to prevent spectral crosstalk and therefore any leak of elastically scattering light at 355 nm, a band-pass filter with high transmittance ($T > 50\%$) from 405 nm up to 435 nm was placed in front of the image intensifier. Mie-scattered light passing the 50/50 beam splitter was directed onto a second 12 bit CCD camera (Hamamatsu C8484-05C, 1,344 × 1,024 pixels, pixel size 6.45 μm × 6.45 μm). This camera was equipped with another Nikon lens ($f_N = 60$ mm, f-number of 2.8) and a yellow glass filter (Schott GG475). For the detection of only vertically polarized light a polarization filter was used (CVI, extinction ratio 10,000:1).

Both detectors were operated in their linear regime. The working distance was approximately 620 mm resulting in a magnification of approximately 10:1 for both LIF- and Mie-signals. The intensified CCD was gated with 600 ns. Both detection units were synchronized capturing signals from the temporally coincident light pulses. The master trigger was the injection pulse. The instant of combined LIF/Mie-images was delayed relative to start of injection.

In a preparative experiment the extinction of both wavelengths was measured along the propagation direction of the laser sheet intersecting the spray. Both wavelengths were weakened notably along the beam propagation direction. However, locally the intensity ratio was nearly constant indicating that extinction losses for both wavelengths did not differ significantly.

15.4 Image Post-processing

The fields of view for both cameras need to be matched. This was achieved by imaging a calibration target that was placed accurately in the laser light sheet plane. This allowed for correction of slightly differing magnifications and for correction of any distortions in the respective imaging systems. Subsequently the images were corrected for inhomogeneities across the profile of the laser sheets. For this purpose ensemble-averaged laser profiles were recorded by flooding the chamber with a homogeneously distributed aerosol. Contours of the spray were deduced by means of a histogram-based algorithm in combination with a triangle thresholding method

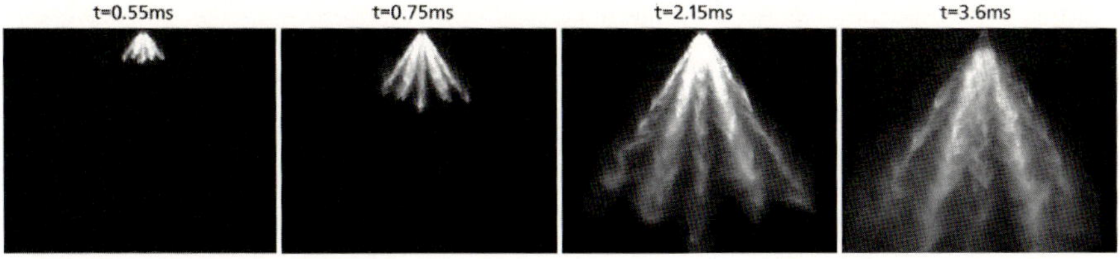

Fig. 15.4 Snapshots of Mie scattering recorded after 0.55, 0.75, 2.15 and 3.6 ms after start of energizing. For drop sizing purposes the central cone and the cone directly right of the central cone are used

described by Sivasubramanian et al. [23]. This method provides the advantage that there are no used-defined parameters to be adjusted.

For each individual Mie-image a threshold was defined and subsequently used for separating spray from surrounding gas. Simultaneously recorded LIF-images were masked accordingly.

The local SMD $(D_{32})_{i,j}$ was deduced from the ratio of LIF- and Mie-signals according to Eq. (15.3):

$$D_{32,i,j} = C \cdot M_{PF,i,j} \frac{\sum_{k=1}^{n} M_{i,j,k} \cdot I_{LIF,i,j,k}}{\sum_{k=1}^{n} M_{i,j,k} \cdot I_{Mie\perp,i,j,k}} \tag{15.3}$$

$M_{\mathrm{i,j,k}}$ is the mask obtained from the thresholding method mentioned above. If no spray is observed at position (i,j) in frame k then $M_{\mathrm{i,j,k}} = 0$, otherwise $M_{\mathrm{i,j,k}} = 1$. Summing over all frames k is necessary as the SMD is a mean quantity. Notice, that each pixel comprehends a multitude of individual droplets that are on purpose not resolved. Spatial locations with low probability to find the spray were excluded from the analysis. Typically areas were considered only if for more than 75% of the images Mie scattering above the threshold was detected. This was achieved by the factor $\mathrm{M_{PF}}$ in Eq. (15.3) that masked areas of low probability of spray occurrence. By this procedure comparable statistical significance for all locations was ensured. The calibration constant C was determined by use of PDA data. Here calibration was performed at an axial height of 25 mm downstream the injector exit.

15.5 Results and Discussion

15.5.1 Spray Visualization by Mie Scattering

In automotive applications sprays are transient. Figure 15.4 shows Mie scattering recorded at different instants after start of energizing the injector (SOE).

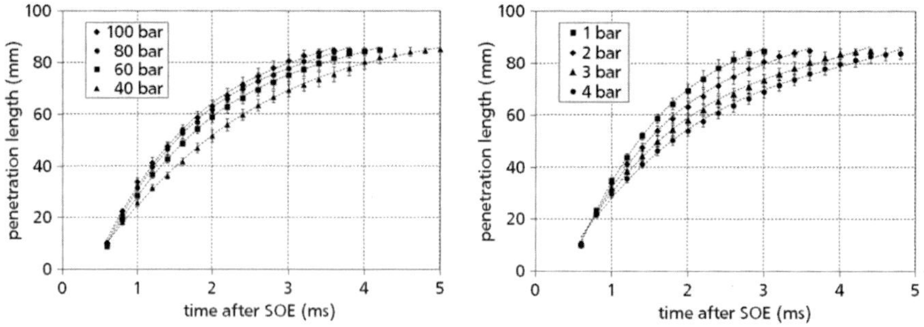

Fig. 15.5 Penetration depth of the spray for various times after start of energizing. *Left*: variation of the injection pressure for a constant chamber pressure of 2 bar. *Right*: variation of the chamber pressure for a constant injection pressure of 100 bar

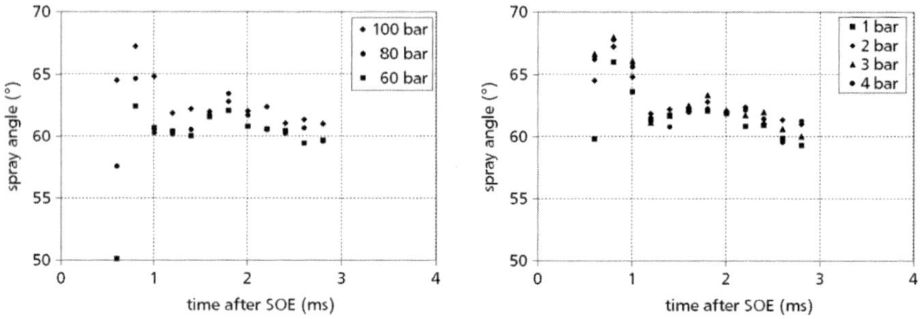

Fig. 15.6 Spray angle during injection. Times are relative to start of energizing (SOE). *Left*: Variation of injection pressure for fixed chamber pressure (2 bar). *Right*: Variation of chamber pressure for fixed injection pressure (100 bar)

Using the contour determination discussed in Sect. 15.4, the leading edge of the central spray cone and the angle of the entire spray can be easily extracted. For various injection and chamber pressures, respectively, in Fig. 15.5 the penetration of the spray is presented as function of time after SOE. As expected the spray penetrates faster for increasing pressure drops across the injector. Due to momentum exchange the spray propagation velocity constantly decreases. The offset from the spray origin presented in Fig. 15.5 is due to the temporal lag caused by the electronics controlling the injector and the inertia of mass of the needle.

Figure 15.6 shows the spray angle as function of time after SOE. During the initial phase of approximately 0.5 ms the spray angle is wider by nearly 10% compared to the second phase. Within the second phase the cone angle remains approximately constant until end of injection. The sensitivity of the spray angle with regard to the injection conditions is rather low.

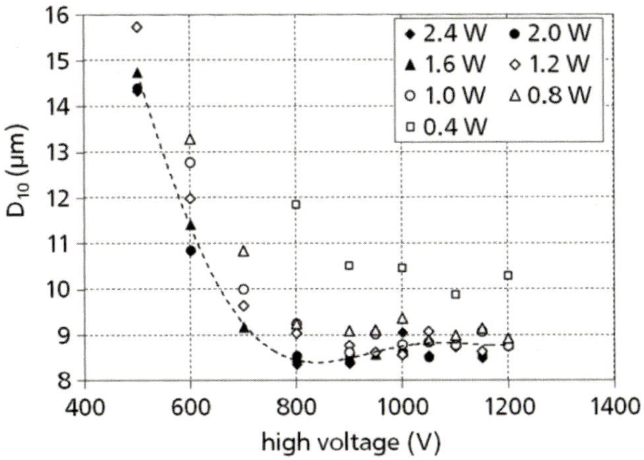

Fig. 15.7 D_{10} measured 2 ms after SOE in the central part of the spray 20 mm downstream the injector exit. The injector was operated at a repetition rate of 5 Hz

15.5.2 Phase Doppler

In the present approach the phase-Doppler method was used for calibration purposes. Quality and reliability of PDA data strongly depend on the parameter-settings described in Sect. 15.3 but also on the specific spray conditions. Considering a transient spray as in this study the droplet distribution may change with time after SOE. Analyzing this characteristic property requires phase-locked PDA-measurements. For 2 ms after SOE Fig. 15.7 presents D_{10} for various laser powers and high voltage settings of the photomultiplier tubes. Obviously there is a parametric sensitivity. For the present configuration a high voltage of 1,000 V and a laser power of 1.0 W served as suitable parameters that exhibit a relatively small sensitivity of D_{10} with regard to the user-defined experimental parameters. Based on this selection PDA-results served as calibration values for the Mie/LIF-drop sizing.

15.5.3 Comparison of Mie/LIF Dropsizing Versus Phase Doppler

In the following drop sizes are discussed that were measured for the central cone and one of the outer cones as can be seen from Fig. 15.4. For an instant of 2 ms after SOE Fig. 15.8 shows Sauter mean diameters (SMD) of these two selected cones. Data at the left hand side show results from PDA-measurements, the corresponding result from the Mie/LIF drop sizing are displayed at the right hand side. Using two slightly overlapping field of views data from two subsequent

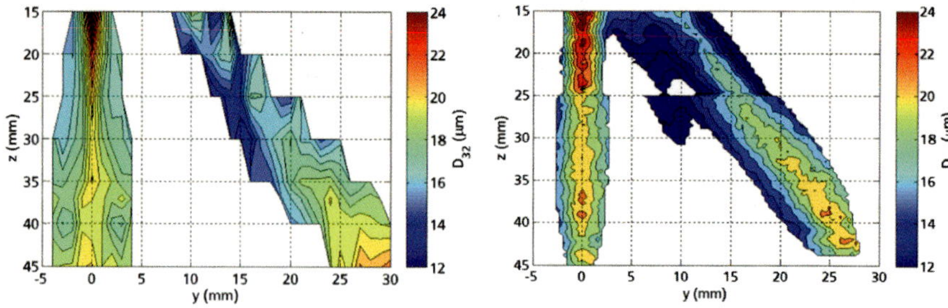

Fig. 15.8 Sauter mean diameters (SMD) for the central and one outer cone. *Left*: Results of PD-measurements. The field has been reconstructed from a coarse spatial grid as discussed in the text. *Right*: Results from Mie/LIF drop sizing. Measurements are conducted in two axially displaced but slightly overlapping field of views. At an axial distance of 25 mm PD-data are used for calibration purpose as outlined above. Results are averaged from 100 single shot exposures

Mie/LIF-measurements are plotted such that one yields connected spray cones. However, due to unavoidable discrepancies slight variations cause the obvious minor mismatch in the reconstruction of the whole spray cones.

PDA measurements were conducted on a relative coarse spatial grid recorded at axial heights of 15, 20, 25, 30, 35, 40 and 45 mm downstream of the nozzle exit. In radial direction the step width was 1 mm. Hence, fine structures below this spatial resolution are not captured and appear naturally different compared to the results from the Mie/LIF approach that is recorded with a higher spatial resolution of 250 μm. For the central cone within the first 10 mm the PDA data indicate a decrease in SMD downstream of the injector. This trend is less obvious in the outer cone. Further downstream an increase of SMD is observed that is more pronounced in the outer than in the central cone. It is speculated that this observation is due to evaporation that is consuming small droplets faster than shrinking of the larger droplets and thereby causing – despite the presence of evaporation – an increase of the apparent SMD. Also collision effects supporting larger droplets downstream cannot be precluded. In close agreement to the PDA data the results from the planar drop sizing by the Mie/LIF approach show very similar trends of the SMD distribution in both cones to those obtained from the PDA-method. More pronounced for the central cone, the SMD decreases within the first 10 mm but – especially for the outer cone – increases further downstream. In both cases the range of diameter observed varies from 12 to 24 μm.

For a more quantitative comparison Fig. 15.9 displays a radial cross section through the central and the right cones at an axial height of 25 mm. Mie/LIF data are shown as line and are directly plotted along with the PDA data. Discrepancies at the order of 10–20% in the SMD are observed. This is overall a good agreement, especially in the view of measurement times that are a factor of 5 shorter for the Mie/LIF measurements compared to the point-wise PDA-measurements.

Moreover, particularly at the edges of the spray, PDA-measurements can be biased as larger droplets tend to be validated with a higher probability than small

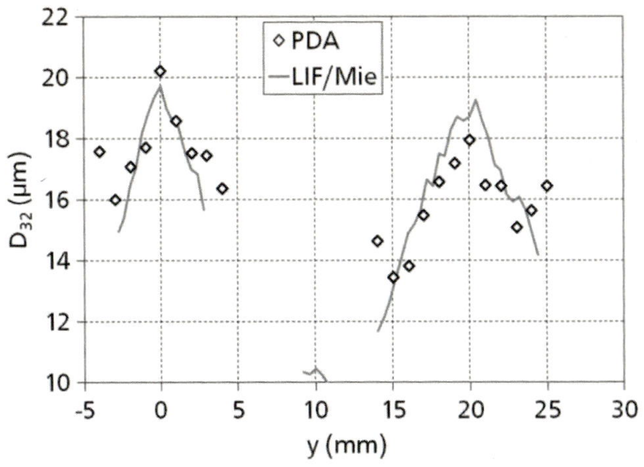

Fig. 15.9 Radial cross section at an axial height of 30 mm comparing results from the Mie/LIF to the PDA-measurements. Overall a good agreement is observed with deviation in the order of 10–20%

droplets. In addition with reduced statistical significance due to fewer PDA-events, this may cause a bias that is contributing to deviations from the results obtained from the Mie/LIF-measurements. However, a more thorough investigation of these type of systematic errors especially in the PDA-measurements are beyond the scope of this contribution.

15.6 Conclusions

Droplet sizes were measured by means of Phase-Doppler-Anemometry and the combined LIF/Mie technique in an automotive gasoline spray. The measurements were carried out without any additional fluorescence dye or tracer. The natural fluorescence was used to obtain a fluorescence signal which then was combined with the signal from the Mie-scattered light to provide two-dimensional information of the spatial distribution of the Sauter-mean-diameter (SMD). Due to the non-stationary behavior of an automotive spray regarding the droplet diameters it was concluded that calibration by PDA measurements require phase-locked measurements within the quasi-stationary spray regime. Here an instant of 2 ms after start of energizing (SOE) was selected where detailed PDA-measurements revealed only very minor temporal variations of the spray.

Exemplary results shown in this contribution show that SMDs obtained by the Mie/LIF approach deviate by less than 20% from the PDA-results. Larger deviations may be observed at the edges of the spray but at these locations poor statistical accuracy of the PDA-results and higher validation rates for lager droplets by trend may contribute to the mismatch as well. With both techniques very similar trends of the axial progression of the SMD in downstream direction have been observed. Measurement time of the combined Mie/LIF approach are by a factor 5 shorter than corresponding PDA-measurements that are conducted on a much coarser grid. In this sense the combined Mie/LIF-technique indeed is an appropriate tool for a significantly accelerated characterization of automotive sprays. However, a thorough calibration by another independent drop sizing technique such as phase–Doppler is required. Therefore Mie/LIF drop sizing can be used as a complementary technique to well-established phase-Doppler measurements.

Acknowledgments Financial support by Deutsche Forschungsgemeinschaft (SFB568, project T3) are gratefully acknowledged. The authors would like to thank BMW Motorsport for test rig maintenance and support, in person Mr. Paul Summerer, Mr. Dietmar Wagner.

References

1. Albrecht, H.E., Borys, M., Damaschke, N., Tropea, C.: Laser Doppler and Phase Doppler Measurement Techniques. Springer-Verlag, Berlin/Heidelberg/New York (2003)
2. Albrecht, H. E., Damaschke, N., Bech, H. FLMT: program for calculation of light scattering of shaped beams on spherical homogeneous particles. Program Documentation (2001)
3. Araújo, M.A., Silva, R., Lima, E., Pereira, D.P., Oliveira, P.C.: Measurement of Gaussian laser beam radius using the knife-edge technique: improvement on data analysis. Appl. Opt. **48**(2), 393–396 (2009)
4. Bakić, S.: Time integrated detection and applications of fs-laserpulses scattered by small particles. PhD thesis, Technical University Darmstadt (2009)
5. Bareiss, S., Bork, B., Bakić, S., Tropea, C., Irsig, R., Tiggesbäumker, J., Dreizler, A.: Application of femtosecond lasers to the polarization ratio technique for droplet sizing. Submitted for publication to Measurement Science and Technology (2012)
6. Berrocal, E., Meglinski, I.: New model for light propagation in highly inhomogeneous polydisperse turbid media with applications in spray diagnostics. Opt. Express **13**(23), 9181–9195 (2005)
7. Berrocal, E., Kristensson, E., Richter, M., Linne, M., Aldén, M.: Application of structured illumination for multiple scattering suppression in planar laser imaging of dense sprays. Opt. Express **16**(22), 17870–17881 (2008)
8. Bohren, G.F., Huffman, D.R.: Absorption and Scattering of Light by Small Particles. Wiley, New York (1983)
9. Domann, R., Hardalupas, Y.: A study of parameters that influence the accuracy of the planar droplet sizing (PDS) technique. Part. Part. Syst. Charact. **18**, 3–11 (2001)
10. Domann, R., Hardalupas, Y., Jones, A.R.: A study of the influence of absorption on the spatial distribution of fluorescence intensity within large droplets using Mie theory, geometrical optics and imaging experiments. Meas. Sci. Technol. **13**(S.), 280–291 (2002)
11. Düwel, I., Kunzelmann, T., Schorr, J., Schulz, C., Wolfrum, J.: Application of fuel tracers with different volatilities for planar LIF/Mie drop sizing in evaporating systems. In: 9th International Conference on Liquid Atomization and Spray Systems (2003)
12. Frackowiak, B., Tropea, C.: Numerical analysis of diameter influence on droplet fluorescence. Appl. Opt. **49**(12), 2363–2370 (2010)
13. Hofeldt, D.L.: Full-field measurements of droplet size distributions: I. Theoretical limitations of the polarization ratio method. Appl. Opt. **32**(36), 7551–7558 (1993); Hofeldt, D.L.: Full-field measurements of droplet size distributions: II. Experimental comparison of the polarization ratio and scattered intensity methods. Appl. Opt. **32**(36), 7559–7567 (1993)

14. Jermy, M.C., Greenhalgh, D.A.: Planar dropsizing by elastic and fluorescence scattering in sprays too dense for phase Doppler measurement. Appl. Phys. B **71**, 703–710 (2000)
15. Kamimoto, T.: Diagnostics of transient sprays by means of laser sheet techniques. In: International Symposium on COMODIA, Yokohama, Japan (1994)
16. Kapulla, R., Najera, S.B.: Operation conditions of a phase Doppler anemometer: droplet size measurements with laser beam power, photomultiplier voltage, signal gain and signal-to-noise ratio as parameters. Meas. Sci. Technol. **17**, 221–227 (2006)
17. Le Gal, P., Farrugia, N., Greenhalgh, D.A.: Laser sheet dropsizing in dense sprays. Opt. Laser Technol. **31**, 286–291 (1999)
18. Lettieri, T.R., Jenkins, W.D., Swyt, D.A.: Sizing of individual optically levitated evaporating droplets by measurement of resonances in the polarization ratio. Appl. Opt. **20**(16), 2799–2805 (1981)
19. Malarski, A., Schürer, B., Schmitz, I., Zigan, L., Flügel, A., Leipertz, A.: Laser sheet dropsizing based on two-dimensional Raman and Mie scattering. Appl. Opt. **48**(10), 1853–1860 (2009)
20. Park, S., Cho, H., Yoon, I., Min, K.: Measurement of droplet size distribution of gasoline direct injection spray by droplet generator and planar image technique. Meas. Sci. Technol. **13**, 859–864 (2002)
21. Sankar, S.V., Maher, K.E., Robart, D.M., Bachalo, W.D.: Rapid characterization of fuel atomizers using an optical patternator. J. Eng. Gas Turbines Power **121**, 409–414 (1999)
22. Schäfer, W., Tropea, C., Elsäßer, W.: Determination of size and refractive index of a single water droplet by using a light source with short coherence length (LED). In: 15th International Symposium on Applications of Laser Techniques to Fluid Mechanics, Lisbon (2010)
23. Sivasubramanian, N.A., Saylor, J.R.: Application of a histogram modification algorithm to the processing of raindrop images. Opt. Eng. **47**(3), 037011 (2008)
24. Zaller, M., Locke, R.J., Anderson, R.C.: Comparison of techniques for non-intrusive fuel drop size measurements in a subscale gas turbine combustor. J. Vis. **2**(3/4), 301–308 (2000)
25. Zimmer, L., Domann, R., Hardalupas, Y.: Simultaneous laser-induces fluorescence and Mie scattering for droplet cluster measurements. Am. Inst. Aeronaut. Astronaut. J. **41**(11), 2170–2178 (2003)

Chapter 16
High-Speed Laser Diagnostics for the Investigation of Cycle-to-Cycle Variations of IC Engine Processes

S.H.R. Müller, B. Böhm, and A. Dreizler

Abstract The work presented in this report was conducted within the Collaborative Research Center 568 funded by the Deutsche Forschungsgemeinschaft over a period of eleven years. The aim of project T4 was the transfer of laser based measurement techniques, which were used and improved in the context of stationary gas turbine combustion, to the investigation of intermittent processes in internal combustion (IC) engines. The focus was on cycle-to-cycle fluctuations as they appear in recent direct injection IC engines. High-speed measurement techniques were applied to investigate the temporal evolution of the in-cylinder flow, fuel distribution and flame propagation. Charge motion was investigated by particle image velocimetry (PIV) and spray by imaging of Mie-scattering. Mixture distribution was captured qualitatively by means of planar laser induced fluorescence (PLIF) of a fluorescing fuel. OH-PLIF was used to investigate the development of the early flame kernel and turbulent flame propagation.

Keywords High-speed laser diagnostics • Particle image velocimetry (PIV) • Laser induced fluorescence (PLIF) • Direct injection engine • Cycle-to-cycle variation

S.H.R. Müller • A. Dreizler
Center of Smart Interface, Technische Universität Darmstadt, Petersenstr. 32, 64287 Darmstadt, Germany
e-mail: mueller@csi.tu-darmstadt.de; dreizler@csi.tu-darmstadt.de

B. Böhm (✉)
Institute for Energy and Powerplant Technology, Department of Mechanical and Processing Engineering, Technische Universität Darmstadt, Petersenstr. 30, 64287 Darmstadt, Germany
e-mail: bboehm@ekt.tu-darmstadt.de

J. Janicka et al., *Flow and Combustion in Advanced Gas Turbine Combustors*, Fluid Mechanics and Its Applications 102, DOI 10.1007/978-94-007-5320-4_16,

16.1 Introduction

To improve fuel efficiency and engine performance direct injection (DI) combustion concepts are widely used in modern gasoline IC engines. Fuel and air are introduced separately and mixture preparation takes place inside the cylinder. For homogenous operation mode, which is commonly used in state-of-the art engines, fuel is injected early during the intake stroke. Charge cooling from the evaporating fuel spray and a reduced knock propensity leads to gains in fuel economy. Further improvement of fuel efficiency can be obtained by stratified charge operation at part load by reducing pumping losses caused by the throttle. Fuel is injected late in the compression stroke resulting in an ignitable mixture around the spark plug while maintaining a globally lean mixture. The stratified-charge mode requires a much tighter control of the in-cylinder processes. Combustion efficiency and pollutant emissions are strongly coupled with mixture preparation, ignition and the subsequent development of a turbulent flame. The stratified charge DI gasoline engine combines the advantages of a gasoline engine at full load with fuel economy advantages of a diesel engine in part load.

In spray guided DI gasoline engines the fuel is injected by a centrally mounted injector. Compact fuel clouds can be generated through the close proximity of the fuel spray with the spark plug enabling stratified combustion over a large load range. This design leads to high velocities of air and fuel droplets in the vicinity of the spark electrodes and to high spatial and temporal gradients of velocity and mixture. This can lead to conditions that are unfavorable for a reliable ignition and flame propagation resulting in partially burned cycle or even misfires [1]. Fuel injection and charge motion need to be adjusted carefully to avoid those stochastic events which cause strong cycle-to-cycle variations. There is a need to investigate processes as the spray development, fuel evaporation, turbulent mixing of fuel and air, the influence of large scale charge motion (e.g. tumble or swirl) on the mixture preparation, early flame kernel development and turbulent flame propagation in inhomogeneous mixtures.

The investigation of the cycle-to-cycle variability (ccv) is a key challenge of DI gasoline engine development and is subject of numerous investigations [2, 3]. To investigate the origin of ccv instantaneous recording of the in-cylinder conditions is required. Although measurements using pressure transducers reveal ccv, planar laser based diagnostics are preferable for a more detailed understanding of the underlying physical processes due to their inherent high spatial and temporal resolution. Those techniques require optical access to observe mixture preparation, charge motion and ignition in a sufficiently large volume. Experimental investigations have been performed regarding fuel distribution using laser induced exciplex fluorescence [4], the influence of ccv of the spray [5], dynamics of ignition conditions in the vicinity of the spark plug using high-speed LIF [6] or the determination of the air/fuel ratio at the spark location from its spectra [7].

Until recently, repetition rates of flow and scalar field measurements were restricted to one exposure per cycle resulting in observations of statistically uncorrelated instantaneous realizations. To characterize ccv, an adequate temporal resolution is required to comprehend the property of interest and its variability in time and space. Recent developments in DPSS-laser (diode pumped solid state laser) and CMOS-camera (complementary metal oxide semiconductor camera) technology facilitate increased pulse and frame rates of several kilohertz [8]. This allows studying the temporal evolution of in-cylinder processes over sequential cycles. Additionally it allows taking large datasets which are required to determine correlations between poor burn and misfire with specific features of the flow or the scalar field. Fansler et al. [1] for example demonstrate correlations between outlier and the early flame kernel development using pressure measurements and time-resolved imaging of the early flame kernel development. Recently a number of investigations have been performed using high-speed measurement techniques for in-cylinder measurements using PIV [9–13], tracer PLIF [14] or even both simultaneously [15].

The aim of this project was the adaption of high speed PLIF and PIV measurement techniques to the instationary processes of DI engines in order to study ccv. From high speed PIV measurements ccv of the in-cylinder charge motion were identified. The reproducibility of the spray injection as well as the interaction with the flow field was investigated by simultaneous PIV and imaging of Mie-scattering of the spray. Tracer PLIF was used to investigate mixture distribution and reproducibility of fuel injection. The focus was on abnormal injection events and variations of fuel distribution to identify regions of overly lean or rich mixtures. This allows identifying main influencing parameter, as for example fuel pressure, injector geometry and charge motion, for a reliable mixture preparation. Further on high speed OH-PLIF was used to investigate flame propagation of the early flame kernel. Abnormal cycles were identified and flame propagation of the early flame kernel was clearly correlated with pressure curves.

16.2 Experimental Setup

This section gives an overview on the optical engine and the high speed diagnostics which were transferred from an unconfined quasi stationary environment to a strongly intermittent process under a harsh environment. More details on the PIV setup can be found in [16, 17], on spray imaging in [18], on tracer PLIF for equivalence ratio measurements in [19] and on OH-PLIF for flame front detection in [20].

16.2.1 Optically Accessible Engine

The experimental setup consisted of an optically accessible spray-guided, spark-ignition, direct injection internal combustion engine (Fig. 16.1). More details on the engine are given in [16–20].

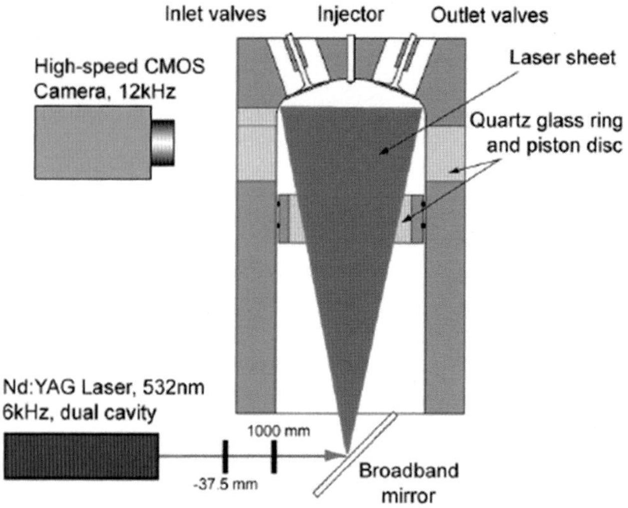

Fig. 16.1 Schematic of the optical engine together with the PIV setup

The single cylinder engine had an 82 mm bore and 86 mm stroke resulting in a displacement of 450 cm^2. The compression ratio was varied from 8 to 10. The engine had a pent-roof cylinder head and was designed for stratified combustion with a central mounted injector and the spark plug mounted in between the exhaust valves at an angle of 30° with respect to the injector.

The in-cylinder charge motion was varied by the horizontally divided dual inlet manifold. A tumble motion was induced by closing the lower half of the twin inlet manifold while a swirl motion was generated by closing one side of the inlet manifold. The optically accessibility was given by a 30 mm quartz glass ring, two additional pent-roof windows and a window embedded into the flat piston crown.

16.2.2 Diagnostics

16.2.2.1 Time-Resolved PIV for Flow-Field Measurements and Spray Imaging

High speed PIV was used to capture the evolution of the instantaneous flow field [16, 17]. A frequency-doubled, dual-cavity Nd:YVO_4 slab laser (Edgewave, INNOSLAB) was operated at 532 nm with a pulse energy of 0.7 mJ/pulse. The repetition rate was set to 6 kHz with a pulse to pulse separation of 10–60 μs according to engine speed. The laser beam was formed into a light sheet and then guided through the piston window into the cylinder. The light sheet width was ~45 mm with a thickness of ~1 mm. Seeding droplet (~1 μm) were generated by an aerosol generator (Palas) using hydraulic oil (Liquimoly) and led into the intake air resulting in a homogenous droplet distribution. The scattered light was recorded at right angles through the quartz glass ring and the pent roof window by a CMOS camera (La Vision, HSS6). The field of view was 43×44 mm^2. Data processing was based on a PIV-algorithm presented in [21]. A multi-pass interrogation with window shifting was used with a final grid consisting of 16×16 pixels, resulting in a resolution of 1×1 mm^2. This setup was also used for simultaneous spray imaging from Mie-scattering of the liquid fuel [18].

16.2.2.2 Time-Resolved Tracer PLIF for Equivalence Ratio Measurements

Planar laser-induced fluorescence (PLIF) of a fluorescing reference fuel was used to measure the equivalence ratio within the cylinder [19]. A frequency-doubled DPSS-laser (Edgewave CX16II-E 80 W) was frequency quadrupled with a single pass external crystal to 266 nm. Pulse energies of 0.5 mJ were achieved at a repetition rate of 6 kHz and were monitored by a high-speed energy meter (LaVision) and a redundant photodiode (Thorlabs). A light sheet was formed 35 mm wide with a thickness of 0.5 mm. A high-speed CMOS camera (Photron FASTCAMSA1) equipped with a dual-stage intensifier (LaVision) and a UV-coated lens (Bernhard Halle Nachfl.) was used to record the fluorescence signals. A set of ultraviolet transmitting filter (Schott UG11) were used to suppress the excitation wavelength. All measurements were performed in nitrogen/fuel mixtures to avoid fluorescence quenching through the presence of oxygen. The fluorescence signals were calibrated with known homogenous air/fuel mixtures obtained through a port fuel injection. A lambda sensor was used to determine the respective equivalence ratio. Then air was replaced by nitrogen and a calibration matrix for each CAD of interest was generated. This procedure attributes the fluorescence signal to an equivalence ratio.

16.2.2.3 Time-Resolved OH-PLIF for Flame Front Detection

Planar laser-induced fluorescence (PLIF) was used to image the combustion radical hydroxyl (OH) [20]. A frequency-doubled dye laser (Sirah Credo, with Rhodamine 6 G) was pumped at 532 nm by a frequency-doubled, Q-switched DPSS-laser (Edgewave IS8II-E). Pulse energies of 360 μJ were achieved at a repetition rate of 6 kHz. The lasers were tuned to 283 nm to excite the Q1(6) line of the A-X (1–0) transition of OH. Fluorescence emission at 308 nm was imaged through the piston window onto a CMOS camera (LaVision HSS6) equipped with a two-stage, lens-coupled intensifier (LaVision, HS-IRO) and a UV-achromat lens (Halle, f = 150 mm, f/2). A band-pass interference filter with 80% transmission at 308 nm (mso Jena) was used to reduce spuriously scattered light (Laser Components UV-B). The field of view of each PLIF system was 53×54 mm^2. To prevent overexposure of the intensifier spark duration was limited to 0.85 μs. A fuel with low aromatic fraction was used to avoid superposition from fluorescence of other species.

16.3 Results and Discussion

Following in-cylinder processes were addressed in this project: The charge motion captured by time-resolved PIV [16, 17], evolution of the spray imaged by Mie scattering of the fuel droplets [18], mixture distribution from PLIF of the fuel [19] and early flame kernel development and turbulent flame propagation captured by time-resolved OH-PLIF [20]. A survey on these measurements is given below. More details can be found in the respective publications [16–20].

16.3.1 Charge Motion

The main focus of the temporally resolved velocity measurements was on the compression stroke, being relevant for cyclic variations due to impact on fuel/air mixing, ignition and combustion. The engine was motored at 500, 1,000 and 2,000 rpm without firing. The compression stroke was tracked temporally from 80° BTDC (before top dead center) up to TDC in the symmetry plane. The flow structure was investigated by individual flow fields and an automated analysis of ensemble averaged flow fields. To characterize the tumble motion and its progress the vortex center was detected for instantaneous flow fields by an automated analysis based on a velocity weighted vorticity map. Additionally kinetic energy and turbulent kinetic energy was determined based on the two measured components of velocity.

Figure 16.2 shows the temporal evolution of the instantaneous flow field for two individual cycles. The high quality of PIV results is demonstrated by showing validated vectors only (no interpolation was performed). The influence of evaporating seeding droplets towards TDC can be observed by the decreasing quality of the vector field. The 11th cycle represents a typical cycle with one large vortex center at 80° BTDC. During compression the vortex is squeezed and moves diagonally towards the upper right corner. In the lower right corner the flow is accelerated by the upward moving piston leading to regions of high velocity. The vortex center is distorted and vertically compressed. From 60° BTDC onwards the large vortex center begins to break up into smaller vortices. At the cylinder head the flow points from the injection nozzle towards the spark plug and continues moving down towards the piston where the fuel cloud is found under stratified conditions. This flow pattern is the standard case. The 51st cycle shows an outlier with a different flow pattern. Two vortices are found at 80° BTDC which are already compressed vertically due to a larger region of high velocity at the piston. This results as well in a different flow pattern in the vicinity of the spark plug. Several crank angle degrees later the vortices seem to have merged.

The large scale structure was characterized in terms of the tumble vortex center location. The temporal evolution of the large scale vortex induced by the tumble motion revealed a substantial variability in the horizontal direction while the vertical location was mainly determined by the upwards moving piston resulting in far lower variability. Figure 16.3 shows the histogram of the vortex centers location at two

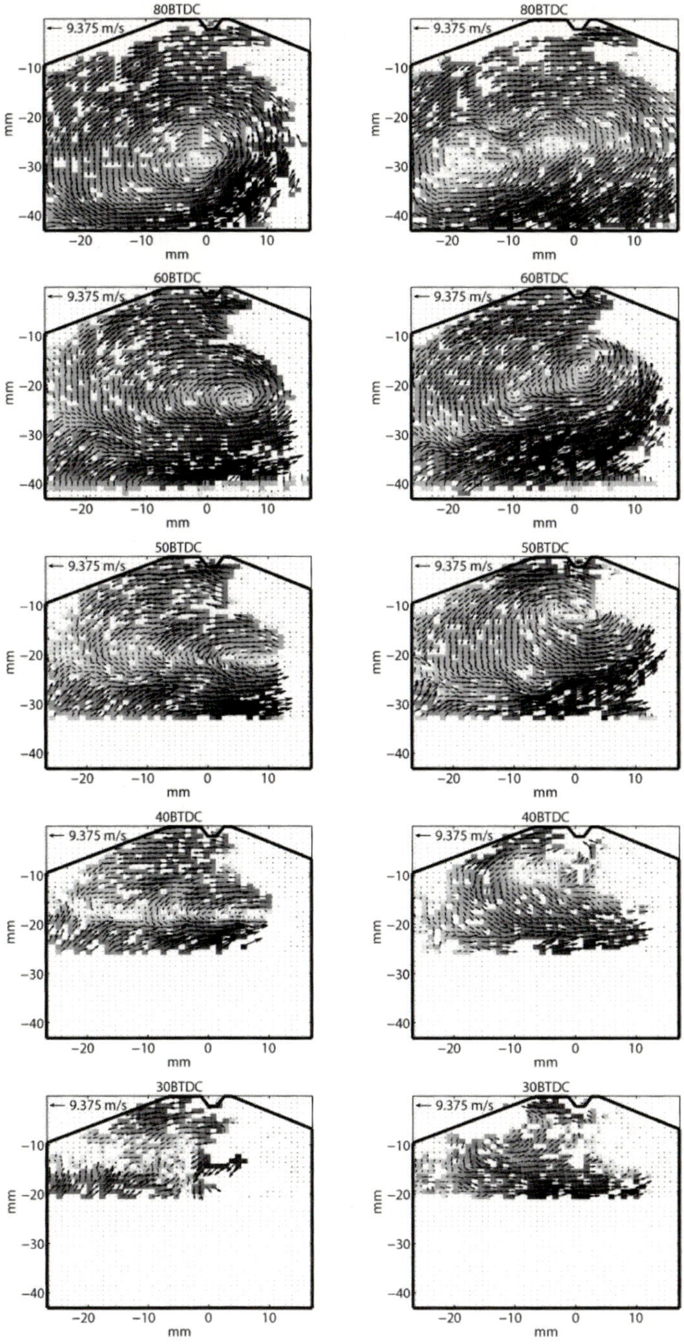

Fig. 16.2 Time sequence of individual flow fields of the 11th (*left*) and 51st (*right*) cycle at 1,000 rpm highlighting ccv of the flow field (Reprinted from [16])

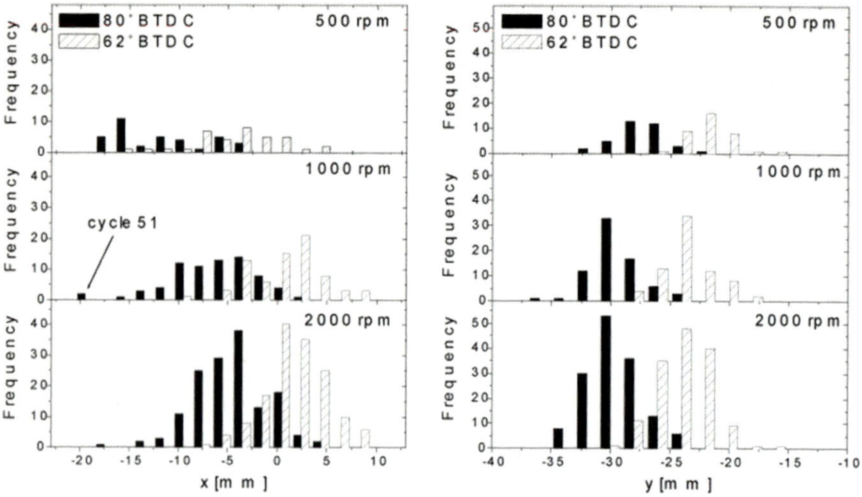

Fig. 16.3 Histograms of vortex center locations for 80° and 62° BTDC for the horizontal coordinate x (*left*) and the vertical coordinate y (*right*), for 500, 1,000 and 2,000 rpm (Reprinted from [16])

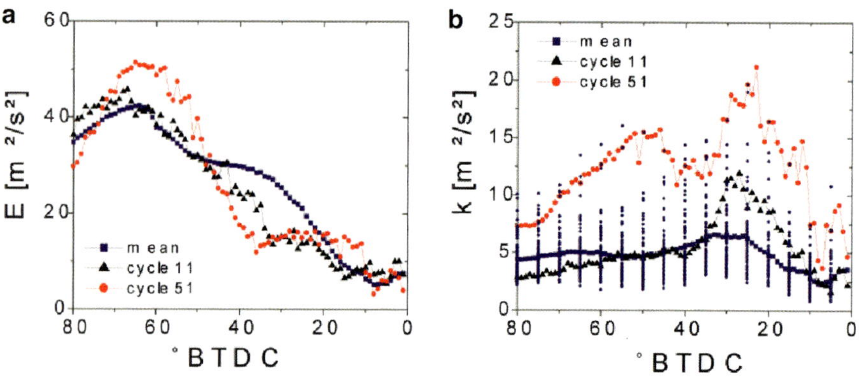

Fig. 16.4 Kinetic energy E (*left*) and turbulent kinetic energy k (*right*) for 1,000 rpm (Reprinted from [16])

selected crank angle (80° and 62° BTDC). The distribution for 500 rpm is broad with a random structure. It becomes narrower for higher engine speeds. Still some isolated outliers are found (51st cycle) were the vortex center is found ~14 mm off the average location.

Flame propagation is not only influenced by the flows direction and the flow structure size but also by its energy content. Therefore kinetic energy (E) as well as the turbulent kinetic energy (k) was determined and spatially averaged over the whole field of view. Figure 16.4 shows averaged kinetic energy E for 1,000 rpm with tumble charge motion. E increases until the piston moves into the image (65° BTDC) due to regions of high velocity in the vicinity of the piston crown. The decay of E over the next 20°CA corresponds with the breakup of the tumble vortex into smaller vortices. Further on production of E balances dissipation for several CAD before it decreases again rapidly. To investigate ccv additionally traces of E and k were investigated for individual cycles. Although kinetic energy levels for the individual cycle are consistent with the instantaneous flow fields (see Fig. 16.2) turbulent kinetic energy was found to be a more sensitive value to identify cycles with a change in their flow structure. While k for the 11th cycle is close to the mean, the outlier cycle 51 presents values of k which are up to three times larger than the mean.

16.3.2 Spray Structure

The influence of charge motion on the developing spray structure was studied with the same high-speed PIV setup. Figure 16.5 shows an example of late injection into the compression stroke (EOI (end of injection) at 50° BTDC) for stratified combustion. The 8-hole injector features a symmetric geometry and spray pattern for undisturbed quiescent environments. During in-cylinder operation with tumble charge motion, the plumes observed in the field of view are increasingly asymmetric. Only 2°CA after end of injection an asymmetric pattern has formed. The flow field shows the tumble vortex in the symmetry plane moving upwards to the left corner. By the superposition of the in-cylinder flow pattern (as seen in Fig. 16.5) and spray momentum the left plume is retarded while the right one is accelerated. This distorts the plumes and leads to an increased penetration depth for the right plume causing piston wetting. It demonstrates the importance of spray investigations under engine conditions together with charge motion. This is even of greater significance for multiple injection strategies, which are used to reduce particulate emissions in DI engines. There the flows influence on the spray events is more prominent due to a lower momentum of the single injection pulses. A further statistical analysis of the spray structure in terms of penetration depth and spray drift revealed that large scale and directed flow structures distort the spray plume while small scale structures have nearly no effect. The variation of injector types (multi-hole, swirl, annular orifice injector) as well as injection parameter (fuel pressure, orientation, multiple injections etc.) can now be studied in detail in order to select the appropriate injector and injection strategy for specific engines.

16.3.3 Mixture Preparation

The next essential step for combustion is a reliable mixture preparation depending on the combustion strategy: well-mixed for maximum combustion efficiency and

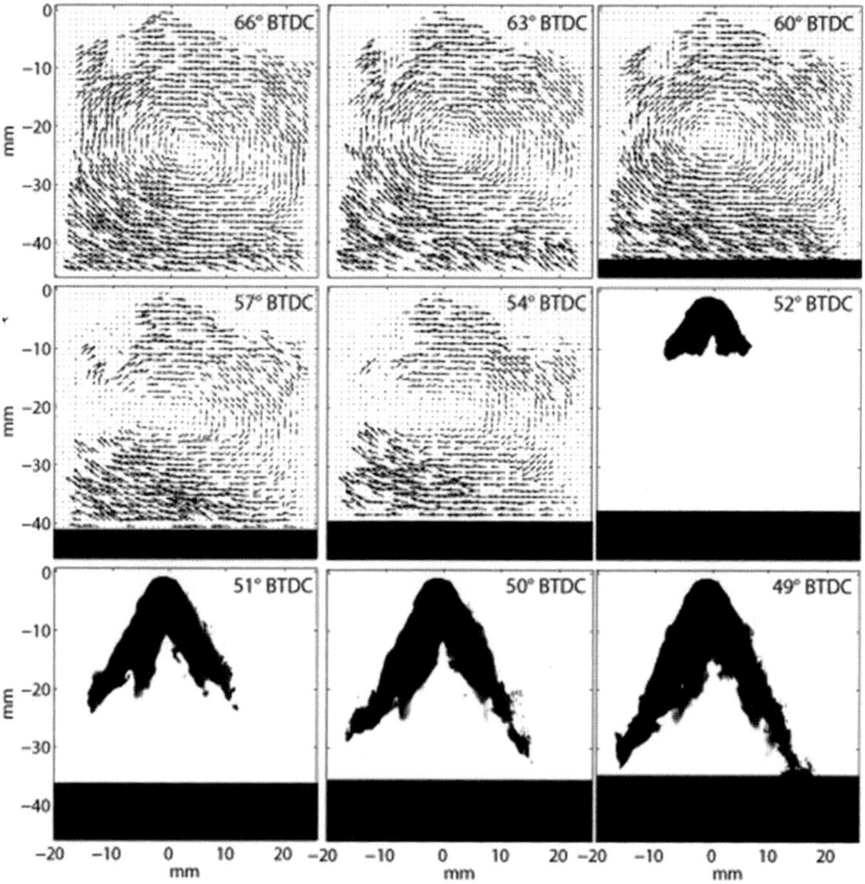

Fig. 16.5 Sequence at 1,000 rpm of individual flow field and injection event for an 8-hole injector (injection pressure at 10 MPa, EOI at 50° BTDC) (Reprinted from [18])

low emissions (especially particulates) for homogenous combustion mode (DI or port fuel injection (PFI)) or a well-connected cloud of ignitable mixture at the spark plug for stratified combustion mode. Figure 16.6 shows the instantaneous mixture distribution in terms of contour plots of equivalence ratio for PFI (1st row) and a stratified DI (2nd and 3rd row). The local distribution with PFI shows a spatially homogeneous map with an approximately stoichiometric mixture. In the stratified mode stratification of the air/fuel mixture is found within the cylinder. Fuel rich mixtures are observed around the spark plug. Strong ccv of the mixture distribution are found for individual cycle. This work presents the feasibility of investigating the mixture distribution within an internal combustion engine at a temporal repetition rate of 6 kHz using planar laser induced fluorescence. At 1,000 rpm, the repetition rate corresponds to a temporal resolution of one crank angle per image.

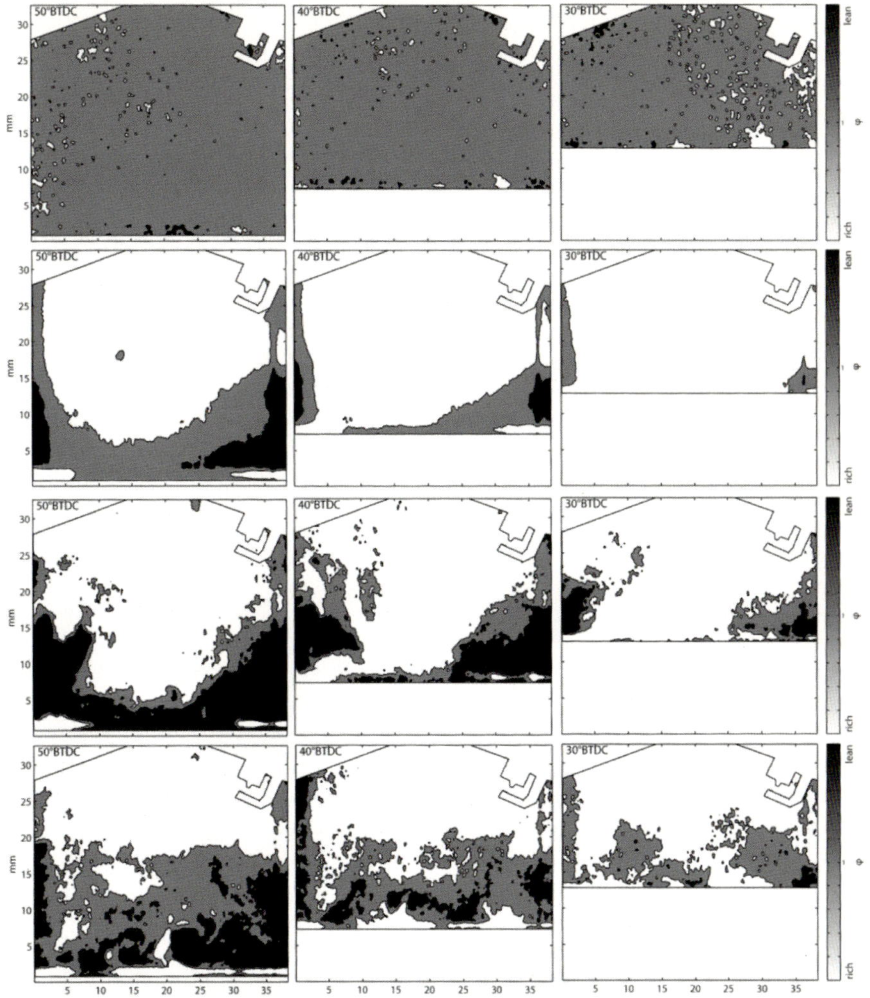

Fig. 16.6 Instantaneous contour plots of equivalence ratio at 1,000 rpm for three individual cycles: PFI (1st row) and DI (3rd and 4th row) with EOI at 300° BTDC and phase averaged images with DI (Reprinted from [19])

16.3.4 Flame Propagation

Transient flame front propagation was resolved within individual cycles using high speed OH-PLIF. The horizontal light sheet entered the cylinder through the quartz glass ring and was aligned horizontally just above TDC position of the piston. A large CMOS camera on-board memory used for detection in combination with continuous UV laser pulse sequences allowed for the recording of hundreds

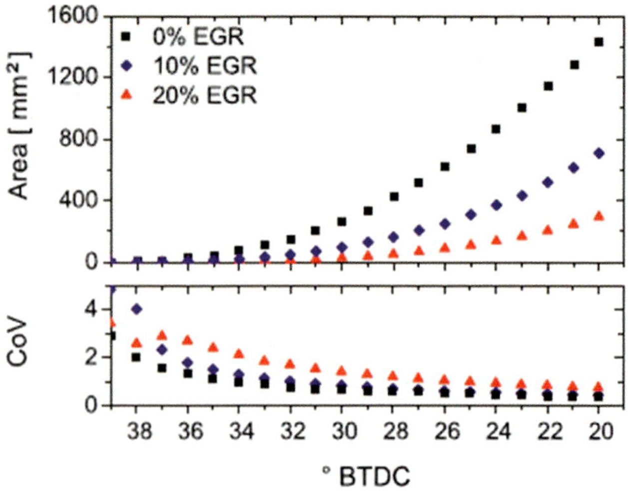

Fig. 16.7 Mean flame area for different EGR rates at 1,000 rpm (*top*) and coefficient of variation (*bottom*) (Reprinted from [20])

of subsequent cycles. The PLIF signals (SNR ~ 7) had a sufficient quality to extract the instantaneous flame front. This gives insights into the flame propagation dynamics. An automated analysis of the flame area growth allowed studying the gas compositions impact on flame propagation as well as ccv.

Figure 16.7 shows an example of the temporal evolution of early flame kernel development for different gas compositions (exhaust gas recirculation (EGR) levels). For operation without EGR, the flame kernel grows into the measurement plane at ~ 37° BTDC. With increasing EGR the flame growth speed is reduced significantly. At a level of 10% the flame growth into the measurement plane is delayed by 1–2° CA and for 20% by 4° CA. For the initial phase up to 25° BTDC, the flame kernel grows only marginally with significant levels of EGR whereas with no EGR the kernel increases constantly and much more rapidly. Mean and the according coefficient of variation are computed from 50 cycles at each EGR level (Fig. 16.7 bottom) demonstrating an increase in ccv for increasing EGR levels.

Although pressure measurements are a powerful tool to characterize combustion they are difficult during early flame kernel development where the pressure-increase is small. In contrast, this presented technique provides high temporal and spatial resolution thus allowing a good characterization of the flame from an early flame kernel to a wrinkled turbulent flame. Even though planar imaging of the flame kernel captures the flame only within the laser sheet and cannot resolve out of plane flame information, it still is a sensitive measure to study ccv. Therefore flame growth was simultaneously recorded with pressure curves. Mean flame area growth at 1,000 rpm is given in Fig. 16.8 together with measurement bars presenting the standard deviation of the dataset. Additionally two individual cycles are shown: a fast (cycle 13) and a slow burning cycle (cycle 32). The respective pressure traces

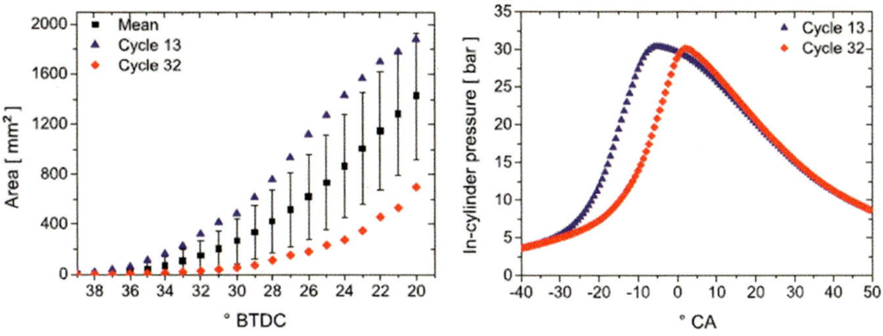

Fig. 16.8 Flame area propagation for mean and 2 individual cycle at 1,000 rpm with no EGR (*left*) with the corresponding pressure traces (*right*) (Reprinted from [20])

correspond with flame area growth rate presenting a steeper gradient for the fast burning cycle 13. The integral area below the pressure trace indicates the shift of 50% mass fraction burned towards later crank angles. This demonstrates the potential of OH-PLIF for the study of ccv because it resolves ccv at earlier CAD with high spatial resolution and allows now to compare the dynamics of the early flame kernel development.

16.4 Conclusions

High-speed measurement techniques have been successfully applied to an optically accessible spray-guided direct injection IC engine. The time evolution of in-cylinder processes, especially important for a stratified combustion mode, is now accessible. This includes charge motion, fuel injection, mixture preparation and finally the development of the early flame kernel and turbulent flame propagation.

The temporal evolution of the large scale tumble vortex tracked by PIV revealed substantial variability in horizontal direction. Individual flow fields at constant CAD showed different flow structures resulting in a change of the flows velocity and directionality in the area of the spray injector and the spark plug. The large scale and directed flow structures were able to distort the spray plume, imaged by means of Mie-scattering. The following mixture preparation was then captured by laser induced fluorescence of a fluorescing fuel. This allowed investigating the mixture distribution in a qualitative manner. It showed homogenous mixtures for PFI while inhomogeneity was severe and varied from cycle-to-cycle for homogenous combustion mode with early injection. Finally OH-PLIF was used to track the early flame kernel and the transient flame propagation. A variation of the in-cylinder gas composition showed a strong impact on the flame dynamics. Increased EGR levels reduced flame growth rates. A correlation of the pressure curves and flame growth was observed. This allows now to study ccv in more detail.

The presented results demonstrate that high-speed measurement techniques are proven to be a powerful tool for the investigation of ccv. These tools are now available and some of them are already used in industry for engine development. In future a combination of these techniques would be of interest to study the interaction of the different in-cylinder processes.

Acknowledgements The authors acknowledge the financial support from Deutsche Forschungsgemeinschaft (DFG) through the SFB 568.

References

1. Fansler, T.D., Drake, M.C., Böhm, B.: High-speed Mie-scattering diagnostics for spray-guided gasoline engine development. In: Proceedings of the 8th International Symposium on Combustion Diagnostics, Baden-Baden, pp. 413–425 (2008)
2. Arcoumanis, C., Whitelaw, J.H.: Fluid mechanics of internal combustion engines – a review. Proc. Inst. Mech. Eng. **201**, 57–74 (1987)
3. Kume, T., Iwamoto, Y., Lida, K., Murakami, M., Akishino, K., Ando, H.: Combustion control technologies for direct injection SI engine. SAE Paper 960600 (1996)
4. Wieske, P., Wissel, S., Grünefeld, G., Graf, M., Pischinger, S.: Experimental investigation of the origin of cyclic fluctuations in a DISI engine by means of advanced laser induced exciplex fluorescence measurements. SAE Paper 2006-01-3378 (2006)
5. Unterlechner, P., Kneer, R.: Experimentelle und numerische Untersuchungen zum Einfluss zyklischer Schwankungen auf die Struktur motorischer Einspritzstrahlen. In: Proceedings in Turbulenz in der Energietechnik, Darmstadt (2005)
6. Sick, V., Smith, J.D.: Laser combustion diagnostics, applications to engines. In: Proceedings der 10. LACSEA (OSA Topical Meeting), paper TuB1 (2006)
7. Fischer, J., Xander, B., Velji, A., Spicher, U.: Cycle resolved determination of local air-fuel ratio at the spark gap using a direct injection gasoline engine. In: Proceedings of the International Symposium on Internal Combustion Diagnostics, pp. 162–173 (2004)
8. Böhm, B., Heeger, C., Gordon, R.L., Dreizler, A.: New perspectives on turbulent combustion: multi-parameter high-speed planar laser diagnostics. Flow Turb. Combust. **86**, 313–341 (2011)
9. Fajardo, C., Sick, V.: Flow field assessment in a fired spray-guided spark-ignition direct-injection engine based on UV particle image velocimetry with sub crank angle resolution. Proc. Combust. Inst. **31**, 3023–3031 (2007)
10. Towers, D.P., Towers, C.E.: Cyclic variability measurements of in-cylinder engine flows using high-speed particle image velocimetry. Meas. Sci. Technol. **15**, 1917–1925 (2004)
11. Justham, T., Jarvis, S., Clarke, A., Garner, C.P., Hargrave, K., Halliwell, N.A.: Simultaneous study of intake and in-cylinder IC engine flow fields to provide an insight into intake induced cyclic variations. J. Phys. Conf. Ser. **45**, 146–153 (2006)
12. Druault, P., Guibert, P., Alizon, F.: Use of proper orthogonal decomposition for time interpolation from PIV data. Exp. Fluids **39**, 1009–1023 (2005)
13. Voisine, M., Thomas, L., Borée, J., Rey, P.: Spatio-temporal structure and cycle to cycle variations of an in-cylinder tumbling flow. Exp. Fluids **50**(5), 1393–1407 (2011)
14. Smith, J.D., Sick, V.: Quantitative, dynamic fuel distribution measurements in combustion-related devices using laser-induced fluorescence imaging of biacetyl in iso-octane. Proc. Combust. Inst. **31**, 747–755 (2007)
15. Peterson, B., Sick, V.: Simultaneous flow field and fuel concentration imaging at 4.8 kHz in an operating engine. Appl. Phys. B **97**(4), 887–895 (2009)

16. Müller, S.H.R., Böhm, B., Gleißner, M., Grzeszik, R., Arndt, S., Dreizler, A.: Flow field measurements in an optically accessible, direct-injection spray-guided internal combustion engine using high-speed PIV. Exp. Fluids **48**, 281–290 (2010)
17. Gleißner, M., Arndt, S., Grzeszik, R., Dreizler, A., Böhm, B., Müller, S.H.R.: Analyse der Brennraumströmung einer Direkteinspritzenden Ottomotors mittels Hochgeschwindigkeits-Particle Image Velocitmetry: Auswirkung auf Gemischbildung und Verbrennung. Presented at 9th Engine Combustion Processes Conference, Munich, Germany (2009)
18. Müller, S., Arndt, S., Dreizler, A.: Analysis of the in-cylinder flow field/spray injection interaction within a DISI IC engine using high-speed PIV. SAE Technical Paper 2011-01-1288 (2011)
19. Müller, S., Arndt, S., Dreizler, A.: Investigation of the air/fuel mixture distribution in an internal combustion engine using high-speed laser induced fluorescence. In: European Combustion Meeting (2011)
20. Müller, S.H.R., Böhm, B., Gleißner, M., Arndt, S., Dreizler, A.: Analysis of the temporal flame kernel development in an optically accessible IC engine using high-speed OH-PLIF. Appl. Phys. B **100**, 447–452 (2010)
21. Böhm, B., et al.: Simultaneous PIV/PTV/OH PLIF imaging: conditional flow field statistics in partially-premixed turbulent opposed jet flames. Proc. Combust. Inst. **31**, 709–718 (2007)

A Projects, Organization, Structure, Members and Participants of the Collaborative Research Center 568

A.1 Research Projects

Project Area A: Injection Systems and Mixture Formation

Project no.	Title	Project leader	Funding period
A1	Aerodynamische Stabilität von verdrallten Airblast-Zerstäubern *Aerodynamic stability of swirled airblast atomizers*	C. Tropea	20012007
A1	Primärzerstäubung eines Gasturbinenzerstäubers *Primary droplet breakup in a gas turbine atomizer*	C. Tropea I. Roisman	20082011
A2	Numerische und experimentelle Untersuchungen der Filmerwärmung und Filmverdampfung in LPP-Kammern *Numerical and experimental investigation of heating and evaporation in a thin liquid film of a LPP chamber*	P. Stephan	20012007
A2	Experimentelle und numerische Untersuchungen der Filmströmung und Filmverdampfung an festen Wänden *Numerical and experimental investigation of fluid flow and evaporation in a thin liquid film on solid walls*	P. Stephan	20082011

J. Janicka et al., *Flow and Combustion in Advanced Gas Turbine Combustors*, Fluid Mechanics and Its Applications 102, DOI 10.1007/978-94-007-5320-4,

Project no.	Title	Project leader	Funding period
A3	Modellierung von Tröpfchenwechselwirkungen	E. Gutheil	20012007
	Modelling of droplet interactions		
A4	Thermodynamisch konsistente Modellierung der Tropfen-Turbulenz-Wechsel-wirkung in Gasturbinen-Sprays	A. Sadiki	20012003
	Thermodynamically consistent modelling of droplet-turbulence interaction in gas turbine sprays	A. Dreizler	
A4	Thermodynamisch konsistente Modellierung von Gastrubinen-Sprays	A. Sadiki	20042011
	Thermodynamically consistent modelling of gas turbine sprays	A. Dreizler	

Project Area B: Combustion

Project no.	Title	Project leader	Funding period
B1	Verlöschen partiell vorgemischter Flammen unter Gasturbinenbedingungen	A. Dreizler	20012003
	Extinction of partially premixed flames under gas turbine conditions	J. Janicka	
B1	Verlöschen vorgemischter und partiell vorgemischter Flammen unter Gasturbinenbedingungen	A. Dreizler	20042007
	Extinction of premixed and partially premixed flames under gas turbine conditions		
B1	Transiente Prozesse in turbulenten Flammen unter Gasturbinenbedingungen	A. Dreizler	20082011
	Transient processes in turbulent flames under gas turbine conditions		
B2	Vereinfachte Reaktionsmodelle für die Verbrennung in Gasturbinenbrennkammern	J. Warnatz	20012011
	Reduced reaction models for combustion in gas turbine combustors	U. Riedel	
B3	Grobstruktursimulation von Verbrennungssystemen unter Gasturbinenbedingungen	J. Janicka	20012011
	Large eddy simulation of combustion systems under gas turbine conditions		

Project no.	Title	Project leader	Funding period
B4	**Entwicklung eines vereinfachten Modells für die Rußbildung in Gasturbinenbrennkammern**	**J. Warnatz**	**20012011**
	Development of a simplified soot model in gas turbine combustors	**U. Riedel**	
B5	Vorgemischte turbulente Verbrennung: Analyse und Modellbildung auf der Basis der Symmetrien der G-Gleichung	M. Oberlack	20012007
	Premixed turbulent combustion: analysis and modelling on the basis of symmetries in the G-equation		
B5	Vorgemischte turbulente Verbrennung: Analyse und Modellierung zu verallgemeinerten Randbedingungen der G-Gleichung	M. Oberlack	20082011
	Premixed turbulent combustion: analysis and modelling in the context of generalized boundary conditions of the G-equation		

Project Area C: Interaction and Fluid-Mechanical Processes

Project no.	Title	Project leader	Funding period
C1	Experimentelle und numerische Untersuchungen zur Wirkung des Verdichters auf die Brennkammereinströmung	B Stoffel	20012003
	Experimental and numerical investigation on the impact of the compressor on the combustion chamber inlet flow		
C1	Experimentelle und numerische Untersuchungen der Wechselwirkung zwischen Verdichter und der Brennkammerströmung	B Stoffel	20042007
	Experimental and numerical investigation on the interaction of the compressor and the combustion chamber flow		
C2	**Strömung und Mischung im Primärzonenbereich von Gasturbinenbrennkam-mern**	**C. Tropea**	**20012007**
	Flow and mixing in the primary zone of gas turbine combustors	**D. Hennecke**	

Project no.	Title	Project leader	Funding period
C3	Numerische Modellierung konvektiver Wärmeübertragung in Gasturbinenbrennkammern unter Berücksichtigung wandnaher Turbulenz	S. Jakirlić	20012011
	Numerical modelling of convective heat transfer in gas turbine combustors considering the influence of near wall turbulence	C. Tropea	

Project Area D: Cross-Sectional Projects

Project no.	Title	Project leader	Funding period
D1	Entwicklung und Analysis numerischer Verfahren für kompressible, reaktive Gleichungen der Strömungsdynamik und Strahlungstransportgleichungen	A. Klar	20012003
	Development and analysis of numerical procedures for compressible reactive equations of fluid dynamics and radiative transport	R. Pinnau	
D1	Entwicklung und Analysis numerischer Verfahren für die Strahlungstransportgleichungen und Kopplung an strömungsdynamische Gleichungen	A. Klar	20042007
	Development and analysis of numerical procedures for radiative transport equations and their coupling to fluid dynamical equations	R. Pinnau	
D2	Effiziente numerische Verfahren zur Berechnung turbulenter reaktiver Strömungen	M. Schäfer	20012007
	Efficient numerical procedures to simulate turbulent reactive flows		
D2	Effiziente numerische Verfahren zur Berechnung und Optimierung turbulenter reaktiver Strömungen	M. Schäfer	20082011
	Efficient numerical procedures to simulate and optimize turbulent reactive flows		

Project no.	Title	Project leader	Funding period
D3	Integrales Modell zur Simulation von Gasturbinenbrennkammern	M. Schäfer	20012011
	Integral model for the simulation of gas turbine combustors	J. Janicka	
D4	Adaptive Fehlerkontrolle bei der Grobstruktursimulation	J. Lang	20042007
	Adaptive error control in the context of large eddy simulation		
D4	Adaptive Qualitätskontrolle bei der Grobstruktursimulation	J. Lang	20082011
	Adaptive quality control in the context of large eddy simulation		
D5	Effiziente numerische Multilevel-Verfahren zur Optimierung von Gasturbinenbrennkammern	S.Ulbrich	20082011
	Efficient numerical multi-level procedures for the optimizatioin of gas turbine combustors		

Project Area T: Transfer Projects

Project no.	Title	Project leader	Funding period
T1	Fortschrittliche Auslegungsgrundlagen für Fluggasturbinenbrennkammern	J. Janicka	20052009
	Advance design principles for aero engine combustors		
T2	Grobstruktursimulation von Zweiphasen-Strömungen und Vormischflammen für Verbrennungsmotoren	J. Janicka	20072011
	Large eddy simulation of two-phase flow and premixed flames for internal combustion engines	A. Sadiki	
T3	Bildgebende laseroptische Messverfahren zur Auslegung von Saugrohreinspritzungen in Ottomotoren	A. Dreizler	20072011
	Laser imaging diagnostics for the design of intake-manifold fuel injection in gasoline engines		

Project no.	Title	Project leader	Funding period
T4	Hochgeschwindigkeits-Laser-Diagnostik zur Untersuchung von Zyklusschwankungen von Verbrennungskraftmaschinen *High-speed laser diagnostics for the investigation of cycle-to-cycle variations of IC engine processes*	A. Dreizler	20082011

Project Area Z: Administrative Tasks

Project no.	Title	Project leader	Funding period
Z1	Zentrale Aufgaben des Sonderforschungsbereichs *Central tasks of the Collaborative Research Center*		20012011

A.2 Scientific Committee

Members

Prof. Dr. rer. nat. Andreas Dreizler
Prof. Dr. rer. nat. Eva Gutheil
Prof. Dietmar K. Hennecke, Ph.D.
Prof. Dr. Ing. Suad Jakirlic
Prof. Dr. Ing. Johannes Janicka
Prof. Dr. rer. nat. Axel Klar
Prof. Dr. Jens Lang
Prof. Dr.-Ing. Martin Oberlack
Prof. Dr. rer. nat. René Pinnau
PD Dr. rer. nat. Uwe Riede
PD Dr.-Ing. Ilias Roisman
PD Dr.-Ing. Tatiana Gambaryan-Roisman
Prof. Dr. rer. nat. Amsini Sadiki
Prof. Dr. rer. nat. Michael Schäfer
Prof. Dr.-Ing. Peter. Stephan
Prof. Dr.-Ing. Bernd Stoffel
Prof. Dr.-Ing. Cameron Tropea
Prof. Dr. rer. nat. Stefan Ulbrich
Prof. Dr. rer. nat. Jürgen Warnatz

Chairman

Prof. Dr. Ing. Johannes Janicka

Vize-Chairman

Prof. Dr. rer. nat. Michael Schäfer

Scientific Coordinator

Prof. Dr. rer. nat. Amsini Sadiki

A.3 Visiting Researchers

Dr. Robert Barlow	2003
Sandia National Laboratories, USA	
Prof. Ismail Celik	2005
West Virginia University, USA	2006
Prof. Satya R. Chakravarthy	2002
IIT Madras, India	2003
	2006
Dr. Mouldi Chrigui	2007
Faculte des Sciences de Gafsa, Tunesia	2008
	2009
	2010
	2011
Prof. P. Coelho	2004
Technical University of Lisbon, Portugal	
Prof. I. S. Ertesvåg	2007
Norwegian University of Science and Technology Trondheim, Norway	
Prof. S. Gogineni	2004
Innovative Scientific Solutionis, Inc., USA	
Prof. K. Hanjalic	2004
Delft University of Technology, The Netherlands	
Prof. T. Ishima	2001
Grunma University, Kiryo City, Japan	
Dr. H. Jasac	2004
Nabla Ltd., UK	
Prof. M. Kameda	2001
Tokio University, Japan	
Prof. W. Kollmann	2001
Davis University of California, USA	2010
	2011
Prof. G. Nathan	2006
University of Adelaide, Australia	

Prof. B. Özdemir	2001
MET University, Ankara, Turkey	2002
Prof. D. Veynante	2005
Ecole Centrale de Paris, France	

A.4 Financial Support by Means of the Deutsche Forschungs-gemeinschaft

The Collaborative Reasearch Centre 568 was supported by grants of the Deutsche Forschungsgemeinschaft totalling EUR 14,763,100.

Year	Grant (EUR)
2001 – 2003	3,195,600
2004	1,571,000
2005	1,322,000
2006	1,272,000
2007	1,494,000
2008	1.796.800
2009	1,366,600
2010	1,421,900
2011	1,298,200
2001–2011	**14,763,100**

B List of Project-Related Publications

B.1 Part I: Injection Systems and Mixture Formation

Ahmad, W., Chrigui, M., Sadiki, A., Ngoma, G.D.: Effect of evaporation on the combustion behaviour of kerosene spray flame. ASME Turbo Expo 2010 (GT2010-22641), Glasgow, Scotland, UK, 14–18 June 2010.

Batarseh, F., Gnirß, M., Roisman, I.V., Tropea, C.: Fluctuations of a spray generated by an airblast atomizer. Exp. Fluids **46**, 1081–1091 (2009)

Chrigui, M., Roisman, I.V., Batarseh, F.Z., Sadiki, A., Tropea, C.: Spray generated by an airblast atomizer under elevated ambient pressures. J. Propuls. Power **26**, 1170–1183 (2009)

Chrigui, M., Sadiki, A., Batarseh, F., Janika, J., Tropea, C.: Numerical and experimental study of spray produced by an airblast atomizer under elevated pressure conditions. In: ASME Conference Proceedings, vol. 3 (2008)

Chrigui, M., Sadiki, A., Janicka, J., Zgahl, A.: Study of n-heptane spray evaporation and dispersion within premixed combustion in complex geometry configuration. In: Accepted to the 32th International Symposium on Combustion, McGill University, Montreal, Canada (2008)

Chrigui, M., Sadiki, A., Janicka, J.: Numerical analysis of spray dispersion, evaporation and combustion in a single gas turbine combustor. In: ASME TURBO-EXPO, GT2008-51253, Berlin, Germany (2008)

Chrigui, M., Batarseh, F.Z., Sadiki, A., Roisman, I., Tropea, C.: Numerical and experimental study of spray produced by an airbalst atomizer under elevated pressure conditions. In: ASME TURBO-EXPO, GT2008-51305, Berlin, Germany (2008)

Chrigui, M., Sadiki, A., Ngoma, G.D.: Unsteady, turbulent, two-phase flow using an Euler/Lagrange approach devoted to two-way coupling conditions. In: International Conference on Multiphase Flow 2010 (ICMF-2010), Floride, USA, 30 May–4 June 2010

Chrigui, M., Hage, M., Sadiki, A., Janicka, J., Dreizler, A.: Experimental and numerical analysis of spray dispersion and evaporation in a combustion chamber. At. Spray **19**, 929–955 (2009)

Chrigui, M., Roisman, I., Batarseh, F., Sadiki, A., Tropea, C.: Spray generated by an airblast atomizer under elevated ambient pressures. J. Propul. Power AIAA **26**(6), 1170–1183 (2010)

Chrigui, M.: N-Hpetane spray evaporation and dispersion in turbulent flow within a complex geometry configuration. J. Comput. Therm. Sci. **2**(1), 55–78 (2010)

Chrigui, M., Sadiki, A., Janicka, J.: Evaporation and dispersion of N heptane droplets within premixed flame. J. Heat Mass Trans. **46**(8–9), 869–880 (2010)

Chrigui, M., Schneider, L., Zghal, A., Sadiki, A., Janicka, J.: Droplet behavior within a LPP ambiance. J. Fluid Dyn. Mater. Process. **6**(4), 399–408 (2010)

J. Janicka et al., *Flow and Combustion in Advanced Gas Turbine Combustors*, Fluid Mechanics and Its Applications 102, DOI 10.1007/978-94-007-5320-4,

Chrigui, M., Moesl, M.K., Ahmadi, W., Sadiki, A., Janicka, J.: Partially premixed prevaporized kerozene spray combustion in turbulent flow. Exp. Therm. Fluid Sci **34**(1), 308–315 (2010)

Chrigui, M., Zghal, A., Sadiki, A., Janicka, J.: Spray evaporation and dispersion of n-heptane droplets within premixed flame. Heat Mass Trans. **46**, 869–880 (2010)

Chrigui, M., Gounder, J., Sadiki, A., Masri, R., Janicka, J.: Partially premixed reacting acetone spray using LES and FGM tabulated chemistry combustion and flame. http://dx.doi.org/10.1016/j.combustflame.2012.03.009 (2012)

Hahn, F., Sadiki, A., Janicka, J.: Large eddy simulation of a particle laden swirling flow based on an Euler-Lagragian approach. In: 6th International Conference on Multiphase Flow (ICMF2007), Leipzig, Germany (2007)

Helbig, K.: Messung zur Hydrodynamik und zum Wärmetransport bei der Filmverdamfung. EPDA Elektronische Publikationen Darmstadt. http://elib.tu-darmstadt.de/diss/000867. Accessed 7 Apr 2011

Helbig, K., Alexeev, A., Gambaryan-Roisman, T., Stephan, P.: Evaporation of falling and shear-driven films on smooth and grooved surfaces. Flow Turbul. Combust. **75**, 85–104 (2005)

Helbig, K., Nasarek, R., Gambaryan-Roisman, T., Stephan, P.: Effect of longitudinal mini-grooves on flow stability and wave characteristics of falling liquid films. ASME J. Heat Trans. **131**, 011601 (2009)

Gambaryan-Roisman, T., Stephan, P.: Analysis of falling film evaporation on grooved surfaces. J. Enhanc. Heat Trans. **10**(4), 445–457 (2003)

Gambaryan-Roisman, T., Alexeev, A., Stephan, P.: Effect of the microscale wall topography on the thermocapillary convection within a heated liquid film. Exp. Therm. Fluid Sci. **29**, 765–772 (2005)

Gambaryan-Roisman, T., Stephan, P.: Flow and stability of rivulets on heated surfaces with topography. ASME J. Heat Trans. **131**(3), 033101 (2009)

Kabova, Y., Alexeev, A., Gambaryan-Roisman, T., Stephan, P.: Marangoni-induced deformation and rupture of a liquid film on a heated microstructured wall. Phys. Fluids **18**, 012104 (2006)

Kunkelmann, C., Ibrahem, K., Schweizer, N., Herbert, S., Stephan, P., Gambaryan-Roisman, T.: The effect of three-phase contact line speed on local evaporative heat transfer: experimental and numerical investigations. Int. J. Heat Mass Trans. **55**(7–8), 1896–1904 (2012)

Löffler, K., Yu, H., Gambaryan-Roisman, T., Stephan, P.: Hydrodynamics and heat transfer of thin films flowing down inclined smooth and structured plates. In: Proceedings of the 4th International Berlin Workshop – IBW4 on Transport Phenomena with Moving Boundaries, Berlin (2007)

Marati, J.R., Budakli, M., Gambaryan-Roisman, T., Stephan, P.: Heat transfer in shear-driven thin liquid film flows. In: Proceedings of CHT-12, ICHMT International Symposium on Advances in Computational Heat Transfer, Bath, England (Accepted) (2012)

Olbricht, C., Hahn, F., Sadiki, A., Janicka, J.: Analysis of subgrid scale mixing using a hybrid LES-Monte-Carlo PDF method. Int. J. Heat Fluid Flow **28**(6), 1215–1226 (2007)

Opfer, L., Roisman, I.V., Tropea, C.: High speed visualization of drop and spray impact on rigid walls with cross-flow, Poster. In: International Conference on Multiphase Flows, Tampa, USA (2010)

Opfer, L., Roisman, I.V., Tropea, C.: Spray impact on walls with cross-flow, Poster. In: Workshop on Near Wall Reactive Flows, Seeheim, Germany (2010)

Opfer, L., Roisman, I.V., Tropea, C.: Spray Impact on Walls with Cross-flow: Experiments and Modeling, ILASS Europe, Estoril, Lissabon (2011)

Opfer, L., Roisman, I.V., Tropea, C.: Laboratory simulations of an airblast atomization: main mechanisms of liquid disintegration and spray characteristics. Exp. Fluids, submitted, March 2012

Pantangi, P., Sadiki, A., Janicka, J., Hage, M., Dreizler, A., van Oijen, J.A., Hassa, C., Heinze, J., Meier, U.: LES of pre-vaporized kerosene combustion at high pressures in a single sector combustor taking advantage of the flamelet generated manifolds method. In: Proceedings of ASME Turbo Expo 2011 (GT2011-45819), Vancouver, Canada, 6–10 June 2011

Roisman, I.V., Batarseh, F.Z., Tropea, C.: Chaotic disintegration of a liquid wall film: a model of an air-blast atomization. Atomiz. Sprays **20**(10), 837–845 (2010)

Roisman, I.V., Batarseh, F.Z., Tropea, C.: Characterization of a spray generated by an airblast atomizer with prefilmer. Atomiz. Sprays **20**(10), 887–903 (2010)

Sadiki, A., Goryntsev, D., Wegner, B., Janicka, J.: Design of technical combustion systems using LES: state of the art and perspectives. In: 7th International ERCOFTAC Symposium on Engineering. Turbulence Modelling and Measurements, (ETMM7), Limassol, Cyprus, 4–6 June 2008

Sadiki, A., Ahmadi, W., Chrigui, M., Janicka, J.: Towards the impact of fuel evaporation-combustion interaction on spray combustion in gas turbine combustion chambers. Part I: effect of partial fuel vaporization on spray combustion. In: Merci, B., Roeckaerts, D., Sadiki, A. (eds.) Proceedings of the 1st International Workshop on Turbulent Spray Combustion, Chap. 3, pp. 111–132, Springer (2011)

Sadiki, A., Ahmadi, W., Chrigui, M., Janicka, J.: Towards the impact of fuel evaporation-combustion interaction on spray combustion in gas turbine combustion chambers. Part II: influence of high combustion temperature on spray droplet evaporation. In: Merci, B., Roeckaerts, D., Sadiki, A. (eds.) Proceedings of the 1st International Workshop on Turbulent Spray Combustion, Chap. 3, pp. 111–132, Springer (2011)

B.2 Part II: Combustion

Böhm, B., et al.: In-Nozzle measurements of a turbulent opposed jet using PIV. Flow Turbul. Combust. **85**, 73–93 (2010)

Böhm, B., et al.: Simultaneous PIV/PTV/OH PLIF imaging: conditional flow field statistics in partially-premixed turbulent opposed jet flames. Proc. Combust. Inst. **31**, 709–718 (2007)

Böhm, B., et al.: New perspectives on turbulent combustion: Multi-parameter high-speed laser diagnostics. Flow Turbul. Combust. **86**, 313–341 (2011)

Böhm, B., et al.: Time-resolved conditioned flow field statistics in extinguishing turbulent opposed jet flames using simultaneous highspeed PIV/OH-PLIF. Proc. Combust. Inst. **32**, 1647–1654 (2009)

Bork, B., et al.: 1D high-speed Rayleigh measurements in turbulent flames. Appl. Phys. B **101**, 487–491 (2010)

Geyer, D., et al.: Finite rate chemistry effects in turbulent opposed flows: comparison of Raman/Rayleigh measurements and Monte Carlo PDF simulation. Proc. Combust. Inst. **30**, 711–718 (2005)

Geyer, D., et al.: Scalar dissipation rates in isothermal and reactive turbulent opposed-jets: 1D-Raman/Rayleigh experiments supported by LES. Proc. Combust. Inst. **30**, 681–689 (2005)

Geyer, D., et al.: Turbulent opposed-jet flames: a critical benchmark experiment for combustion LES. Combust. Flame **143**, 524–548 (2005)

Gregor, M.A., et al.: Multi-scalar measurements in a premixed swirl burner using 1D Raman/Rayleigh scattering. Proc. Combust. Inst. **32**, 1739–1746 (2009)

Ketelheun, A., Olbricht, C., Hahn, F., Janicka, J.: NO prediction in turbulent flames using LES/FGM with additional transport equations. Proc. Combust. Inst. **33**(2), 2975–2982 (2011)

Ketelheun, A., Aschmoneit, K., Janicka, J.: CO prediction in LES of turbulent flames with additional modeling of the chemical source term. In: ASME Turbo Expo, Copenhagen, Denmark, accepted for publication, 11–15 June 2012

Kuehne, J.: Analysis of combustion LES using an Eulerian Monte Carlo PDF method. PhD thesis, TU Darmstadt (2011)

Kuehne, J., Ketelheun, A., Janicka, J.: Analysis of sub-grid PDF of a progress variable approach using a hybrid LES/TPDF method. Proc. Combust. Inst. **33**, 1411–1418 (2011)

Kuenne, G., Ketelheun, A., Janicka, J.: LES modeling of premixed combustion using a thickened flame approach coupled with FGM tabulated chemistry. Combust. Flame **158**, 1750–1767 (2011)

Lebiedz, D.: Computing minimal entropy production trajectories: an approach to model reduction in chemical kinetics. J. Chem. Phys. **120**(15), 6890–6897 (2004)

Lebiedz, D.: Entropy-related extremum principles for model reduction of dynamical systems. Entropy **12**(4), 706–719 (2010)

Lebiedz, D., Reinhardt, V., Kammerer, J.: Novel trajectory based concepts for model and complexity reduction in (bio)chemical kinetics. In: Gorban, A.N., Kazantzis, N., Kevrekidis, I.G., Theodoropoulos, C. (eds.) Model Reduction and Coarse-Graining Approaches for Multiscale Phenomena, pp. 343–364. Springer, Dordrecht (2006)

Lebiedz, D., Reinhardt, V., Siehr, J.: Minimal curvature trajectories: Riemannian geometry concepts for slow manifold computation in chemical kinetics. J. Comput. Phys. **229**(18), 6512–6533 (2010)

Lebiedz, D., Reinhardt, V., Siehr, J., Unger, J.: Geometric criteria for model reduction in chemical kinetics via optimization of trajectories. In: Gorban, A.N., Roose, D. (eds.) Coping with Complexity: Model Reduction and Data Analysis, pp. 241–252. Springer, Berlin (2010)

Lebiedz, D., Siehr, J., Unger, J.: A variational principle for computing slow invariant manifolds in dissipative dynamical systems. SIAM J. Sci. Comput. **33**(2), 703–720 (2011)

Nauert, A., et al.: Experimental analysis of flash back in lean premixed swirling flames: conditions close to flash back. Exp. Fluids **43**, 89–100 (2007)

Nauert, A., Dreizler, A.: Conditional velocity measurements by simulatenously applied laser Doppler velocimetry and planar laser-induced fluorescence in a swirling natural gas/air flame. Z. Phys. Chem. **219**, 635–648 (2005)

Omar, S.K., et al.: Investigation of flame structures in turbulent partially premixed counter-flow flames using laser-induced fluorescence. Prog. Comput. Fluid Dyn. **4**, 241–249 (2004)

Reinhardt, V., Winckler, M., Lebiedz, D.: Approximation of slow attracting manifolds in chemical kinetics by trajectory-based optimization approaches. J. Phys. Chem. A **112**(8), 1712–1718 (2008)

Schneider, C., Dreizler, A., Janicka, J.: Fluid dynamical analysis of atmospheric reacting and isothermal swirling flows. Flow Turbul. Combust. **74**, 103–127 (2005)

Siehr, J., Lebiedz, D.: An optimization approach to kinetic model reduction for combustion chemistry. Submitted (2012)

B.3 Part III: Interaction and Fluid-Mechanical Processes

Jakirlić, S., Hanjalić, K.: A new approach to modelling near-wall turbulence energy and stress dissipation. J. Fluid Mech. **539**, 139–166 (2002)

Jakirlić, S., Jester-Zürker, R., Tropea, C.: Joint effects of geometry confinement and swirling inflow on turbulent mixing in model combustors: a second-moment closure study. J. Prog. CFD **4**(3–5), 198–207 (2004)

Jakirlić, S., Eisfeld, B., Jester-Zürker, R., Kroll, N.: Near-wall, Reynolds-stress model calculations of transonic flow configurations relevant to aircraft aerodynamics. Int. J. Heat Fluid Flow **28**(4), 602–615 (2007)

Jakirlić, S., Kniesner, B., Kadavelil, G., Gnirß, M., Tropea, C.: Experimental and computational investigations of flow and mixing in a single-annular combustor configuration. Flow Turbul. Combust. **83**(3), 425–448 (2009)

Jakirlić, S., Kadavelil, G., Kornhaas, M., Schäfer, M., Sternel, D.C., Tropea, C.: Numerical and physical aspects in LES and hybrid LES/RANS of turbulent flow separation in a 3-D diffuser. Int. J. Heat Fluid Flow **31**(5), 820–832 (2010)

Jakirlić, S., Jovanović, J.: On unified boundary conditions for improved prediction of near-wall turbulence. J. Fluid Mech. **656**, 530–539 (2010)

Jakirlić, S., Kniesner, B.: Near-wall RANS modelling in LES of heat transfer in backward-facing step flows under conditions of constant and variable fluid properties. In: ASME 3rd Joint U.S.-European Fluids Engineering Summer Meeting: Symposium on "DNS, LES and Hybrid RANS/LES Methods", Montreal, Quebec, Canada, Paper No. FEDSM-ICNMM2010-30354, 1–5 August 2010

Jakirlić, S., Jester-Zürker, R.: Convective heat transfer in wall-bounded flows affected by severe fluid properties variation: a second-moment closure study. In: ASME 3rd Joint U.S.-European Fluids Engineering Summer Meeting: "7th Symposium on Fundamental Issues and Perspectives in Fluid Mechanics", Montreal, Quebec, Canada, Paper No. FEDSM-ICNMM2010-30729, 1–5 August 2010

Jakirlić, S., Chang, C.-Y., Kadavelil, G., Kniesner, B., Maduta, R., Sarić, S., Basara, B.: Critical evaluation of some popular hybrid LES/RANS methods by reference to flow separation at a curved wall (invited lecture). In: 6th AIAA Theoretical Fluid Mechanics Conference, Honolulu, HI, USA, Paper No. AIAA-2011-3473, June 27–30 2010

Jakirlić, S., Kniesner, B., Kadavelil, G.: On interface issues in LES/RANS coupling strategies: a method for turbulence forcing. JSME J. Fluid Sci. Technol. **6**(1), 56–72 (2011)

Jester-Zürker, R., Jakirlić, S., Tropea, C.: Computational modelling of turbulent mixing in confined swirling environment under constant and variable density conditions. Flow Turbul. Combust. **75**(1–4), 217–244 (2005)

Jester-Zürker, R.: Reynolds-Spannungsmodellierung des Skalartransports unter Bedingungen variabler Stoffeigenschaften in Drallbrennerkonfigurationen (Second-moment closure modelling of scalar transport in swirl combustors under variable flow property conditions). PhD Thesis, Technische Universität Darmstadt, Shaker Verlag: ISBN 978-3-8322-6742-1, 27–30 June 2006

John-Puthenveettil, G., Jia, L., Reimann, T., Jakirlić, S., Sternel, D.C.: Thermal mixing of flow-crossing streams in a T-shaped junction: a comparative LES, RANS and Hybrid LES/RANS study. In: 7th International Symposium on Turbulence, Heat and Mass Transfer, Palermo, Italy, 24–27 September 2012

John-Puthenveettil, G.: Numerische Modellierung von komplexen Strömungen mittels wirbelauflösender Turbulenzmodelle unter Berücksichtigung der wandnahen Turbulenz (Computational modelling of complex flows using eddy-resolving turbulence models accounting for the near-wall turbulence). PhD Thesis, Technische Universität Darmstadt (2012) (in preparation)

Kniesner, B.: Ein hybrides LES/RANS Verfahren für konjugierte Strömung, Wärme- und Stoffübertragung mit Relevanz zu Drallbrennerkonfigurationen (A hybrid LES/RANS method for conjugated flow, heat and mass transfer with relevance to swirl combustor configurations). PhD Thesis, Technische Universität Darmstadt (2008)

Šarić, S., Jakirlić, S., Čavar, D., Kniesner, B., Altenhöfer, P., Tropea, C.: Computational study of mean flow and turbulence structure in inflow system of a swirl combustor. In: Tropea, et al. (eds.) Notes on Numerical Fluid Mechanics and Multidisciplinary Design, vol. 96, pp. 462–470. Springer, Berlin (2007). ISBN 978-3-540-74458-0

B.4 Part IV: Cross-Sectional Projects

Chrigui, M., Gounder, J., Sadiki, A., Masri, A.R., Janicka, J.: Partially premixed reacting acetone spray using LES and FGM tabulated chemistry. Combust. Flame **159**(8), 2718–2714 (2012)

Erdmann, B., Lang J., Roitzsch, R.: Kardos user's guide. ZIB-Report 02–42, ZIB (2002)

Gauß, F., Schäfer, M., Siegmann, J., Sternel, D.: On the influence of boundary discretization schemes on the accuracy of flow simulation with local refinement. In: Diez, P., Bouillard, Ph. (eds.) In: IV International Conference on Adaptive Mo- delling and Simulation, pp. 97–100.

ECCOMAS, International Center for Numerical Methods in Engineering, May 2009
Hertel, C., Schümichen, M., Löbig, S., Fröhlich, J., Lang, J.: Adaptive large eddy simulation with moving grids. Preprint Technische Universität Dresden, submitted to Theoretical and Computational Fluid Dynamics (2011)
Janicka, J., Sadiki, A.: Large eddy simulation of turbulent combustion systems. Proc. Combust. Inst. **30**, 537–547 (2005)
Janus, B., Dreizler, A., Janicka, J.: Experiments on swirl stabilized non-premixed natural gas flames in a model gas turbine combustor. Proc. Combust. Inst **31**(2), 3091–3098 (2006)
Jakirlic, S., Kadavelil, G., Kornhaas, M., Schäfer, M., Sternel, D.C., Tropea, C.: Numerical and physical aspects in LES and hybrid LES/RANS of turbulent flow separation in a 3-D diffuser. Int. J. Heat Fluid Flow **31**, 820–832 (2010)
Ketelheun, A., Olbricht, C., Hahn, F., Janicka, F.: NO prediction in turbulent flames using LES/FGM with additional transport equations. Proc. Combust. Inst. **33**(2), 2975–2982 (2011)
Kornhaas, M., Sternel, D.C., Schäfer, M.: Influence of time step size and convergence criteria on large eddy simulations with implicit time discretization. In: Meyers, J., Geurts, B., Sagaut, P. (eds.) Quality and Reliability of Large-Eddy Simulations. ERCOFTAC Series 12, pp. 119–130. Springer, Berlin (2008)
Lang, J.: Adaptive incompressible flow computations with linearly implicit time discretization and stabilized finite elements. In: Papailiou, K., Tsahalis, D., Periaux, J., Hirsch, C., Pandolfi, M. (eds.) Computational Fluid Dynamics '98. Wiley, Chichester/New York (1998)
Lang, J., Verwer, J.: ROS3P—an accurate third-order Rosenbrock solver designed for parabolic problems. BIT Numer. Math. **41**, 730–737 (2001)
Lang, J., Cao, W., Huang, W., Russell, R.D.: A two-dimensional moving finite element method with local refinement based on a posteriori error estimates. Appl. Numer. Math. **46**, 75–94 (2003)
Löbig, S., Dörnbrack, A., Fröhlich, J., Hertel, C., Kühnlein, C., Lang, J.: Towards large eddy simulation on moving grids. Proc. Appl. Math. Mech. **9**, 445–446 (2009)
Olbricht, C., Hahn, F., Sadiki, A., Janicka, J.: Analysis of subgrid scale mixing using a hybrid LES-Monte-Carlo PDF method. Int. J. Heat Fluid Flow **28**(6), 1215–1226 (2007)
Pantangi, P., Sadiki, A., Janicka, J., Hage, M., Dreizler, A., van Oijen, J.A., Hassa, C., Heinze, J., Meier, U.: LES of pre-vaporized kerosene combustion at high pressures in a single sector combustor taking advantage of the flamelet generated manifolds method. In Proceedings of ASME Turbo Expo 2011 (GT2011-45819), Vancouver, Canada, 6–10 June 2011
Sternel, D.C., Kornhaas, M., Schäfer, M.: High-performance computing techniques for coupled fluid, structure and acoustics simulations. In Competence in High Performance Computing 2010: Proceedings of an International Conference on Competence in High Performance Computing, June 2010 Springer, Germany, 2011
Sternel, D.C., Schäfer, M., Heck, M., Yigit, S.: Efficiency and accuracy of fluid–structure interaction simulations using an implicit partitioned approach. Comput. Mech. **43**, 103–113 (2008)
Ullmann, S., Lang, J.: A POD-Galerkin reduced model with updated coefficients for Smagorinsky LES. In: Pereira, J.C.F., Sequeira, A., Pereira, J.M.C. (eds.) Proceedings of the V European Conference on Computational Fluid Dynamics ECCOMAS CFD 2010, Lisbon, Portugal (2010)
Wegner, B., Maltsev, A., Schneider, C., Sadiki, A., Dreizler, A., Janicka, J.: Assessment of unsteady RANS in predicting swirl flow instability based on LES and experiments. Int. J. Heat Fluid Flow **25**, 528–536 (2004)
Wegner, B.: A large-eddy simulation technique for the prediction of flow, mixing and combustion in gas turbine combustors. PhD thesis, Technische Universitaet Darmstadt, VDI Verlag GmbH, ISBN 978-3-18-354906-1 (2007)
Wegner, B., Maltsev, A., Schneider, C., Sadiki, A., Dreizler, A., Janicka, J.: Assessment of unsteady RANS in predicting swirl flow instability based on LES and experiments. Int. J. Heat Fluid Flow **25**, 528–536 (2004)

Printed by Printforce, the Netherlands